A. Sølvberg D.C. Kung

Information Systems Engineering

An Introduction

With 263 Figures

Springer-Verlag
Berlin Heidelberg New York
London Paris Tokyo
Hong Kong Barcelona
Budapest

Arne Sølvberg
Faculty of Electrical Engineering
and Computer Science
The Norwegian Institute of Technology
University of Trondheim
N-7034 Trondheim, Norway

David Chenho Kung
Department of Computer Science Engineering
University of Texas at Arlington
P. O. Box 19015
Arlington, TX 76019, USA

ISBN-13:978-3-642-78003-5 e-ISBN-13:978-3-642-78001-1
DOI: 10.1007/978-3-642-78001-1

Softcover reprint of the hardcover 1st edition 1993

Cover design: Konzept & Design, Ilvesheim
Typesetting: Camera-ready copy from the author
45/3140 - 5 4 3 2 1 0 - Printed on acid-free paper

Preface

This book presents a selection of subjects which the authors deem to be important for information systems engineers. The book is intended for introductory teaching. We have tried to write the book in such a way that students with only fragmented knowledge of computers are able to read the book without too many difficulties. Students who have had only an introductory course in computer programming should be able to read most of the book. We have tried to achieve simplicity without compromising on depth in our discussions of the various aspects of information systems engineering. So it is our hope that also those who have deeper knowledge in computing may find pleasure in reading parts of the book.

The writing of a textbook is a major undertaking for its authors. One is quite often forced to reexamine truisms in the subject area, and must be prepared to reevaluate one's opinions and priorities as one learns more. In particular this is so in new fields, where formalisms have been scarcely used, and where consensus has not yet emerged either on what constitutes the subject area or on how practical problems within the field shall be approached. Contemporary practice in computer applications is confronted with an increasingly complex world, both in a technical sense and in the complexity of problems that are solved by computer. There is an increasing need for a coherent framework to be put forward which has a sound theoretical underpinning and which relates the various skills needed by professional system designers. Our text in Information Systems Engineering is intended to support a further professionalization of computer application development.

The selection of subjects in this book is a reflection of the two authors' priorities, within the limit of half-a-thousand pages of text and figures. This is not to say that there are not other important subjects as well. Given more time, and more energy to devote to textbook writing, we would happily expand our text with chapters on document processing, software metrics and cost estimation. We would most of all like to expand our text further in the realm of formal systems modeling, and associated development tools. Many interesting model proposals have been forwarded over the last few years which very well deserved to be treated in a text on Information Systems Engineering.

Several concerns have influenced the selection of material for the book:

- The basic components in an information system must be described.
- The recommended design techniques should be widely accepted in practice.
- Methodological aspects should be discussed.
- Systems evolution problems should be stressed.
- Future methodologies should be discussed.
- Different approaches should be related to each other.

We have tried to satisfy these requirements as follows:

- The first half of the book is based on well known information systems design techniques (Chap. 2–5).
- The next quarter of the book is concerned with systems evolution, project management issues, engineering design, various views of information systems and software (Chap. 6–10).
- The last part of the book covers several formal design approaches, and has more the flavor of a research monograph than the previous chapters (Chap. 11–14).

Furthermore, we have made the following particular choices:

- The so-called structured analysis and design approach forms the basis for our introduction to information systems engineering.
- The object oriented approach is introduced and explained relative to the "structured" approaches.
- Software design, database design and rule modeling are treated separately, and presented as the "basic" design components.
- User interface design is treated as a side issue, partly because of lack of space, but also because of lack of methodological maturity of the field.
- Performance evaluation issues are discussed relative to software design as well as to database design.
- Systems evolution aspects are treated, partly from a software maintenance perspective, and partly from the perspective of the installation of common systems that are built from existing software components.
- Project management techniques have been treated only at an elementary level.
- Information systems engineering is seen as a sub-discipline of engineering in general.
- The presentation in the last part of the book takes as its starting point that future methodologies will be based on the availability of formal, executable models of information systems, so that a full integration becomes possible of the different specification languages that are currently used during the various phases of systems development.
- Only a small selection of formal models should be presented, because of space limitations.

This text has been developed over several years. It would never have been finished unless both authors were given possibilities to concentrate on textbook writing for limited periods of time. These times of concentration have been partly supported by the National Science Foundation (U.S.A.) and by the Royal Norwegian Council for Scientific and Industrial Research. Most of the text has been written in Trondheim, and at the Naval Postgraduate School's Department of Computer Science, Monterey, Ca., where one of the authors spent a sabbatical year. Professors David Hsiao and Vincent Lum are warmly thanked for providing excellent working conditions for bookwriting in California. There has also been excellent help in preparing the final manuscript. Hilde Berg has been particularly helpful in the last phase. Anne Sørvik helped with the preparation of earlier drafts of the manuscript. We are also grateful to Ali Arifoglu for preparing the exercises of the book, as well as the index. Gunnar Brataas has been a great help in drawing a number of the figures. Several other doctoral students in Trondheim have been most helpful in letting us use some of their material prior to their thesis-publication, as well as sharing freely their general mental support for the endeavor. We are particularly grateful to Rudolf Andersen, Ming Wei Yang, Odd Ivar Lindland, Anne Helga Seltveit, Jon Atle Gulla, Geir Willumsen, John Krogstie and Vidar Vetland. Finally, we thank Springer-Verlag for good cooperation, especially J. Andrew Ross for his careful copy-editing.

Trondheim, Norway, and Arlington, Texas, USA, May 1993

Arne Sølvberg and David C. Kung

Contents

Chapter 1

Introduction

A modern society is completely dependent on computers. They are found everywhere. They are used for supporting information processing tasks that range from the most mundane to the most complex tasks imaginable. The production and servicing of computers is one of the largest industries in the world.

Computers were first used for fairly mundane tasks, like doing the enormous number of additions and multiplications necessary in order to find solutions to numerical mathematical problems, for example inverting a matrix. Other examples are statistical analysis of large masses of data, text editing, and so on. As the years have passed, more and more ambitious applications of computers have been realized, from robotics to artificial intelligence and decision support systems for managers, to mention a few. Computers are increasingly becoming integral parts of large systems. They interact with human beings, and with other computers and technical systems. Computers become useful to the extent that they are properly integrated in their environments.

This text is about information systems engineering. It is about the application of computers in large information processing systems. It is about the development of computer assisted information processing systems. Examples of such systems are:

- The passenger booking system of an airline: it has been said that a modern airline consists firstly of a computerized passenger booking system, and secondly of planes, flight crews, and technical maintenance systems.

- The transaction processing system of a bank: can a modern bank exist today unless it is computerized over and over again?

- The flight control system of a modern airplane.

Some of the systems are extremely large. The software development for the space shuttle is reported to have cost approximately 25 000 work years. Most information systems cost only a fraction of this, from one work year and up, depending on the system's complexity. There are, however, similarities among the systems, regardless of size.

Better quality is needed in information system development

Information system development has developed into a big industry. On the one hand, the investment is rapidly increasing, while on the other hand, systems are growing larger and more complex. However, the current situation in system development leaves much to be desired. Projects that fail to meet schedule, cost, and performance objectives are still too common. Some of the most common problems are

- System deliveries are often far behind schedule.
- System development often exceeds projected costs.
- Systems do not satisfy the user information requirements, e.g., they do not provide the right information at the right time to the right user.
- Systems are so unreliable that constant maintenance is required to correct errors during operation.
- Systems are badly structured and poorly documented, making enhancement and maintenance extremely difficult.

Success in system development depends on effective human communication

The solution to these problems lies in the nature of system development. To bring large and complex information systems into existence is not a task for one or two persons. It is a team effort involving the development team and the various user communities. It has to consider the political and economic interests of these parties. In particular, documentation of the information systems requirements is the result of coordinating differing, or even conflicting opinions and needs. This necessitates the use of languages for communication within the team, and for communication between the team and the users. Therefore, the most critical problem in developing information systems is the communication problem. It is particularly evident when developing computer-assisted information systems, where information processing tasks are shared between computers and human beings. It is quite crucial to the success of the system that the users can participate in the development of the information system requirements because the users are the specialists in the application domain where the information system is built.

Information systems support business information needs through effective use of software and computers

Information systems are built in order to provide information services that enable tasks to be performed in other systems and other contexts, e.g., a customer's purchase of a piece of ribeye steak is supported by a sales clerk at a point-of-sales terminal, which is a component in a complex system.

Information systems must be planned in a total context, taking all relevant actors into account. Only in this way can a sufficiently wide framework be provided to enable the development of information systems with enough desirable features. The three major subsystems that have to be considered are the computer (and communication) system, the (application) software system, and the (human) organization that is being served by the two other subsystems (Fig. 1.1).

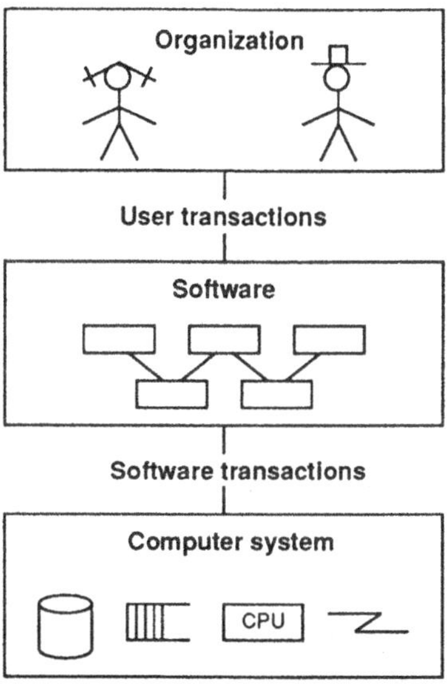

Fig. 1.1. Information systems comprise computer components, application software components and human/organizational components

Information system development methods must to some extent be holistic, because all relevant aspects have to be considered, be they of a technical, sociological or political nature. Many of the relevant issues are fuzzy. There is always a temptation to deal with fuzzy issues in a fuzzy way. This should be avoided. One should not forget that those parts of an information system that are to be computerized must be precisely described. Computers require precise, complete and consistent instructions in order to work reliably. A major objective of information system development methods is to permit the designers to effectively proceed from the fuzzy to the formal.

The functional, technical and operational aspects of information systems

Every information system has a functional aspect, a technical aspect and an operational aspect (Fig. 1.1). WHAT is the system to do? HOW shall things be done? WHO shall do it?

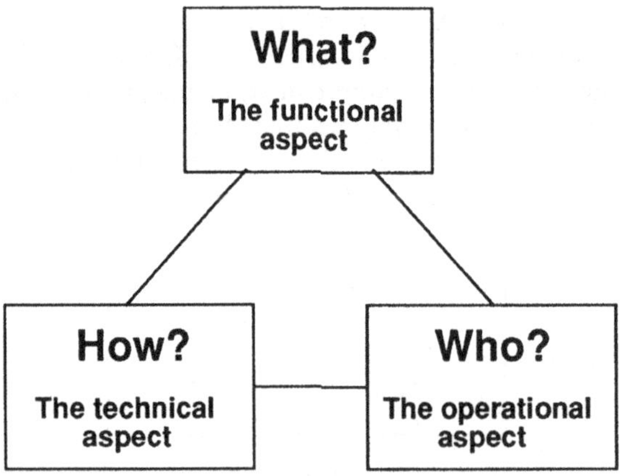

Fig. 1.2. Three aspects of an information system

The WHAT-aspect of an information system can only be sorted out relative to the needs of the system's environment. For example, a passenger booking system for an air carrier and a patient booking system for a hospital will most probably have to satisfy different functional requirements, imposed by the different environments.

The HOW-aspect must be analyzed relative to the available technical solutions. Important design issues are software structures, database structures, interfaces between computers and people, interfaces between computers, etc.

The WHO-aspect includes costs and operational characteristics of computers, datanets, other technical devices, and people. The various technical and human resources have different abilities, capacities, costs, and reliabilities.

The basic system development issue is:

> Can a satisfactory functional design (*what* to do) be realized by an available technical solution (*how* to do it) for an acceptable price (*who* to do it)?

The three issues are, of course, not unrelated to each other. It will do us little good if we propose a functional design for which there is no available technical solution. We also cannot propose technical solutions that are too expensive, or that cannot support the functional requirements. System development projects accordingly have to iterate among the three issues of *what*, *how* and *who*, until a satisfactory solution has been found.

The major phases of a system development project are usually called *system analysis*, *system design* and *system implementation*. These three development phases differ in that the work in each of them concentrate mostly on only one of the *what-how-who* issues. System analysis is mostly concerned with the functional aspects, system design with the technical aspects and system implementation with the operational aspects (Fig. 1.3).

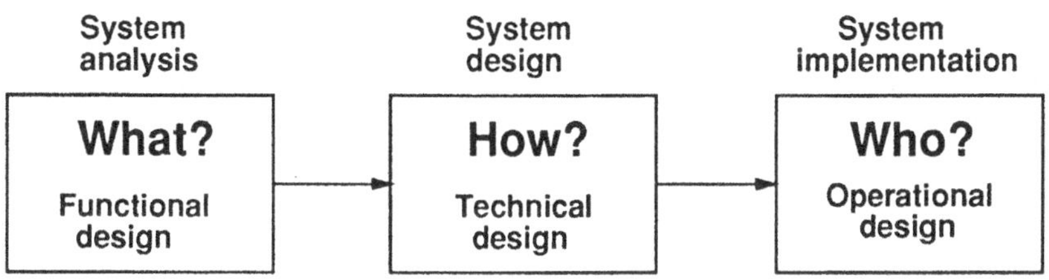

Fig. 1.3. Different development phases concentrate on different system aspects

From an engineering point of view all major project phases are concerned with design. In order to master high system complexity it is advisable to treat different design issues separately whenever possible. This principle is adhered to in the organization of work in the three major phases of systems analysis, systems design and systems implementation.

The organization and management of a development project are issues in their own right. Computerized information systems are among the most complex artifacts that have ever been created. Managing this complexity requires sophisticated organization and tools.

The purpose of project management is to institute a development discipline, so that different design issues are treated in proper order. Furthermore, the development discipline must ensure that design decisions are properly evaluated with respect to system requirements, e.g., cost, functionality. Finally, the development discipline must ensure that proper action is taken whenever previously stated requirements can not be satisfied by the proposed designs, be it designs of functional properties, technical properties or operational properties.

1.1 Two Information System Examples

A good way to understand the rationale behind the introduction of modeling concepts is to examine the examples used to illustrate the concepts. We will therefore present some typical problem solving situations. The first one is the IFIP Conference example [OLLE83]. The other one is a description of a university administration system [WINO79].

1.1.1 The IFIP Conference Example

The IFIP Conference example, developed by T.W. Olle, has been used as a test example for comparing different information system development methods [OLLE83].

The International Federation of Information Processing (IFIP) is a federation of more than 30 national data processing societies. IFIP has a number of Technical Committees and Working Groups, each of which is set up to deal with a specific professional topic area.

An IFIP Working Conference is an international conference designed to bring together experts from all IFIP countries to discuss topics of specific interest to one or more IFIP Working Groups. The usual procedure, and the one to be considered for present purposes, is a conference which is not open to everyone, but only to those invited. For such conferences it is something of a problem to ensure that members of the IFIP Working Group(s) and Technical Committee(s) involved are invited even if they do not come. Furthermore, it is important to ensure that sufficient people attend the conference so that the financial break-even point is reached without exceeding the maximum number dictated by the facilities available.

IFIP policy on Working Conferences suggests the appointment of a Program Committee to deal with the technical content of the conference and an Organizing Committee to handle financial matters, local arrangements, and invitations and/or publicity. These committees need to work together closely with the same current information so that their recorded information is consistent and up to date.

The information system should be designed to support the activities of both a Program Committee and an Organizing Committee in arranging an IFIP Working Conference. The involvement of the two committees is analogous to two organizational entities within a corporate structure using some common information. The following activities of the committees should be supported.

Program Committee:

- Preparing a list of those to whom the call for papers will be sent.
- Registering the letters of intent received in response to the call.
- Registering the contributed papers on receipt.
- Distributing the papers among those undertaking the refereeing.
- Collecting the referees' reports and selecting the papers for inclusion in the program.
- Grouping selected papers into sessions for presentation and selecting a chairman for each session.

Organizing Committee:

- Preparing a list of people to invite to the conference.
- Issuing priority invitations to National Representatives, Working Group members and members of associated working groups.
- Ensuring all authors of each selected paper receive an invitation.
- Ensuring authors of rejected papers receive an invitation.
- Avoiding sending duplicate invitations to any individual.
- Registering acceptance of invitations.
- Generating the final list of attendees.

This example has been used for comparing a number of different system development methods. The results of the exercises were presented at IFIP Working Conferences in 1983 and 1986 [OLLE83], [OLLE86]. It should be noted that budgeting and financial aspects of the Organizing Committee's work, meeting plans of both committees, hotel accommodation for those who

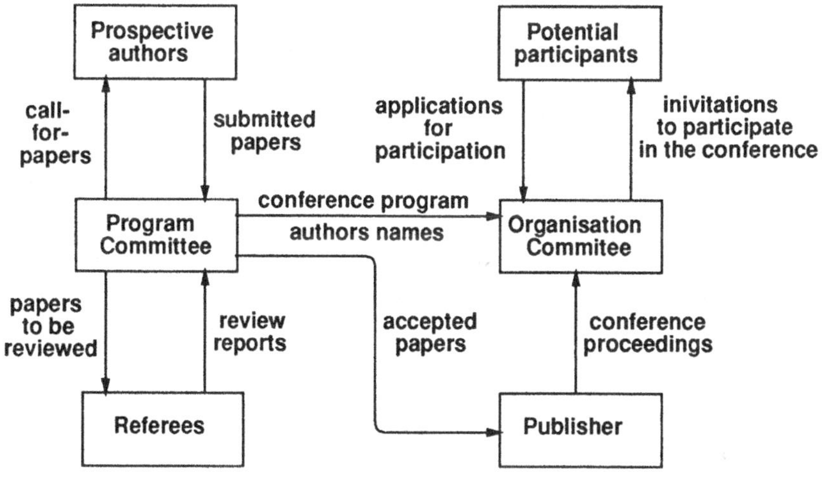

Fig. 1.4. Overview picture of IFIP conference example

attend and the matter of preparing camera-ready copy of the proceedings is omitted from the exercises, although some submissions include several of these additional aspects. The production of conference proceedings from the accepted papers has been added into Fig. 1.4.

The IFIP Conference example is a typical example of a low-technology application of high technology. One very important issue in such applications is the human-computer interface. Can the system be used to the advantage of the (low-technology) human tasks, or is the (high-technology) solution just another pain in the neck for the human operators?

1.1.2 A University Administration System

Winograd has developed the following problem description [WINO79]. Imagine that you have come to work as a system designer and programmer for a large university. The university administration has a computer system for scheduling and planning room use. Users at several sites on campus access the system through workstations that display building floorplans, as well as text and other graphic data.

The system keeps track of the scheduled use of all rooms, such as long-term lab and office assignments, regular class schedules, and special events. It is able to answer questions regarding current scheduled use and availability. Querying is done from the workstations, through structured graphic interaction (menus, standardized forms, pointing, etc.) and a limited natural language interface. The system does not make complex abstract deductions, but can combine information in the database to answer questions like "Is there a conference room for 40 people with a projection screen available near the education building from 3 to 5 on the 27th?" Users with appropriate authorization enter new information, including the scheduling of room use and changes to the facilities (including the interactive drawing of new or modified floorplans). In addition to the current assignments, the system keeps a history of use for analysis. Standard statistical information and data representation (such as tables, bar charts, graphs, etc.) are produced on demand for use in long-range planning.

The university wants the system to provide more help in making up the quarterly classroom assignments. It should be possible to give it a description of the courses scheduled for a future quarter and have it generate a proposed room assignment for all courses. In deciding assignments, the system should consider factors such as expected enrollment (using past data and new information that is available on estimated enrollments), proximity of rooms to the departments and teachers involved, preference for keeping the same location over time, and the nature of any special equipment needed. It should print out notices summarizing the relevant parts of the plan for each teacher, department, dean, and building supervisor. Any of these people should be able to use the normal querying system to find out more about the plan, including the motivations for specific decisions.

Properly authorized representatives of the dean's office would be able to request changes in the plan through an interactive dialogue with the system in which alternatives can be proposed and compared. When a change is made, the system should readjust whatever is necessary and produce a notification for the affected people.

It will take many programmer-years of effort to build a system like this. A development project will be successful only if the project is managed extraordinarily well. However, it is not difficult because of the intrinsic complexity of the tasks the system must carry out. The system combines hardware and software facilities that have been demonstrated in various combinations a number of times.

The problem lies in the difficulties of organizing complex systems. The integration of all of the components of the "initial system" would be a major achievement, calling for the best design tools and methodologies. The idea that a new programmer could come in to such a system and make widespread changes to handle the "assignment" is enough to make an experienced programmer shudder. It is tricky enough to add new types of questions (e.g., asking for explanations for decisions), new information (e.g., estimated enrollments), and new output forms (e.g., schedule summaries for departments). Additionally, we are asked to integrate a new kind of data (e.g., projected plans) into a system that was originally built to handle only a single current set of room assignments and a record of their history. This new data, the projected plans, must be so well integrated with the existing data for all of the facilities (including floorplan drawing, question answering, statistics gathering, etc.) that users can operate on them just as they operate on the initial database.

1.2 Information Systems Modeling

In the development of information systems, there are at least two language levels. One is the programming language level, e.g., COBOL. The other is the application language level, e.g., the set of user-oriented concepts developed for a payroll application. Computer professionals and users communicate in the application level language. Therefore, if the development and definition of the application language is not properly taken care of, communication between software designers and software users will probably be distorted, or even break down.

The key to achieving meaningful communication among human beings is getting the parties to share the relevant conceptual knowledge. This is usually achieved by developing so-called conceptual models.

1.2.1 Conceptual Models Have a Key Role in IS-Design

A conceptual model is not developed all at once, but rather through a process of consecutive iterations. Personal interviews and questionnaires are most often used to collect information about the application. An example conceptual model which is partially developed is shown in Fig. 1.5. This simple model displays that, in a specific application, students and teachers are subsets of persons. Some of the properties of persons are social security number (ssn), name, and address. These properties are inherited by students and teachers because these are both persons. In addition, students have property major which non-students do not have. Teachers have title and salary in addition to the other properties which persons have. In the conceptual model the properties are either **identifiers** or **attributes**. An identifier is a unique name to an entity, while many entities may share some attribute value. In this example the conceptual model displays that persons are to be identified by their social security numbers. This identification mechanism is also inherited by students and teachers.

There is a relationship between students and courses called EXAM, which have an attribute-property called grade. The EXAM relationship represents the examination results for students taking the courses. Another relationship is between teachers and courses that displays which teacher teaches

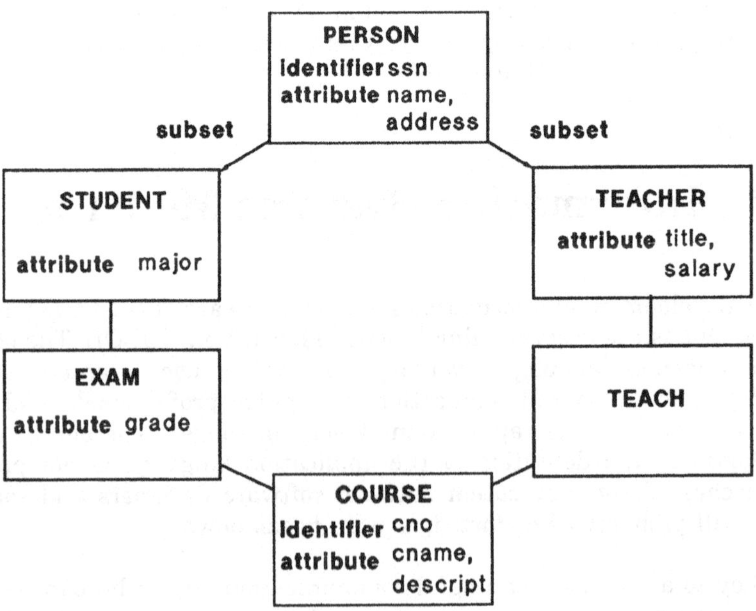

Fig. 1.5. A simple conceptual model

which course(s). This conceptual model, although simple, can be used to facilitate communication and enhance understanding between the system analyst and the users. It lays down a basis upon which additional information can be gathered.

A conceptual model is an abstract representation of the real world phenomena that are of interest to the application in question. The conceptual model is usually constructed during the system analysis phase. The role of such a conceptual model includes the following:

- It serves as a common reference framework used during the system analysis phase to communicate with the future users of the system.
- It serves as a model of reality offering insight into the application domain. In other words, the construction of the conceptual model enables the system analysts to have a better understanding of the application and the users needs.
- It serves as a basis upon which the design and implementation of the database can be carried out and against which the design and implementation can be tested.
- It provides documentation of the system which can be used during the maintenance phase to facilitate modification and enhancement.

1.2.2 Four Different Modeling Approaches

The most common modeling approaches may be classified into approaches that are either *process-oriented, data-oriented, rule-oriented* or *object-oriented.*

Process-oriented approaches take as a starting point the description of the processes of the system. This approach was first adopted in the information systems area, where the specification of the information processing functions within the organization was emphasized. The descriptions of the processes are usually supported by a separate data model that contains a description of the data items and data structures that are manipulated by the processes.

Data-oriented approaches take as their starting point the description of data structures and data semantics. The approach stems from the database area, where the specification of the data items, records, and their relationships were focused. A representative modeling tool belonging to this category is the Entity-Relationship or ER data model [CHEN76].

In a *rule-oriented approach,* knowledge about the application is represented by a set of formal assertions, such as logical formulas. Events and operations that affect the knowledge base are specified as derivation rules. A derivation rule states that when a set of conditions is true, then the consequences must be true. That is, it states WHAT is to be done when the conditions are true. In this sense, a rule-oriented approach aims at specifying WHAT the information system is going to do rather than HOW to do it. Whereas in a process-oriented approach, the emphasis is on HOW the information system is to process the information rather than WHAT it is supposed to do. Therefore, the rule-oriented approach has a higher degree of data processing independence. However, rule-oriented approaches are more difficult to implement and the implementation efficiency is usually low.

Object-oriented approaches take as their starting point the operational aspects of an information system. The system is seen as consisting of interacting objects, each object having particular skills, resource requirements, and potential for result production. Object-oriented approaches have their origin in simulation methods.

1.2.3 Modeling Approaches May Be Classified According to Their Time Perspective

The conceptual model depicted in Fig. 1.5 describes only a snapshot of the application. It does not reflect the evolution of the application such as the assignment of courses to teachers which results in the establishment of the TEACH relation. In this model, it is assumed that the events of assigning courses to teachers are considered outside the model. In particular, the model displays only the static properties of the application; hence, it is called a static model. Static models are adequate for applications involving many complex objects and relationships and a small number of events and operations. Question-answering systems or subsystems are examples of such an application.

Static models are easy to construct, understand and check, and their development costs are relatively lower. However, many applications today require considering the dynamic aspect of applications, where the transition from one state of the system to another needs to be modelled. For example, a retail company information system supporting automated replenishment of parts may require modeling the transactions which affect the stock levels of parts. In some applications, for example office information systems, temporal properties involving sequences of states are required to be modelled as well as time points and intervals.

In Sect. 1.2.2, we considered the classification of modeling approaches in terms of the way in which the model is specified. Another dimension along which the modeling approaches can be classified is called the temporal dimension, where we consider how well the temporal aspects of an application can be modelled. Along the temporal dimension, the existing modeling approaches fall into four classes (Fig. 1.6):

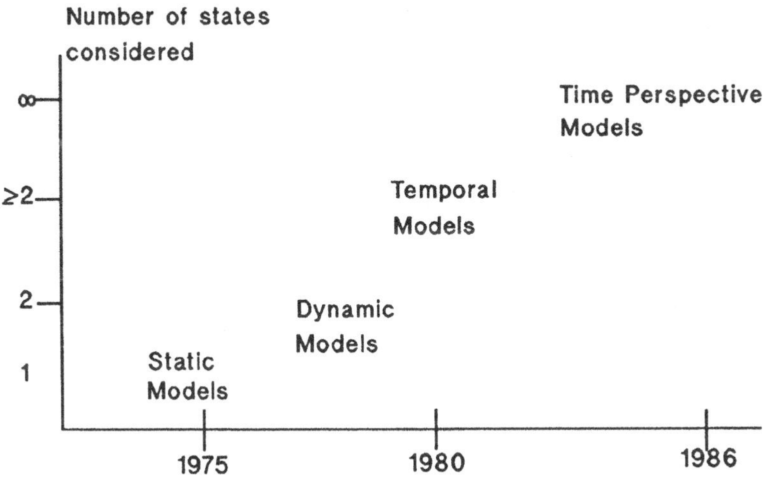

Fig. 1.6. Four types of modeling approaches

1) *Static approaches*: These provide facilities for describing only a snapshot of the application. Variants of this type may include process models which can be interpreted as computer instructions. The imperative style implies a prescription for the software design. In this approach only one state of reality is explicitly considered at a time. Static approaches were proposed and focused by the mid-1970s.

2) *Dynamic approaches*: These provide facilities for modeling the state transitions without considering the mechanisms that achieve them in full detail. For example, an event or operation can be specified by using a precondition and a postcondition. When a system state satisfies the precondition, the event can take place, or the operation can be performed. In the resulting state the postcondition is true. In this approach, two states are explicitly considered at a time, i.e., the prestate and the poststate. Dynamic approaches began during the late 1970s.

3) *Temporal approaches*: These allow the specification of time dependent constraints such as "age must not decrease". In general, sequences of states are explicitly considered in this type of approach. Temporal approaches began in the 1980s.

4) *Full-time perspective approaches*: These emphasize the important role and particular treatment of time in modeling. A full-time perspective approach eliminates notions such as states, operations, processes, transactions, etc. The number of states that are explicitly considered at a time is infinite. This approach was also introduced in the 1980s [OLIV82].

1.2.4 Desirable Features of a System Specification Model

The roles of conceptual models as they have been described earlier in this chapter suggest a number of desirable features that an information systems specification model should have. The model should be easy to understand, powerful in expression, processing independent, checkable and testable, and easy to change.

A specification should be easy to understand

A conceptual model is used as a common reference framework for communication between the user community and the design team. Therefore, comprehensibility should be the most important feature of a conceptual model. A conceptual model is easy to understand if:

- It uses user-oriented concepts and constructs,
- Both informal and formal descriptions are readable,
- It is unambiguous,
- It has high clarity, and
- It is intuitive.

First, in order to be understood by the users, the conceptual model should describe things in terms of the user's language. This suggests that concepts and constructs of a modeling approach should be as user-oriented as possible.

The model description should be easy to read. A conceptual model can be described formally or informally and requires that both descriptions are readable. Usually, highly informal language employs more natural language constructs and the readability is high. However, the possibility of ambiguity is also high.

In contrast, high formality usually implies less use of natural language constructs and has low readability. However, the clarity is high. Thus, unambiguity acts as a constraint upon informality. Unambiguity is required; otherwise, no-one will be able to understand properly what is expressed in the description.

In addition to these "semantical" concerns, we need to consider other facets stemming from the "syntactical" realm, e.g., clarity and intuitivity. Clarity means that redundancy of expressions is allowed only for enhancing the understanding of the application. Intuitivity assumes some means of representation, e.g., diagrams, graphs, or tables for improving the understanding of the business.

The specification language must be powerful in its expressions

Power in expression includes three aspects: capability to express everything that needs to be expressed, good resolution in describing the "small" aspects of the real-world phenomena and support for modeling the temporal dimension.

Good resolution and unambiguity are different. Unambiguity guarantees that there is a unique interpretation. Resolution is concerned with the accuracy of the detailed descriptions. Resolution refers to the primitive constructs. Unambiguity refers to the way formality and informality are used.

Processing independence is desirable in a specification language

Before introducing the conceptual model, the designers have a large number of design alternatives. The use of a conceptual model should not cause drastic reduction of the design space. This implies that the conceptual model should be as free as possible from data processing considerations. Research in achieving data processing independence indicates that a declarative rather than a procedural approach is desired. Such an approach tends to specify what to do rather than how to do it.

Recent development has revealed that a complete separation of *what* and *how* is almost impossible. The reason is that most real world application problems are "wicked" problems, for which no complete formulation can be found. One of the properties of wicked problems is that the formulation of the problem is the solution to the problem and vice versa. Therefore, an evolutionary or prototyping approach is needed.

The specifications should be checkable and testable

It is desirable to check for the validity and consistency of a conceptual model; that is, to ensure that the model describes exactly the relevant part of the application and that the model is free from conflict. Moreover, it should be possible to test whether an implementation of the system satisfies the requirements described in the conceptual model.

Specifications should be easy to change

Incompleteness of specifications and ever-changing reality suggest that a conceptual model should be easy to change. This is required during model formulation, system development, and system maintenance phases. Changes are easy to make if a few pieces of the model need to be modified. Pieces can then be added or removed easily, and the model structure can be readjusted automatically.

1.3 Contemporary Changes in System Development Practices

In the past, information systems have mostly been developed in a "tailor-made" fashion. Companies have developed their own in-house systems from scratch. This practice has led to increasing maintenance burdens on their DP-departments, and to embarrassingly high information systems expenses.

Some of the changes that are taking place in system development practices may be characterized as follows:

- The software profession is slowly maturing from garage ventures into an industry.
- An appropriate management culture and methodology will evolve for DP-departments, as well as for software houses.
- Software development is becoming increasingly capital intensive.
- Software engineering environments will become a costly must.
- A number of prescriptive software engineering standards will emerge.
- Software management and maintenance will increasingly become major problem areas, and will necessitate the use of rigorous software design methods.
- The cost for a company to enter the professional software market will increase greatly.

There is currently a clear trend towards increased sharing of the costs of information systems software

Some years ago, when there were not so many software products available in the marketplace, the usual way of implementing a system was to program all of the needed software oneself. Each implementation tended to consist of special purpose software. This was very expensive because there were no economy-of-scale savings to be gained from this implementation philosophy. The maintenance costs and reliability costs, that is, the quality costs for each software component, had to be carried by one user system alone.

Systems are no longer constructed from scratch. They are increasingly assembled from available software components, which are interfaced and integrated into systems. Vendor-supplied application platforms will appear in the marketplace for more and more application areas.

Systems designers must have skills in all relevant areas of information systems engineering

An information system may be viewed as consisting of three layers (Fig. 1.1):

- the organization layer,
- the application software layer,
- the computer system layer.

Information system design consequently comprises organizational design, application software design, and computer system configuration, plus appropriate interface design between the layers.

An information system designer will have to participate in designing all three layers. Even if appropriate design skills relevant to all three levels are needed, specialization within the different levels should be expected. Some of the relevant skill areas are

- business analysis,
- socio-technical design,
- human-computer interface design,
- functional analysis,
- software design,
- database design,
- performance evaluation,
- communication system design,
- computer system configuration,
- system administration,
- project management, and so on.

A trend in information system development is towards increased reliance on higher-level software platforms

Some of the most important developments that will influence information system development techniques in the years to come are found in the realms of:

- application platforms,
- system development environments,
- integrated specification models.

An application platform contains basic software functions for a particular application domain, plus possibilities for the platform user to add on company-specific software. An application platform is a vendor-supplied "common system". It is intended to comprise 40–80 % of the software that is required in an information system within some particular application area. The remaining 20–60 % of the application software defines additional behavioral properties of those information systems that are based on the platform. The add-ons therefore define the competitive edge of the respective organizations. The basic idea of a common application platform is indeed an old one. The new development is that finally the idea seems to be able to fly.

Computer aided design environments for information systems engineering are becoming the preferred development platforms

A system development environment is a particular type of platform, which provides an infrastructure for the integration of available system development tools, beyond the conventional level of compilers and database systems. A most important component of system development platforms is the system encyclopedia (data dictionary, repository, specification database), which makes it possible to integrate the various application systems of a company, so that it may become possible to obtain control over the evolution of a company's software.

An integrated specification model makes it possible to formally express a system's properties on every level of abstraction, from the business policy level to the operational data processing level, such that the specifications may be formally massaged and analyzed. "Automatic programming" based on integrated systems specifications may soon be within reach.

Developments like those indicated above will contribute to promoting programming from being an art to becoming an engineering discipline. The preferred talents and skills of the programmer may be different in the future from what they have been until now.

1.4 System Development Activities

Let us distinguish between *the development object* and *its environment, the development team* and *its environment, the development project* and *the development methods*.

The development object is the information system that is being built. It might be a banking system, a payroll system, an air traffic control system, a telephone switching system, or any other computerized information processing system.

The system's environment is the environment in which the development object is being embedded. The environment of a banking system is the bank, its employees, its decision making procedures, and its customers. The environment of an air traffic control system is the airplanes, the controllers, the airport procedures, neighborhood airports and so on. The information system (i.e., the development object) and its environment are non-overlapping systems. The term *total system* is used to denote the information system together with its environment.

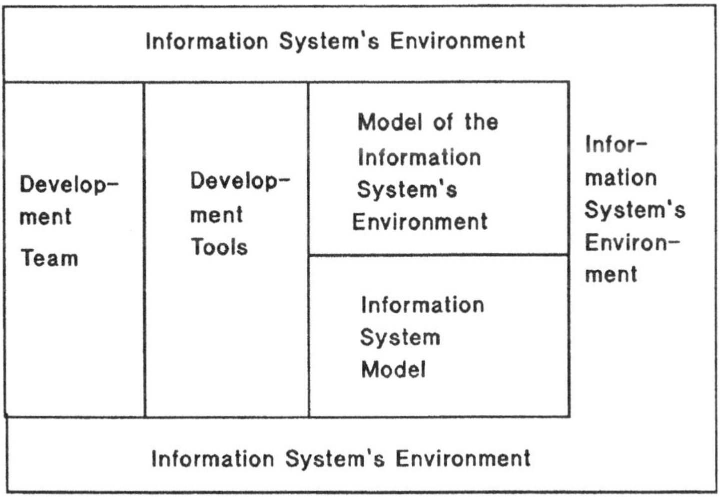

Fig. 1.7. Relations between developers and their environment

The development team consists of persons who build the development object. The size of the team differs with the size and complexity of the development object. The team must be organized, which is the responsibility of project management.

The development environment is the working environment of the development team. It includes the project organization, and its design and implementation procedures. It also includes the available development tools and design methods, the physical work conditions and so on.

A development project is usually divided into several project phases. Typical phases include: the problem definition phase, the design phase, and the implementation phase. Each phase usually consists of several development steps. Typical steps are: the man-machine interface design step, the database design step, and so on.

Development methods provide ways of solving systems problems. The methods differ depending on the nature of the problems being solved. In most branches of engineering the intermediate system products are specified in natural languages, tables, text, programming languages, drawings, diagrams, mathematical formulas, or in combinations of these main forms.

There is a complex interaction between the objects in these domains. Some of the relationships are indicated in Fig. 1.7. The development team is seen to interact with the development tools of the development environment, as well as with the organization at large. The information system (that is, the

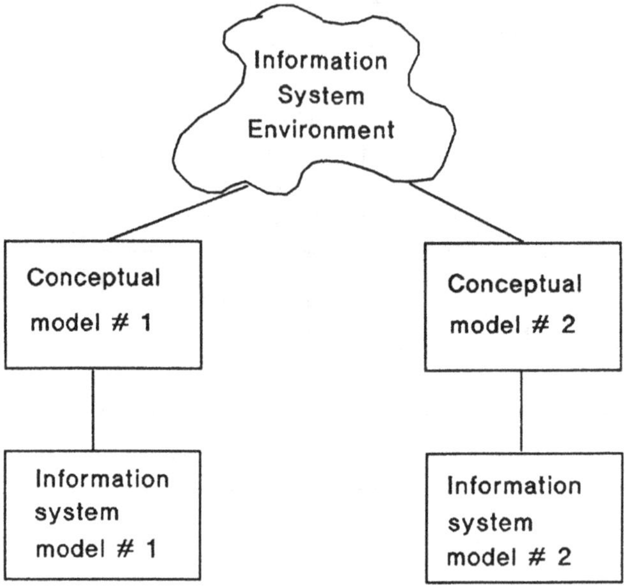

Fig. 1.8. Information systems models are expressed in terms of the conceptual models of the system's environment

development object) is related to its environment. The development team is seen to interact with the development object (the information system model and the environment model) through the development tools.

It is worthwhile to note that a model of the information system's environment must contain a definition of the concepts that are used for talking about the environment. These concepts are also used for specifying the information system model (Fig. 1.8). This means that it is impossible to check during the specification stage whether two information system models are equivalent (i.e., exhibit the same behavior) unless they have the same conceptual model.

Two independent development teams will most probably develop different models of a system and its environment. Unless the conceptual structures of these models are harmonized, it will be impossible to compare (in a formal sense) the information system specifications that the two teams develop. This simple argument shows the importance of conceptual modeling.

Project management controls the development process

The management of a development project must provide a structured and controllable path for transforming the development object from a conceptual to an operational state. Each project management method provides its own structure of phases and steps. The different methods are quite similar in the phase and step structures that are recommended. The methods differ on details rather than on principle. Usually, each step and/or phase has to result in some tangible system product, such as a problem definition document, a tested piece of software, or an approved proposal for a change in the database structure.

During a project period, the development object proceeds from a highly conceptual, abstract state to a concrete, well-defined, operational state. During the process, the development object undergoes a number of transformations, each transformation incorporating results of decisions made about future properties of the development object. As an example, consider the process of building a house. In the first project phase, the architect and the customer will work out the concept of the house. The tangible systems product is the architect's sketches of the house. In the next stage, the sketches are given to the civil engineering branch, where the engineering specifications are incorporated into the plans. Here, decisions about the future system (house) are made and the detailed design is determined. The tangible system product consists of detailed house drawings which are the basis for carpenters, plumbers, electricians, and other workers, who have the job of implementing the design.

Development methods in information systems engineering are heavily oriented towards software engineering problems

Information systems engineering has traditionally used natural languages, diagrams and programming languages for systems specification purposes. Diagrams have mostly been used for making quick sketches of data flows and data structures. The widespread use of mathematical formulas was not introduced until recently (1975–80), when programming languages based on mathematical logic, e.g. PROLOG, began to impact on software development methods.

When using software-oriented development methods, the development of an information system can be viewed as an orderly sequence of transformations of texts, starting with a natural language statement describing the purpose of the system and ending with a program text which can be interpreted by computers. Nowhere in the textual transformation process has there been any chance of formally deriving the properties of the information system. When designing a bridge it is customary at some point in the design process to perform some calculation to assure oneself and one's client that the bridge will not fall down in a severe storm. During the information system design phase, similar reasoning requires significant effort and cost because the systems specifications are mostly in the form of natural language.

Contemporary specification methods do not support quality assessment of information system specifications

When natural language is used for specifying system properties, the quality of the specifications can only be verified by inspecting the specifications. No wonder that so-called Quality Reviews are integral parts of every project management system for information systems design. The quality review meetings are absolutely necessary when contemporary systems specification languages are used; there is no other way of evaluating design quality. The situation can be improved only if we learn how to use modeling techniques in information system development. Such modeling techniques should be similar in spirit to modeling techniques that are used in other engineering disciplines, like electrical and civil engineering.

In the implementation phase, resources for the realization of the specified system have to be found and organized so that the systems requirements are satisfied

The major task of the implementation design is to assign appropriate organizational and operational resources in an orderly and reliable manner, so as to satisfy the stated requirements for information processing. The manner in which implementation design is done is of course dependent upon the availability of suitable resources. In information systems engineering the most important resources are people, computers, printing facilities, telecommunication systems, and various software packages, e.g. for text editing, or document processing. If appropriate software systems are not available, suitable software and associated databases have to be created. Sometimes it is possible to buy software which is close to what is needed. In such cases it must be determined whether the cost of necessary modifications in the system structure in order to use the available software will outweigh the costs of building special purpose software.

Lower costs may be achieved through software reuse

To lower implementation and maintenance costs it is necessary to learn how to utilize software components which may have been written for other purposes, that is, to reuse software components. To do this it must be easier than it is now to understand what a piece of software is really doing, what information processing it is supporting, which rules it is employing, how reliable it is, and so on. The current state of the art does not permit this to be easily done. It has never been easy to read a computer program and really understand what it is doing. Even if the program is exceptionally well documented, it is difficult to comprehend a program text, which takes time and much intellectual energy to read and understand.

We are far from solving the problems of building information systems through the integration of already available software components. It seems that a solution to this problem requires novel ways of programming so that comparisons can be more easily made between component properties and the required properties than can be achieved with contemporary implementation models. In the meantime, we have to live with the old programming languages and make the best of them.

1.5 The Methods Problem in Information Systems Engineering

The major motivation for looking at information systems engineering methods has been the desire to reduce the cost of producing and maintaining the software components of the systems. Additional motivations include: developing effective man-machine interfaces, developing reliable programs to support non-stop computer systems, and designing the software to effectively support its organizational environment. Although these motivations have become increasingly important, they have been secondary to the software cost problem; their influence on information systems engineering methods has been secondary to the need to lower software costs.

Contemporary information systems engineering methods are based on a project-oriented approach to system development

The prevailing view is that a problem analysis phase comes first, followed by phases for defining user requirements, and for doing the overall information systems design. The next phase comprises the derivation of the software requirements, which forms the basis for software design, eventually ending in the implementation, operation, and modification phases. Consequently, there must be tools and techniques for problem analysis, user requirements analysis, information systems design, software design, database design, implementation, and modification.

In large projects there must be effective control of costs and achievements. While costs are fairly easily measured, it is a lot harder to measure achievements. Project managers need to have documents that reflect the achievements in one phase before authorizing further work on subsequent phases. This need leads managers to wish that the development techniques produced complete specifications within each phase, so that it is easier to test if the work in one phase satisfies the previous phase. The techniques and tools should preferably have the property that what is produced by one tool in one phase can be used as input to the next tool in the next phase.

Contemporary information systems engineering methods fall short of satisfying the demanding requirement for methodological integration

The situation is characterized by a plethora of tools, techniques, and methods. They can be roughly grouped according to which project development phase they claim to support. Within the method groups there are many similarities among the various alternatives. Between groups there are neither many similarities, nor many connections. In practice this is manifested by the severe difficulties encountered when trying to use the systems specifications developed by one method as input to the methods supporting the succeeding development phase.

Contemporary information system development methods originate from administrative data processing. This was the first area where computers were applied on a large scale. Therefore it was the first computer application area where a need was felt for systematic methods in the development of application software. Contemporary methods are strongly influenced by this origin. For example, in some of the method proposals it is difficult to distinguish the issue of organization design from the issue of information systems design.

Every system development technique has a descriptive aspect as well as a prescriptive aspect

The descriptive aspect is related to the project management's need to know where the design is headed, and the need for easily understandable system specifications in the design team's environment. Project management needs are manifested in checklists of tasks that should be performed and in standard report forms for each predefined task. The ease-of-understanding needs are manifested in presentation methods that suppress system detail and simultaneously try to depict major structural and behavioral properties of the systems.

The prescriptive methods are directed at providing a technique to guide the analyst through the design of an information system. There are prescriptive methods for program design, as well as for database design. Some of the methods can be formalized to such an extent that they can be supported by computer tools. The most widely known methods for software structure design are based on dataflow analysis and data structure analysis. These two methods are often called "structured design" methods. Because they enjoy great popularity, they will be discussed in more detail in the next few chapters.

Many recommended development methods mainly address the descriptive aspects

Most of the early techniques were mainly aimed at giving support to project management, and were consequently driven by checklists and forms. A technique of this type involves a standard procedure that specifies in detail the sequence and the content of the steps to be followed by the designer during each stage of a project. There is usually no detailed recommendation for doing the individual project development step. The actual design work is left to the creative talent of the designer. Particular methods for the individual design steps may be amended by each company.

Approaches driven by checklists and forms provide a framework for system development which in principle is open-ended with respect to the detailed prescriptions for how to perform the individual steps. Most organizations have a standard that specifies the sequence and content of the system lifecycle to be used, and the reporting guidelines and milestones. In fact this approach is so common today that one would hesitate to call it a method, reserving that word for the prescriptive methods and techniques. Many of the project management oriented techniques rely heavily on standard documentation forms to guide the designer in collecting and analyzing data and in specifying design details.

The prescriptive aspects reflect the particularities of a development method

It is unavoidable that the appearance of individual forms is based on certain presumptions about the systems analysis and design techniques to be supported. Even if in principle, the project management-oriented techniques should be open for the application of various prescriptive methods, limitations concerning choice of method are frequently designed into the technique, as manifested in the structure of the forms that the individual technique recommends. Two well-known examples are data-structure- and dataflow-based approaches to software design.

The data-structure-based approach proposes detailed prescriptions for designing programs based on analysis of the structure of the input and output data. This is a problem oriented approach with the problem expressed in the mapping of the input data structure to the output data structure. The solution emerges from the application of the method. This approach assumes that the program function is implicit in the data-structures. The most widely known method of this type was conceived by Michael Jackson, known as Jackson Structured Programming, or abbreviated JSP [JACK75]. Other examples are proposed by Warnier [WARN76] and Orr [ORRK77].

In the dataflow-based approach, the program design emerges from a top-down decomposition of the dataflow transformations in an information system. The system's components are grouped into procedures and modules guided by prescriptive rules of thumb. The dataflow-oriented approach originated in the mid-1970s, and has been widely promoted over the years [STEV74], [YOUR75], [MYER75].

1.6 Information Systems Analysis Approaches

The purpose of information systems analysis is to identify the information flow within the organization and to determine the different information needs, ranging from operation of the production equipment to effectively implementing management decisions. The information system and the socio-technical organization which it supports are usually closely intertwined. Therefore, the information systems analyst must know the organization well in order to competently propose solutions that will serve the organization satisfactorily. We find that the need for acquiring knowledge about the information systems environment is reflected in the recommended analysis techniques.

In studying the information system there are several approaches [HART69]. They can be classified into methods for

- Departmental study
- Stimulus-response analysis
- Action-initiation analysis
- Object description
- Information needs analysis

Various information systems analysis techniques can be associated with one or more of these classes of methods.

Departmental study methods provide a complete understanding of the range of operations within the departments being investigated in the organization. An obvious guideline is to perform the study by adapting to the existing departmental structure. However, care must be taken in observing the manner in which information passes between departments. The information interfaces are not always obvious. Different parts of a task are frequently found to be performed by different departments. Persons working in one department often lack the systems overview that is necessary in seeing how their work fits in with the work in other departments.

Stimulus-response analysis methods recommend starting with a stimulus, or trigger, to which the organization must respond (e.g., a customer order), and following the information flow associated with the response until the train of action comes to an end. One must be cautious in examining the response to each kind of stimulus to which the organization is required to respond. There are internal as well as external stimuli to an organization and both types must be investigated.

Action-initiation analysis methods determine which information is required to initiate or control certain specific actions, e.g., preparation of a purchase order, delivery of goods, or landing an aircraft. Action-initiation methods are not complete methods for the study of complex information flows, but in combination with other methods they may be very useful.

Object-description methods concentrate on the analysis and description of properties and behavior of the objects that are produced and controlled by, or otherwise associated with, the organization. Examples of such objects are products manufactured, customers served, etc. This method is also not in itself a complete method for studying complex information systems. Nevertheless object-oriented specification methods can be used advantageously to support any type of systems analysis method.

Information-needs analysis methods start the analysis with the investigation of system outputs. First, the information needs of the organization must be established. This may be done by analyzing the operational needs, by some action-initiation analysis method. By starting with an analysis of decision processes, etc., one determines which outputs the information system should render to its environment. Next, by working backwards through information archive updating, intermediate processing, etc., to the creation of the information sources, an understanding of the structure of the information flow will be obtained.

All of these approaches are useful when used appropriately. None of the approaches are complete in the sense that they can be used successfully alone. The last three methods – action-initiation analysis, object description, and information-needs analysis – are partial methods only, because they require that the analyst already has acquired an overall understanding of the system. For example, a prerequisite for analyzing initiation conditions of actions is that the actions have already been defined. Similarly, objects must be defined prior to their detailed description, and the information users must be determined prior to an analysis of their information needs.

The departmental method and the stimulus-response method are best suited to the early systems analysis and design stage. The stimulus-response method stresses the relationship between processes and gives a dynamic picture of the information flow. The departmental method is excellent for making the systems analyst familiar with an existing system. For a more detailed analysis, each of the five approaches is suitable; and best results are obtained if they are applied in combination.

Chapter 2

Structured Analysis and Design

The Structured Analysis and Design method is the currently most widespread and popular approach to information system development. It is a simple method. It is easy to understand, and it captures most of the relevant features of conventional, transaction-oriented data processing.

There are two major project phases: Structured Analysis and Structured Design. The analysis phase is often seen as developing so-called functional requirements specifications. The design phase is seen as developing software and database solutions that satisfy the functional requirements, as well as non-functional requirements concerning, for example, performance and security.

Structured analysis and design is a process-oriented method, because the starting point of the information system development is in the analysis of the information processes of the system. These processes are described by the use of so-called Data Flow Diagrams (DFDs). From this starting point the other major systems components of data, rules, and program objects are discovered and designed. There are different description tools to be used for each of these major systems components.

The method is mainly based on the use of diagrammatic tools. The structured analysis and design method was developed during the 1970s, before workstations and PCs were commonplace. The diagrammatic tools were designed for paper and pencil environments. These modest origins may be easily spotted in the otherwise powerful CASE tools which are used today for supporting this widespread method (CASE – Computer Aided Software Engineering).

In the original method there is no explicit design step for developing user interfaces. This can be interpreted as a clear indication of that user interfaces were not seen as very important parts of the design in the early 1970s. Other design problems took preference. The user interface design issue is today of much greater importance, and is supported by most CASE tools. User interfaces are shortly discussed in Sect. 2.3.

2.1 Structured Analysis

A functional specification is intended to describe everything the user sees of the software and the interaction between the software and its environment. The functional description must be given in a terminology that is familiar to the user. That is, the system's properties must be described by using the terminology of the particular computer application area being investigated. If the application area is banking, the functional specification must be phrased in bankers' terminology; if it is widget making then that particular widget-community's terminology must be used. The functional description should also refrain from describing the implementation details that the user does not see. That is the implementer's business. The implementation specification must necessarily be phrased in computer language and contain a number of additional details. These details are probably of little interest to the user and are also difficult to understand in the limited time the user is willing to spend on them.

Structured Analysis is one particular approach to abstract away implementation details and concentrate on describing those features of an information system that are particularly important to the user. The functional specification that results from applying this approach must also form an effective basis for the subsequent design and implementation of the software.

So the specification documents must make sense both to the users and to the software designers. Experience shows that this is a very difficult task that in the current state of the art is not performed satisfactorily.

The tools of Structured Analysis are:

- Data flow diagrams (DFD);
- Tools to describe processing rules;
- Data access diagrams.

These tools provide the systems analyst with the means to carry out a disciplined analysis, and provide the end-product of analysis – the functional specification. The objective of using these tools is the preparation of a functional specification that

- is well understood by the users,
- sets out the functional requirements of the system without dictating its implementation,
- expresses users' preferences and system trade-offs.

We will use an example to support the description of the Structured Analysis approach. The example is a variant of an example from [GANE79]. We will analyze a company that sells widgets, Widget Incorporated. The mode of operation is a standard one. The WINC company accepts orders from their customers by telephone, by mail, or directly through WINC salespersons. If WINC has the desired quantity in stock, they prepare a shipment to the customer and send it by mail when necessary. Sales over-the-counter are handled elsewhere and are not considered in this example. The WINC company buys its widgets directly from the large producers and also from dealers who serve the small producers. Purchase orders have to be prepared and the received goods inspected. For WINC to survive, payments need to be received on time so that the vendors can be paid. The systems analysis task is to prepare an effective information system for WINC.

2.1.1 The Process Aspect – Dataflow Diagrams

The modeling constructs of data flow diagrams are

- External entity: source or destination of data,
- Flow of data,
- Process that transforms flows of data,
- Store of data.

Concept	Gane & Sarson symbol	de Marco symbol
External entity	Double square	Square
Flow of data	Arrow →	Arrow →
Process	Rounded rectangle	Circle
Data store	Open-end rectangle	Parallell bars

Fig. 2.1. Two different standards for writing DFD symbols

There are two widely accepted standards for drawing the graphic symbols that represent these concepts (Fig. 2.1). One was developed by deMarco [DEMA78], and the other by Gane and Sarson [GANE79]. The two drawing techniques can be compared in Figs. 2.2 and 2.3.

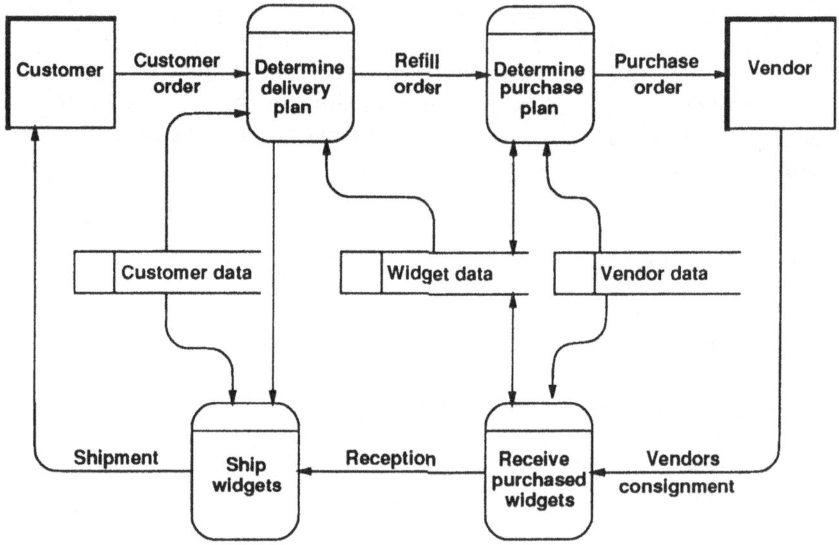

Fig. 2.2. Widgets Inc., dataflow overview, Gane and Sarson style

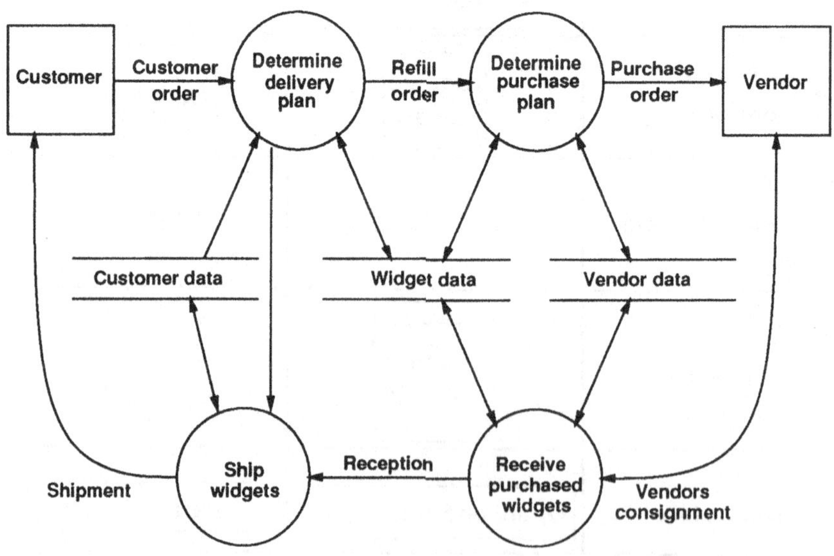

Fig. 2.3. Widgets Inc., dataflow overview, de Marco style

There is no difference between the two diagrams except in layout style. We prefer to use the Gane and Sarson drawing style primarily because it is awkward to write text within a deMarco circle. The text always seem to cross the border of the circle. Text within rectangles looks tidier.

Proponents of dataflow diagrams claim that one of the best, most valuable properties of the technique is that the control flow of the system is not shown in the diagrams. Traditional flowcharts, in contrast, show both dataflow and control flow. Unfortunately, dataflows and control flows in conventional flowcharts are so intertwined that they are difficult to separate. Consequently, implementation issues, as modeled by control flows, and functional issues, as modeled by dataflows, are inseparable in the flowcharts. This is the main reason that conventional flowcharts are not considered to be a good specification technique.

Dataflow diagramming explained through an example

In the dataflow diagramming technique the emphasis is on determining the dataflows in the system. Control flows have a second-order status. The rationale behind this is that one wants to direct the system designer's attention to the functional aspects of the information system, and to discourage the designer from getting involved with software and implementation issues from the start.

The weak point of the dataflow technique is that it is not possible to easily express dynamic system features, like user-system dialogues. Such dynamic interaction features can only be modeled if there are suitable modeling concepts available, like the control flow concept. The dataflow diagram technique falls short of being able to effectively model interactive systems, e.g. real time systems.

The DFD technique applies hierarchical decomposition of processes until the component processes are either easily understood by users and designers or are available as products, prototypes, or design proposals. Processes that are not otherwise available must be further specified in terms of detailed processing rules.

We will elaborate on the application of the dataflow diagramming technique by using the Widgets Inc. example. In Fig. 2.2 we show a proposal for a so-called "level 0" DFD of the WINC company. Already in this diagram we find a symbol that is not shown among the DFD symbols in Fig. 2.1.

The double-directed arrows related to the WIDGET_DATA data store indicate that some of the processes update the data store as well as use data from the data store. Local "DFD-dialects" can easily be created by making company-internal standards for drawing the diagrams.

In the DFD "level 0" proposal for WINC's dataflow, we have not considered the money flow. If money is not coming from customers, then WINC will soon be broke. If WINC is not paying its vendor, soon there will be no more widgets to sell. In both cases WINC will be out of business; and since we do not want that to happen, we propose an extension of the "level 0" DFD (Fig. 2.4) to incorporate the money-management functions.

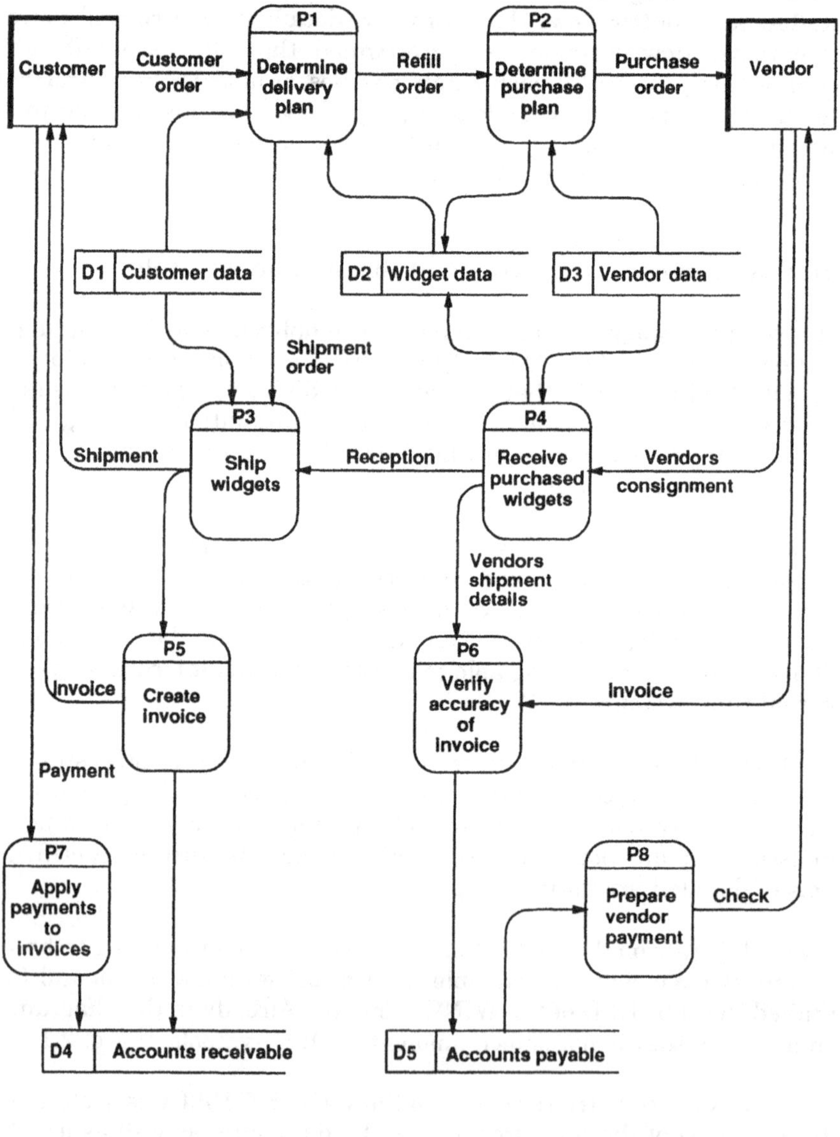

Fig. 2.4. Widgets Inc., completed level 0 DFD

Local DFD "dialects" may be designed to capture more specification details

It is useful to make room for attaching short identifiers to both process-symbols and data store-symbols (Fig. 2.5). It is also useful to be able to refer to the "processor" performing the process. This is usually done by indicating the name of the processor at the bottom of the process-symbol. In Fig. 2.5 it is indicated that process P23 Edit order is to be performed by the Sales Department, and that process P72 Verify Credit is to be performed by Program R27.

Fig. 2.5. DFD symbols, refinements, Gane and Sarson style

Dataflow diagrams support hierarchical decomposition

Each of the processes, and the data stores, in the level 0 DFD may be hierarchically decomposed. A part of the decomposed structure is depicted in Figs. 2.6–2.9. We have decomposed process P1 `Determine delivery plan` of level 0 (Fig. 2.4) into three processes: one for order verification, one for working out a delivery plan in case of conflicting demands on a limited supply of widgets, and one process for determining when to refill the widget supply (Fig 2.6).

Note that the data store D2 Widget data of level 0 (Fig. 2.4) has been decomposed into four component data stores

D2.1 Widget descriptions
D2.2 Inventory status
D2.3 Back orders
D2.4 Purchase orders

which appear in different parts of the decomposition structure. We have furthermore decomposed the level 2 process P1.1 `Verify order is valid` into three processes: `Accept order`, `Edit order`, and `Verify Credit` (Fig. 2.7).

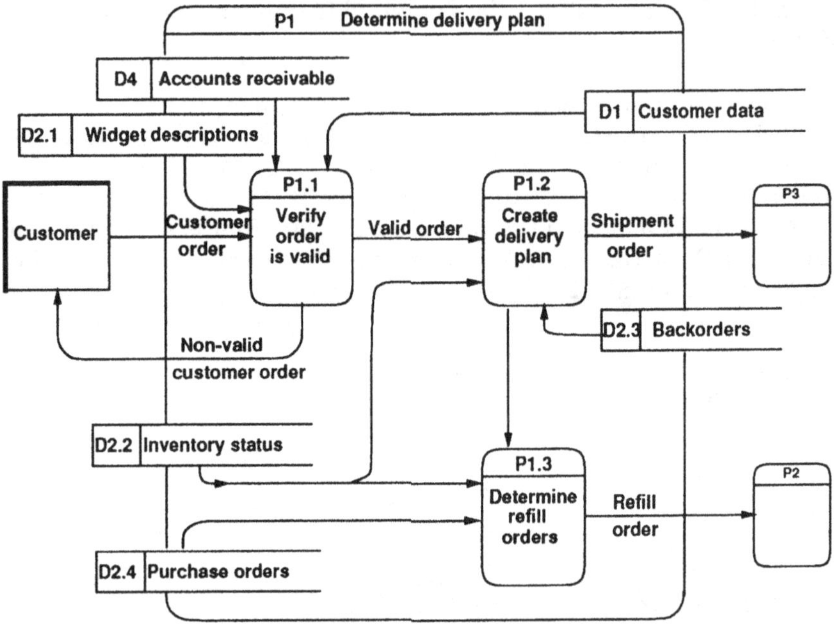

Fig. 2.6. Widgets Inc., level 1, DFD for P1 `Determine delivery plan`

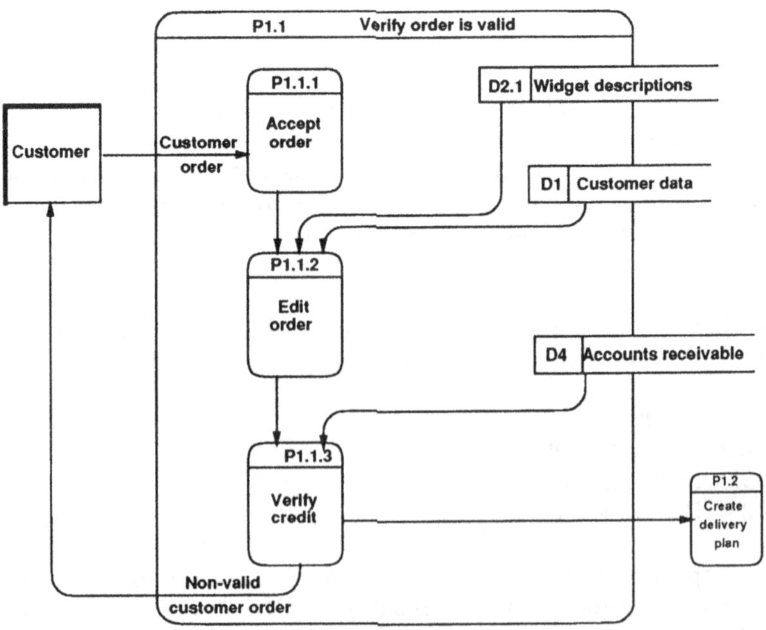

Fig. 2.7. Widgets Inc., level 2, DFD for P1.1 `Determine delivery plan`

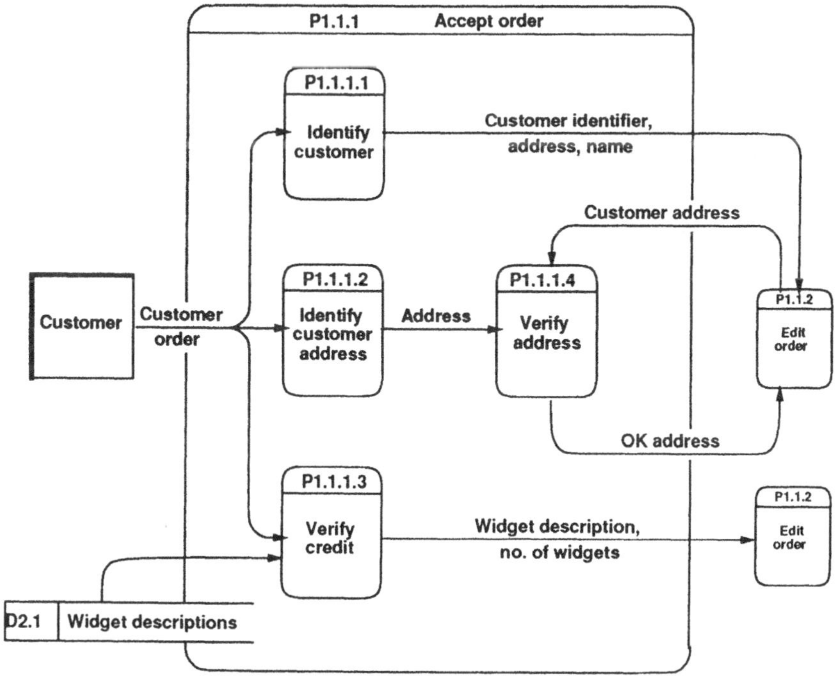

Fig. 2.8. Widgets Inc., level 3, DFD for P1.1.1 `Accept order`

Through the decomposition process we are gradually approaching a detailed level of specification. An even more detailed level is reached in Figs. 2.8 and 2.9, where `Accept order` and `Edit order` have been decomposed. It is not advisable to decompose further from this level of detail by using DFDs. Different specification tools should be chosen for expressing the detailed rules in the next step of refinement, that lead us to the computer program level of detail.

We see that by analyzing process P1 `Determine delivery plan` we have introduced changes on the process' borders, relative to the starting point as given by the level 0 diagram of Fig. 2.4. We have introduced the D4 data store Accounts receivable to be accessed by process P1. We consequently have to correct the level 0 DFD in order to keep internal consistency in the specification. The result is shown in Fig. 2.10. Note that D4 appears twice in the DFD, in order to simplify the graph. That D4 appears twice in the DFD is indicated by the double vertical bar to the left in the datastore symbol.

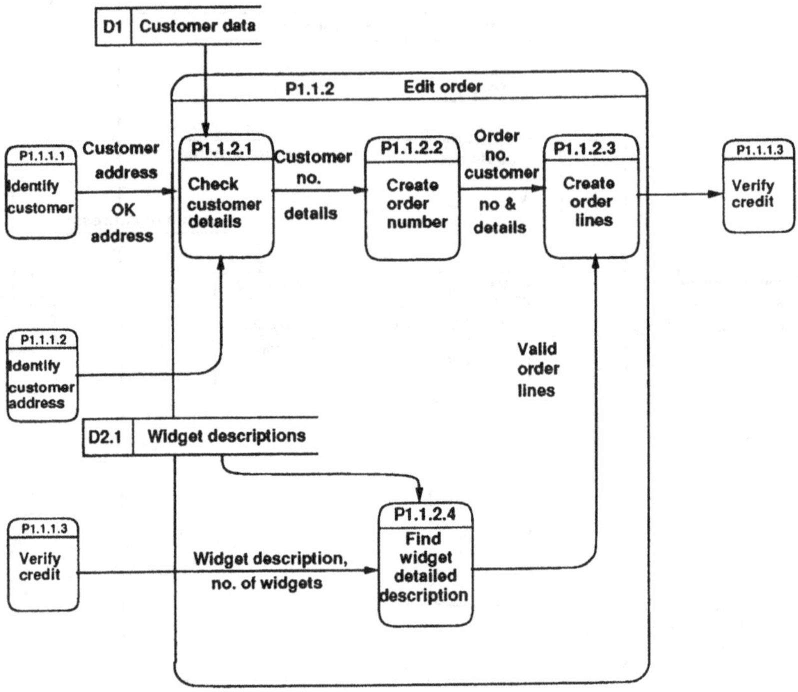

Fig. 2.9. Widgets Inc., level 3, DFD for P1.1.2 `Edit order`

This exercise in DFD decomposition shows the technique's strengths as well as its weaknesses.

The strong points are:

- The DFDs give the reader a good conceptual understanding of the problem.
- The graphic notation is simple and easy to learn.
- The concepts of dataflow and process are essential for understanding the function of an information system.

The weaknesses are:

- DFDs are clumsy when used for detailed specifications on the implementation level.
- DFDs do not show all of the essential specification detail, e.g. control flow.
- It is difficult to know when a problem has been decomposed enough.
- DFDs have weak modeling concepts for representing user interaction problems.
- DFDs have weak modeling concepts for data descriptions, databases, and data access descriptions.

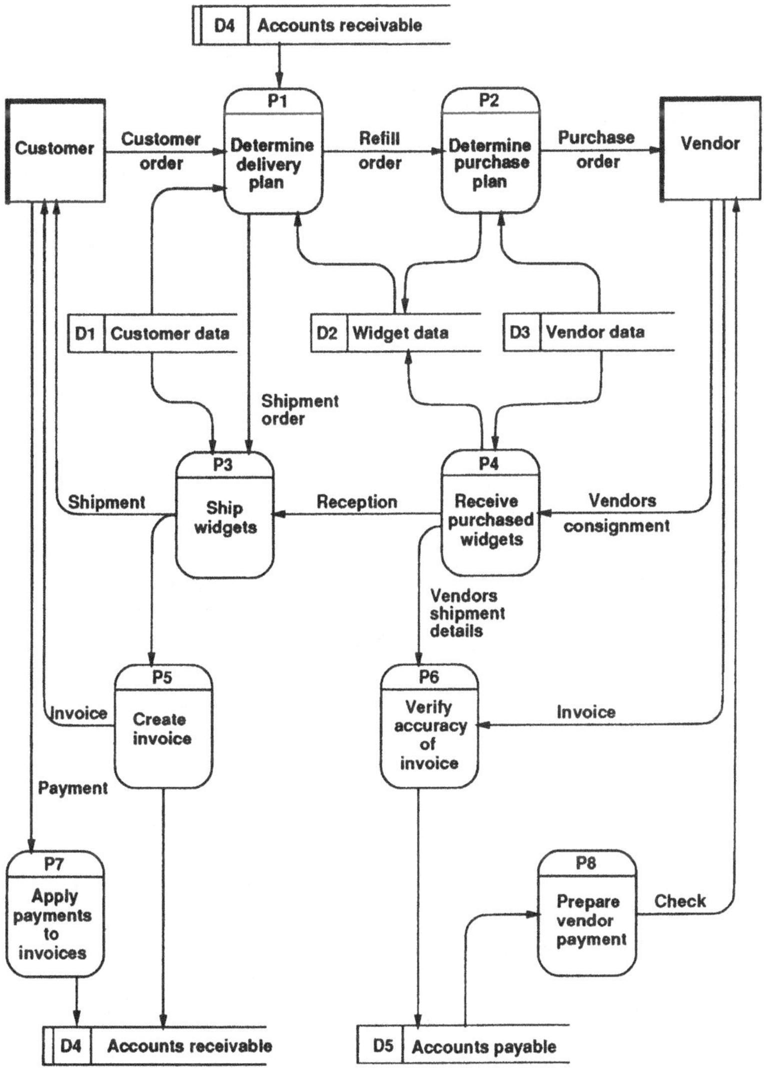

Fig. 2.10. Widgets Inc., revised level 0, DFD

In general, DFDs contain too few specificational details to make it possible to map them directly into completed software design. This means that it is necessary to add further design information to the DFDs to achieve a completed systems design.

2.1.2 Defining the Processing Rules

It was pointed out when introducing the data flow diagrams that the technique of process decomposition can only be used until we reach a level of specificational detail where the algorithmic relationships among data elements come into the focus of interest. This happens when rules have to be formulated for handling specific cases. For example, rules for verifying customers' credit have to determine exactly which treatment each customer shall receive. The customers' properties are represented by data elements in dataflows and data stores. So the rules have to be formulated as relationships among the relevant data elements. The rules have to be formulated as precise algorithms. The algorithms determine what is going on inside the processes of the dataflow diagrams.

Business rules are most often formulated in an imprecise fashion

Natural language is often used to formulate business rules, for obvious reasons. Natural language is easy to understand for humans, and is therefore effective in transmitting knowledge about policy and procedures. But natural language is not precise enough for a computer to follow. Ambiguities which humans would be able to resolve, because of their knowledge about the context in which the ambiguities occur, leave the computer utterly bewildered and useless. Ambiguities which may never be discovered in a non-computerized information system will leave a computerized system paralyzed and non-functioning. Rules may be of very different complexity. Consider the two cases described in the sequel.

Assume that Widgets Inc. are trading both with retailers as well as with ordinary customers. To keep the retailers happy the company must avoid competing for the same customers. Some discount policy must be worked out so that only high-volume customers are attracted to deal directly with the company. The discounts that are offered must not put the retailers in an uncomfortable competing position with the company. The discount policy may well say something like

> "Retailers discount is 35 %. For direct customers, 5 % discount will be given for orders of more than $200, 10 % on orders for $1000 or more, or if the order is for a quantity of more than 50 of only one widget type, 20 % will be given for orders of $5000 or more. An additional 10 % will be given on all orders for more than 100 of a single widget type."

The next example is taken from the tax laws:

> "When a company has a controlling owner interest (see § 3) in a building being used for industrial purposes (see § 4) of this company or another company, then the company that owns the controlling interest, for the accounting period (see § 5), shall be given a discount in the capital tax, calculated according to the capital that is invested in the building (§ 6)."

The latter example is considerably more complicated than the former. Only a few years ago the complexity of the latter rule was considered to be too high for giving it any practical formalization, so that computerized systems could be built that would reason routinely in such complex terms. Not so today. Advances in the application of mathematical logic for the formalization of complex rules have made possible a higher level of ambition than before.

Even if the rule of the first example seems very simple, this kind of policy rules can quickly get confusing. Exceptions and policy changes are usually dealt with by writing memos rather than rewriting the original policy document. The people who change the policies have to have the rules expressed in a way that they can readily understand. Computer language will not do for this purpose. More powerful tools are necessary.

Rule specification tools must support the human understanding of process logic

When rules are expressed in non-formalized languages they may well carry ambiguities that will not be discovered, even after close and thorough inspection. The analyst therefore needs tools that are both effective in detecting ambiguities, and effective in communicating ideas among the users' representatives, who are supposed to decide upon rules and business policy. If one looks closer at the first example above, it is easy to see that the level of precision of the rule leaves much to be desired. Do we give 10% discount for orders in excess of $1000 regardless of the widget type ordered? Do we give 10% for orders over $1000 as well as for orders of more than 50 of one widget type regardless of the money amount? Or, do we give 20% for orders of more than 50 of one widget type only if the ordered money amount is $5000 or more? There are plenty of possibilities. We see that we need methods for expressing the rules in a way which lends itself to formal analysis, in order to be able to pinpoint the ambiguities and improve the precision of the rule.

There are additional requirements on tools for specifying process logic. It should be possible to express the rules in such a way that they are easily understood by lay-people. It should also be possible to execute the rules directly by a computer. This cannot be done if the rules are expressed in natural language, just to mention one obvious example of a specification tool.

At the end of the day the rules will have to be processed by computers. If we are forced to express the rules in different languages, one for human understanding and one for computer understanding, we are bound to get into difficulties. If we are forced to keep the rules in two forms, we shall have to keep the two specifications consistent with each other. It is obvious that this will be both difficult and expensive. In order to avoid a translation from rule expressions in human-understandable language into computer-processible language, it would be preferable if humans and computers could easily understand the same language. This is not yet fully achievable by any known specification method.

There are several commonly used rule specification methods

The most common rule specification methods are *decision trees, decision tables, stylized natural languages, programming languages, diagrammatic languages, mathematical logic*. None of the rule specification methods are perfect, in the way that one method is superior to the others in every situation. Each of them have their stronger and weaker properties. The various methods will be explained in detail in later chapters. For the remainder of the discussion in this chapter it is sufficient to know that rules may be expressed in ways that are intelligible to humans and/or computers. One may therefore envisage the interior of the DFD-processes to be specified in the style of programming language, natural language or mathematical language. Any one will do.

However, to give the flavour of some of the methods already at this stage, we shall show a formulation of the same problem in natural language, decision tables, decision trees and stylized natural language. The example is taken from [VESS86]. It is a recipe for the preparation of vegetables:

> Vegetables that are both leafy and crispy should be fried, while those that are crispy but not leafy should be boiled. Prior to cooking, all vegetables that are not crispy should be chopped. Then, those that are green and hard should be boiled, while those that are hard but not green are steamed; those that are not hard are grilled.

This natural language formulation of the recipe example is reformulated in a stylized natural language form called structured English, in a decision table, and in a decision tree, as shown in Figs 2.11–2.13.

```
if crispy
   if leafy then fry
   otherwise boil
otherwise chop
   if hard
      if green then boil
      otherwise steam
   otherwise grill
```

Fig. 2.11. Recipe example in structured English formulation

	R_1	R_2	R_3	R_4	R_5
Crispy	Y	Y	N	N	N
Hard	—	—	Y	Y	N
Leafy	Y	N	—	—	—
Green	—	—	Y	N	—
Fry	X				
Chop			X	X	X
Boil		X	X		
Steam				X	
Grill					X

Legend:

Y – yes

N – no

— – indifferent

X – action marker

Columns are marked R_1, R_2...,R_5 and indicate rules (e.g **if** crispy **and** leafy **then do action** fry)

Fig. 2.12. Recipe example in decision table formulation

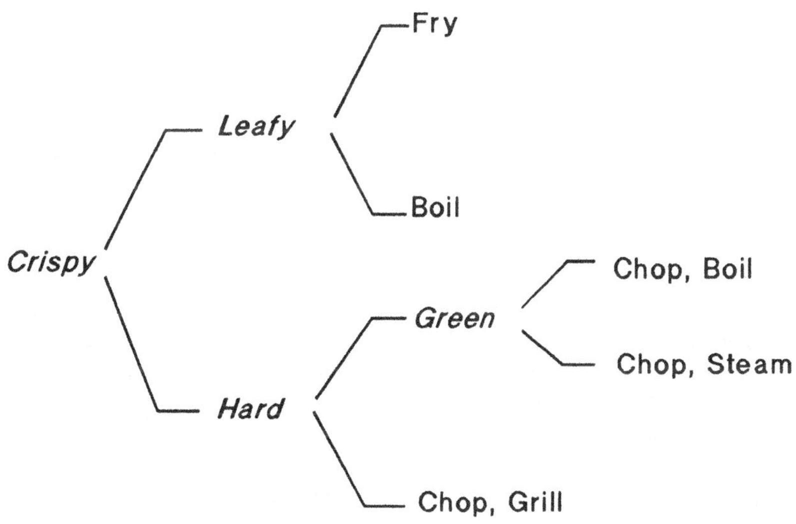

Fig. 2.13. Recipe example in decision tree formulation

2.1.3 The Data Aspect – Flows and Stores

We have so far shown how a detailed specification of the functional features of an information system can be worked out, even if the details of the data flows and the data stores have not yet been determined. This may be achieved because the names chosen to identify the data flows and the data stores usually have good *associative power*. For a name to have a good associative power means that all those persons who use the name will associate the name with the same system feature when exposed to the name in its appropriate context.

To rely on the associative power of the names of data flows and data stores is, of course, only possible as long as the amount of specificational detail is limited. As soon as system complexity grows beyond a limit, one can no longer rely on human association and human memory. In order to stay in control of the creation of new names, and the associations that those names induce in the minds of those who use the names, one has to be precise when describing the content of the data flows and data stores which are being named. Only then can one also start to be precise concerning the operations that are to be performed on the data flows. That is, only then can one precisely specify the algorithms of the processes that use and produce the data flows and the data stores.

Data flows may be thought of as pipes, down which parcels of data are sent

In a manual, non-computerized, information system, the data flows may very well be flows of paper-documents. The obvious way of specifying document data is to specify the document structure in a way that reflects how the data is seen by the users of the system. In most cases this would lead us to use conventional hierarchical data structures. There is no reason to make things more complicated than necessary so the use of hierarchical data structures for describing data flow contents may be readily recommended. After all, the main purpose of specifying the data flows is that the description is easy to understand. Therefore it is advisable to use a notation that is well known and easy to learn.

Let us exemplify this by describing the data flow "Customer order" which flows from Customer to P1 `Determine delivery plan`, see DFD level 0. Let us assume the following breakdown:

```
CUSTOMER_ORDER
   Order_identification
   Customer_details
   Widgets_ordered
```

We can explain these further, as follows:

```
CUSTOMER_ORDER
   Order_identification
      Order_date
      Customers_order_number  ....Optional, usually present
   Customer_details
      Customer_name
      Phone
      Shipping_address
      Billing_address
   Widgets_ordered    ..........This group of data items to be
      Widget_name               repeated one or more times
      Widget_number
      Quantity_ordered
```

The names that we choose for data items and other objects must follow some formal rule for their composition at this stage of the detailing of the system. Unless some naming standard is imposed on the analyst we shall end up with a plethora of names, each analyst following his own naming convention, and nobody being able to associate anything with names introduced by somebody else. A naming standard is a prerequisite for obtaining high associative power of the names, thereby making it easier to exchange ideas and systems descriptions among the interested parties.

All of the data flows that are candidates for any kind of formalization must be described at the level of detail which is indicated above. Note, however, that these data structures tell us nothing about format, layout data representation, editing characteristics and so on. The physical detail is omitted. Only the content is of interest at this stage of the system development process.

Defining the contents of data stores: what goes out must come in

The next task is to propose the content of the data stores. Luckily for us, it is quite easy to determine the preferred content of a data store provided that we know the contents of the data flows coming out of the store. For a data element to be extracted, it must be put into the data store in the first place. We can therefore examine the details of the data flows coming out from the store, and compare that with the details of the data flows going into the store, to check that all the necessary data elements are being stored and to see whether anything is being stored that never gets used.

Let us apply this approach to determining the content of data store D1 Customer data, in the Widgets Inc. example. From the data flow diagrams we see that there are two outcoming flows from D1, one to P1 `Determine delivery plan` and one to P3 `Ship widgets`. There are no ingoing flows to D1 Customer data. This in itself indicates that there are serious shortcomings in the specification. The future system has to have facilities both for accepting new customers, and for updating the company's data store with new knowledge about old customers. The ingoing data flows are usually determined in response to needs manifested by the outcoming flows. Therefore we shall do best to concentrate on the outcoming flows first.

Retrieval requirements on datastores come from the information processes

The data flows that are associated with D1 Customer data are depicted in Fig. 2.14, based on extracts from the DFDs of the previous sections of this chapter. We see immediately that we cannot determine all of the details of D1 until all of the processing details have been decided upon. The best we can do is to guess a first approximation of the content of D1, and modify that approximation as further analysis forces more decisions to be made about the properties of the information system.

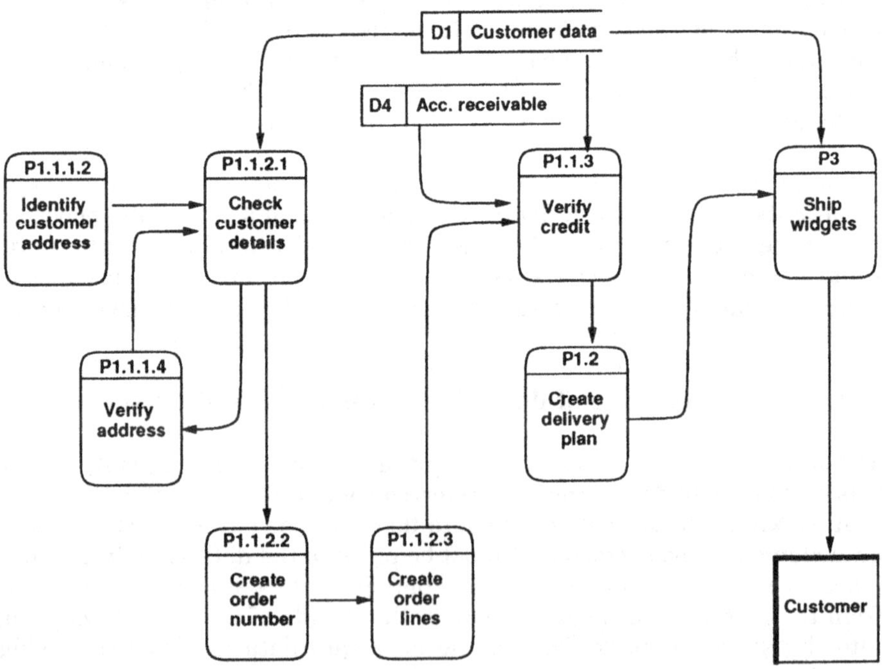

Fig. 2.14. Outcoming data flows from D1 CUSTOMER_DATA

Let us inspect the three processes. First, we see that process P1.1.2.1 **Check customer details** needs to check D1 for a match to the data coming from P1.1.1.2 **Identify customer**. We will need to have the data elements Customer_identifier, Customer_name, and Customer_address from D1 to P1.1.2.1 **Check customer details**.

Further, we must reason about process P1.1.3 **Verify credit**. The credit verification may be a very simple one, whereby it is only checked if the customers who have ordered new deliveries already owe us so much money for previous deliveries that we do not want to honor new orders. One may also envisage complex credit verification routines where individual customers' credit is determined relative to the size of their order, for example by running credit checks with outside agencies instead of collecting all of the necessary information in the company's own data stores. The point being made is that the content of the D1 data store cannot be determined unless policy decisions are made about how to run the company's credit verification.

One may give every customer a credit limit for a start, and increase or decrease this according to the willingness of the customer to pay his bills on time. This would be a very simple policy, indeed, and would only require that a credit limit be added to the content of the D1 data store. In addition, one must of course also specify the details of the processes which change the credit limits for the customers. Let us assume for the present that we need to have Customer_identifier and Credit_limit from the D1 datastore to the process P1.1.3 **Verify credit**.

Furthermore, the requirements from process P3 **Ship widgets** may add new data elements into the data store. This process must be analyzed in much greater detail before its effect on the content of D1 can be determined. To perform a detailed analysis of the shipment of widgets only to find that a lack of policy decisions will prevent us from determining which additional data elements that D1 should contain, is probably to go too far for the purposes of this little exercise. On the other hand, it seems quite possible that there will be some use for the data elements which were listed in the Customer_details group of Customer_order. Especially it seems to be a good idea to let the customer have two addresses, one billing address and one shipping address. We shall therefore be content at this stage to propose only that the data elements

Customer_identifier
Customer_name
Shipping_address
Billing_address
Phone
Telefax

be retrieved from D1 by P3 **Ship widgets**.

Requirements on datastore content may uncover a need for having additional information processes

Now we know that D1 CUSTOMER_DATA must at least contain the data which is necessary to satisfy the requirements from the three DFD-processes that have been inspected above. We shall therefore propose that the first approximation for the content of D1 is

CUSTOMER_DATA
  Customer_identifier
  Customer_name
  Customer_address
  Shipping_address
  Billing_address
  Phone
  Telefax
  Credit_limit

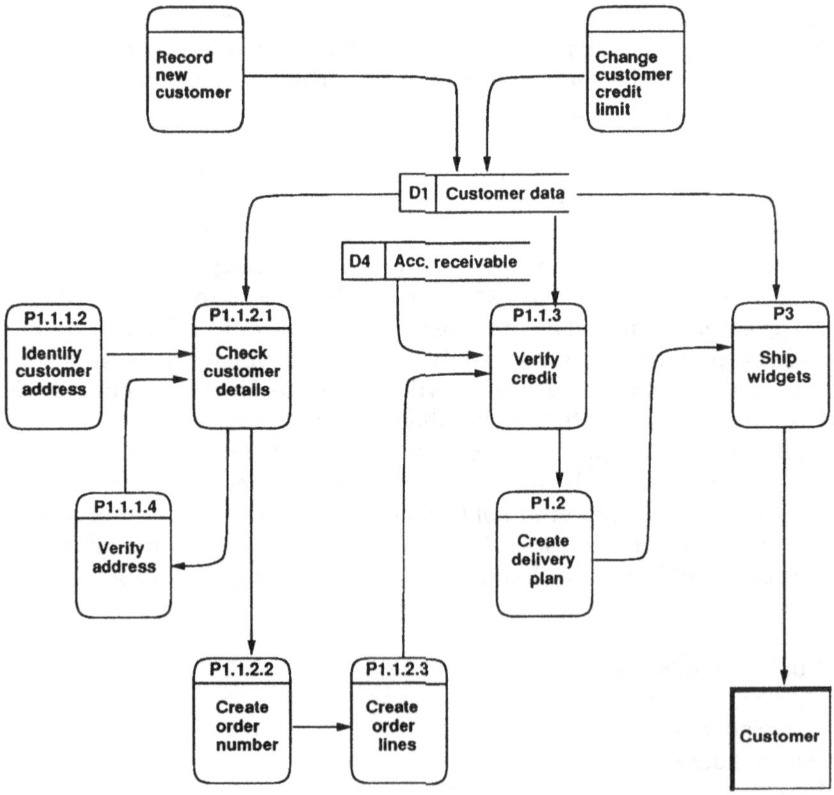

Fig. 2.15. Ingoing and outgoing data flows from D1 CUSTOMER_DATA

with the understanding that neither P3 `Ship widgets` nor the other relevant processes have been properly analyzed. Even this modest proposal requires that the system be modified with additional processes for introducing new customer data into the data store, and for modifying the customer data which has already been stored. System modifications are indicated in Fig. 2.15.

We have not tried to modify the data flow diagrams that have already been drawn up earlier with the exception of the revision of the level 0 diagram. The number of diagrams is already large enough for our limited purpose of introducing the techniques and tools of Structured Analysis. Without having access to suitable tools it takes too much time to modify the DFDs for every new system modification. The name of the game is continuous modification of the system specifications until everybody agrees on everything, and the system is finished and implemented. We shall discuss tools in a later section.

Overlap between data stores must be detected, and possibly removed

Since data can only get into a data store via some data flow, and cannot get out unless it has been put in, the contents of a data store can be determined by analyzing the specifications of the incoming and outgoing data flows, as has just been shown. This type of analysis should be done for all of the data stores which are introduced during the data flow analysis stage. We would thus get a first approximation of the contents of the data stores of the system.

When comparing the data stores one will usually find a great deal of overlap between their contents. This is because similar data may be needed for various purposes in different parts of the system and therefore be defined to be parts of local data stores. It would be too good to be true if the need for customer data only appeared in that part of the DFDs that had to do with customer orders. A much more common situation is that two or more analysts, who are not aware of each others' detailed work, all need some of the same data. Customer data may therefore appear in several data stores, as a result of the independent work of different persons.

Problems may also be introduced if the analysts in the course of their work choose different names for the same concepts, that is, if the analysts create synonyms. The same name may also be found to denote different concepts, when introduced in different parts of an information system, by different people. This is of course to be expected when people are free to choose their own local names for the new concepts that they introduce at various stages during the systems analysis.

The data store proposals have to be harmonized. Overlaps have to be removed. Differences in naming of concepts have to be resolved. This is not an easy process. Assume that one of the data stores contains a data element named Customer_address, and another one contains Billing_address and Shipping_address. Are these three addresses really different, or is Customer_address always one of the other two? One may easily think of much more complicated situations where the names chosen for similar concepts are so different that nobody will guess there is any relationship among them at first glance.

One technique that may be used for pinpointing overlaps between data stores is to first decompose each data store into independent chunks, each one having a unique key. Examples are

Customer_identifier, Customer_address
Customer_identifier, Customer_name
Customer_identifier, Billing_address
Customer_identifier, Phone

The next step is to inspect all of the "elementary" data stores and divide them into groups that share a common identifier. Further inspection of each group will most probably lead to most of the similarities and overlaps being identified and removed. This simple and straightforward analysis technique is quite effective, even if it is based on inspection alone.

Our purpose is to decide on the definition of a set of non-overlapping data stores that will be practical to work with when continuing the analysis. We may choose to group all of the elementary data stores that share a common identifier into one data store only, or we may choose to form several groups. For example, if we choose to form only one data store from the four elementary stores above, we get

Customer_identifier
Customer_name
Customer_address
Billing_address
Phone

We may alternatively form two data stores with contents

Customer_identifier	and	Customer_identifier
Customer_name		Billing_address
Customer_address		Phone

The choice between the different alternatives has to be taken on entirely pragmatic grounds, based on an educated guess on which of the data store structures will lead to the simplest systems description.

2.1.4 Resolution of Data Access Conflicts

Much harder than specifying contents is deciding what immediate access is needed to each data store and to provide a storage structure for the data that satisfy the access requirements. The meaning of "immediate" in this context is that data must be put into or extracted from the data store so fast that one cannot wait for the entire data store to be passed (and possibly sorted) in order to produce the required results. If the needed response is faster than we can pass or sort the data store, the designer will have to create some type of index or some other structure to provide access to the data store with the desired search arguments.

A recommended approach [GANE79] is to first analyze the access requirements to each individual datastore, and subsequently create a structure for the datastore that resolves access conflicts. The next step is to merge all of the individual datastores into one data map. Data accesses that involve more than one datastore are analyzed, and structures are created in order to resolve access conflicts. Access frequencies are analyzed and added to the picture. The final result is a system-wide access map, which may serve as a first proposal for the database of the information system.

The various access needs are often in conflict with each other. The multiple access conflict is typical. An example is shown in Fig. 2.16. The data store CUSTOMER_DATA, as introduced in the previous section, is supposed to be represented as a sequential file. The key to the CUSTOMER_DATA file is Customer_identifier. The file is subject to access in three different ways.

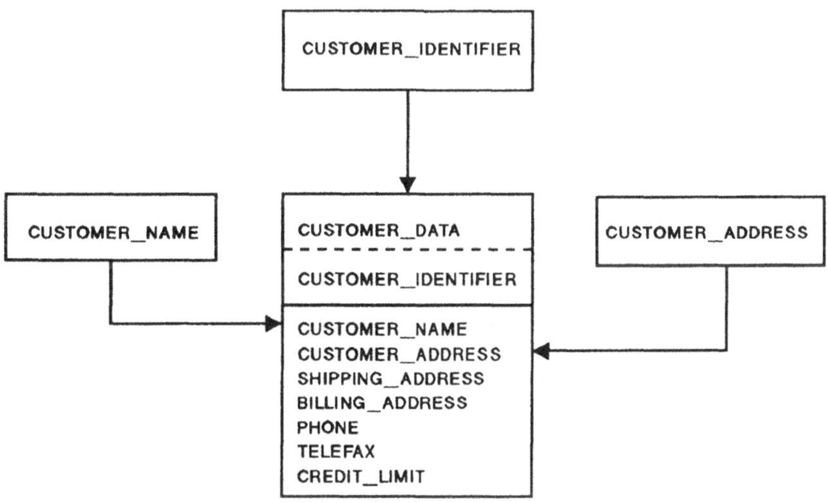

Fig. 2.16. Multiple access conflict of a data store

In the first case access is wanted to the customers' data provided that the customers' names are known. It is implicit in the situation that there may be several customers that have the same name, because Customer_name is not specified to be a unique identifier of the data store. In the second case somebody wants access to the data store through Customer_identifier which is the key to the file. This case is of course not problematic in itself. There is exactly one record in the file for every value of the key. Files are usually organized in increasing or decreasing order of their key. The third case is similar to the first case. Access is wanted in order to get data about customers when their Customer_address is known. In principle, several customers may share the same address, so we may retrieve more than one record from the file for each address.

There is no problem involved in treating each of these access requirements satisfactorily on an individual basis. The problem is, however, that not all of the queries can be answered immediately if the data store is to be kept as an independent sequential file. The file cannot be sorted on three different criteria at the same time. We have an access conflict. To resolve access conflicts is one of the main purposes of database design. There are several techniques and tools for achieving databases which effectively support multiple access. Some of them will be presented in a later chapter. What is of interest right now is to show how to gather information about the access needs of the various interest groups, and how to present this information in a consistent manner.

There are typically two kinds of access to the data stores. First, we have those where the content of the data that is going to be inserted or extracted from a data store can be predicted ahead of performing the actual data manipulation. Such accesses are related to the well-prepared processing of routine transactions. The other kind of access is related to the unprepared queries from users who want to extract data from the data stores in combinations which are not necessarily known beforehand.

In both cases the access needs must be expressed in such a way that they give sufficient information to the database designer, whose job it is to provide a structure for the database that will satisfy all of the access needs. A consequence of this is that the unprepared queries must also be prepared, so that the limits for the formulation of queries are clearly stated. Unless this is done, it will be impossible to prepare a physical structure for the database which can accommodate "unprepared" queries. After all, one can only expect to have an answer for those queries that request data which are present in the database. One can also expect to have an answer within a reasonable time only if this has been prepared by setting up an appropriate access path for that type of query.

Data access may be specified in a data manipulation language, and possibly also in a graphical formalism

An access specification must define

- which data store is to be accessed,
- how the records in the data store are to be restricted,
- which data is to be retrieved or inserted.

It turns out that it is difficult to give an accurate access specification in graphical form, for example in a dataflow diagram. An example of a DFD specification [GANE79] is shown in Fig. 2.17.

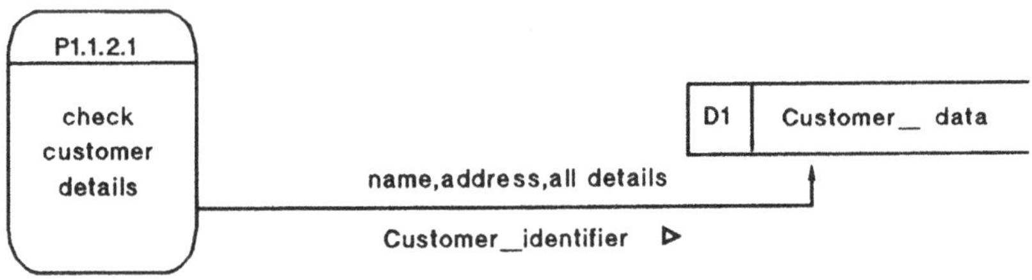

Fig. 2.17. Access specification in a dataflow diagram style

We see that the search argument, Customer_identifier, is marked with the special symbol ▷, pointing in the direction of the data store which is to be retrieved or updated. In this example we have only indicated informally which data elements are to be retrieved or updated. Such informal specifications are common, and have to be accepted in order not to overload the diagrams. If everything were spelt out in detail, then the diagrams would become so overloaded that they would hardly be readable.

The other alternative for access specification is to use a data manipulation language, for example SEQUEL. Every database management system is equipped with one Data Definition Language – a DDL, and one Data Manipulation Language – a DML. The languages reflect the particular data model that is used by the database management system. SEQUEL is particularly well suited for expressing manipulation on "flat" files, that is, on files whose records are non-hierarchical in the sense that they do not contain any repeating groups. A typical SEQUEL statement would look as follows:

```
select  Customer_name, Customer_address, Phone, Telefax
from    Customer_data
where   Customer_identifier = <some value>
```

We see that the SEQUEL statement contains the necessary access specification components, the data store: the **from** statement, the data that is searched for: the **select** statement, and the arguments that restrict the search: the **where** statement. To use an available data manipulation language for the data store access specification may also be rewarding because the specification then is directly computer executable. This means that one does not have to translate the access specification over into computer programs because the specification is already one.

The data access map provides an overview of how the different datastores relate on a system-wide scale

The users' needs for immediate access to data emerge over the whole period of study and systems analysis, as part of the iterative process of gathering information, proposing trial solutions, evaluating the proposals, developing new proposals and so on. There is a need for tools both for the detailed, formal specification of data store accesses, like SEQUEL, and for the overview-oriented presentation of how the access paths interplay and relate to each other. Methods for the latter will be introduced through an example which is based on [GANE79].

Assume that Widgets Inc. decides to expand the company's business by resorting to direct marketing of whole systems of widgets that have been tailored to the needs of selected market segments. A daughter company called WIDSYPE is established. WIDSYPE is an abbreviation for WIDget SYstems for the PEople. A sales force is hired. Standard widget system proposals are worked out. The sales force travel to the people and try to persuade them to buy as many and as expensive widget systems as possible, in this way doing their best to simultaneously keep the people happy and the company prosperous.

WIDSYPE salespersons contact prospective customers and offer them standard proposals for widget systems. The customers may place their orders with the salesperson, who reports the order to the company, where the order is assembled and shipped to the customer. All in all, a straightforward and quite common type of business. We have the very usual entities of salespersons, customers, orders and proposals. Each of them must be kept track of in the normal course of business, so that one may get information about outstanding orders, outstanding proposals, prospective customers, successful salespersons, and so on, as the needs of the business determine. The information system to be developed will keep track of the proposals, the orders, the deliveries, the accounts, and whatever more is needed in order to keep WIDSYPE operational and effective.

Let us assume that we have performed a data flow analysis of the type described earlier in the chapter, and that we have determined the need for four data stores CUSTOMER_DATA, SALESPERSON_DATA, PROPOSALS and ORDERS. Let us further assume that we have determined (a first approximation of) the content of the four data stores, as shown below.

```
CUSTOMER_DATA
  #Cust_no
    Cust_name
    Cust_address
    Phone
```

```
SALESPERSON_DATA
  #Emp_no
    Emp_name
    Emp_address
    Region
```

```
PROPOSALS
  #Proposal_no
    Cust_no
    Proposal_date
    System_type
  *Wid_system
    #Widget_no
      Quantity
      Unit_price
      Part_amount
    Money_value
    Order_date
```

```
ORDERS
  #Order_no
    Cust_no
    Order_date
    System_type
    *Order lines
      #Widget_no
        Quantity
        Unit_price
        Line_price
      Order_value
      Plan_del_date
      Actual_del_date
```

The keys are marked with # in front of the key-name. Repeating groups are marked with * in front of the group name.

Let us assume the following access requirements:

(1) Management wants to know orders and outstanding proposals for any customer and for any salesperson.

(2) One needs the ability to calculate the value of orders delivered in any specific future period, in order to be able to predict the future cash-flow situation.

First, we note that there does not appear to be any obvious way to access either ORDERS or PROPOSALS from SALESPERSON_DATA. The same observation is valid for CUSTOMER_DATA. We may cure this shortcoming by creating access paths from SALESPERSON_DATA and from CUSTOMER_DATA to ORDERS and PROPOSALS respectively (Fig. 2.18). This will take care of access requirement (1) above.

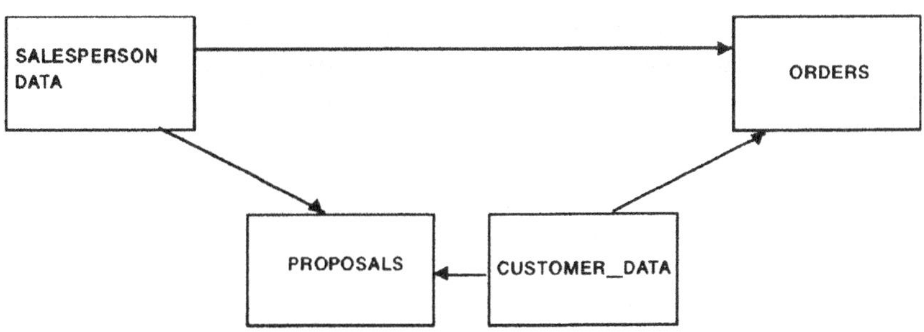

Fig. 2.18. A diagram indicating direct access paths

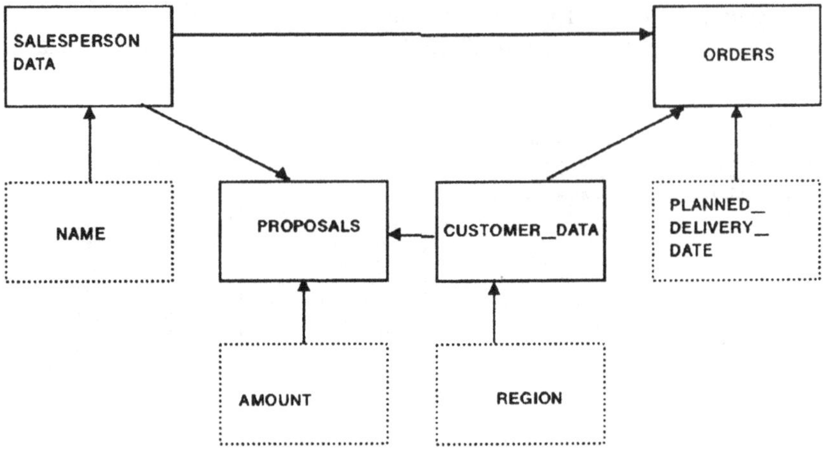

Fig. 2.19. Direct access paths with several secondary indices

Next, let us satisfy access requirement (2). We may get to the data that is needed by adding an extra so-called secondary index to ORDERS, namely PLANNED_DELIVERY_DATE as shown in Fig. 2.19. Secondary indices may be added on freely to represent additional access requirements. An example was shown earlier in this section (Fig. 2.16), whereby one could retrieve CUSTOMER_DATA from knowledge only of non-key attributes such as customers' names, addresses and so on. Let us add the following access requirements to our list:

(3) We want to find data about salespersons given just their names.
(4) We want to be able to retrieve data about customers in specific geographical regions.
(5) It should be possible to retrieve proposals based on money value.

The resulting secondary indices are added on in Fig. 2.19.

The data access map incorporates non-functional requirements, e.g., authorization restrictions on access to stored data, as well as numerical characterizations of access frequency

The direct access paths that have been specified so far are only part of the whole story. A complete specification must also state the identity of those who are authorized to read and to update the different parts of the database. Further, the maximum response times to be permitted for the various queries to the database must be stated in greater detail.

The question of security of data against theft, destruction or alteration has become more and more important as more of our data is entrusted to computerized systems. Confidential data in a data store may, unless we are careful, be read by unauthorized persons, leaving no trace of their theft.

Most commercial database management systems offer some provision for protection against unauthorized access to data. As usual, the better the facilities the higher the price. Consequently, the analyst should establish the data security requirements and the authorization system fairly early in the game, both in order to develop a basis for choosing an appropriate database management system, and in order to constrain the number of marginally necessary access paths at an early stage.

The various access requirements are usually in internal conflict, if the system is of some size. It may therefore not be possible to simultaneously satisfy all of the requirements completely. Design alternatives have to be evaluated with respect to how well they satisfy the access requirements. Each access path has to be supported by a physical data structure in the database. Each extra data structure that is added on for the support of an additional access path is associated with an extra operational cost for maintaining the appropriate links during database update. The extra cost has to be compared to the value of having available the extra, possibly marginal access path.

To enable the database designers to estimate the cost versus benefit ratio for the various access requirements, additional information is needed about the interaction between the processes and the data stores. Typical information required by the designers concerns access frequencies, estimates of acceptable response times for the various user transactions, and so on. The latter is usually fairly straightforward. The response times are either required to be short, that is, for interactive systems maximum a couple of seconds, or they may be in the order of minutes, hours or even days.

For the calculation of the accumulated effect of all of the user transactions on the data store access frequencies, one has to choose a reference period, say a day, and normalize all of the user transactions to this time period. Next, one has to estimate the number of logical accesses on each path in the data store structure. This estimate is called an access map, and is a most useful piece of information for the database designer. It is useful to have different access maps for read-only transactions and update/insert/delete transactions, because of their different effect on the database. It is also useful to distinguish between time-critical transactions with short response time requirements, and other non-critical transactions.

An access map is illustrated in Fig. 2.20. The access map was created by adding the following requirement to our previous list:

(6) We want to know, for each proposal over a given amount, the name of the salesperson for that proposal, and the names and credit limits of the customers of the orders that this salesperson has already placed.

For illustrative purposes the whole access path necessary for supporting the query in requirement (6) is plotted into the figure. This is of course not the

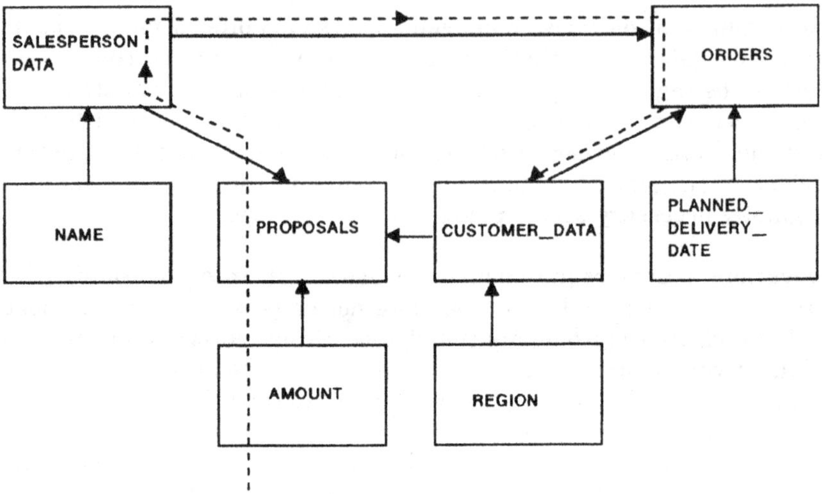

Fig. 2.20. Access map for WIDSYPE, with the access path for requirement (6) plotted in

usual way of presenting database access paths. The figures would soon look like a bird's nest if such practices were encouraged. Numerical characterizations of access frequencies have been suppressed for the same reason.

Please note that the access map does not necessarily reflect the way that the data are physically stored in the database. The access map gives a picture of the users' need for data from the database. The actual storage of the data has to be determined so that, for example, performance requirements are satisfied, as well as usage requirements. Database design issues are treated in Chap. 4.

2.2 Structured Design

The next design step is to create software structures which satisfy the functional requirements as they are expressed in the dataflow diagrams. Subsequent steps may include design changes to satisfy performance requirements and other non-functional requirements, for example, security and reliability requirements. We shall only discuss the functional design issues here.

The central issue is: How does a systems designer transform an information system model in a dataflow diagram into structures of programs and subprograms? What are the principles for packaging programs into implementation units? There are usually many different software solutions that may satisfy a set of functional requirements. Each of the solution alternatives will have different nonfunctional properties. Some will perform better than others, giving shorter response times. Some solutions may be easier to maintain than others. We shall discuss aspects of satisfying nonfunctional requirements in subsequent chapters. In this chapter we shall only treat the problem of creating one example of a subprogram structure that satisfies the functional requirements as stated in a dataflow diagram.

A design strategy called *transform analysis* provides the main link between the dataflow diagrams and the subprogram structure diagrams. There are also other techniques, e.g. transaction analysis, but transform analysis seems to be the recommended design technique. There is no particular database design method that is recommended in the literature on structured design. The structured design method should therefore be regarded as a particular way of designing subprogram structures, and must be supplemented with a database design method.

Starting out with a system description in the form of dataflow diagrams supplemented with a description of the logical database, as explained in Sect. 2.1.4, the recommended design steps are:

(1) Determine the systems architecture,
(2) Design the (physical) database,
(3) Choose subprogram structure,
(4) Refine the subprogram design,
(5) Package the subprograms into implementation units ("modules").

These design steps are not necessarily to be performed sequentially, in the sense that each step is to be completed prior to performing the next design step. It is perfectly reasonable to iterate over the design steps, and gradually proceed towards a full design by filling in design details as they are discovered.

The database design step will be treated later in the text, and will not be elaborated on here. The database design step more often than not leads to restructured data stores. Some of the original data stores may become decomposed into new ones, and others may be concatenated. Some of the original retrieval/update requirements may be found to be too expensive, and replaced by new ones. All of these changes should be incorporated in the dataflow diagrams, so that the subsequent program design is based on the most recent decisions concerning the system's properties.

The step of refining the subprogram design involves designing program objects that are easily maintainable and that satisfy the other nonfunctional requirements. Some of the design aspects are treated in the next chapter.

2.2.1 Software Structure Specification

The simplest form of connection between two subprograms is the call-relation. In structured design call-relations are represented in a graphical specification language. The pictures are called *structure charts* or *structure diagrams*. The term structured design is often abbreviated to SD.

The arrow in the graphic representation of Fig. 2.21 a represents a normal subroutine call where the direction of the arrow shows which program calls which. Note that there is no way of showing how many times A calls B, i.e., if B is called only once from A's code, or if there are several calls from different parts of A's code, etc. It is only stated that A calls B one or more times. However, it is possible in the graphic language to indicate that the B-call is not mandatory. The diamond symbol at the start of the call-arrow of Fig. 2.21 b indicates that it is possible that B sometimes is not called by A. So the interpretation of the diagram is that B is called none, one, or several times by A. The diamond symbol represents selection. The diagram in Fig. 2.22 shows that subprogram A calls either B, C, or D.

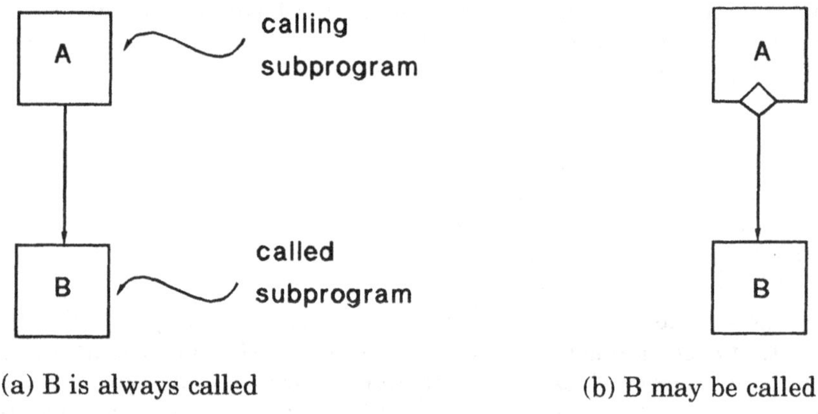

Fig. 2.21. Call relations between subprograms

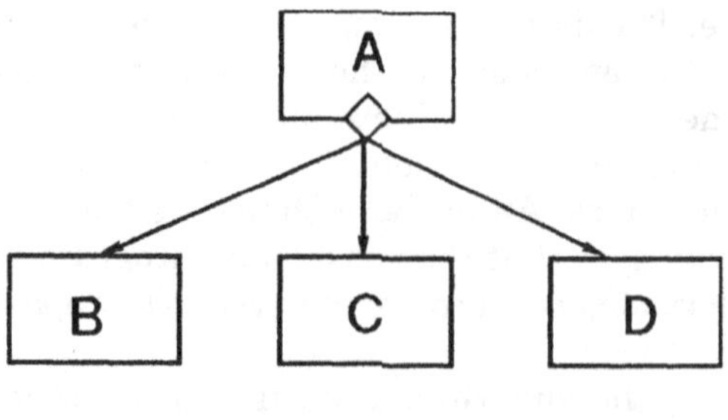

Fig. 2.22. The subprogram A calls either B, C, or D

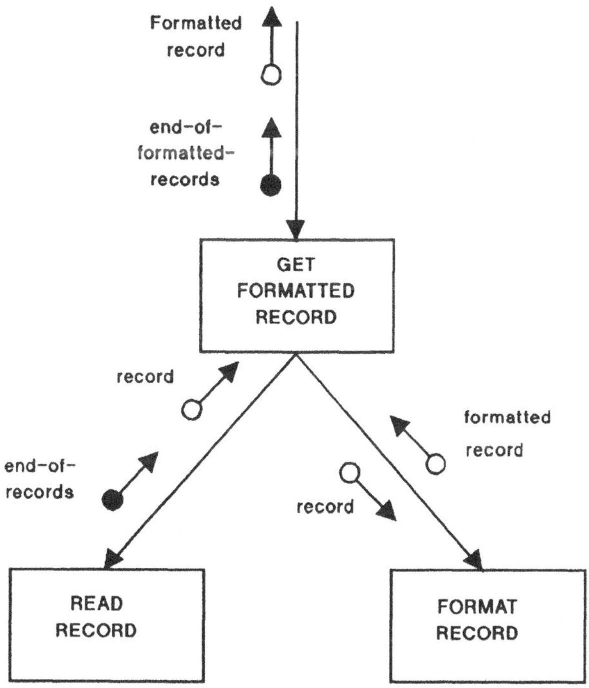

Fig. 2.23. Communication between subprograms

The communication between subprograms is indicated by putting small "communication" arrows with the call-arrows. The direction of a "communication" arrow is from sender to receiver (Fig. 2.23). The "communication" arrows indicate which information is transmitted between the subprograms. From Fig. 2.23 we read that the subprogram **GET FORMATTED RECORD** receives the data element record and the control element end-of-records after the **READ RECORD** subprogram is invoked. We can also see from the figure that when subprogram **FORMAT RECORD** is invoked, and given the record input data, it will return formatted record data to the invoking subprogram. It should be noted that there is no concept of sequence for the invocations of subprograms. It cannot be inferred from the diagram that **READ RECORD** is invoked prior to **FORMAT RECORD**. However, it is recommended that early in the design process, in order to indicate the most probable course of the call sequence, diagrams be made in which subprograms appear to the left of those that are invoked later in the sequence.

Transform-Centered Software Structures

Transform-centered design is appropriate for systems whose primary task it is to receive a flow of similar input transactions and turn them into an output flow of data, after having taken some action on each of the input

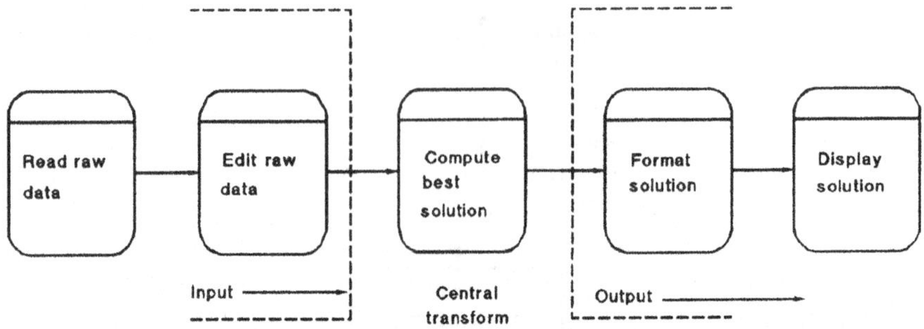

Fig. 2.24. Central transform of hypothetical dataflow

transactions, e.g. updating a file, or accumulating a sum. The recommended procedure [GANE77] is to follow the major data flows, from input to output, and determine where a flow changes character from being an input flow to becoming an output flow (Fig. 2.24).

One may also do the analysis starting from the output and working towards the input. Once this has been done, we can create a transform-centered design as shown in Fig. 2.25. Note that we try to develop a standard subprogram structures by adhering to standard design rules. If such standard structures can be found for the problem at hand, it is of very great advantage with respect to later program modifications, which are almost always bound to come some time in the future. If a standard program structure can be recognized in an otherwise unknown program, this will be of great help for the (unfortunate) maintenance programmer who is fighting his or her way through pages of unfamiliar code.

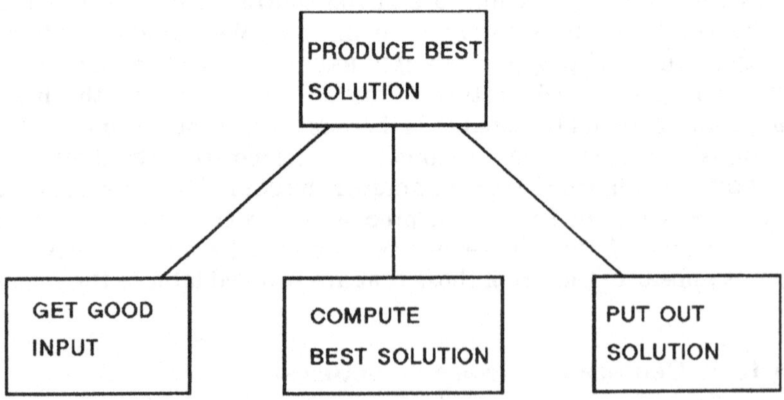

Fig. 2.25. The most abstract level of a transform-centered systems structure

Not all information systems conform to the central transform concept. For such systems there is little consolation to find in the literature which champions the structured design method. The only supplementary technique that is widely recommended in the literature is called *transaction analysis*. In transaction analysis the principle is to separate the various transactions by type and not by any common processing requirements. Therefore, a separate subprogram should be created for the processing of each type of transaction. The standard transaction-centered systems structure is discussed in Chap. 3.

After the different transactions have been separated from each other, each of them may be made subject to transform-centered design. The subprogram structures for the various transactions may in a next design step be examined for similarities to see if they can be assembled into one larger structure chart, that can be made flexible to modifications by using appropriate design techniques.

Transform analysis is a major strategy for converting a transform-centered DFD into a structure chart. Transform analysis consists of the following steps:

(i) Identify the central transforms of the system, and locate the main input and output dataflows (Fig. 2.26);

(ii) Define the top level main program of the system, which will control the system's overall processing;

(iii) Define one subprogram for each of the central transforms and each of the major dataflows, and put these immediately subordinate to the main program;

(iv) Factor all processes and dataflows in the input leg and the output leg, until all elements in the DFD are represented in the structure chart.

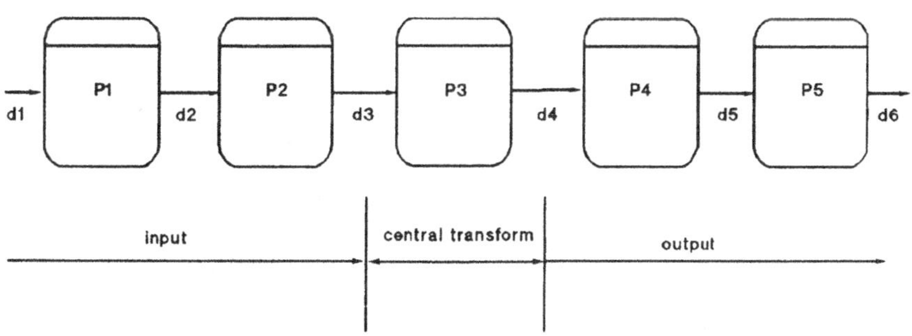

Fig. 2.26. Transform-centered DFD

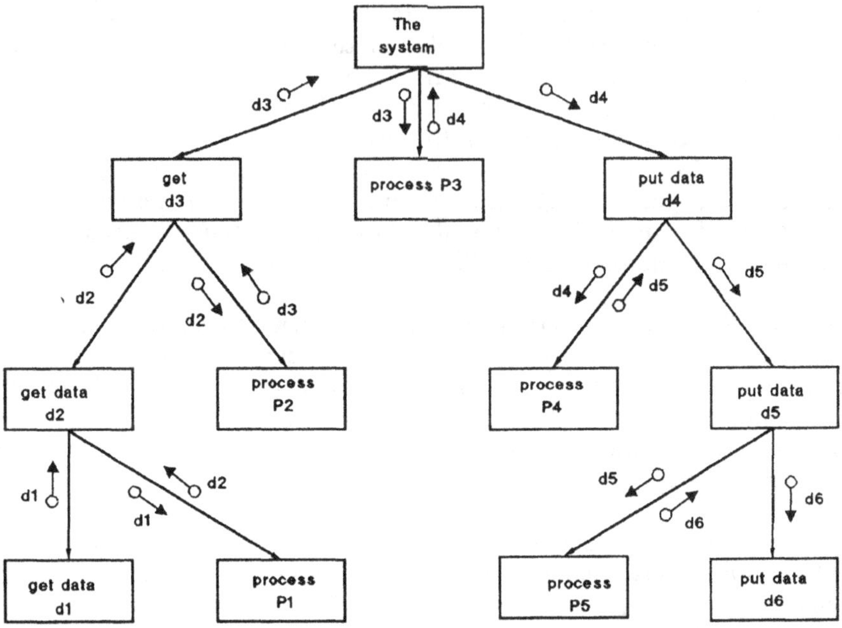

Fig. 2.27. Initial structure chart proposal

At the end of the transform analysis, an initial structure chart (Fig. 2.27) will have been produced which contains all the structural information of the DFD.

It should be pointed out that the transform analysis technique does not produce a polished final design, but merely provides an initial structure chart quickly which can then be modified and improved.

2.2.2 The System's Architecture

The first step of structured design is to determine the system's architecture. This includes to decide on the automation boundaries of the information system. An example is shown in Fig. 2.28. For the widget system that was analyzed previously, it is suggested that delivery planning is automated as well as the money handling (invoices and payments), but that purchasing and shipping of widgets are to be done by people.

The consequences for the proposal are that user interfaces must be designed for the altogether 11 dataflows that cross the automation boundary. The dataflows have to be formatted, and appropriate manual procedures have to be designed both for the preparation of the data input to the automated system, and for treating the data that is produced by the

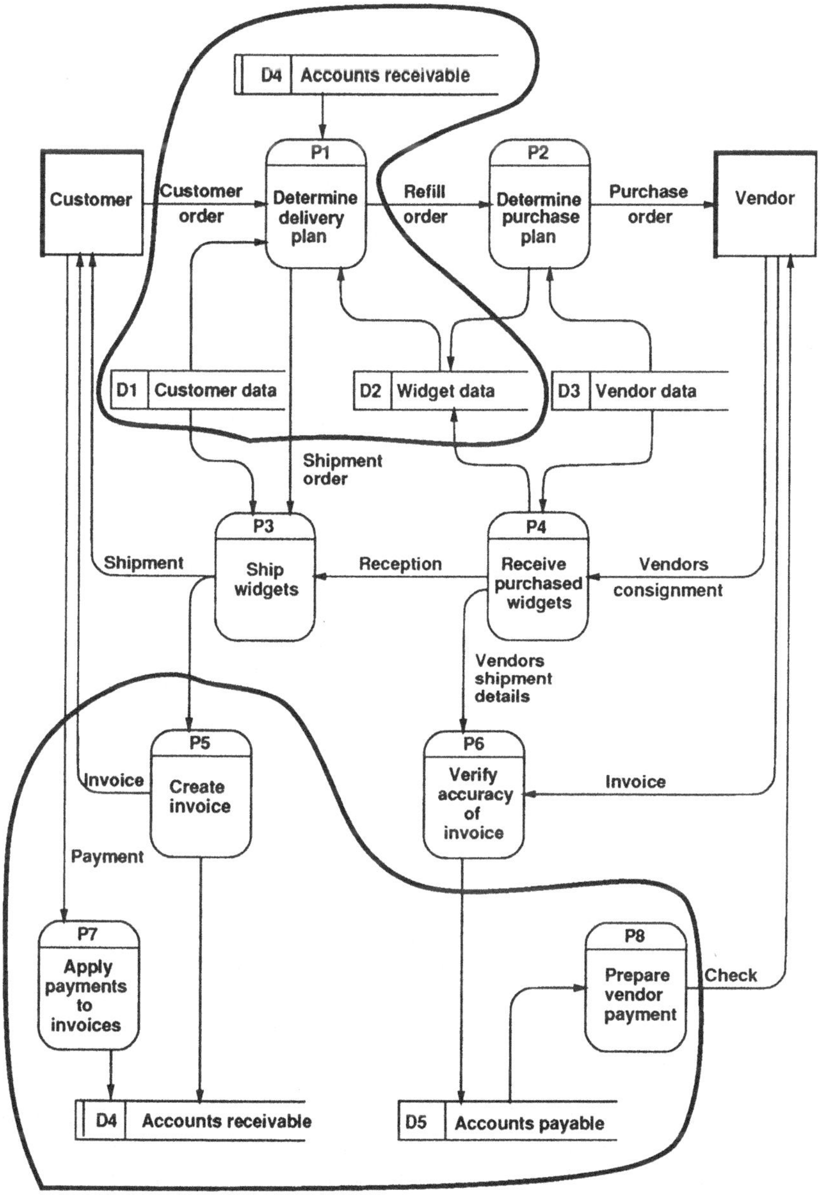

Fig. 2.28. Widgets Inc., automation boundaries

system and presented to its users. Further, the proposal includes the computerization of the three data stores Customer data (D1), Widget data (D2), Accounts receivable (D4), and Accounts payable (D5), while Vendor data (D3) is to be kept as an old fashioned paper archive.

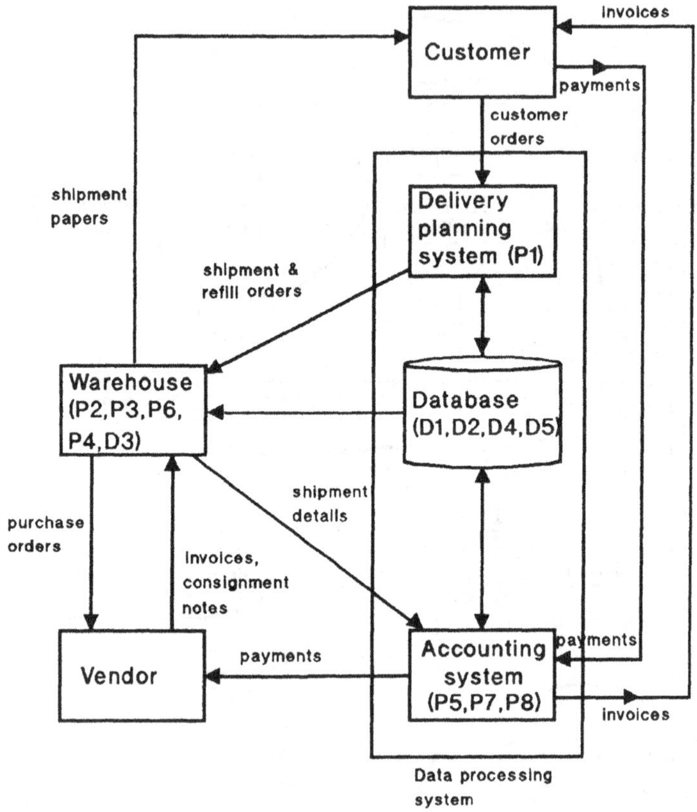

Fig. 2.29. System Architecture for Widgets Inc.

The automation boundaries determine the main features of the system's architecture (Fig. 2.29). In the automated data processing system we have decided on having two subsystems for `Delivery planning` and for `Accounting`. The non-automated parts of Widgets Inc. are to be performed by the organizational unit called Warehouse. It is indicated in the figure which DFD-processes and datastores are to be realized by which subsystem. The next step is to develop the major features of the software architecture. This has to be done already at this stage because it is necessary to evaluate the operational consequences of the proposed system boundary.

Two alternative solutions for the automation of the DFD-process P1 `Determine Delivery plan` are depicted in Figs. 2.30 and 2.31. A batch processing solution is indicated in Fig. 2.30. New customer orders are supposed to be manually entered, verified and stored in intermediate storage, as they arrive at the company. The validated orders are furthermore supposed to be periodically processed (in batches), rendering shipment orders, and orders for refilling the widget inventory whenever the inventory level drops below a predetermined level.

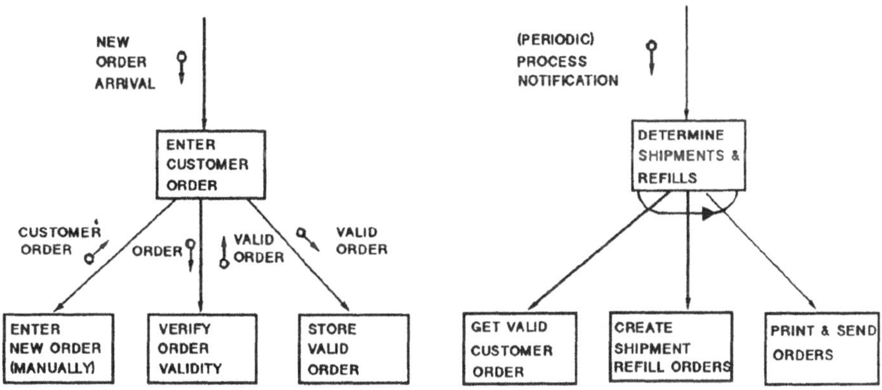

Fig. 2.30. Batch processing solution for P1 Determine delivery plan

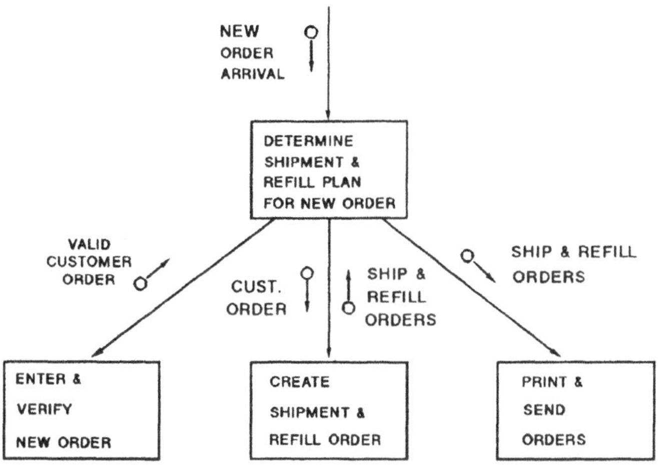

Fig. 2.31. On-line processing solution for P1 Determine delivery plan

An alternative on-line solution is depicted in Fig. 2.31. The software system is supposed to be activated whenever a new customer order arrives. All of the processing that is necessary in order to create a shipment order (for valid orders), and possibly to create a refill order, is to be done on an order-by-order basis, as indicated by the "loop"-symbol on the bottom side of **DETERMINE SHIPMENTS & REFILLS**.

The two alternative solutions have different operational properties. They will interact differently with the other automated parts of the information system. In both cases, printed shipment orders and printed refill orders are supposed to be sent to the respective information processing functions of the warehouse (Fig. 2.28 and 2.29, processes P2 and P3).

It can be inferred from the dataflow diagram of Fig 2.28 that most of the data in the shipment order will have to be re-entered into the computer, in order to produce the invoices of P5 `Create invoice`. The proposed solutions seem to hold great potential for improvement, to put it mildly.

The investigations and analysis that provide the basis for decisions about which functions shall be part of the data processing system, and which shall not, should be thorough and might therefore become quite costly. For each alternative a preliminary design has to be carried out, and development costs and operational costs have to be estimated, as well as the expected benefits.

From a technical point of view the important point is that whatever decision is made concerning the automation boundary, this is always the starting point for the data system design. The automated functions plus the manual functions must always add up to the same logical dataflow diagram for all of the various implementation alternatives.

The interaction between the data processing system and the manual system that follows from the determined automation boundary must be analyzed in detail. It may very well happen that the manual procedures that are found to be necessary will require additions and modifications both in the data stores and in the process structure of the dataflow diagrams that have already been developed for the information system. Furthermore, it may also happen that some of the process components of those processes that are to be automated cannot be performed satisfactorily by computers, but have to be performed by people. In such cases there will be a need for designing additional man-machine interfaces following a detailed analysis of the dataflow details within those processes.

Note that although we have been talking about only one computer system serving the functions within the automation boundary, there is no problem with defining many automation boundaries, each cluster of functions being served by different autonomous, interacting computer systems.

2.2.3 Refining the Subprogram Design

The dataflow diagrams usually do not contain enough detail to permit the direct design of a complete subprogram structure. Details most often have to be added. The subprogram structure will usually not mirror the relevant DFDs directly, e.g. for efficiency reasons. One therefore has to be careful when laying out the design of the subprogram structure, in order to ensure that the design is kept consistent with what has already been specified in the DFDs. An example of a subprogram design that is directly derived from its relevant DFDs is shown in Fig. 2.32.

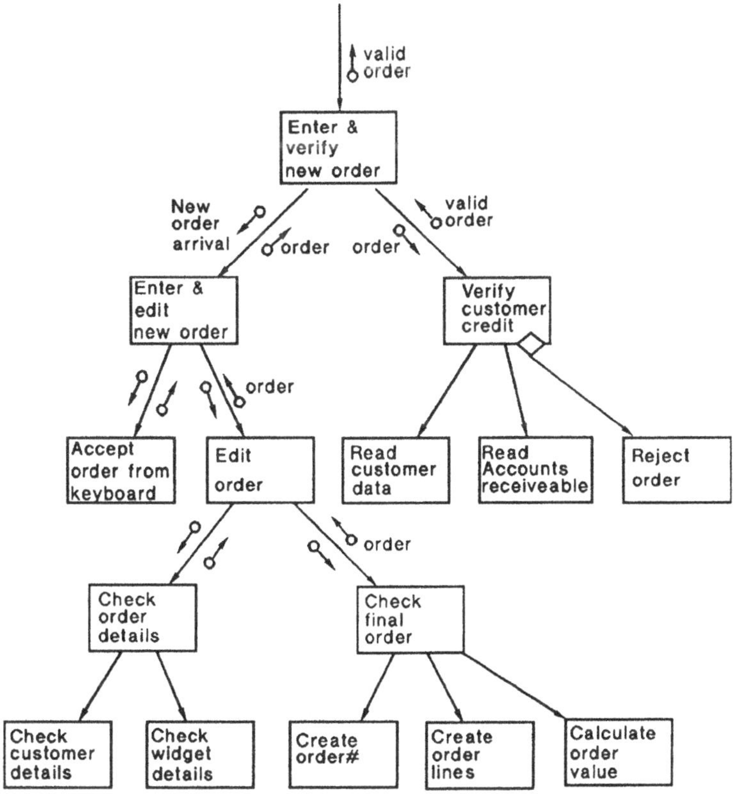

Fig. 2.32. Details of the "Enter and verify orders" of the on-line solution proposal for process P1 "Determine delivery plan"

The starting point for this subprogram design is the on-line solution for P1 `Determine delivery plan` of Fig. 2.31. Only one of the three subprograms, the "Enter and verify new order" subprogram, is being subjected to refinement. By comparing the subprogram structure of Fig. 2.32 with the DFDs of Figs. 2.7 – 2.9, we may see how the functional properties that are set down in the DFDs are supported by the chosen subprogram structure.

2.2.4 Packaging of Subprograms into Implementation Units

Basic to the structured design method is the idea that the development of a software system should proceed from a logical representation to a concrete implementation, but that the implementation issues should be postponed for as long as possible. We are now at the programming stage, and implementation issues can be postponed no longer. One central task when developing large systems is to develop a plan for packaging subprograms into implementation units, and to do this before programming starts.

Some of the most common implementation units are [PAGE81]:

- *Program*: a hierarchy of one or more subprograms,
- *Load unit*: a hierarchy of one or more programs whose calls are linked before execution,
- *Job step*: one main program with (optionally) a hierarchy of subprograms,
- *Job*: a sequence of one or more job steps,
- *Program system*: one or more application-related jobs.

The two major stages of packaging are [PAGE81]:

- By the end of the structured analysis stage, to package the system into separate jobs and job steps,

- By the end of the structured design stage, to package each job step into possible further job steps, load units and programs, in order to improve the operational characteristics of the program system.

The packaging at the end of the analysis stage is done in conjunction with the determining of the automation boundaries, as explained earlier in this chapter, and may be seen as an extension of this task. There are three types of physical boundaries that have to be drawn on the dataflow diagram, in addition to the automation boundary. They are [PAGE81]:

- Hardware boundaries, when different parts of the program system are to be implemented on different computers,
- Batch/online/real-time boundaries,
- Periodicity boundaries, when different programs are to be run at different intervals of time.

In this phase one should also consider the splitting of jobs into job steps. There are three considerations that are of special importance for this:

- Commercial software packages may be used to implement some of the functions of the system,

- Safety requirements may call for the introduction of otherwise unnecessary job steps and/or intermediate files,

- Computational resources are too restricted to permit the job to be executed all at once.

By the end of the structured design stage the time has come to package the subprograms into programs and load units. How to do this is dependent on the possibilities given by the chosen programming language. Of the general advice to be offered, the best is to repeat that one should strive to obtain high cohesion and low coupling on all levels of packaging.

The principles of object-oriented software may lead to a quite different packaging discipline (see Sect. 3.4).

2.3 User Interfaces

Most computer systems have an interface through which human users and computers interact. This interface is often the single most important factor in determining the success or failure of an information system. It is also one of the most expensive parts of a system. Yet, despite its importance, the user interface is one of the less well understood aspects of a system. Its success or failure is determined by a complex range of poorly understood and subtly interrelated issues, including whether the system is congenial or hostile, easy or difficult to learn and to use, responsive or sluggish, forgiving or intolerant to human error [BAEC87].

Effective interface development is a multidisciplinary process requiring a holistic view of the design problem. Skills in graphic and industrial design are appropriate, as well as a deep understanding of human cognitive, perceptual and physical skills. An understanding of organizational dynamics and processes is required, as well as knowledge about display technology, input devices, interaction techniques and design methods. Good interface designers should also have an aptitude for elegance in design. No single individual is likely to possess all of the required skills. Team work becomes a necessity.

Because of its complexity we can only give a short introduction to user interfaces. In the sequel we shall briefly discuss the relationship between the user and the usage of interactive systems, and we shall discuss some features of interaction styles and techniques.

2.3.1 The User and the Usage of Interactive Systems

In order to discuss users and usage of interactive systems we need a theoretical orientation to the modeling of the processes that take place during the interactions. *Information processing psychology* develops such models [LIND77], [GARD87]. It views man as an information processing system which is a component of a larger system consisting of human beings and physical entities. The human information processor is in turn viewed as consisting of four subsystems for sensing (perceptual), processing (cognitive), memorizing and responding (physical). Knowledge about the basic characteristics of these four aspects of human behavior is essential to the understanding of human-computer interaction and to interface design. To avoid misunderstanding we state that the information processing metaphor is used for convenience alone. We do not mean to imply that the brain's function is analogous to that of a computer.

Perception is intimately related to cognition

Perception and cognition are considered by many to be separate processes. The perceptual mechanisms are viewed as a kind of preprocessor that captures signals from the outside world and passes them on for cognitive processing. This view is simplistic and misleading. Perception is an active process that involves cognition. In viewing a scene, for example, there is a great deal of processing involved. One has to look, postulate hypothesis, and use real-world knowledge to interpret a scene. So it is, to a greater or lesser degree, in all perception. This fact has important design implications, especially as concerns audio and visual stimuli.

Cognitive resources are in short supply, and there is a high demand

Cognitive activities can be discussed in terms of the resources that they consume. The resource issue has to do with human memory, processing "cycles", and internal communication channels. Assume that a conversation is taking place in a noisy environment. One would expect that more resources are being used to support the intended communication than when the conversation is taking place in quiet surroundings. A good measure of the difficulty of a task is the number of resources it uses. This measure is known as the *cognitive load*, which relates directly to factors like learning time, fatigue, and stress.

There are two concepts from information processing that are of particular importance in understanding cognition. Those are the concepts of critical and limited resources [BAEC87]. Critical resources are resources that are required to perform a particular task. Limited resources are resources that are in short supply relative to their need. In cognition, as in operating systems, problems arise when critical resources are limited, and supply cannot meet the demand.

Problem solving resources should be directed to developing solutions for the primary tasks

In working with a computer there are two classes of problems that confront a user: operational and functional. Operational problems have to do with the means of performing work, which usually refers to a sequence of operations needed to accomplish a taste. Functional problems have to do with the content of that work. One objective of interface design should be to minimize operational problem solving, so that as many as possible of the available human resources can be used for the primary application for which the computer was adopted in the first place. The more cognitive resources that can be saved, say by making the perceptual processes more effective, the more resources are made available for solving the primary functional problems.

User interfaces must conform to the fact that users have limited knowledge about the system they are interacting with

Every system can be characterized by the skills that are required to utilize its full functionality. Users of complex systems usually do not possess the full range of appropriate skills. The typical situation is that users' understanding of a system is limited, both with respect to its operation and its function.

Take as an example a computer game flight simulator. The functionality of the simulator reflects the aerodynamic behavior of a small aircraft. The operation of the simulator is reflected by the commands that a user can give to the simulator. Most users will have a less than complete understanding of the relevant aerodynamics. Some will even have a false view of aircraft behavior, for example that the aircraft will immediately fall like a stone to the surface of the earth and crash if the engine of the aircraft suddenly stops. Similar discrepancies between system capability and users' knowledge is found for operational aspects as well as for functional aspects. Most users are inexperienced pilots and can be expected to have only rudimentary knowledge on the art of flying. So they will have difficulties in understanding and remembering the diversity of commands that the flight simulator will accept as well as having a less than desirable feeling for the aircraft's expected responses to various commands.

A user's model of a piece of reality is typically a subset of the system's model of the same reality. Furthermore a user will normally have a less than complete understanding of the information processing capabilities of an information system. Just think of an inexperienced user trying to find his way through a text processing system. A major objective of information system engineering is to ensure reliable operation of systems in spite of these model discrepancies. The user interface is central in this respect.

User interfaces should be designed so that they accelerate the process whereby novices begin to perform like experts

What characterize experts is that they are skilled in their particular fields of expertise. Unlike the attentive nature of problem solving, skilled task performance is automatic. Skilled task performance consumes negligible resources compared to problem solving.

Achieving a skilled level of proficiency in any task is difficult. Skill acquisition is usually expensive. It is therefore advisable to design user interfaces so as to exploit existing skills in the user community. It is recommended [BAEC87] to

- build upon the users' existing skills,
- keep the set of required skills to a minimum,
- use the same skill wherever possible in similar circumstances,
- use feedback to reinforce similar contexts and distinguish those that are dissimilar.

The idea is that by keeping the repertoire of skills small, the skills are used more often. The novice gets more practice and will sooner reach a desired level of proficiency. Using the same skill in similar circumstances is critical. When a user is confronted with a new situation which is similar to a familiar situation, he should be able to transfer what he has previously learnt to the new situation.

The main communication channels between humans and computers are the visual channel, the haptic channel and the audio channel

The main vehicle for delivery of information from a computer to a human being is the *visual communication* channel. The relevant technology is advancing at breakneck speed, and includes devices for making three-dimensional images as well as well-known devices for visualizing text, charts, graphs, maps and diagrams.

Haptic communication involves physical contact between the user and the computer (the word *haptic* evolves from a greek word meaning *touch* or *contact*). The haptic channel is mostly used for manual input, e.g., by using a joy-stick, keyboard or mouse. Haptic output devices are rare, one example being Braille-printers.

While sound provides the primary means for communication between humans, the *audio channel* is rarely used between human and computer. Yet this is an area in transition, and audio is a useful option in many design situations. The audio channel is not restricted to speech. It can support a wide variety of non-speech messages, such as alarm signals.

It goes far beyond the scope of this text to treat any of these communication channels in any detail, and we refer to the many available texts on the subject, such as [BAEC87] and [SHNE92].

2.3.2 Interaction Styles and Techniques

A diverse set of interactive techniques are used in various computer systems. These include command languages, menus, icons, windows, help buttons, and many others. One way of organizing these techniques is in terms of dialogue style. An interface with a unified and consistent set of interaction techniques will impose a certain dialogue style on the user, e.g., the interface of a Macintosh computer is very different from the interface of the UNIX operating system. They most certainly impose very different dialogue styles on the users. The major categories of interactive styles are based on four basic dialogue techniques:

- *menu selection*, in which a user issues commands by choosing from among a menu of displayed alternatives,
- *form filling*, in which a user issues commands by filling in fields in one or more forms displayed on the screen,
- *command languages*, in which a user types instructions to the computer in a formally defined command language,
- *direct manipulation*, in which a user manipulates a graphic representation of the underlying data.

The major types of interaction are

- *natural language* interaction, in which the user's command language is a well-defined subset of a natural language, and/or system feedback is expressed similarly,
- *iconic interaction*, in which user commands and system feedback are expressed in graphical symbols or pictograms instead of words,
- *graphical interaction*, in which a user defines and modifies sketches, diagrams and other images and pictures,
- *audio interaction*, in which spoken language, or pre-defined sound signals, are used for communication.

Window systems are interaction environments in which the user's screen is divided into a number of possibly overlapping rectangular areas, each of which handles a specific function. The general problem for many computer users is the need to consult multiple sources rapidly, while minimally disturbing their concentration on their task. In a *multi-window strategy* separate windows are assigned to each information source. A user can move between the different windows, and move data from one window to another.

The *desktop* metaphor has proved to be particular useful. The user's screen is viewed as a desktop which contains documents. Each document on the desktop is associated with a window, and may be moved around on the screen. Documents may be archived in folders, they may be cut and pasted. The desktop metaphor is supported by a direct manipulation strategy: the user interacts directly with the documents.

Menu selection is attractive because it eliminates training and memorization of complex command sequences

The primary goal for menu designers is to create a comprehensible and memorable semantic organization of commands that are relevant to the users' tasks. Menu items should fit logically into categories and have easily understood meanings. Users should have a clear idea of what will happen when they make a selection.

Menu selection applications range from trivial binary choices to complex choices between thousands of items. When alternative choices are few, a single menu may suffice. But when alternatives are many it becomes necessary to organize alternatives in groups, and to create menus to help in choosing among the groups prior to selecting among the alternatives within each group. When groups are many it becomes necessary to organize groups of groups, and we may end up by creating a hierarchy of menus in many levels.

Whenever hierarchical menu organization is attempted one often finds that one and the same item may be classified in more than one way. It may belong to more than one group, and therefore should be reachable by more than one path through the menu tree. A network organization of menus may become desirable. One should nevertheless keep in mind that a network organization may reflect confusion and disagreement on the basic classification scheme.

Classification and indexing are complex tasks, and in many cases there is no single solution that is acceptable to everyone. If the groupings at each level are natural and comprehensible to users, and if users know what they are looking for, then menu traversals may be accomplished in a matter of a few seconds. But if groupings are unfamiliar and the users have only a vague notion of what they are looking for, then they may get lost in the tree of menus for hours.

Form-filling dialogue is attractive when many fields of related data are to be transferred to a computer

A menu is a display of alternatives, in which one option is selected in each interaction. A form is a display of requirements, in which various options and values are specified and integrated in a single display screen [PERL85]. A form may also be viewed as a continuous single menu from which multiple selections are made rapidly, and which cover many fields of data [SHNE92]. The form-filling approach is attractive because the full complement of the desired information is visible. The approach resembles familiar paper forms, and the user is given a feeling of being in control of the dialogue.

Many companies offer tools for creating form filling dialogues. But even with excellent tools, the designer must still make many complex decisions. The elements of form filling design include the following [SHNE92]:

- meaningful title of the form,
- comprehensible user instructions,
- logical grouping and sequencing of fields (related fields should be adjacent),
- visually appealing layout of the form,
- familiar field labels: common terms should be used,
- consistent terminology and abbreviations,
- visible space and boundaries for data-entry fields,
- convenient cursor movement,
- error correction for individual characters and entire fields,
- error messages for unacceptable values,
- optional fields clearly marked,
- explanatory messages for fields,
- completion signal.

These considerations may seem obvious, but often designers have omitted one or the other, or have rushed through without doing a serious enough job.

Command languages are created especially for precise and concise communication within a limited domain

Command languages originated with operating systems, and are traditionally directed to controlling very complex systems of software and hardware. Command language designers usually *know* their systems in spite of their complexity, and they design their languages for other people who are also supposed to know the inner workings of the systems.

Many command languages seem to have been designed in an ad hoc manner, like UNIX. They contain a plethora of commands which are poorly related to each other. Users of many command languages seem to be expected to be able to recall notation and initiate action without much help from the system. Users are often called upon to accomplish remarkable feats of memorization and typing.

Command languages are distinguished by their immediacy and by their impact on devices or information. Users issue a command and watch what happens. If the result is accepted, the next command is issued. Commands are brief and their existence is transitory. The goal of the user is more to create a result than to create a program.

In general, people use a computer system if it gives them power that is not otherwise available. The critical determinant of success is the functionality of the system. Next, the command language must be designed so that the functionality is made available to users. It is often seen that only a small part of the available functionality of a system is used because cryptic and poorly structured command languages prevent users from easy access to useful functions.

Few guidelines exist in the literature for designing good command languages. Most of them seem to boil down to recommendations for providing structure and order:

- simplicity,
- each command should carry out a single task,
- the full set of commands should be organized into a tree structure, like a menu tree,
- arguments should be consistently ordered.

The subject of command language design is too large to be treated in a short survey like here. The reader should consult the specialist literature, e.g., [SHNE92]. We shall nevertheless give a taste of what can be found in this literature. Shneiderman [SHNE92] refers to an experiment by Carroll [CARR82], where two design variables were altered to produce four versions of a 16-command language for controlling a robot. Commands could be hierarchical (verb-object-qualifier, e.g., **move robot forward**), or they could be non-hierarchical (verb only, e.g., **advance**, **straighten** or **push**). Commands could be symmetric (e.g., in pairs like **advance/retreat** or **right/left**) or nonsymmetric (**go/back** or **turn/left**).

Students studied the command sets in manuals, gave subjective ratings, and then carried out paper-and-pencil tasks. Subjective ratings prior to performing tasks showed that the students disapproved of the non-hierarchical non-symmetric form of, say, **go**, **turn**, **back**, **sweep**, **reach**, and gave the highest rating for the non-hierarchical symmetric form of **advance**, **retreat**, **raise**, **lower**. An analysis of the problem solving tasks showed that the symmetric forms were clearly superior, and that the hierarchical forms were superior for several independent measures. Error rates were dramatically down for the symmetric hierarchical forms, like **move robot forward, move robot backward**. Symmetry helped users to remember natural pairs of concepts. The hierarchical structure helped users to master 16 commands with only one rule of formation and 12 different keywords, such as **forward**, **open**, **left**.

Direct manipulation dialogues

The central ideas of direct manipulation dialogues is the visibility of the objects and actions of interest, and the replacement of command language with direct manipulation of the objects and actions through the use of pointing devices. The designer is encouraged to minimize the use of computer concepts, and deal directly in a model of the reality of the task domain. The WYSIWYG (*what you see is what you get*) approach to document processing and the desktop metaphor are typical examples of direct manipulation approaches.

The critical part of creating a direct manipulation system is to come up with an appropriate representation of reality. Some designers may find it difficult to think about information processing problems in a visual form. For many applications the jump to visual language may be difficult. This may be so particularly when abstract problems of a theoretical nature are treated, for example, visual programming languages have so far met with only limited success.

Exercises

1. What are the basic steps of structured analysis and design? Explain each step briefly.

2. Data flow diagrams do not show the control flows. Discuss the advantages and disadvantages of this modeling approach.

3. Explain the term "hierarchical decomposition" and describe how data flow diagrams support hierarchical decomposition on an example.

4. What are the weak and the strong points of
   (a) Structured English ?
   (b) decision tables ?
   (c) decision trees ?

5. What are the methods for resolving data access conflicts in data flow diagrams?

6. The car rental agency, CRA, wants to use computers to increase productivity in its operations. The basic activities of the CRA are:

   - reserving a car,
   - the post-rental process, including charging the customer and returning the car to the available stock,
   - the maintenance process,
   - the accounting process, including the financial reports,
   - various inquiries about cars and customers.

   Use the structured analysis approach to information systems development to:

   (a) draw data flow diagrams,
   (b) identify data stores, flows and entities,

(c) show the access paths (both direct and secondary) on data stores,
(d) derive the access map,
(e) show the automation boundaries, and finally the system architecture.

Make your own assumptions where needed.

7. What are the criteria for packaging subprograms into implementation units?

8. Explain the fundamentals of user interface design.

9. Describe the user interfaces in a software package you are familiar with (e.g., a wordprocessor or a spreadsheet).

10. What type of interaction combinations do you use when you are developing

    (a) an information system for a small bank,
    (b) an operating system for mini-computers.

    Give examples of the interaction types to be used for each task above.

11. Explain how structured analysis and design provides abstractions of information systems.

Chapter 3

Software Design

Data system design is the process of taking a logical model of an information system, together with objectives for that system, and producing the specification of a data processing system and its external interactions that will meet those objectives.

It is useful to distinguish between three different kinds of system objectives:

- *Organizational objectives*, e.g., increased revenue, lower operational costs, improved service;

- *Information system objectives*, e.g., providing more up-to-date and accurate information, providing more relevant information quicker than before;

- *Data system objectives*, e.g., cost less than $ 100 000, response time less than 2.5 sec for 95 % of the on-line transactions, mean-time-between-failures more than 3000 hours operational time.

These three different kinds of objectives are supposed to be consistent with each other, so that the data systems objectives support the information systems objectives, which in turn support the organizational objectives. An overall good result should therefore be expected if the data systems design meets the data systems objectives, provided that the objectives are sensible.

Software designs must be evaluated with respect to performance, reliability and maintainability

The overriding objective of data systems design, of course, is to deliver the functions required by the user. There are usually many alternatives for doing that. The three kinds of objectives with respect to which a physical design must be evaluated are:

- *Performance*, e.g., how fast and at what cost the data system will be able to do the work required;

- *Reliability*, e.g., how reliable the operation of the data system will be with respect to human errors, computer breakdown, or sabotage;

- *Maintainability*, e.g., how easy it is to modify the data system after it has been put in operation, say, in order to meet users' requirements for new or modified functions.

These three factors may work against one another. A reliable system is usually achieved by employing tight controls, and this will most often tend to degrade its performance. After all, if high reliability came for free, everybody would like to have it, and it would not be a design issue. It is also quite common to find that data systems that are designed for very high performance may not be easily modified. It is the designers' task to clarify how the different design alternatives will fare with respect to these design objectives as early as possible in the system development process, in order to provide the basis for choosing among the alternatives.

Of the three kinds of objectives, the reliability and maintainability objectives cannot be as easily quantified as can the performance objectives

Reliability is usually expressed in *mean-time-between-failures*, e.g. 1200 hours. It is, however, hardly possible to derive a reliability measure for a software system through a formal analysis of the program text of that system. The *mean-time-between-failure* measure has to be evaluated on the basis of judgements of how reliable the individual software components are, and on judgements of how software components interact in the systems structure.

Maintainability measures are even harder to operationalize, in the sense that one can say that one data system design is 20 % more maintainable than another data system design. It is possible to set as an objective that each systems modification should not cost more than 20 hours to implement, but it is a lot harder to evaluate a design with respect to this requirement.

On the other hand, experience shows that some program structures are easier to modify than others, and that some security measures prove to be more effective than others. Even so, neither reliability issues nor maintainability issues are well enough understood to make it possible to deal with them except in a superficial manner. It is possible to formulate guidelines to follow for achieving reliable and maintainable data processing systems, but it is hardly possible to formulate detailed step-by-step procedures for doing the design.

Software system design is heavily influenced by the priorities given to the different design objectives

The major design tasks in data system development are:

- physical data base design,
- program structure design,
- communication systems design,
- systems environment design,
- interface design.

All of the design tasks are interrelated. How much emphasis should be put on the different tasks depends on the characteristics of the data system to be designed. Some systems have a large database and very little computation, while others have very strict requirements for timely computations but are void of anything resembling a data base. Some data systems are embedded in other technical systems, while others interact extensively with human beings. For some systems the ease of modification is an overriding issue, while others are built as solutions to well defined problems, so that the designs will seldom or never have to be modified. Because of this variety of data systems profiles, it seems pretty futile to propose general "best" solutions to the data system design problem.

In the sequel we shall look at three different aspects of software design: maintainability, structural standardization, and performance. We shall first discuss the maintainability issue, and will start our discussion with a review of programming terminology.

3.1 A Review of Software Terminology

The aim of structured design techniques is to achieve control of the software design process, so that design costs and design quality can be predicted and the course of events can be influenced. For this to be possible, one must be able to talk about those properties of the software one aims at influencing. It is necessary to have a model of software, and to define those properties of software that make the design process effective.

3.1.1 A Standard Software Terminology is Lacking

Structured design distinguishes three levels of software: *code* level, *module* level, and *package* level, where a *package* is an implementation unit. Unfortunately, various authors employ different terminologies, as do the different programming languages. For example, Wirth distinguishes the procedure level from the module level [WIRT76], [WIRT81]. Procedures are resources which can process data, while modules are syntactical units which define groups of resources based on some measure of similarity among the resources, e.g., similarity of access rights. So a Wirth module is a syntactical structure which controls the transparency of its resources with respect to its environment. A Wirth module cannot be executed. It is the procedure resources of the module that process data. This view of procedure and data is implemented in the Modula-2 programming language.

Wirth's view differs from the way similar concepts are defined in the programming language Ada. Early Ada proposals defined a module to be a collection of resources [ICHB79]. In a later proposal of an Ada standard, it was suggested that Ada modules and Ada program units are represented by the same concept [HONE82]. An Ada program unit is either an executable resource, a subprogram, or a collection of subprograms, to name a few. That is, Ada modules are executable programs as compared to Wirth modules.

No wonder some authors seem to avoid using the word *module* because it has so many different meanings in the literature and in the various programming languages. Instead of *module* the word *system component* is sometimes used. A *system component* may consist of program code, documentation, an interface description, and references to other system components. One should also note that the older programming languages like FORTRAN, COBOL, PL/1, and Pascal do not support any module concept.

A study of the literature can create a confusing picture. There is no agreement in the software engineering community on terminology and conceptual structure, either in theory or in practice. It is evident, however, that the way that the word *module* is used in structured design is different from the way the word is used in the other subfields of computer science. A *structured design module* is similar to a procedure, a subroutine, and a subprogram in programming languages [PAGE80]. On the other hand, a *"Wirth" module* can be viewed as the basic implementation unit. It is the Wirth module that is bought, sold, and maintained. A Wirth module is similar to the package concept in structured design.

The terminological picture is so confused that it is very difficult to choose words and concepts in a way that pleases everybody. *Module* in structured design means something different from *module* in software engineering. The software engineering use of the word is the most widespread. Therefore we shall use the word *subprogram* instead of *module* in structured design. We will use *module* in the software engineering sense for software units that comprise other software and data resources.

Subprogram denotes program units on the level of procedure, subroutine, and function. We will sometimes substitute the words *program* and *routine* for *subprogram*, where there is minimal risk of confusion.

3.1.2 Software Component Independence

One overriding objective of software design is to structure the software in such a way as to achieve independence among the software's components. This is important because functions may be compartmentalized and interfaces simplified. Independent components are easier to maintain because secondary effects caused by design and code modification are limited, error propagation is reduced, and *plug-in* components are made possible.

The two most important design principles are called information hiding and abstraction

But how do we decompose a software solution to obtain the best set of components? The principle of *information hiding* [PARN72] suggests that software can be modularized by aiming to keep the information (procedure and data) contained within one software component inaccessible to other components that do not need such information. Independence is achieved by designing the software components to have a "single-minded" function and to limit their interaction with other components.

The principle of *abstraction* is another central principle in software design. This principle suggests that properties of entities are classified in levels of increasing detail. For example, vehicles can be crudely described by properties like speed and type (e.g., car, boat). Detailed descriptions may consider properties of the motor or the electrical ignition system. By classifying the properties into several levels, it is possible to choose a level of detail on which to communicate. When discussing vehicles in general, we don't want to be bothered with details about the form of the lightbulb in the driver's cabin. We say that we abstract away non-essential details.

The principles of abstraction and information hiding must be translated into operational features

The principles of abstraction and information hiding are used to design structures of highly independent software components. Both of these attributes must be translated into operational features which are characterized by time history of incorporation, activation mechanisms, and pattern of control of the software components [PRES82].

Time history of incorporation refers to the time at which a software component is included within a source language description of the software. Configuration control tools are required to keep track of the various compile-time macros, subroutines, and modules, when they are introduced and where they are linked in to the software system.

Activation mechanisms are of two kinds. Either a software component is invoked by reference (e.g., a call-statement) or by interrupt; that is, an outside event causes a discontinuity in processing that results in passage of control to another software component.

Pattern of control of a software component describes the manner in which it is executed internally. Conventional subroutines have a single entry and single exit and are executed sequentially. Reentrant code is designed such that it does not in any way modify itself or the local addresses that it references. Therefore, the code may be used by more than one task concurrently.

Subprograms can be characterized as follows:

- *Sequential subprograms* are referenced and executed without apparent interruptions by the application software (compile-time macros, subroutines, and procedures).

- *Incremental subprograms* can be interrupted prior to completion by applications software and subsequently restarted at the point of interruption (co-routines).

- *Parallel subprograms* execute simultaneously with other subprograms in concurrent multiprocessor environments (con-routines).

Let it be clear that the detailed techniques of structured design focus on supporting development of software that consists of sequential subprograms. This means not that structured design concepts should not be used when designing real-time systems, but that these detailed techniques may be improved.

3.1.3 Code-Level Concepts

Important concepts in discussing code-level quality are *abstraction*, *control flow constructs*, and *readability*.

The concept of abstraction ranks as one of the most important advances in programming

The concept of abstraction is the basis for high-level languages, virtual machines, data abstractions, and many other concepts. The whole concept of bottom-up design consists of building up layers of abstract machines which get increasingly powerful until only one instruction is needed to solve the problem (Fig. 3.1). In most cases we stop short of defining the one superpowerful instruction, because otherwise we would have built a very, very special purpose machine that could only be used in this one way. We want to build devices with a broader range of application.

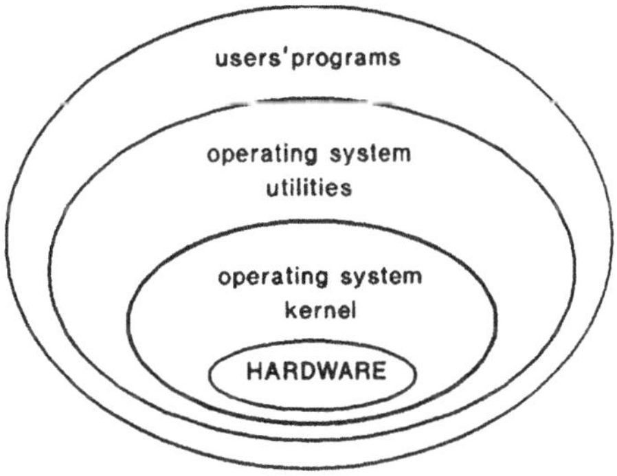

Fig. 3.1. "Onion" layers of abstract machines

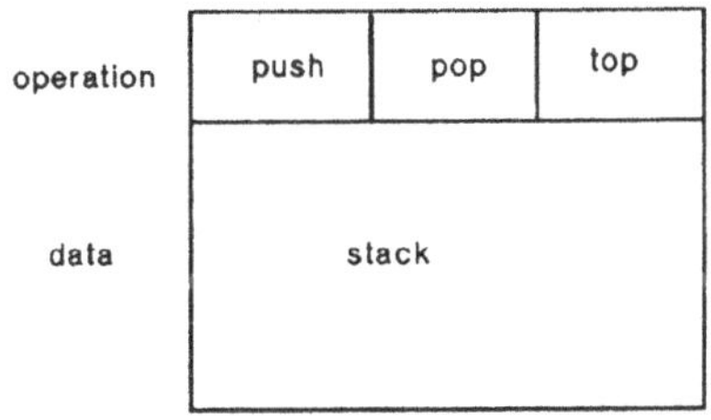

Fig. 3.2. Abstract data type

In many cases the objective is to abstract out many of the complicated interactions which can occur when many users or user programs are sharing the same machine. In other cases a virtual machine is created to hide the idiosyncrasies of a particular machine from the user so that the resulting program will be more portable.

This is often the case when defining *abstract datatypes* (Fig. 3.2). Generally speaking, an abstract datatype is a datatype plus the operations on that datatype. The structure of the datatype is not transparent from outside, and the data content can only be reached and manipulated through the operations. We have in effect created a virtual machine which contains instructions that are the operations of the abstract datatype.

This means that we can concentrate on a problem at some level of generalization without regard to irrelevant low-level details. Use of abstraction also permits us to work with concepts and terms that are familiar in the application environment without having to transform them to an unfamiliar structure.

The number and type of control flow constructs is limited in order to achieve structural simplicity

The basic control flow constructs are

- sequence
- selection (**if then else** and **case** constructs)
- repetition (**repeat** and **while** constructs)

The flowchart notation that corresponds to the three constructs is depicted in Fig. 3.3. For comparison we also show the graphic notation used in Jackson's data-structure-oriented design method [JACK75]. The main advantage of using these three constructs is that the number of possible

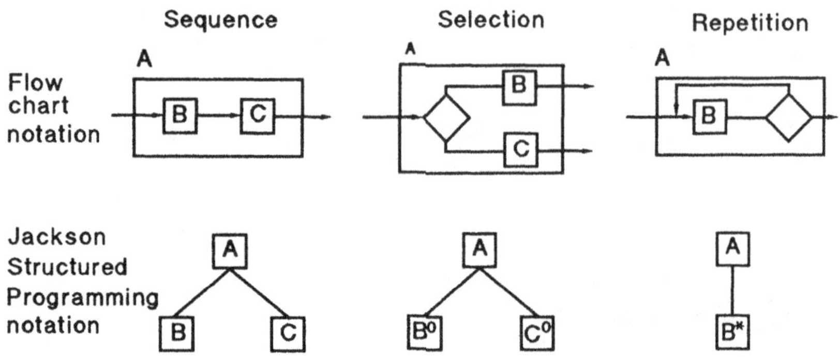

Fig. 3.3. The three control flow constructs of structured programming

program paths as a function of the number of instructions will not grow as fast as in programs that employ more liberal constructs, like the infamous **go to** construct.

Readability of programs determines their lifecycle cost

A program communicates with both people and machines. The life cycle costs of operating a program depend in most cases far more on how well it communicates with people than on how much it was optimized. Unfortunately most textbooks on software design do little more than point out the importance of writing readable code.

It has never been an easy task to explain complicated matters in a simple way. In programming it is made even more difficult because of the imperative nature of programming languages. The solution of a problem is a long list of instructions for a machine to perform. The form of the solution may be rather different from the form of the problem formulation. Therefore, it is a good practice to make frequent references from the program text to the relevant problem statement.

Another way of obtaining readability is to enforce a standard programming style within a company. This is difficult because a standard programming style is not the most efficient for every programmer in every situation. The programmers will in many cases have to be persuaded to use a less efficient programming style than they would otherwise use, for the benefit of the people who are going to do the future maintenance. It is difficult to persuade people to change their procedures when they are simultaneously being pressed by management to deliver their programs yesterday.

3.2 Software Design for Maintainability

As mentioned in the introduction to this section, the literature about structured design uses the word "module" for a concept that is similar to a procedure. Page-Jones [PAGE80] defines a module to be a collection of program statements with four basic attributes:

input	–	what it gets from its invoker
output	–	what it returns to its invoker
function	–	what it does to its input to produce its output
mechanics	–	the code by which it carries out its function
internal data	–	the data to which it alone refers

What may be considered *modules* in structured design are

in PL/1 or ALGOL:	**Procedure**
in COBOL:	**Program, section, paragraph**
in FORTRAN:	**Subroutine, function**
in APL:	**Function**

We have a splendid basis for using the word *subprogram* for this concept. There are three aspects of subprograms that are of special interest:

structure	–	connections between subprograms
coupling	–	communication between subprograms
cohesion	–	association between the components (i.e., program statements) of a subprogram

One of the objectives of software design is to create subprograms that require little interaction with programs that appear in other parts of the software. It is quite possible to get rid of much of a subprogram's external interaction by lumping together most of the interacting code into one big program. In this way one gets low external interaction at the expense of highly complex internal interactions among the program components. One objective of structured design is to create program structures that lead to a better balance between the complexity of internal and external interactions.

To be able to discuss internal interaction phenomena intelligently, we need concepts that characterize various types of internal interaction. For this we use the cohesion concept. A subprogram has high cohesion if it does just one, restricted thing. It has low cohesion if the code for a number of non-related tasks is lumped together without much thought on the part of the program designer. Coupling and cohesion are two of several metrics for characterizing programs. Others are information flow [HENR84] and design complexity [MCCA89] metrics.

3.2.1 Subprogram Cohesion

Cohesion is a measure of the strength of functional association of the code-components of a subprogram. The concept was first proposed by Stevens, Myers, and Constantine [STEV74]. Cohesion is the "glue" that holds the program sentences of a subprogram together. Generally, one wants the highest level of cohesion possible. A highly cohesive subprogram performs a single task within the software structure requiring little interaction with subprograms being performed elsewhere. While no quantitative measure of cohesion exists, a qualitative set of levels was first suggested by Yourdon and Constantine [YOUR75].

The levels of cohesion listed from high to low cohesion are

- *functional*
- *sequential*
- *communicational*
- *procedural*
- *temporal*
- *logical*
- *coincidental*

Coincidental cohesion is the lowest level of cohesion. At this level the subprograms are determined by chance. When one organization introduced a standard that programs are modular if no module contains more than 50 statements, one programmer took a listing of a 2000-statement program and a pair of scissors, cutting the listing every 50 lines and making subprograms out of the pieces. Those subprograms were coincidentally cohesive.

Logical cohesion is present when several similar, but slightly different functions are combined, making a more compact subprogram than if each function had been programmed separately. An example might be a subprogram composed of ten different print routines. The routines do not work together or pass work to each other, but logically perform the same function of printing. Nevertheless, it is tempting to use common sections of code where appropriate, and branch around other sections of code when required for a particular print routine. Subprograms of this type are often very difficult to change because the logic paths through them are so complex. They should be replaced by special-purpose subprograms, one per function.

Temporal cohesion is present when a subprogram performs a set of functions that are related in time, such as "initialization", "house-keeping", and "wrap-up". An example might be a subprogram for PANIC-ABORT containing code for closing-of-files, ringing-the-warning-bell, producing-of-error-messages, saving of check-pointed data, and so on. The only connection between these operations is that they are performed within the same limited time-span. The ability to change such subprograms is improved by isolating the code for each separate function into its own subprogram.

Procedural cohesion is found when the processing elements of a subprogram must be executed in a specific order. Within each subprogram several functions are carried out, which are unrelated except for the flow of control that relates them in this particular case. Functions that can be represented together well on a flowchart are often grouped together in subprograms of procedural strength. Because changes of procedural sequence are frequent when the processing components are not otherwise closely related, procedural cohesive subprograms might become quite difficult to maintain after several modifications have taken place.

Communicational cohesion results when functions that operate on common data are grouped together. Suppose that we wish to determine some facts about a book. For instance, we may wish to find the title of the book, its price, code number, and author. These four activities are related because they operate on common data – the data about the book – which makes the corresponding subprogram communication cohesive. Communication-cohesive subprograms are quite maintainable, although there can be problems stemming from the temptation to share code among the activities within it. This should be avoided because it can make it difficult to change the code of one activity without destroying another.

Sequential cohesion is found when the code-segments of a subprogram are such that output data from one segment serves as input data to the next; thus the subprogram does not perform a complete function, but only participates in doing part of the function. Typically these subprograms accept data from one subprogram, modify or transform it, and then pass it on to another program.

Functional cohesion results when every code-segment within the subprogram contributes to performing one single function, and when the function is completely performed by the subprogram. The subprogram often transforms a single input into a single output. Functionally cohesive subprograms can usually be described by single phrases, with an active verb and a single object, such as "compute square root", "find best solution", "edit inquiry", and "display answer". This is the highest level of cohesion and is desirable whenever it can be achieved.

In practice it is unnecessary to determine the precise level of cohesion. Rather, it is important to strive for high cohesion and recognize low cohesion so that software design can be modified to achieve greater independence among the software components. Note that code-segments within the same subprogram may be cohesion-related in several ways. A subprogram may in general be classified to the lowest level of cohesion present.

Page-Jones proposes a guideline for determining the level of cohesion in a subprogram [PAGE80]. A decision-tree formulation of the guideline is depicted in Fig. 3.4.

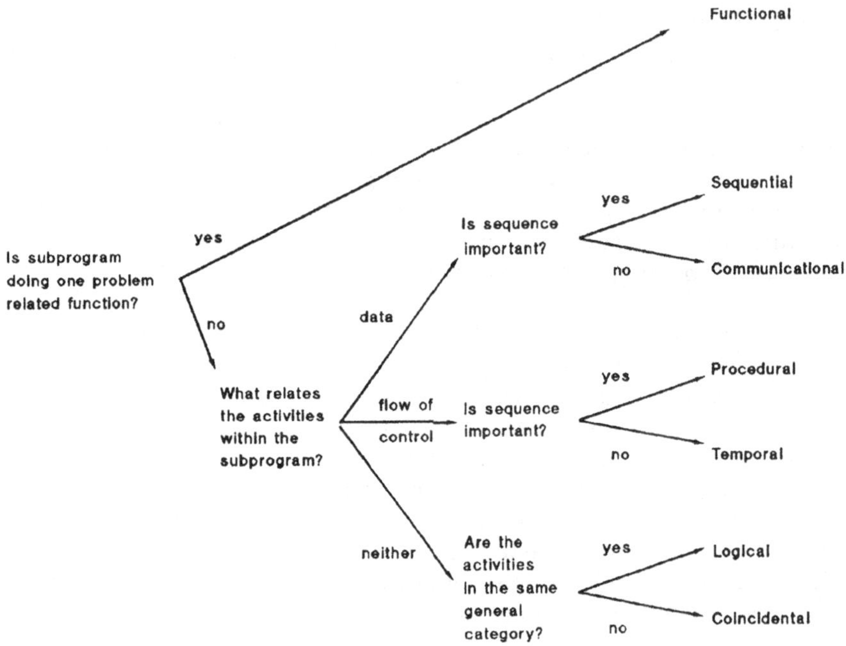

Fig. 3.4. Scheme for determining subprogram's cohesion (from [PAGE80])

3.2.2 Subprogram Coupling

Coupling is a measure of interconnection among subprograms in a software structure. Coupling depends on the interface complexity between subprograms, the point at which entry or reference is made to a module, and which data pass across the interface.

Subprograms may be tightly coupled or loosely coupled (high coupling and low coupling). We usually strive for the lowest possible coupling. The objective is to obtain a subprogram structure such that no subprogram has to know about the particular internal structure of any other subprogram. There are several types of coupling:

- *data coupling*
- *stamp coupling*
- *control coupling*
- *external coupling*
- *common coupling*
- *content coupling*

Data coupling and stamp coupling are considered to be low couplings, while common coupling and content coupling are high couplings.

Data coupling is present when two subprograms communicate by parameters, each parameter being either a single field or a homogeneous table. Data coupling is graphically depicted by an arrow with an open circle at the tail of the arrow. The direction of the arrow points from the sender program to the receiver (Fig. 3.5).

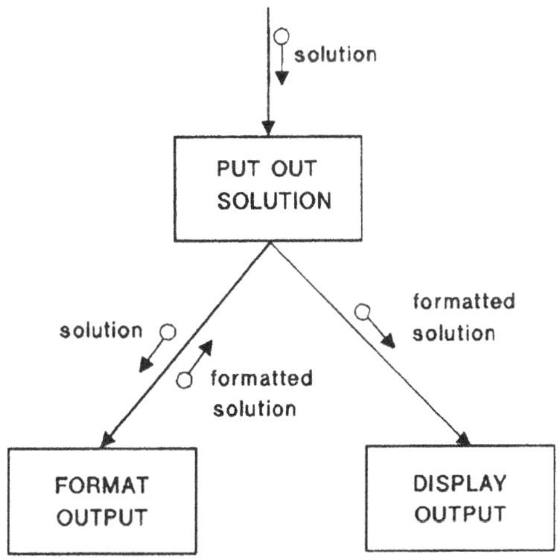

Fig. 3.5. Data coupling among subprograms

Stamp coupling is a variation of data coupling, and is found when two subprograms refer to the same data structure, but use different parts of the data structure. In this case, a change to this data structure will affect all of the subprograms that refer to it, even those subprograms that don't refer to the actual part of the data structure that has changed. The problem inherent in stamp coupling is that it creates dependencies between otherwise unrelated subprograms. Stamp coupling leads to so-called *tramp data*, which is transferred from subprogram to subprogram without being read or modified. Another situation to be avoided in stamp coupling is known as *bundling*. Otherwise unrelated data and/or control items are bundled into artificial data structures and used for communication purposes.

Control coupling occurs when data that characterize the state of the system (e.g., state variables) are passed between subprograms and are intended to control the internal logic of the receiving subprogram. Control coupling is very common in most software designs. In its simplest form, control is passed by means of a "flag" on which decisions are made in a subordinate or superordinate module. The graphic symbol is an arrow with a solid circle at the tail point of the arrow (Fig. 3.6). Control coupling has a more serious effect on changeability than data coupling and should be kept to the minimum necessary for the system to function. The more switches and flags, the more complex the maintenance programmer's job. Passing control

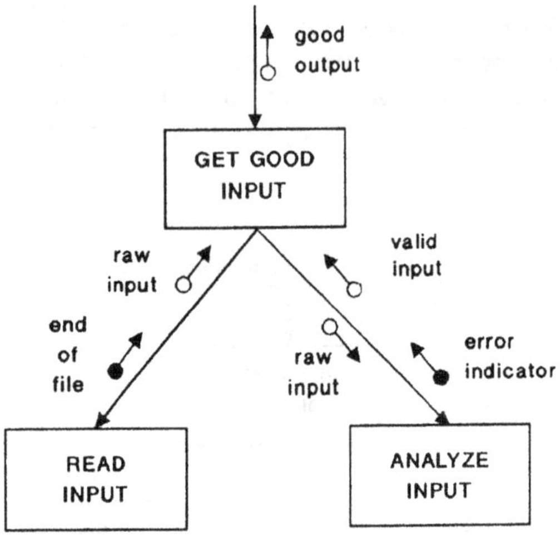

Fig. 3.6. Control coupling among subprograms

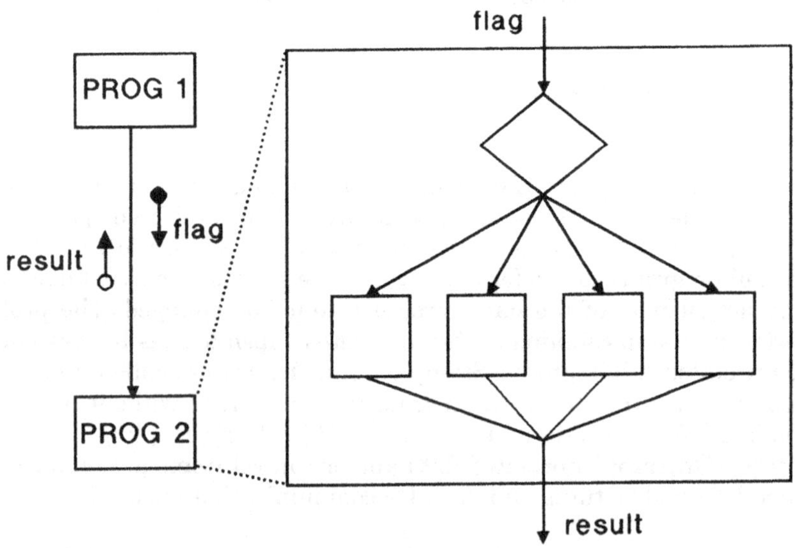

Fig. 3.7. Control coupling down the hierarchy

flags down the hierarchy is undesirable because it implies that the invoked subprogram contains a mixture of functions, and will execute in different ways depending on the control flag (Fig. 3.7).

External coupling occurs when subprograms are tied to an environment external to the software being designed. Examples are couplings to I/O devices, special I/O formats, and communication protocols. External coupling is essential, but should be limited to a few subprograms.

Common coupling is present when subprograms refer to the same global data area. The term "common" is taken from FORTRAN's COMMON. However, globally accessible data areas can be created in most programming languages. The danger inherent in common coupling is that one subprogram will use some data which have been erroneously produced by some other subprograms a long time ago. The resulting faulty result may be very difficult to diagnose. This does not mean that the use of global data is necessarily "bad". It does mean that a software designer must be aware of potential consequences of common coupling.

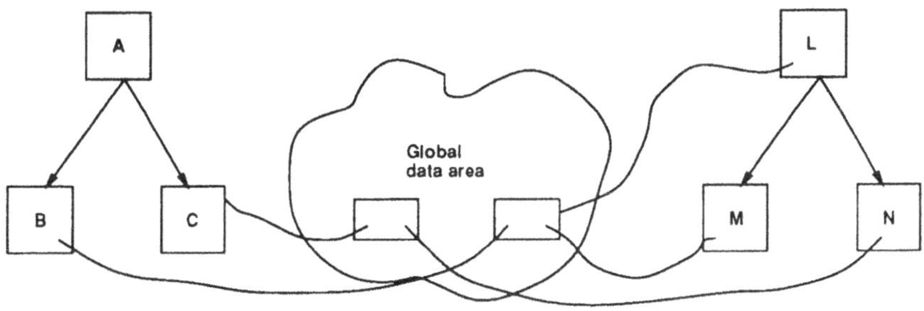

Fig. 3.8. Common coupling: B, L and M are common coupled, C and N are common coupled

Content coupling is present if one subprogram refers to the inside of another subprogram in any way, by either extracting some data defined within the second program, branching control to the inside of the second program, or modifying the way the second subprogram executes. This mode of coupling can and should be avoided. If, for example, a transaction counter is updated in one subprogram and referred to from another subprogram without having been passed up and down the hierarchy, a maintenance programmer may easily modify the first subprogram without realizing the damage done to the second program (Fig. 3.9). This is where the endless "ripple effects" begin, when errors occurring at one location propagate through a system creating unexplained errors and much havoc along their unpredictable propagation path.

The coupling modes discussed above occur because of design decisions made when the structure was developed. A good rule of thumb for designing the way a subprogram gets and returns its information is to imagine it as a library program. How would the program be easiest to understand? How would it be most usable to other people in the organization?

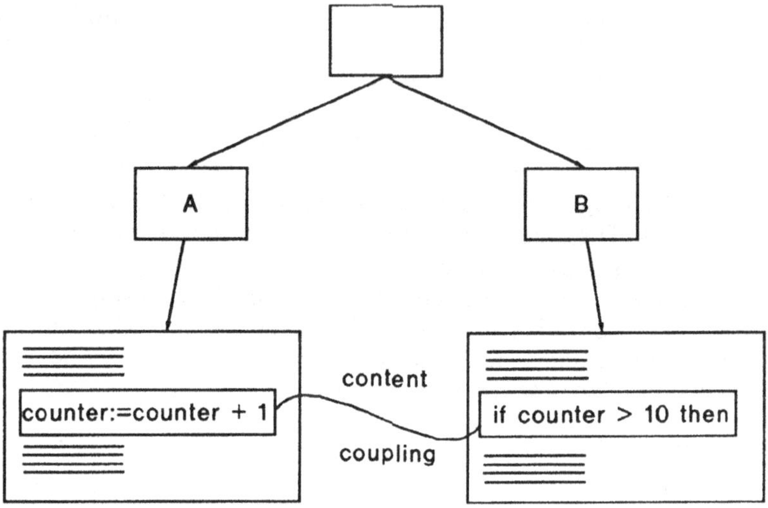

Fig. 3.9. Content coupling

3.2.3 Subprogram Structures

We need more than the criteria of cohesion and coupling to design maintainable software systems. We need to consider other guidelines in forming good and easily maintainable structures.

Experience with structured design and examination of systems known to be changeable suggest that the subprograms in a changeable system often resemble the units of a military organization [GANE79]. Each subprogram has its own job that it performs only when given orders from above; it communicates only with its superior officer and with its subordinate to whom it will in turn issue orders. Figure 3.10 illustrates this. In the military analogy the subprograms do not communicate directly with one another, but only with their commanders. This is one way of simplifying intermodule coupling and making it easier to understand the behavior of the system.

Systems with this type of structure in which one input leg handles all input functions, a transform leg takes input and produces a result, and an output leg handles all output of that result, are called transform-centered systems. They handle input transactions that follow the same or closely similar paths through the subprogram structure. The standard structure for transform-centered systems is depicted in Fig. 3.11 [GANE79].

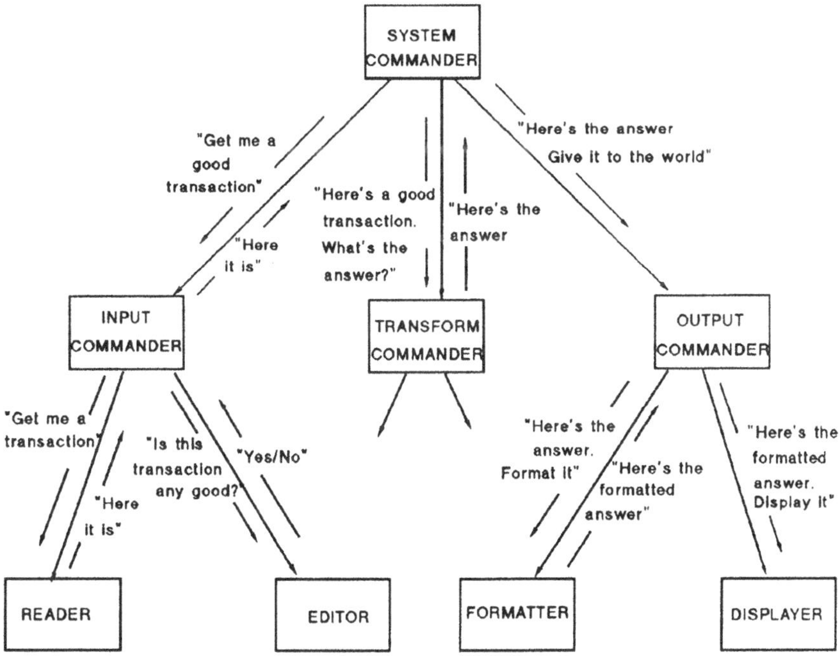

Fig. 3.10. Transform-centered system structure (from [GANE79])

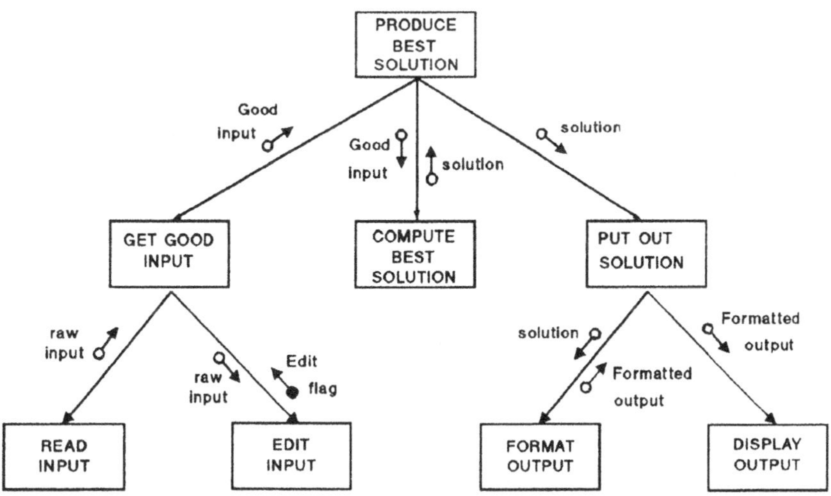

Fig. 3.11. Transform-centered system (from [GANE79])

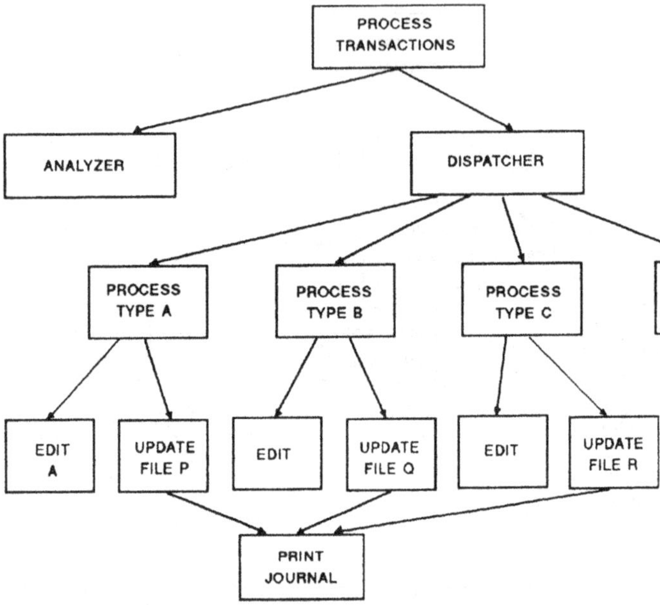

Fig. 3.12. Transaction-centered system (from [GANE79])

Often, especially in commercial systems, this is not the case; the input data consist of several different types of transactions, each of which has to be handled in a separate way, e.g., editing, file updating, etc. The standard transaction-centered subprogram structure is depicted in Fig. 3.12 [GANE79]. In this type of system the main executive program first invokes an analysis-program that reads the transaction and determines its type. The main executive then invokes a dispatcher-program which routes the transaction to an appropriate subsystem set up for each transaction type. Clearly, these two standard structures can only be used as starting points for designing maintainable software structures.

Guidelines for creating "good" subprogram structures can be formulated

The most important guidelines for obtaining high quality software structures are [PRES82], [PAGE80]:

- reduce coupling and improve cohesion,
- minimize fan-out, strive for fan-in,
- keep scope-of-effect within scope-of-control,
- reduce complexity of subprogram interfaces,
- strive for single-entry, single-exit subprograms,
- avoid subprograms with "internal" memory,
- strive for moderation in the application of guidelines for the structuring of software.

One of the most effective techniques is called factoring which is the separation of a function contained as code in one subprogram into a new subprogram of its own. It may be done for any of the following reasons:

- to reduce module size,
- to make the system easier to understand,
- to avoid having the same function carried out by more than one subprogram,
- to separate work (calculations) from management (calls),
- to provide more generally useful subprograms,
- to simplify implementation.

The process of factoring leads to the subprogram structures that will eventually be implemented.

There is no single measure of "structural quality" for a subprogram structure. Nevertheless, it is possible to formulate a few guidelines as listed above. The guidelines are explained in the following.

(i) *Reduce coupling and improve cohesion* can be done by either factoring out new subprograms, or by grouping subprograms together. Factoring involves the examination of subprograms to determine if common process components can be found, and if factoring would indeed improve coupling and cohesion. In cases where high coupling is present, grouping of subprograms can sometimes reduce passage of control, reference to global data, and interface complexity in general (Fig. 3.13).

(ii) *Minimize fan-out, strive for fan-in* as the depth of the structure-tree increases. High fan-out means that few subprograms are ever used more than once in the subprogram structure. High fan-in to a subprogram means that the subprogram is used by many other subprograms. We shall always strive to create reusable code. High fan-in structures indicate that this objective has been achieved. High fan-out structures indicate the opposite. (Fig. 3.14).

(iii) *Keep scope-of-effect within scope-of-control* for all of the subprograms. The scope-of-control of a subprogram is all of its subordinates, that is, all of the programs that can be invoked down the call-hierarchy. The scope-of-effect of a subprogram is all of the subprograms that are affected by decisions concerning computational state in the first subprogram. A standard situation is exemplified in Fig. 3.15. Assume that juveniles and adults are treated differently when buying car insurance with regard to premium and to payment plan. If we apply a structure where we first calculate the premium and then determine the payment plan we may get the following situation:

In calculating the premium we classify the customers as juveniles or adults, and we determine the status variable to be **true** or **false**. The status variable is passed to the payment-plan part of the structure via the system commander. Next, the juvenile/adult classification is made on basis of the value of the status variable. So the behavior of the right leg of the structure depends on classifications made in the left leg of the structure.

The right leg is by no means within the scope of control of the left leg because none of the subprograms of the right leg can be invoked from the subprograms of the left leg. Therefore the scope-of-effect of **COMPUTE-PREMIUM**, which includes **DETERMINE-PAYMENT-PLAN**, is outside the scope-of-control of **COMPUTE-PREMIUM**, which does not include **DETERMINE-PAYMENT-PLAN**.

Scope-of-effect/scope-of-control conflicts indicate the presence of control coupling. Conflicts may be cured by moving the classifications higher up in the hierarchy to avoid multiple decisions. In Fig. 3.20 it is shown how the scope-of-effect/scope-of-control conflict can be removed by a simple restructuring of the subprograms.

(iv) *Reduce complexity of subprogram interfaces* so that the data that passes through the interface is simple and consistent with the function of the subprogram. Seemingly unrelated data passes through the interface via an argument list or other techniques as an indication of low cohesion. The subprograms in question should be reevaluated.

(v) *Strive for single-entry, single-exit subprograms* and avoid content coupling.

(vi) *Avoid subprograms with "internal" memory* such that the subprogram's behavior depends on how the situation was when the subprogram was previously invoked.

(vii) *Strive for moderation in the application of guidelines for the structuring of software*. There is often a trade-off between performance and "good" software structures. Structured design guidelines aim to design easily maintainable software. This is, among other things, achieved by designing a system as a set of small changeable subprograms, rather than a few large programs. However, this may also involve much more overhead caused by modules calling other modules. In some cases this may prove fatal to the performance of the planned system. We then have to trade maintainability for performance. One should always be prepared for this situation and be ready to make trade-offs.

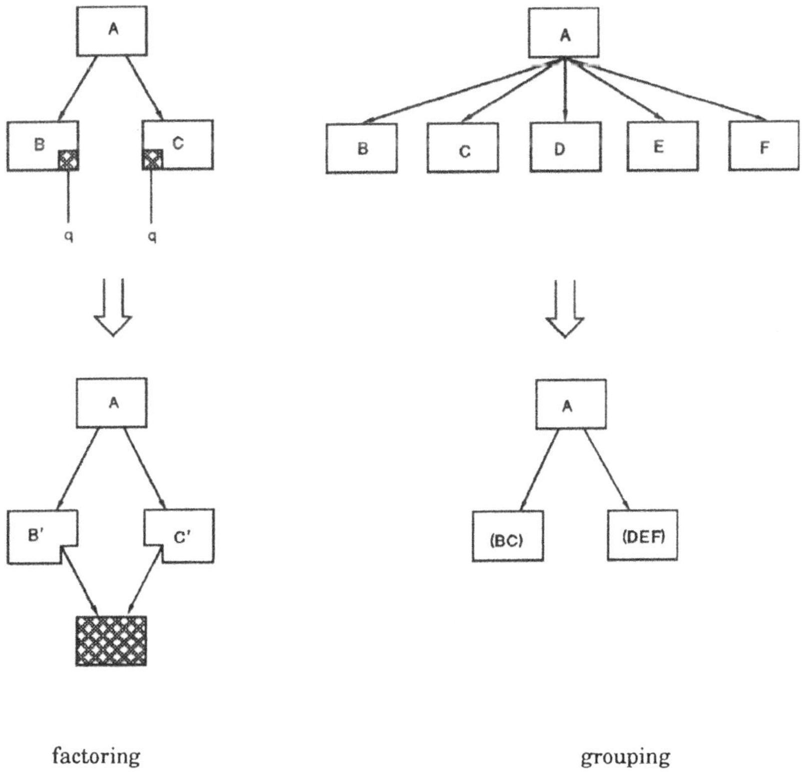

Fig. 3.13. Improving coupling and cohesion by factoring and grouping

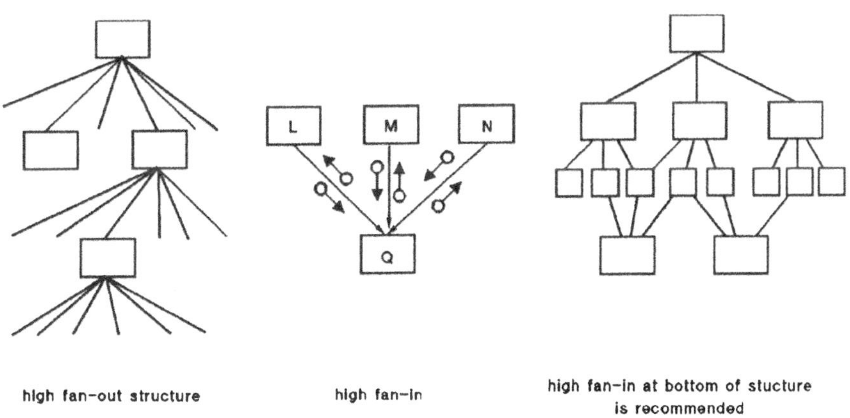

Fig. 3.14. Fan-in and fan-out in subprogram structures

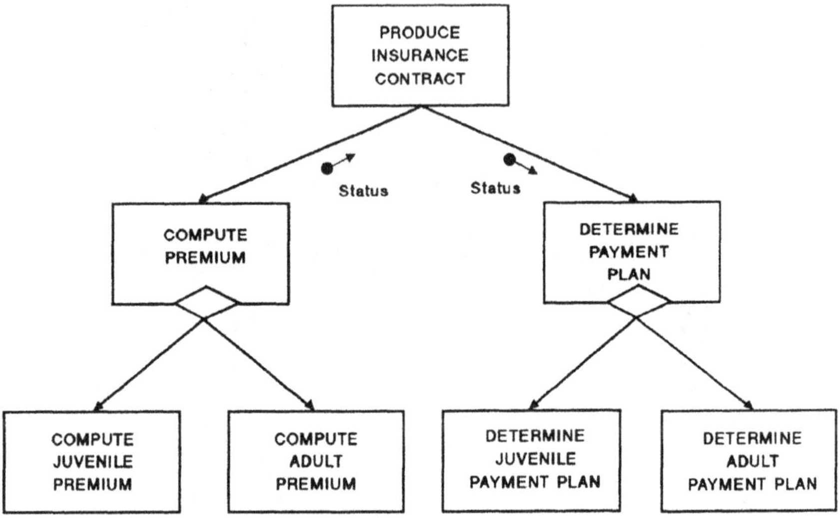

Fig. 3.15. Subprogram structure with duplicate decisions, leading to scope-of-effect/scope-of-control problem (from [GANE79])

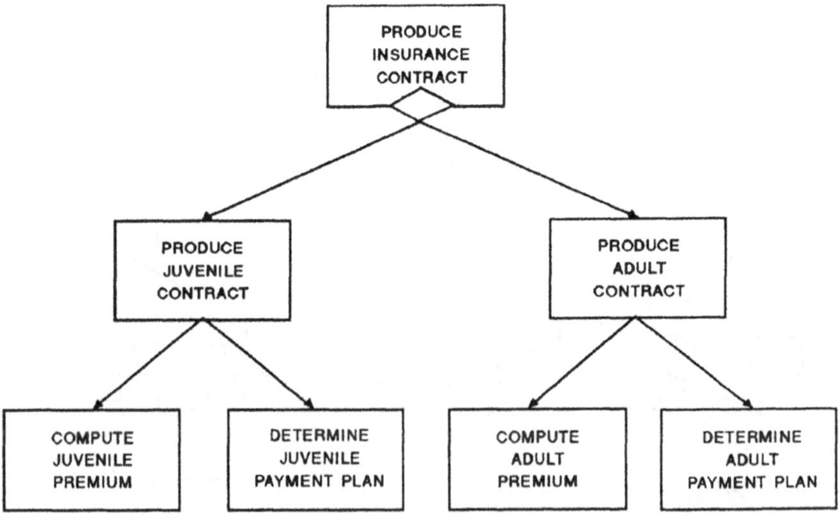

Fig. 3.16. Scope-of-effect included in scope-of-control (from [GANE79])

3.3 Program Structures for Hierarchical Files

The logical model of an information system may take different forms, depending on which specification method has been used. We have so far only looked at the structured analysis method. When this method is used the logical design is expressed in dataflow diagrams, data stores, access maps, and process logic, together with a statement of which processes and data stores are to be realized by computerized means, and which are to be realized by the use of other technical resources, or by human resources, e.g., clerks, decision makers, or managers.

A design method that is based on a general system development method such as structured analysis will become very general itself. It will therefore become difficult to make the method effective for every conceivable design situation. Design methods that are developed to support special situations will usually be more effective than the general method. One method that is especially effective for the design of programs that support hierarchical file processing is Jackson's structured programming (JSP). We shall apply JSP to an example in order to explain the approach. We shall also describe how the structured design method fares when applied to the hierarchical file design problem.

The processing of hierarchical files is most common in conventional administrative data processing. Two usual situations are:

- Transactions are entered via terminals, a transaction log is collected over a period of time, the transaction log is analyzed and reports of the analysis results are produced.

- A master file containing data records is updated with transactions containing modifications to the content of the master file, producing a new version of the masterfile, and reports of the modifications made.

Assume that we want to analyze the daily movements in the widget warehouse. For each kind of widget, we want to add up the number of widgets received from vendors during a day of business, and subtract from that the number of widgets that are delivered from the warehouse to customers. The input data consists of sorted transaction records of the form <widgetnumber, widgetname, quantity>. We are required to produce a report that shows the net change of widgets in store during the reporting period (Fig. 3.17). The corresponding hierarchical data structure charts are shown in Fig. 3.18.

WIDNO	WIDNAME	QUANTITY
W1	FRUITS	-15
W1	FRUITS	+57
W1	FRUITS	+22
W2	NUTS	-387
W2	NUTS	+530
W2	NUTS	...
...	...	...
...	...	...

WIDGET WAREHOUSE REPORT		
NUMBER	NAME	NET-CHANGE
W1	FRUITS	-32
W2	NUTS	+180
W3	SCREWS	-10
W4	BOLTS	+5
W5	...	...
Total charge (units)		+163

Fig. 3.17. Inventory transactions and inventory report

This seems like a straightforward, easy little job. And it is, provided that one knows how to do it. On the other hand, there is no end to the difficulties that one may encounter if one plunges into the programming of a straightforward solution to this problem without preparing for all of the unexpected things that may happen during operation of the data system. How will the program react if somebody tries to execute it without providing it with input data? What is the reaction if there are transactions only for one kind of widget, say for NUTS only? What about executing the system with one transaction only? An amusing exposé of all of the horrid things that may happen to a simple program like this can be found in [BERG81].

3.3.1 Jackson's Structured Programming (JSP)

Jackson has proposed a standard way of dealing with the processing of hierarchical data [JACK75]. The method is known as Jackson Structured Programming (JSP). The essence of the method is that the programs that handle hierarchical data should also have a hierarchical structure. We mention this method here, even though it is a programming method and not a system development method, because JSP has later been expanded to become Jackson's System Development method (JSD), and because the method enjoys considerable popularity and is being used in the grey area between systems analysis and programming.

JSP uses the three basic procedural constructs of structured programming: sequence, selection and repetition. In addition the method proposes a set of transformation procedures in order to deal with unexpected and non-standard situations. It is through the application of these procedures that the method may be adapted to variations in data structures, and to differences in structure between input and output data.

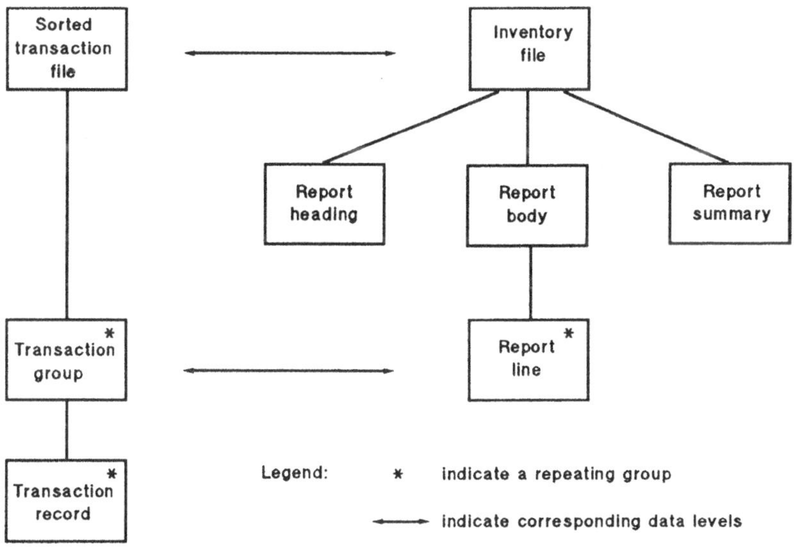

Fig. 3.18. JSP data structures for a transaction processing problem

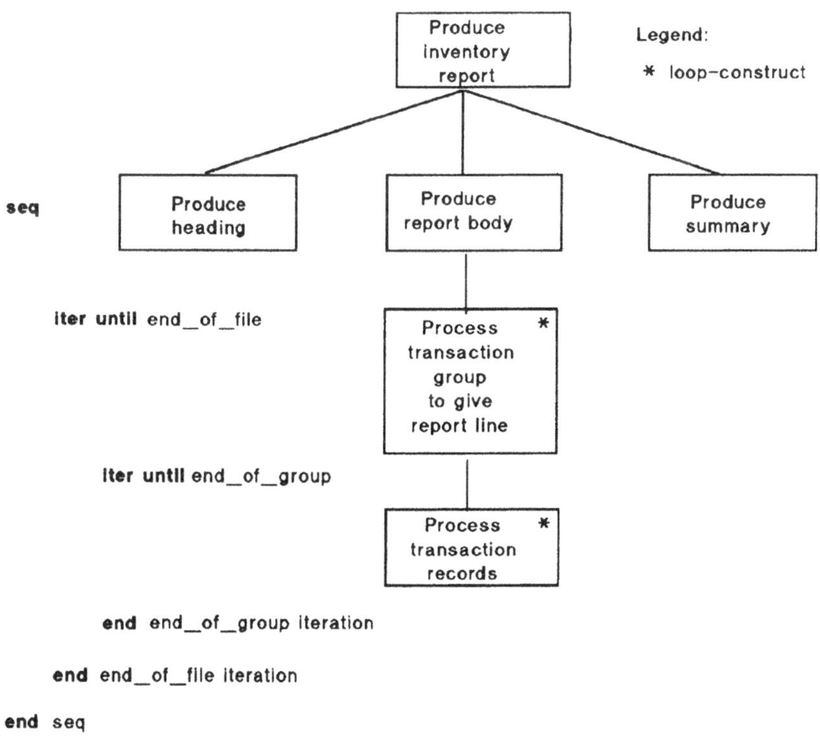

Fig. 3.19. JSP program structure for the transaction processing problem

The JSP rules have been applied to the data structures of Fig. 3.18. Input transactions in a sorted transaction file are to be processed against a master file that contains data about an inventory. The resulting program structure is depicted in Fig. 3.19. We see that each repeating data group corresponds to a loop construct in the JSP program. In this example we have not shown the select construct, which corresponds to a case construct in a computer program. JSP derives a processing hierarchy as the primary program structure. The program skeleton of a possible procedural representation of the design is indicated in the Fig. 3.19.

The corresponding program structure is shown below, in pseudocode.

```
Produce_inventory_report seq;
   open Inventory_file;
   open Transaction_file;
   Produce_heading;
   Produce_report_body
   iter until End_of_transaction_file;
          < Initialize processing>
          Produce_report_line
          iter until End_of_transaction_group;
             Process_transaction_records;
          end End_of_transaction_group;
          < write report line>
   end End_of_transaction_file;
   Produce_summary;
   close Inventory_file;
   close Transaction_file;
end Produce_inventory_report;
```

JSP is an example of a data structure oriented design method. The form of the data structures dictates the form of the corresponding programs. Therefore one can characterize JSP programs in terms of being *right* or *wrong*, that is, a JSP program may very well be considered *wrong* even if it performs correctly, if it violates the JSP rules for program structures.

JSP is a method for detailed program design. A complete discussion of the technique is beyond the scope of this text, and the interested reader is referred to more detailed and thorough literature on the subject, e.g., [JACK75]. However, an overview of a few of the more important supplementary techniques is presented in the following.

Program designs are usually based on the assumption that every conceivable combination of situations has been analyzed and prepared for. This is of course seldom so in practice. Program modifications are to be expected as soon as unexpected situations appear. JSP offers three program constructs to make it easier to deal with the unexpected. They are called **posit**, **admit**, and **quit**:

- **posit** indicates that processing is to occur on the basis of a hypothesis of correctness, e.g., input data to a program is assumed correct;
- **admit** provides alternative processing rules for the abnormal situation;
- **quit** indicates that if the hypothesis fails, then control is to be passed to an **admit** construct.

These three constructs make it possible to start up processing based on an assumption of correctness (**posit**), and leave the normal processing path (**quit**) in a controlled manner (**admit**) as soon as abnormal situations are detected. In JSP terminology this is called backtracking.

The problem of dealing with erroneous data is difficult in data structure oriented design. The reason for this is that erroneous data are not represented in the data structures. Therefore, only the correct, *normal* data will get a corresponding program structure. Unlike dataflow oriented approaches where error processing can be analyzed and presented in the dataflow diagrams, such additional design characteristics can best be treated in JSP by using supplementary "backtracking" constructs.

An example of a typical program structure for dealing with error processing is shown below:

```
PROCESS_TRANSACTION_RECORDS posit
   GOOD_RECORD seq
   quit GOOD_RECORD if error E1;
   do P1;
   quit GOOD_RECORD if error E2;
   do P2;
   end GOOD_RECORD
PROCESS_TRANSACTION_RECORDS admit do P_ERROR;
end PROCESS_TRANSACTION_RECORDS
```

The program is to be read as follows: If a transaction record is found to contain the error E1, then the error routine `P_ERROR` is executed, else the P1-routine is executed. Following the execution of `P1`, if the transaction is found to contain the E2-error, then the `P_ERROR` routine is executed because of the E2 error, else the P2 routine is executed. We see that for any transaction the error routine is executed once if the transaction contains either error E1 or E2. Both `P1` and `P2` will be executed when transactions are free from errors. `P1` will always be executed when the E1-error is absent.

In many software applications the input data has little or no structural commonality to the output data. For example, the sorting order of two files may not correspond to each other. In JSP such situations are called *structure clashes*. They are usually dealt with by introducing some transient intermediate files into the program structure. This can of course add substantially to processing overhead. In order to avoid this, there are additional design procedures called program inversion and multithreading that may be applied to eliminate the use of intermediate data structures.

3.3.2 Structured Design for Hierarchical Files

One of the major weaknesses of dataflow diagrams is their lack of ability to capture control information. This is felt especially when the dataflow diagrams are to be converted into program structures. Because the control information is lacking, the designer has to add it to the dataflows. There are of course many possible alternatives in every case, so it may require inventive designers to come up with good solutions.

We shall show the situation through the same example that was used in the previous section. An incoming flow of inventory transactions are to be analyzed, to produce an inventory report. Assume that the incoming transactions are sorted on the widget identifier *widno*. A possible dataflow diagram is depicted in Fig. 3.20.

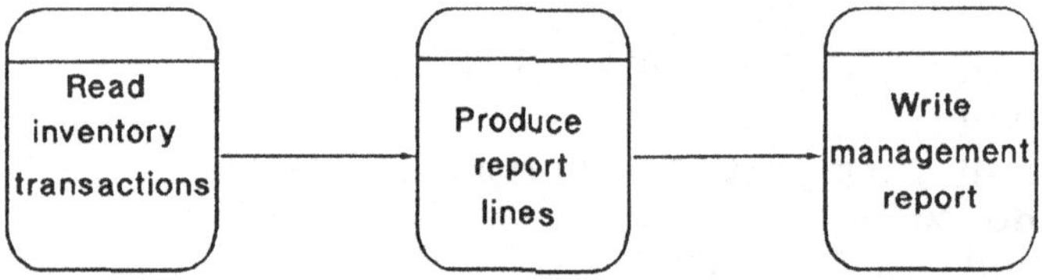

Fig. 3.20. Dataflow diagram for producing an inventory report

A corresponding subprogram structure is depicted in the diagram of Fig. 3.21. We can see that much detail has had to be added to the dataflow specifications in order to reach a specificational level that is close to an executable computer program. There are still more details that must be specified to reach the program code level. These details are, however, easily filled in, as shown in the sequel.

The structures of the data flows may be found from the tables of Fig. 3.17:

structure(nt) : widno, widname, quantity
structure(prl): widno, widname, net_change
structure(prl) = **structure**(nt)
structure(rline)= **structure**(prl)
structure(eof): **boolean**

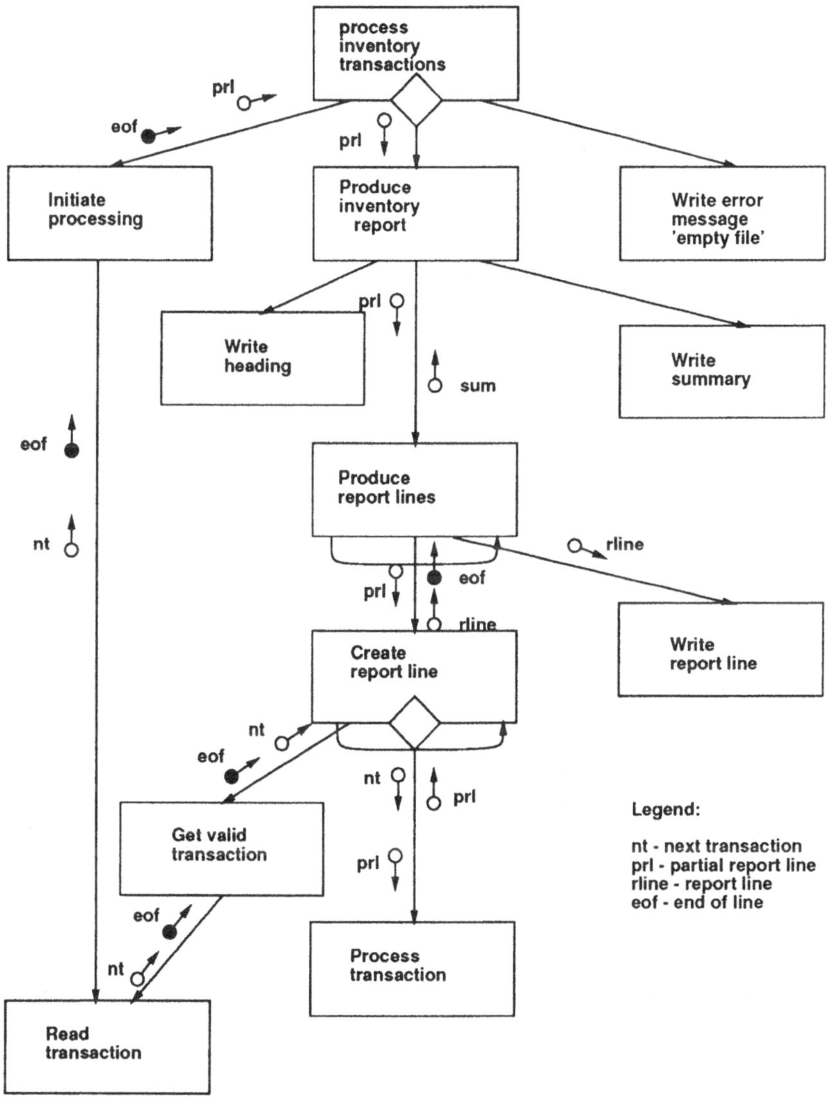

Fig. 3.21. Subprogram structure for inventory report example

The internal details of the various subprograms in pseudocode form:

```
program Process_inventory_transactions;
   Initiate_processing(output prl,eof);
  if eof then write('empty file')
        else Produce_inventory_report(input prl);
end of Process_inventory_transactions;

subprogram Initiate_processing(output prl,eof);
  eof:=false;
  read(nt,eof);
  prl:=nt;
end of Initiate_processing;

subprogram Produce_inventory_report(input prl);
  write('WIDGET WAREHOUSE REPORT
        NUMBER NAME NET-CHANGE')
  Produce_report_lines(input prl, output sum);
  write("Total charge', sum);
end of Produce_inventory_report;

subprogram Produce_report_lines(input prl, output sum);
  sum:=0;
  while(not eof)
      begin
        Create_report_line(prl, output rline,eof);
        write(rline);
        sum:=sum+rline.net_change;
      end;
end of Produce_report_lines;

subprogram Create_report_line(prl, output rline,eof);
  Get_valid_transaction(output nt,eof);
  while( (not eof) and nt.widno=prl.widno)
       begin
         Process_transaction(prl, input nt);
         Get_valid_transaction(output nt,eof);
       end;
  rline := prl;
  prl := nt;
end of Create_report_line;

subprogram Get_valid_transaction(output nt,eof);
  read(nt,eof);
end of Get_valid_transaction;

subprogram Process_transaction(prl, input nt);
  prl.net_change := prl.net_change + nt.quantity
end of Process_transaction;
```

If the various subprograms are put together into one main program for the processing of these inventory transactions, we would get a program as shown below. Even if this is not a complicated program, by any measure, it is large enough for one to appreciate the usefulness of documentation that is contained in the structure charts of Fig. 3.21.

```
program Process_inventory_transactions;
comment Initiate_processing;
eof:=false;
read(nt,eof);
prl:=nt;
comment end of Initiate_processing;
if eof then write('empty file')
   else
   begin
      comment Produce_inventory_report;
      write('WIDGET WAREHOUSE REPORT
             NUMBER NAME NET-CHANGE')
      comment Produce_report_lines;
      sum:=0;
      while(not eof)
      begin
            comment Create_report_line;
            read(nt,eof);
            while( (not eof) and nt.widno=prl.widno)
              begin
              prl.net_change := prl.net_change + nt.quantity
              read(nt,eof);
              end;
            rline := prl;
            prl := nt;
            comment end of Create_report_line;
            write(rline);
            sum:=sum+rline.net_change;
        end;
      comment end of Produce_report_lines;
      write('Total change', sum);
   comment end of Produce_inventory_report;
   end;
comment end of Process_inventory_transactions;
end;
```

3.4 The Object Oriented Approach

The basic idea of the object oriented approach to software design is to organize software in such a way that it supports the direct manipulation of concepts as they exist in the minds of the users of the software. In the object oriented approach a software model is designed to directly reflect on the properties and behavior of, say, a person. Relevant properties, e.g., *name, address and age*, are encoded in a data structure for that person. Models of the assumed skills of the person are represented by software units, which are called *methods*. For example, the skill of a person of being able to move from one location to another would have to be reflected in a method. An obvious consequence of applying a `person-changing-location` method to the person-object would be to modify the content of the address field in the person-object's data structure. Models of a person's various behaviors may be composed by applying different combinations of pre-defined methods.

The invocation mechanisms are different for software objects and abstract datatypes

Object oriented software is similar to abstract datatypes. They share the principle of information hiding, but differ in invocation method. The data structure of an abstract datatype is not transparent from outside, and the data content can only be reached and manipulated through the operations that belong to that datatype. The data structure of a software object is also not transparent from outside. It can only be reached and manipulated through the *methods* of the software object. Abstract datatypes and object oriented software apply the principle of information hiding in a similar way.

Different invocation methods are used for abstract datatypes and software objects. Abstract datatypes are invoked by reference, e.g., a call-statement. Software objects are invoked by interrupt, e.g., when receiving a message from some other object. In the object oriented view the world is seen as consisting of autonomous units which communicate with one another. Object oriented software is made to directly reflect on this view of the world. Object oriented software is therefore well suited to represent dynamic behavior in systems of interacting components.

3.4.1 Object Orientation and Structured Analysis/Design

The object oriented approach is in contrast to the structured analysis approach of building a data model for, say, a person and his dealings with other persons. The structured approach leads us to designing systems consisting of data stores, and of software that is appropriate for dealing with the stored data. This has been described in previous chapters. The object oriented approach leads us to software systems which more directly reflect the objects of the real world and their behavior.

Software objects have a more complex internal structure than subprograms that are created through structured design. The overriding objective of structured design is to create program structures that are maintainable. To this end it is recommended that subprograms have high cohesion and low coupling. Object oriented design strives to satisfy the additional requirement that the software units should match the designer's mental model of the appropriate part of the world.

In the object oriented approach, maintainability seems to be seen as less important than the real-world match. This may be because a conceptual model of the world is seen to be more stable over time than are functional requirements for an information system. Chances for re-using software objects are perceived to be higher than for functionally oriented software, because of this higher stability. So maintainability is traded for re-usability. The practical problem is that the claim of higher re-usability for object oriented software has so far not been persuasively proven.

The behavior of a software object depends on the object's internal state when invoked

One of the major rules of structured design is that process and data should be separated. When executing a subprogram the same input data should always result in the same output data, that is, processes should behave in a deterministic way. For object oriented software this is different. The behavior of a software object depends on the method being applied, on the content of the incoming message, and on the *internal state* of the object.

Take as an example a bank customer who wants to withdraw money from his account. In the structured analysis approach the corresponding information system will consist of one credit control process and one money withdrawal process, where both of the processes interact with a customer account data store (Fig. 3.22). The customer credit control process is shown to return a result based on knowledge about the amount requested and the balance of the appropriate account.

The details of the credit control process are not shown in the diagram. Common sense knowledge about banking would nevertheless make us expect that if the amount of money requested is less than the amount on the account, then the request will be granted. We would also expect that the credit control process would always return this verdict. The result would not depend on when we ask for a credit validation, or where we are located when we ask, but only on the input data which is given to the process from the outside. Analogous reasoning may be applied to the money withdrawal process.

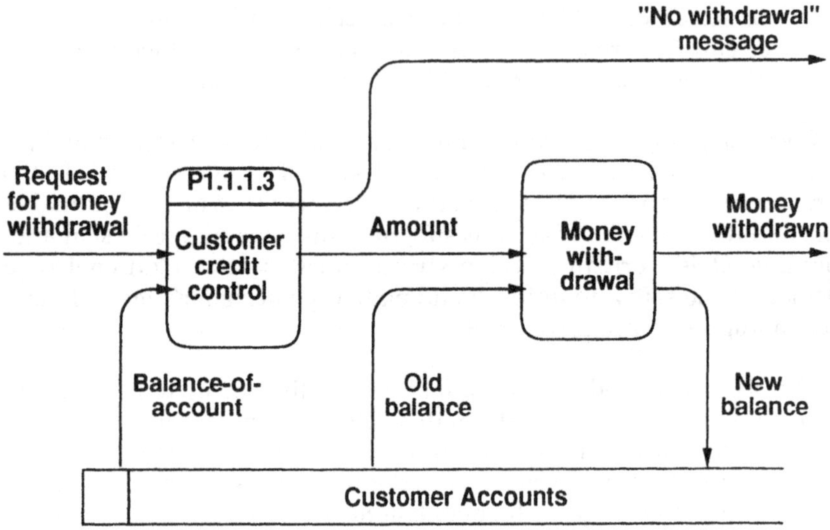

Fig. 3.22. DFD diagram for a money withdrawal situation

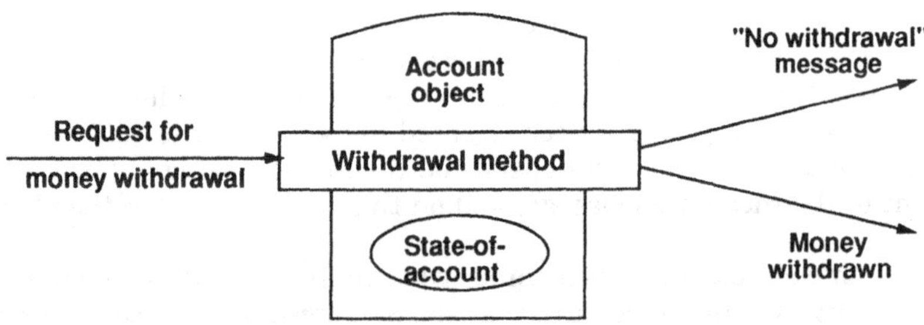

Fig. 3.23. A software object for supporting money withdrawal

When using the object oriented approach for solving the same problem, we suggest establishing an Account object to directly reflect on the common sense world view (Fig. 3.23). A Withdrawal method would have to be defined in order to reflect the account's behavior when money withdrawal is requested. The application of the Withdrawal method would result either in refusing the money withdrawal, or in granting the requested money. The chosen alternative would depend on the state of the customer's account. If the account contains less money than requested, the money request will be refused, otherwise it will be granted.

The customer's account is kept internal to the Account object, and is not transparent from outside. Consequently the behavior of the Account object appears to be non-deterministic. No matter how large a customer's account is, sooner or later he will run out of money if he continues to withdraw without making deposits. A request for the withdrawal of, say, $100 may therefore be refused even if a previous request for the same amount was granted.

The internal state space of a software object may become too complex to be easily understood

The result of applying a method to a software object depends not only on the input values given to the method, but also on the time history of the software object, e.g., the sequence of deposits and withdrawals. This information is hidden from outside. The possibility for human beings to understand the properties of a software object depends on how well its structure is presented. The literature on the object oriented approach to software design gives few recommendations on how to achieve improved understanding of software objects, beyond general recommendations on how to design readable programs.

In our example of a bank account the internal state of the Account object appears to be simple. The internal state may be represented by the accounts balance. In the practical case the internal state space of an Account object may be designed to become considerably more complex (Fig. 3.24). Withdrawals may, for example, be accepted even if they lead to a negative balance of the account, say if the customer has money in other accounts in the bank. Accounts may also become temporarily closed, for example if the customer has lost the check book or the credit card.

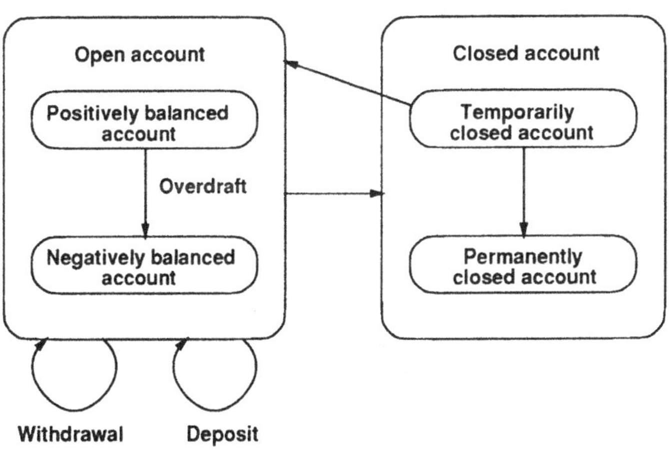

Fig. 3.24. A possible state transition diagram for the Account object

In general there is no limit to the complexity of the internal state space of a software object. While this improves the possibilities for representing real world complexity in software objects it may also make software specifications difficult to read and to understand.

The WIDSYPE example revisited with object orientation leads to a description which reflects the real-world situation

In the WIDSYPE example (Chap. 2.4) we had a sales force which proposed to customers that they buy widget systems. We developed a proposal for an information system to support the WIDSYPE organization. We discussed customer data and sales-person data, and we gave data descriptions of proposals and orders. We discussed the processing of the WIDSYPE data.

When using the object oriented approach to find a solution to the same problem we would have defined a model consisting of customers, sales-persons, proposals and sales-orders (Fig. 3.25). The very simple picture of Fig. 3.25 uses a notation proposed by Booch in [BOOC91]. The notation is informal and is used to depict message passing between objects.

When detailing the behavior of the various objects we would have designed *methods* for modifying the descriptive data of the objects that take part in an interaction. For example, as part of its pre-defined behavior, a sales-person object might have been given the skill of sending a proposal object to a customer object. If in the simulation of an encounter between a sales-person and customer, a particular sales-person object was made to actually send a particular proposal object to a particular customer object, all of the involved objects would have had their internal data modified in order to reflect on the changes in their situation from before to after the encounter.

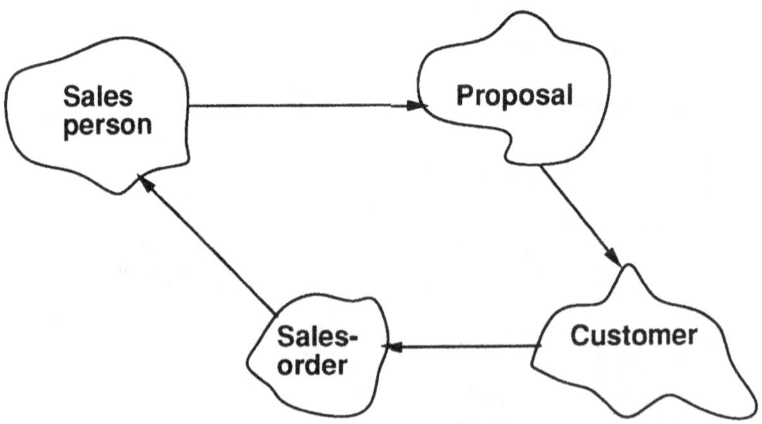

Fig. 3.25. WIDSYPE in an object-oriented view

3.4.2 Properties of Software Objects

The term *object* originated in programming. Object oriented programming was first discussed in the late 1960s in the Simula language [BIRT73]. Simula was developed for simulation of dynamic systems. A primary motivation was to match the software representation of a dynamic system closely to the conceptual view of the real world. The Simula approach to simulation proved to be a powerful approach to programming in general. It was later taken over to the Smalltalk language in the 1970s [GOLD83]. In Smalltalk everything is based on the object concept. Integers are objects, strings are objects, and even the language syntax itself consists of objects. All things that can be done to an object, say, an integer, is represented as methods associated with the object.

The concept of *class* is fundamental to object orientation

In object oriented programming all objects that share a common implementation are abstracted into one class [BIRT73]. Instead of defining individual software objects, e.g., for persons, we give one common definition for a class of objects, e.g., the class of all persons. Because classes originally were defined by their implementations much knowledge of detail was needed in order to understand their behavior. In order to avoid dealing with implementation detail when defining object classes, the concept of *abstract class* was introduced, and was later extended to a concept of *object type* [CARD85].

The introduction of the concept of class has also necessitated a slight change in terminology, in order to avoid confusion. The word *class* is used to mean a collection of things with common attributes and/or behavior. The word *instance* is used to mean one of the things in a class. The word *object* is used to mean either a class or an instance. This coincides with common-sense use of the words. In particular, because a class may be a member of another class (the set-of-sets situation), the word object must be available for class-things as well as for other things.

The most fundamental properties of objects are related to the concepts of *encapsulation, interaction* and *inheritance*

We have already discussed encapsulation and interaction in the previous section. The major points are:

- Software objects are designed to *encapsulate* their data and their code to be reached from outside through the application of predefined methods. This is in contrast to structured design which encapsulates code into procedural units of software, and which keeps data and process separated.

- Software objects *interact* through communication with messages. Software objects are activated when they receive messages from other software objects.
- Software objects may *inherit* properties of other software objects, e.g., a woman-object may inherit the common properties of persons from a person-object where the relevant properties have already been described.

The principle of inheritance is used to explicitly express commonality among object classes

Inheritance allows us to specify common methods and data structures *once* for all of the object classes that carry the shared properties, as well as extend and specialize methods and data structure into specific cases. Thus, we may recognize men and women as special cases of persons, as well as recognizing managers as special cases of employees, which in turn are special cases of persons.

Classification structures depict how some classes are included in other classes, e.g., every man-object is also a person-object, so the class of man-objects is included in the class of person-objects. Every class inherits the properties of its superclasses, e.g., every man-object inherits the properties of person-objects. For each subclass special properties may be defined that are not shared by their superclasses, e.g., the length of beard is a shared property of man-objects, but is not a property shared by all person-objects.

3.4.3 Object Oriented Analysis and Design

Object oriented programming has been around since the late 1960s. The concepts were taken over in systems analysis and design in the late 1980s. Most of the relevant literature has been published since 1987. Many approaches are little more than new words applied to old concepts. Others, e.g., [RUBI92], involves new notations and approaches that are different from the more traditional function- and data-oriented techniques such as structured analysis and design. There are also proposals for organizational approaches that are different from the traditional way of organizing development projects [HEND91]. But as yet there is little agreement among the different proponents of the variety of object-oriented development techniques.

The object-oriented approaches to systems analysis and design emerge from two very different areas of study: object-oriented programming languages and information modeling. From programming languages comes the idea of encapsulation of data and method. From information modeling comes the idea of semantic modeling of problem-domain objects. Most of the various proposals for analysis and design techniques have grown out of either the one or the other of these two starting points.

Software objects may be distinguished in problem domain objects or in solution domain objects

Software objects may be classified as *semantic objects, base objects, utility objects, application objects* and *interface objects* [MONA92]. The semantic objects belong to the problem domain. They represent things or concepts used in describing the problem at hand. Examples are object classes like Customer, Order, and Employee.

Object-oriented analysis (OOA) techniques are mostly concerned with problem domain objects. During object oriented design (OOD) the emphasis is on finding a solution to the problem at hand. The semantic classes may be extended during design, as useful abstractions are discovered, for example, a semantic object class Person may be created as the superclass of Customer and Employee. The distinction between problem domain objects and solution domain objects may consequently become blurred.

Objects which represent software concepts belong to the solution domain. They include interface objects, application objects, base objects and utility objects. *Interface objects* represent users' views of the semantic objects. Icons are interface objects. *Application objects* are the drivers or control programs for the software that represent the problem solution. This corresponds to main programs in procedural languages. Examples of *base objects* are strings and numbers, e.g., the class of integers. An example of a *utility object* is a window in the workstation's screen. Base objects and utility objects are consequently independent of the applications as well as of the problem domains.

There is no clear distinction between analysis and design in the object oriented approach

The object oriented analysis process (OOA) comprises the identification and description of semantic object classes, their attributes and associated data structures, their behavior and their relationships to other object classes. The object oriented design process (OOD) is more of the same, but mainly applied to solution domain objects of the base/utility kind, or to interface objects or application objects [MONA92].

The main objective of design, as well as of analysis, is to identify and describe objects, their attributes and methods. The boundary between the two is so far not clearly defined in the literature. Some processes used by one author during OOA may be included in another author's OOD technique. Most authors blend their methodological proposals from techniques that are already well known in other contexts. Many of the relevant techniques are mentioned and described in various parts of this book, even if they do not appear under the object oriented banner. We shall return to the object oriented approach to systems analysis and design in a later chapter.

3.5 Principles for Creating Software with Acceptable Response Times

The performance of a data processing system refers to its response time or throughput as seen by the users. The structure of software and data together with their frequency of use determine the workload on the available computer systems. Developers routinely sought performance in the early days of computing. Computers were expensive. It was necessary to choose designs that would make large programs fit on small computers. Most information systems had a performance aspect.

As hardware became more plentiful, more systems appeared that did not come with critical performance requirements. A "fix-it-later" attitude was adopted. Performance considerations were deferred to later phases of the system life cycle. Problems were corrected with additional hardware, with systems and software "tuning", or both. Most system designers nevertheless had an intuitive "feel" that led them to create systems with acceptable performance, even when they were told not to worry about performance. Today's software developers have grown up with a "fix-it-later" attitude. No wonder we experience growing performance problems as we try to fit ever larger and more complex software into our computer systems, to satisfy our appetite for increased information systems functionality.

Increased hardware capacity is the traditional remedy for performance problems. In many cases this is also the most cost-effective solution. If so, it should be explicitly chosen early in the life cycle to increase hardware resources when needed, not as a last resource when everything else has failed. There are many examples of performance problems that no amount of hardware will ever correct, because of ill-chosen software structure and/or data base structure.

Tuning can always improve performance, but not as much as appropriate design can. The fix-it-later attitude will introduce modifications to the code that may lead to increased software complexity, and thus to increased costs for software maintenance. The rationale of the fix-it-later tuning approach was to save development time, expense and maintenance costs. The reality is that the additional time and expense for correcting performance problems are often greater than the cost of built-in performance.

It is increasingly dangerous to neglect the performance issues. All kinds of systems that provide customer-service functions are particularly vulnerable to performance problems. Customers are not willing to wait in line because of slow computers. They may choose to do their business with competitors that have better systems, or they may choose to use their money for other goods or services. Real-time systems must often meet critical response-time requirements to prevent disasters. It is indeed dangerous to apply a fix-it-later policy when building these types of systems.

Seven major design principles for creating responsive software have been established by Connie U. Smith [SMIT90]. They are called: *fixing-point*, *locality-design*, *processing-versus-frequency-tradeoff*, *shared-resource*, *parallel-processing*, *centering*, *instrumentation*. Each principle is briefly reviewed. Readers are referred to [SMIT90] for details.

Fixing-point principle

For responsiveness, fixing should establish connections at the earliest feasible point in time, such that retaining the connection is cost-effective.

Fixing connects the desired action to the instructions that accomplish the action. Fixing also connects the desired result to the data used to produce it. The "fixing point" is a point in time. The latest fixing point is during execution immediately before the instructions or data are required. Connections which are established at compile-time are examples of early fixing, while connections which are interpretatively established during run-time are examples of late fixing. Note that "binding" is a subtype of fixing.

Assume that a set of data is required in sorted order. Late fixing sorts the data at the time that the ordered results are requested. Early fixing establishes that data are kept sorted at all times, with all additions preserving the order.

The fixing-point principle says that early fixing leads to better responsiveness when it is cost-effective to retain the fixing connections. In the example above that means that early fixing is preferable if it costs less to keep the data set sorted during updates than it cost to sort it every time the data are requested. This obviously depends on the frequency of update compared to the frequency of requests, as well as the costs of updating and requesting.

Flexibility is influenced by the choice of fixing point. Late fixing may lead to more flexible and general designs than early fixing. Consider the difference between a design which permits ad-hoc queries to a data base, compared to a design which only permits predefined queries. The latter can be supported by early fixing of standard queries to the data base structure, while ad-hoc queries are treated interpretatively through late fixing.

Locality-design principle

Create actions, functions, and results that are "close" to physical computer resources.

Locality refers to the closeness of desired actions and results to physical resources such as processors and storage devices. Examples of applying the principle:

- Group data that are used together and store them adjacent to each other,
- Physically group together procedures that reference one another and do not mix those that are unrelated,
- Create large blocks of file records when sequential access is frequent,

If improved memory locality can be achieved, the software's memory requirement is reduced, and more concurrent users may be allowed. The software may also execute faster because it spends less time loading code and data from external storage.

Processing-versus-frequency-tradeoff principle

Minimize the product of processing times frequency.

This principle addresses the amount of work done per processing request and its impact on the number of requests made. The principle is mostly applied by trying to reduce processing overhead costs. Assume a situation where users are in the habit of doing several additional functions after they have done their first function. For example, when a salesperson requests some data about a customer, he may also want to get hold of data about the inventory status of the widgets that this customer has ordered. If all requests are treated independently, access rights have to be obtained for each individual request. One way of saving computational resources is to let the operating system give all of the access rights to the user for all of the resources he may acquire whenever he enters the system. This may save work if the cost of obtaining access rights for all resources is small compared to obtaining the access rights for one resource.

Another common situation is when transactions can be grouped into batches to fulfill multiple requests for each overhead penalty. This can be done when the batching delay is small compared to required response time, e.g., if the throughput time of an individual request is 250 msec of which 200 msec are due to processing overhead, one may process 6 requests in 500 msec if the requests are grouped, even if they are processed in sequence, while saving (6 – 1)*200 msec = 1000 msec of overhead work.

Shared-resource principle

Share resources when possible. When exclusive access is required, minimize the sum of the holding time and the scheduling time.

Computer resources are limited. Processes compete for their use. Some resources may be shared. Other resources require exclusive use: processes take turns – each process has exclusive use of the resource, one at a time. Exclusive use affects performance in two ways. There is a possible waiting time to get access to the resource, and there is additional processing overhead to schedule the resource.

If some process holds one particular resource for a long time, this means that the other processes must wait until the first process releases the resource. It is therefore important that holding times be short. There are several ways of achieving this:

- Minimize the processing time (by using the other principles),
- Hold a resource only when needed, e.g., request a resource just before it is used, and release it immediately afterwards,
- Request smaller resource units, e.g., don't request exclusive access to the entire data base when only a limited access is needed,
- Fragment the resource requests, by partitioning one request into multiple requests, each having a shorter holding time.

Distribution of smaller resource units may lead to increased scheduling overhead. There will be a net improvement only if the additional scheduling processing can be done while the other processes are waiting for resources. If the scheduling overhead becomes so large that processes have to wait for the scheduling to take place, there is of course no gain in making the resource units even smaller. Wait time is difficult to quantify. To evaluate this tradeoff, one will need quite sophisticated performance models.

Parallel-processing principle

Execute processing in parallel (only) when the processing speedup offsets communication overhead and resource contention delays.

Processing time can sometimes be reduced by partitioning a process into multiple concurrent processes. We have real concurrency when the processes execute at the same time on different processors. We have apparent concurrency when the processes are multiplexed on a single processor. Benefits derived through apparent concurrency are usually not significant compared to those achievable using other principles. Real concurrency will be effective if processing time reductions outweigh the additional overhead for communication and coordination of the concurrent processes.

Centering principle

Identify the dominant workload functions and minimize their processing.

The centering principle focuses attention on the parts of a large software system that have the greatest impact on the performance. The principle proposes that one should create special, streamlined execution paths for the dominant workload functions.

Centering is based on the 80 – 20 rule for the execution of code within programs, which claims that approximately 20 % of a program's code accounts for approximately 80 % of its computer resource usage. Performance enhancements made to these key areas of the software system thus greatly affect the overall responsiveness of the system. It is generally not profitable to worry about low-frequency functions, unless they are either vital functions or they have such large resource requirements that they influence the performance of the more frequent functions.

The centering principle applies to all systems. What one centers on depends on the type of system and on its performance goal. Note that most general purpose software systems, e.g., database systems, have dominant workloads which may not be known by those who develop software on top of these systems. Unless the "bottlenecks" of such systems are known by the software developers, they may center on the wrong performance issues, and gain nothing.

Instrumenting principle

Instrument systems as you build them to enable measurement and analysis of workload scenarios, resource requirements, and performance goal achievement.

Instrumentation is essential to controlling performance. Probes should be inserted to collect the number of times software components execute, and their resource requirements. Instrumentation technology is limited. It is difficult to add software probes after implementation. Instrumentation should therefore be designed into the software.

3.6 Workload Analysis of Software Design Specifications

The minimal requirements that a software system must satisfy are that users' information needs are supported, and that desired information can be retrieved and updated in combinations specified by the system's users. Such requirements are called functional requirements. Additional non-functional requirements are that response time and performance cost constraints are met. Furthermore, the system's performance should not be unacceptably degraded when users' information needs and retrieval needs change over time. Otherwise one would experience frequent rewrites of the software, as the functional requirements change over time.

In realistic project situations, the totality of the users' requirements is so complex that it is necessary to develop the systems design in phases, each new phase producing a description of the future system at some increased level of detail. In most cases, several "next-level" design alternatives may satisfy the users' functional requirements on the level above. Each functional design decision contributes to determining the basis for the final design of software and data base structures. The final design must satisfy both the functional and non-functional requirements.

Each selection of a functional design alternative, on every level of specificational detail, may influence the performance characteristics of the future software system. Experienced designers will therefore try to make crude estimates of the expected performance for each of the design alternatives, before they settle for one of them. The centering principle recommends that one should try to identify those parts of a system that create most of the workload on the computer resources. However, this is not a straightforward matter.

To be able to predict performance, it is necessary to have quantitative knowledge of application software characteristics and data base characteristics. Examples of relevant system parameters are transaction frequencies, branching probabilities in the algorithms, number of characters of data item types, etc. Numerical values for such parameters are usually not known by the designers during design.

Lack of time, money and manpower limit the possibilities of getting detailed quantitative knowledge of relevant system parameters. In practical design situations, we shall therefore never have complete knowledge of all of the relevant design parameters. But we still have to design. So we need a design method which will work even if our knowledge of relevant parameters is weak. The design method should follow the centering principle in that it should direct the designer to give preference to estimating those design parameters that are most important for determining the system's performance.

Relevant design parameters may be obtained by

- Measurement of the performance of existing software components,
- Intelligent guesses by experienced software designers,
- Analysis of design specifications,

or by a combination of all of these three methods.

We shall here give an outline of a method for formal analysis of design specifications, for finding relevant parameters for the estimation of information systems performance [OFTE81]. The method shows the limitations of applying formal analysis techniques for performance evaluation, during the design stage of information system development. Because of this it may be of value to look at the method, and because it can give us valuable hints as to when one has to resort to measurement or to guesswork in order to estimate the performance of a design proposal.

The method is based on the application of sensitivity analysis for identifying parameters which have a major impact on the predictions of a system's performance. It was originally proposed for database design. The design procedure consists of:

- Specification of users' transactions and their use of system resources, including their use of the data stores of the system, e.g., files and archives.

- Quantification of appropriate system parameters.

- Calculation of the usage of system resources, including the system's data stores.

- Identification of the crucial design parameters, according to the centering principle.

- Reestimation of the crucial design parameters, and recalculation of the resource usage.

The database design aspect comprises an additional design step of

- Restructuring of the system's data stores in order to obtain a database with satisfying performance.

The last design step is described in Chap. 4. In the present section we restrict ourselves to viewing the data stores as similar to other computational resources like memory, CPU, etc., relative to the application software. When the need for data store resources has been determined, there will be a next step of estimating the subsequent need for data storage devices to support the accesses to the data stores. So in the first part of the workload estimation procedure we shall restrict ourselves to determining the access frequencies of the access map of a system's data stores, based on a formal analysis of the performance properties of the users' transactions with the system.

3.6.1 An Example of Transactions on Data Stores

The outcome of the functional design, e.g., structured analysis, is a specification of processes, data flows, and data stores of an information system. For our performance estimation purposes it is necessary that the functional specification comprises the branching points of the users' transactions. This is needed for determining the relative use of resources by the different parts of the system. For the simple example that we are going to use in the following, the flowchart level gives the appropriate specificational detail.

A typical example of a specification is shown in Figs. 3.26 and 3.27. The transactions RETURN and LOAN provide operating information for a library.

The LOAN transaction records requests for loans of library items, e.g., books, magazines, gramophone records. After the request has been analyzed, the date of return of the item is determined. One rule of the library is that magazines cannot be reserved. So only books and records are checked for reservation. Finally, the loan is recorded.

The RETURN transaction records the return of items to the library. First, the input is checked for errors. If the returned item has been reserved by others the recorded reservation is detected and an output message is written. Then the recorded loan information is deleted for the returned library item.

Figures 3.26 and 3.27 illustrate the transactions and their references to the systems data stores. Keys are written in *italics*. The subprocesses of each transaction are denoted $\alpha_1, \alpha_2, \ldots$

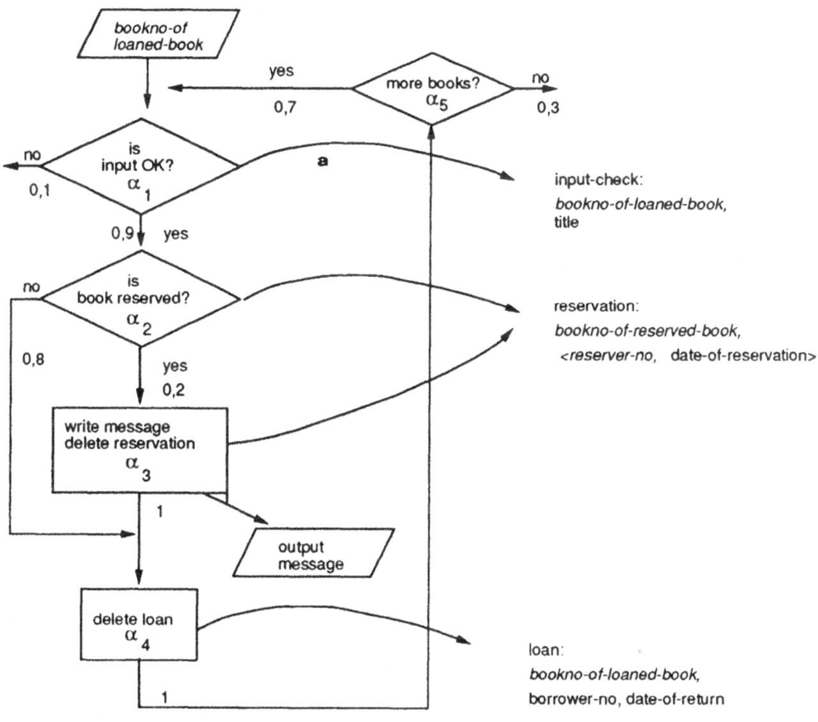

Fig. 3.26. Library transaction RETURN (from [OFTE81])

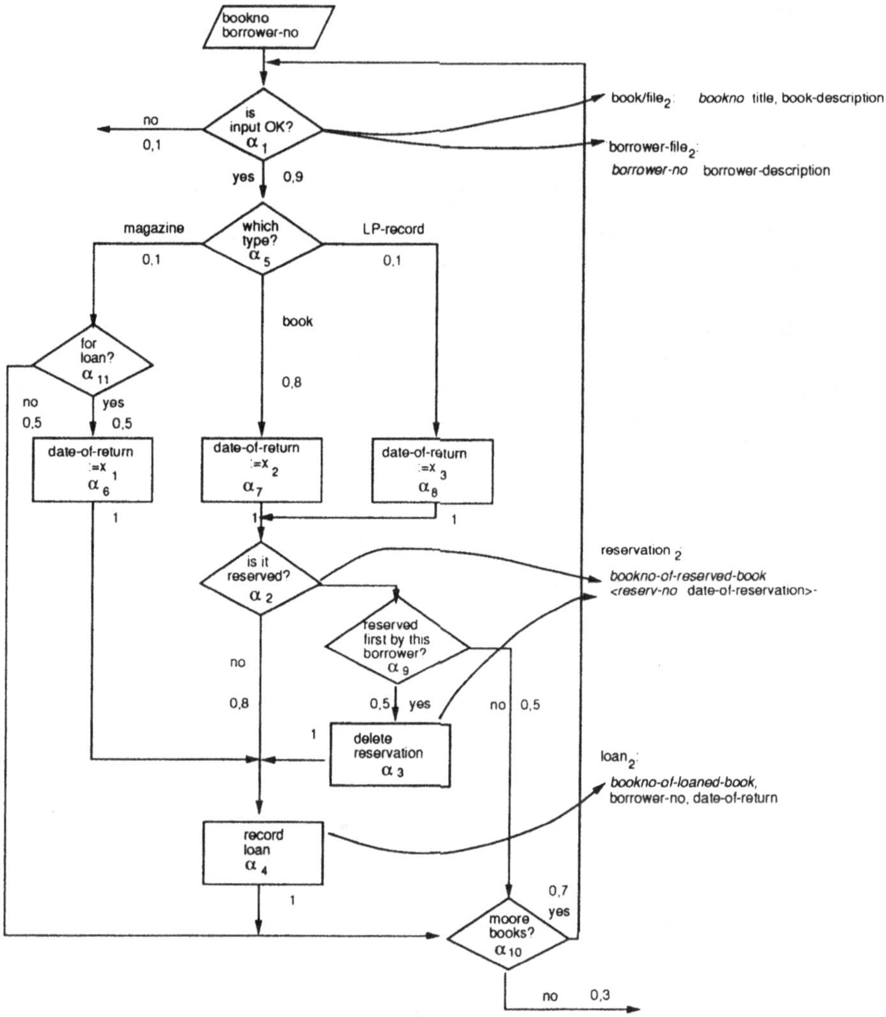

Fig. 3.27. Library transaction LOAN (from [OFTE81])

The computation in a transaction is specified in terms of next activity probabilities, e.g., there is a 10 % probability of an input error in the RETURN transaction, and there is a 90 % probability that the activity following the input check will be to check if the returned book is reserved.

It is unlikely that branching probabilities will be recorded during the requirement definition phase, unless the systems analysts are specifically asked to do so. By looking at the RETURN and LOAN transactions, we realize that most of the branching probabilities can be estimated quite easily. One problem is, however, that in a reasonably sized system there are so many of them. Most of them are also irrelevant, provided that the 20 – 80 rule holds. We shall nonetheless assume that some of the branching probabilities always can be assessed by the designers. We shall use the branching probabilities for predicting the traffic load on the system resources, including the traffic load on the access map of the data stores.

The data stores are usually structured in a way which is convenient for the transactions using the data. It is often convenient to describe the data stores in terms of separate files and/or tables, as it was done for the RETURN and LOAN transactions.

Each of the data stores is a part of the system's total information repository. The data stores as they are defined during structured analysis represent the users' views on the content of the system's data base. To be able to estimate the transaction traffic on the access map of the future data base, we have to integrate all of the data stores into one global information structure.

The global information structure is characterized by a triplet (**A**, **W**, n) where **A** is the global information set, **W** is a vector containing characteristic properties of the elements of the set, and n is the number of elements in the set. Thus,

$\mathbf{A} = [a_u; u=1,2,\ldots,n]$ and $\mathbf{W} = [w_u; u=1,2,\ldots,n]$.

Depending on which property of the information structure we want to study, w_u may be the space needed to store a_u or the data transfer cost of a_u, etc.

The global information set of the library example is:

input-check:	*bookno-of-loaned-book*, title	(a_1)
reservation:	*bookno-of-reserved-book*, <*reserver-no*, date-of-reservation>	(a_2)
loan:	*bookno-of-loaned-book*, *borrower-no* date of return	(a_3)
book-file:	*bookno*, book-description	(a_4)
borrower-file:	*borrower-no*, borrower-description	(a_5)

where a_1, a_2 and a_3 are requested by both the LOAN and the RETURN transactions.

3.6.2 Estimation of Traffic Load

A transaction is represented by a 6-tuple (α, **V**, **E**, **P**, f, m) in which

α	=	$\{\alpha_i\}$ is the set of processes in the transaction,
V	=	$\{v_i\}$ is a vector containing characteristic properties of the processes (such as processing time or number of data base calls performed by the processes),
E	=	$\{e_i\}$ is a vector of entrance probabilities to the processes,
P	=	$\{p_{ij}\}$ is the (Markov) matrix of transitions between the processes,
f		is the number of times that the transaction is invoked over some period of time, and
m		is the number of processes in the transaction.

E and **P** can be used to derive additional knowledge of the behavior of the transaction. The expected number of times that each process will become active during the lifetime of the transaction is given by the formula

$$\Gamma^T = \{\gamma_i\} = \mathbf{E}^T(\mathbf{I} - \mathbf{P})^{-1} \tag{1}$$

The formulas are derived as follows. Given that p_{ij} is the probability of the process α_j being activated immediately after the termination of process α_i, then we know that the probability of α_k being activated immediately following the termination of α_j, because of the original α_i-activity, is equal to $p_{ij} \cdot p_{jk}$ (Fig. 3.28). The probabilities are additive, so that if α_k is being activated because of the previous activity of α_i either via α_j or via α_r, then the probability is $p_{ij} \cdot p_{jk} + p_{ir} \cdot p_{rk}$. We may then calculate the accumulated number of process activations

$\mathbf{D} = \{d_{ij}\}$ where d_{ij} = the expected number of times that α_j becomes active because of that α_i was active at time t=0

We assume that all processes are active at time t = 0, and represent this by the unit matrix $I = \{i_{ij}\}$, where the diagonal elements $i_{ii} = 1$ and all other elements $i_{ij} = 0$. The probability for the next processes being activated after the first termination of all of the first process activations is given by $\mathbf{P}\cdot\mathbf{I} = \mathbf{P}$, and for the next process activation after that the activation probability is given by $\mathbf{P}\cdot\mathbf{P}$, and so on. The accumulated number of process activations thus becomes

$$\mathbf{D} = \mathbf{I} + \mathbf{P} + \mathbf{P}^2 + \mathbf{P}^3 + \ldots + \mathbf{P}^k + \ldots$$

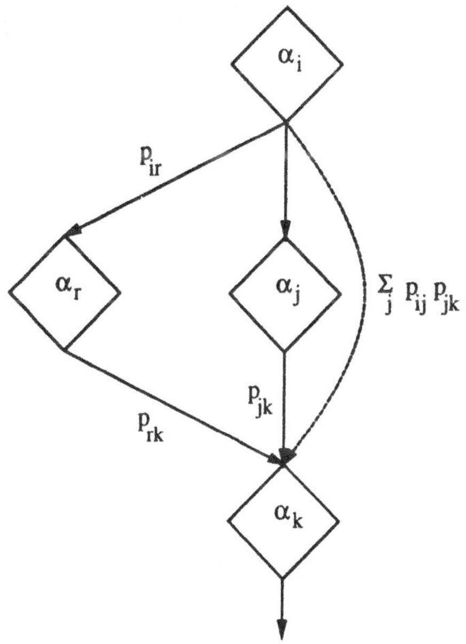

Fig. 3.28. Probability of α_k being activated by processes that have first been activated by α_i

By subtracting

$$\mathbf{P}\bullet\mathbf{D} = \quad \mathbf{P} + \mathbf{P}^2 + \mathbf{P}^3 + ... + \mathbf{P}^k + \mathbf{P}^{(k+1)} + ...$$

we get $(\mathbf{I} - \mathbf{P})\bullet\mathbf{D} = \mathbf{I}$ and $\mathbf{D} = (\mathbf{I} - \mathbf{P})^{-1}$ provided that $\mathbf{P}^k \rightarrow 0$ for large k. This assumption means that there must be no tight loops in the transaction program, and the program must always terminate.

If we now multiply **D** by the entrance vector, we select those processes of a transaction program that are active at time t = 0, and we get the expected number of times each process component will become active during one execution of the transaction, as expressed by the vector (see equation (1))

$$\Gamma^T = \{\gamma_i\} = \mathbf{E}^T(\mathbf{I} - \mathbf{P})^{-1}$$

The traffic within a transaction is expressed by

$$\tau = \{\tau_{ij}\} \qquad (2)$$

where $\tau_{ij} = \gamma_i p_{ij}$ is the expected number of control transfers from α_i to α_j.

3.6.3 Estimation of Penalties Because of One Transaction Activation

The next step is to estimate the penalties for activating the process components of a transaction. There are two types of processing costs involved. The first type is associated with the number of times each process α_i is activated. If the penalty for activating α_j is v_i, then the total activation cost for the transaction is

$$C = \Gamma^T \mathbf{V} \tag{3}$$

The second type of cost is associated with the transfer of control from one process α_i to another process α_j. If the programs associated with the two processes are situated on different computers, there will be a penalty associated with the transfer. The total cost of the control transfers, as expressed by τ, depends on the associated number of boundary crossings between the computers where the programs are situated.

We shall apply the calculation of control transfer penalties to the estimation of access frequencies of a data store access map. One reference from a process to a data store involves a reference to the data base. The coupling between an arbitrary transaction structure and the global information structure is formally expressed by the m x n matrix

$$\mathbf{Q} = \{q_{iu};\ i = 1,2,\ldots,m;\ u = 1,2,\ldots,n\},$$

where $q_{iu} = 1$ if there is a reference from a process $\alpha_i \in \alpha$ to a data store $a_u \in A$, and
$q_{iu} = 0$ if there is no such reference.

By means of **Q** the behavior of the transaction structures can be transformed to access frequencies of the global information structure. One may choose to build one **Q**-matrix for each type of reference (retrieval, update, insertion and deletion), e.g., for updates $q_{iu} = 1$, only if α_i updates a_u.

The traffic within transactions can be mapped into traffic between the different parts of the global information structure. Let $\mathbf{T} = \{t_{uv}\}$ describe the traffic within the global information structure induced by one transaction. Then t_{uv} is the expected number of transitions from a_u to a_v for one invocation of this transaction. The penalty associated with transitions from a_u to a_v might be avoided by implementing a_u and a_v in the same record class (file, data base set). The expected number of transitions is given by

$$t_{uv} = \sum_{i,j} q_{iu} \tau_{ij} q_{jv} \tag{4}$$

or alternatively

$$T = Q^T \tau Q \tag{5}$$

The number of references to a_u is

$$g_u = \sum_i \gamma_i q_{iu} \tag{6}$$

that is,

$$G = \{g_u\} = Q^T \Gamma \tag{7}$$

which is called the activity vector.

3.6.4 Estimation of Penalties for Systems of Many Transaction Types

Most systems comprise many transaction types. We introduce an extra index k to distinguish between transactions, e.g., $p_{ij,k}$ denotes the transition probability from $\alpha_{i,k}$ to $\alpha_{j,k}$ within transaction k. For simplicity, we adopt the notation that, if transactions are considered in isolation, the index k is left out.

The traffic loads on the global information structure induced by different transactions are additive. The expected traffic within the structure is obtained by adding the effects of all transactions:

$$T' = \{t_{uv}\} = \sum_k f_k T_k \tag{8}$$

where f_k is the number of invocations of transaction k over some period of time. **T′** is useful for logical record definition and schema design.

Additivity of the activity vectors implies that the total number of references to the elements of the global information structure is

$$G' = \{g_u\} = \sum_k f_k G_k \tag{9}$$

Special attention should be given to those parts of the structure which are referenced frequently. We define

$$C = G^T \cdot W \tag{10}$$

which is a measure of the cost of referencing the information structure. The impact of high activity (large g'_u), may be compensated by limiting the corresponding **W**-value to a minimum (e.g., by reducing record length), thereby reducing the cost C.

Another important quantity is the entrance vector of the global information structure. It is defined by

$$\boldsymbol{E}' = \sum_k f_k \boldsymbol{Q}_k^T \boldsymbol{E}_k \tag{11}$$

Since $\mathbf{E}_k$ is the entrance probability vector of transaction k, $\mathbf{E}'$ represents the expected number of entrances to the different parts of the global information structure. It may indicate which parts of the structure should be directly accessible and how fast the access method should be (indexed sequential, randomized or direct).

3.6.5 An Example of Traffic Load Analysis

The analysis procedure will be illustrated by the library example of Sect. 3.6.1. The data base is required to support the two transactions RETURN and LOAN.

Based on the estimates of branching probabilities we shall derive the total number of references $\mathbf{G}'$ to the global information structure and the traffic $\mathbf{T}'$ within the structure. The global information structure of the example consists of the 5 data stores a_1, a_2, a_3, a_4, a_5.

The RETURN transaction

There is no reference cost associated with the process α_5. For the sake of simplification we therefore remove this process, so that $\{\alpha_1, \alpha_2, \alpha_3, \alpha_4\}$ constitutes the set of relevant processes. We accordingly let

$$\mathbf{E}^T = [1 \quad 0 \quad 0 \quad 0]$$

$$\mathbf{V}^T = [1 \quad 1 \quad 1 \quad 1]$$

where **V** represents unit cost of information structure references.

The modified probability matrix becomes

$$P = \begin{bmatrix} 0 & 0.9 & 0 & 0 \\ 0 & 0 & 0.2 & 0.8 \\ 0 & 0 & 0 & 1 \\ 0.7 & 0 & 0 & 0 \end{bmatrix}$$

Using (1) and (3) the expected cost of information structure references becomes C=8. The coupling between the processing structure and the global information structure is given by

$$Q = \begin{bmatrix} 1 & 0 & 0 & 0 & 0 \\ 0 & 1 & 0 & 0 & 0 \\ 0 & 1 & 0 & 0 & 0 \\ 0 & 0 & 1 & 0 & 0 \end{bmatrix}$$

The activity of, and the traffic within, the information structure induced by RETURN then becomes

$$\mathbf{G}^{\mathrm{T}} = [2.7 \quad 2.9 \quad 2.4 \quad 0 \quad 0]$$

and

$$T = \begin{bmatrix} 0 & 2.4 & 0 & 0 & 0 \\ 0 & 0.5 & 2.4 & 0 & 0 \\ 1.7 & 0 & 0 & 0 & 0 \\ 0 & 0 & 0 & 0 & 0 \\ 0 & 0 & 0 & 0 & 0 \end{bmatrix}$$

The LOAN transaction

After elimination of the processes that do not reference the data stores, the remaining set of processes becomes $\{\alpha_1, \alpha_2, \alpha_3, \alpha_4\}$ (Fig. 3.29).

In the figure it is assumed that bookno-of-loaned-book is contained in bookno of book-file$_2$. We let

$$\mathbf{E}^{\mathrm{T}} = [1 \quad 0 \quad 0 \quad 0]$$

$$\mathbf{V}^{\mathrm{T}} = [3 \quad 1 \quad 1 \quad 1]$$

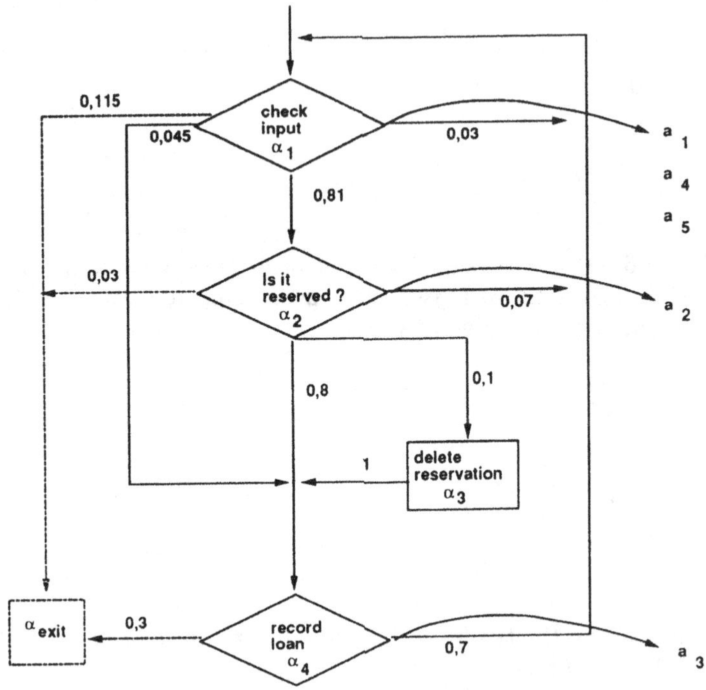

Fig. 3.29. LOAN-modified transaction structure (from [OFTE81])

α_1 has three references to the global information set (a_1, a_4, a_5). If the sequence of the references is of importance we should split α_1 into three processes, each with one reference. The modified probability matrix becomes

$$\boldsymbol{P} = \begin{bmatrix} 0.03 & 0.81 & 0 & 0.045 \\ 0.07 & 0 & 0.1 & 0.8 \\ 0 & 0 & 0 & 1 \\ 0.7 & 0 & 0 & 0 \end{bmatrix}$$

The cost is C = 12.6.

The coupling is given by

$$\boldsymbol{Q} = \begin{bmatrix} 1 & 0 & 0 & 1 & 1 \\ 0 & 1 & 0 & 0 & 0 \\ 0 & 1 & 0 & 0 & 0 \\ 0 & 0 & 1 & 0 & 0 \end{bmatrix}$$

We now find that

$$G^T = [2.7 \;\; 2.4 \;\; 2.1 \;\; 2.7 \;\; 2.7]$$

and

$$\boldsymbol{T} = \begin{bmatrix} 0.1 & 2.2 & 0.1 & 0.1 & 0.1 \\ 0.2 & 0.2 & 2.0 & 0.2 & 0.2 \\ 1.5 & 0 & 0 & 1.5 & 1.5 \\ 0.1 & 2.2 & 0.1 & 0.1 & 0.1 \\ 0.1 & 2.2 & 0.1 & 0.1 & 0.1 \end{bmatrix}$$

The combined effect of RETURN and LOAN

The total traffic load on the global information structure can be calculated from (8) and (9) if we know the transaction frequencies of RETURN and LOAN.

Let us assume that the transaction frequency of LOAN is 150 times per time-period, and 100 times per time-period for RETURN.

The total traffic load on the global information structure becomes

$$G'^T = [676 \;\; 653 \;\; 557 \;\; 405 \;\; 405]$$

and

$$\boldsymbol{T}' = \begin{bmatrix} 13 & 571 & 18 & 13 & 13 \\ 23 & 82 & 539 & 23 & 23 \\ 390 & 0 & 0 & 220 & 220 \\ 13 & 328 & 18 & 13 & 13 \\ 13 & 328 & 18 & 13 & 13 \end{bmatrix}$$

The calculations show that the reference activity is fairly equally distributed on all information sets.

3.6.6 On the Estimation of the Design Parameters

We have shown that it is possible to calculate the transaction load on the computational resources that are needed for the operation of information systems, provided that we can estimate numerical values for the appropriate design parameters. If the appropriate parameters for some reason cannot be estimated we have to resort to measuring or to guessing. The various measures for the transaction load that have been developed previously in this section may also be estimated directly, without any previous calculations. That is, it is quite appropriate to estimate, for example, the activity vector **G** for some transaction if the parameters are missing for making a calculation of **G** possible. Estimates may be based on measurements as well as the designer's experience.

Estimation of each parameter value is associated with a specification cost. There will usually be so many branch points in a system specification that it will be too expensive to estimate all of the branching probabilities. There will be neither manpower, time nor patience available for such an effort. So we have to find some way of directing the designer's attention to those parameters that are critical to the estimation of the performance.

The initial parameter estimation will usually only comprise those parameters that are easily available. In order to be able to perform a first calculation of the transaction traffic, the remaining parameters must be given default values. The following rules may be applied:

- All unknown outcomes from a branch point are considered equally probable.
- All unknown resource demands are given some preset value.

The next step in the quantification procedure is to identify the parameters that are most important for the workload calculation. For example, if it is found that an estimation error in one particular branch probability is dominant for the workload calculation, it seems to be reasonable to concentrate on refining that particular estimate instead of putting a lot of work into estimating some other less important parameter.

We may identify the critical parameters by calculating the sensitivity of the transaction cost with respect to estimation errors in the branching probabilities, or with respect to the transaction frequencies, or the property vectors **V** and **W**. The parameters that are most critical with respect to estimation errors are the primary candidates for re-estimation.

What we have to do in order to determine the critical parameters is to find the differential of the transaction cost with respect to the branching probabilities, or the various penalties, or the transaction frequencies. The

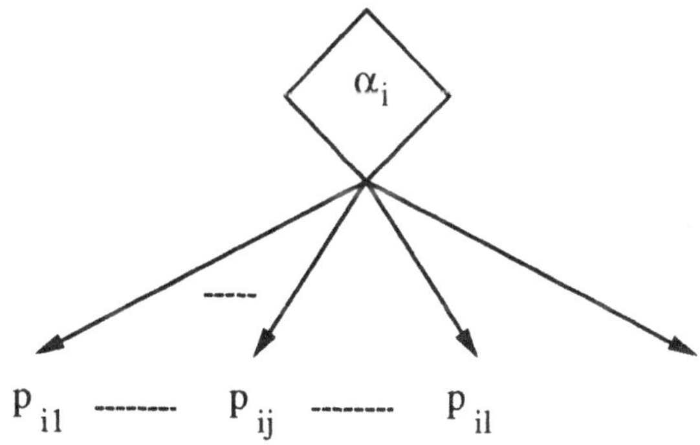

Fig. 3.30. Branchpoint α_i with transition probabilities

highest differential determines the most critical parameter. We shall sketch the highlights of the calculations in the sequel, without giving extensive details, which can be found in [OFTE81].

Recall that the transaction cost C = C(**P**,**V**,**E**). When analyzing the transaction cost's sensitivity relative to changes in transition probabilities we regard **V** and **E** as constants and consider transaction cost as a function of **P** alone, i.e., C = C(**P**). We now want to find the derivative of the cost C with respect to the transition probabilities as represented by **P**. That is, we want to find the transition probability estimate p_{ij} where an estimation error has the largest impact on the calculation of the cost C.

The total differential of the transaction cost with respect to the transition probability p_{ij} is

$$\frac{dC}{dp_{ij}} = \sum_{l} \frac{\partial C}{\partial p_{il}} \cdot \frac{dp_{il}}{dp_{ij}} \qquad (12)$$

where the terms p_{il} represent the other transition probabilities out of the branch point α_i (Fig. 3.30).

The differential dC/dp_{ij} is a measure of the impact on the transaction cost of an infinitesimal change in a transition probability p_{ij}. The first of the two factors of this expression, $\partial C/\partial p_{il}$, is given by

$$\frac{\partial C}{\partial \boldsymbol{P}} = \boldsymbol{D}^T \boldsymbol{E} \boldsymbol{V}^T \boldsymbol{D}^T \qquad (13)$$

where $\mathbf{D} = (\mathbf{I} - \mathbf{P})^{-1}$.

This formula is derived as follows.

We have $C(\mathbf{P}) = \Gamma^T \mathbf{V}$.

A change in **P**, **dP**, implies that

$$\begin{aligned} C(\mathbf{P}+d\mathbf{P}) &= \mathbf{E}^T(\mathbf{I} - \mathbf{P} - d\mathbf{P})^{-1}\,\mathbf{V} \\ &= \mathbf{E}^T[(\mathbf{I} - \mathbf{P})(\mathbf{I} - (\mathbf{I} - \mathbf{P})^{-1}d\mathbf{P})]^{-1}\,\mathbf{V} \\ &= \mathbf{E}^T[(\mathbf{I} - (\mathbf{I} - \mathbf{P})^{-1}d\mathbf{P})]^{-1}\,(\mathbf{I} - \mathbf{P})^{-1}\,\mathbf{V} \end{aligned}$$

We set $\mathbf{D} = (\mathbf{I} - \mathbf{P})^{-1}$

and get
$$\begin{aligned} C(\mathbf{P}+d\mathbf{P}) &= \mathbf{E}^T(\mathbf{I} - \mathbf{D}d\mathbf{P})^{-1}\bullet\mathbf{D}\bullet\mathbf{V} \\ &= \mathbf{E}^T[\mathbf{I} - \mathbf{D}d\mathbf{P} - (\mathbf{D}d\mathbf{P})^2 + \dots]\,\mathbf{D}\bullet\mathbf{V} \\ &= \mathbf{E}^T\bullet\mathbf{D}\bullet\mathbf{V} + \mathbf{E}^T\bullet\mathbf{D}\bullet d\mathbf{P}\bullet\mathbf{D}\bullet\mathbf{V} + \dots \\ &= C(\mathbf{P}) + \mathbf{E}^T\bullet\mathbf{D}\bullet d\mathbf{P}\bullet\mathbf{D}\bullet\mathbf{V} + \dots \end{aligned}$$

By setting $\mathbf{B} = \mathbf{D}\,\mathbf{V}$ and remembering that $\Gamma^T = \mathbf{E}^T\bullet\mathbf{D}$

we get

$$C(\mathbf{P}+d\mathbf{P}-C(\mathbf{P}) = \Gamma^T\, d\mathbf{P}\bullet\mathbf{B} = \sum_{ij}\sum \gamma_i\, dp_{ij}\, b_j$$

and consequently $\dfrac{\partial C}{\partial p_{ij}} = \gamma_i\, b_j$ or, alternatively,

$$\frac{\partial C}{\partial \boldsymbol{P}} = \Gamma\,\boldsymbol{B}^T = \boldsymbol{D}^T\,\boldsymbol{E}\,\boldsymbol{V}^T\boldsymbol{D}^T$$

as proposed above.

In order to calculate the total differential $dC/d\mathbf{P}$ we need to find dp_{il}/dp_{ij} for all i, j, l. For any process α_i we have that $\sum_i p_{ij} = 1$ because there is always a next action, until the transaction is deactivated. Let p_{ij} be increased by dp_{ij}. Because $\Sigma p_{ij} = 1$ both before and after the change we have $\sum_i dp_{il} = 0$. The value of the second factor dp_{il}/dp_{ij} depends on how we want to satisfy the requirement $\Sigma p_{ij} = 1$. For binary decisions there is no ambiguity, because $dp_{i1} = -dp_{i2}$. For decision processes with more than two outcomes, $-dp_{ij}$ has to be distributed over the remaining outgoing branches. A discussion of the distribution problem is found in [OFTE81], where a distribution proportional to the uncertainty of the probabilities is chosen. For simplicity, it may be assumed that the uncertainty is given by $p_{ij} = 0.5 - |0.5 - p_{ij}|$, which is the assumption used in the example calculations in the sequel.

The most critical decision probability is the probability with $\max_{i,j} \left| \frac{dC}{dp_{ij}} \right|$.

This identifies the parameter p_{ij} where an estimation error has the largest consequences with regard to the transaction cost C. Similar calculations can be made with respect to other parameters, but they will not be shown here.

The criteria which have been derived are useful for collecting additional information to support the cost calculations. It is desirable to obtain good estimates of the most critical parameters while less good estimates are sufficient for less critical parameters.

The motivation for ranking parameters in this way is that inexact estimates, for example default values, have a lower "intellectual" specification cost than more exact estimates. Concentration on important issues alone might decrease the total specification cost.

3.6.7 An Example of Sensitivity Analysis

In Sect. 3.6.5 we calculated the activity vector **G′** and the transition matrix **T′** of the library example of Sect. 3.6.2. **G′** and **T′** were derived from initial estimates of the system parameters, and from default values of branching probabilities, e.g., in Fig. 3.27, $p_{11,10} = p_{11,6} = 0.5$ are default values.

We shall use this example to show how critical parameters are determined. We shall show how to use the formulae derived earlier. For illustration purposes it is sufficient to study the modified transaction structures. The transaction frequency of LOAN is more critical than that of RETURN because

$$f_{loan}\, C_{loan} > f_{return} \cdot C_{return}.$$

Let us determine which of the transition probabilities of LOAN we should consider reestimating first. We introduce an extra column and row in the probability matrix to represent the dummy exit-process of Fig. 3.29 and get

$$\boldsymbol{P} = \begin{bmatrix} 0.03 & 0.08 & 0 & 0.045 & 0.115 \\ 0.07 & 0 & 0.1 & 0.8 & 0.03 \\ 0 & 0 & 0 & 1 & 0 \\ 0.7 & 0 & 0 & 0 & 0.3 \\ 0 & 0 & 0 & 0 & 0 \end{bmatrix}$$

Then we may evaluate equation (12) to give

$$\frac{dC}{d\boldsymbol{P}} = \begin{bmatrix} 14.4 & 17.1 & x & 6.6 & -29.1 \\ 7.3 & x & 2.9 & 0.0 & -23.1 \\ x & x & x & x & x \\ 26.2 & x & x & x & -26.2 \\ x & x & x & x & x \end{bmatrix}$$

Matrix elements to which no uncertainty is attached are indicated by an x. We note that even though $p_{24} \neq 0$ the corresponding derivative is equal to zero. Thus, the cost is insensitive to a small change of p_{24}. The most critical probabilities are p_{15} and p_{41} (and, consequently, p_{45}). This means that better knowledge is required of the frequency of typing errors and of how many books are loaned at a time. We also note that all elements of the last column are negative. This means that the earlier the transaction terminates, the lower is the cost, as expected. Note also that the paths $\alpha_2 \rightarrow \alpha_9 \rightarrow \alpha_{10} \rightarrow \alpha_1$ and $\alpha_2 \rightarrow \alpha_9 \rightarrow \alpha_3$ (with the default values 0.5 from α_9) are not very critical. This is seen indirectly from dC/dp_{21} and dC/dp_{23} being small.

Exercises

1. What are the differences between structured analysis and design and object oriented analysis and design?

2. XYZ Travel agency has as its main activity to provide transport for passengers between cities. It is a medium size company. It started its business in 1970 with three buses. Today, the company runs 70 buses. All reservations are made manually, hence causing many mistakes. Automation of the agency's information system is urgently needed.

   The main operations of the company can be summarized into the following groups:

   - selling a ticket,
   - reservation,
   - canceling a reservation,
   - abandoning a reservation,
   - scheduling the journeys,
   - preparation of daily and monthly sales reports,
   - getting journey statistics for future planning.

There are several sales points, and there are several PCs available in the company. Hence the information system software should be available on a PC network.

(a) Prepare a solution by using the structured analysis and design approach.
(b) Prepare a solution by object-oriented analysis and design.

3. Give an example of a module having the "information hiding" property.

4. Discuss the relationships between the concepts of "software component independence" and "software re-usability".

5. Discuss how you can provide readability for C programs.

6. The measures of coupling and cohesion are very well suited for systems developed by the structured analysis and design approach. Explain how you can apply those measures to the object oriented approach.

7. Decide upon the cohesion level of the following modules and explain your reasoning:

(a) DISPLAY-MAIN-MENU : Displays the main menu of an application package
(b) PRINT-FORMS : A system utility which reads data from various resources and prints them in a desired predefined form.
(c) GET-EMPLOYEE-INFO: Using employee number as key, firstly it retrieves the EMPLOYEE master file to read the dynamic information for an employee, then using the same key it reads the family information of the employee from EMPLOYEE-FAMILY file.

8. Explain basic characteristics of responsive software briefly.

9. Apply the traffic load analysis technique given in this chapter to Exercise 2. Use your own assumptions about the frequencies.

Chapter 4

Database Design

The overall objective of database design is to enable all users to obtain the data they require in an efficient and timely manner. This objective breaks down into two areas:

(1) The database should be able to satisfy all of the data needs that have been specified, and it should be sufficiently flexible to meet those that will be required in the future.

(2) The database must perform efficiently, i.e., allowing data to be accessed fast enough for application systems to meet their objectives while using a minimum of computer resources.

The design results of the major database design steps are (Fig. 4.1)
- the logical database structure
- the database access map
- the physical database model
- the database definition (the database schema)

The starting point in the design is modeling the real-world situation of the organization's data. The database model adopted is an abstraction of this reality which suits the information needs of the organization, the constraints of the available database management systems, and the computing resources that are to be used.

The first two steps of database design are partly contained in the structured analysis technique. The structured analysis task of defining the data store content is the same as the first database design step of creating a logical database structure. The access path definition of structured analysis is contained in the second database design step of creating an access map, based on an analysis of the processing needs of the organization. The third step of designing a physical database model is not contained in the structured analysis technique.

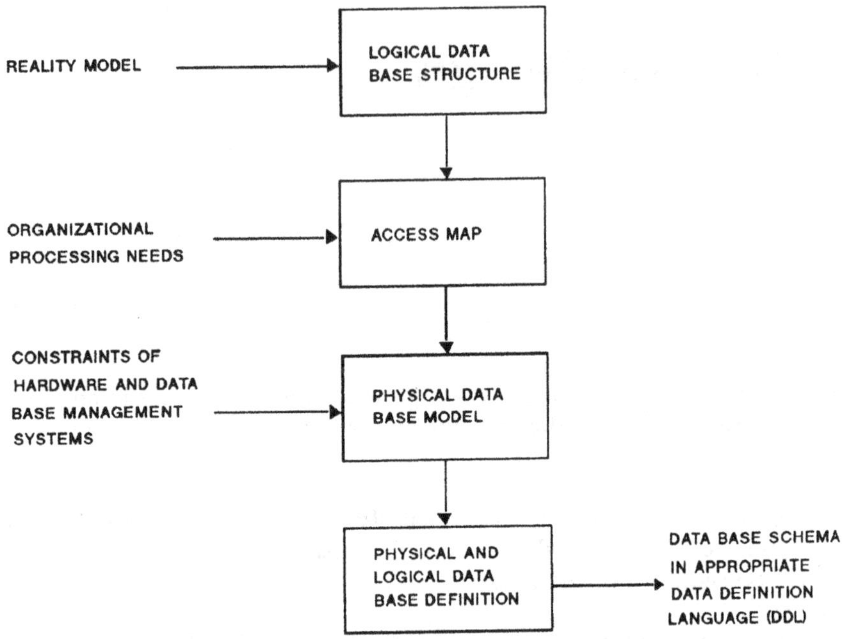

Fig. 4.1. Steps in the design of a database

Performance issues have to be treated at the physical level

The physical database model describes how data is actually stored in the mass storage devices of the computer systems. Hence the name *physical* database model. However, the actual allocation of data to the physical storage devices is in most cases dependent on the implementation of the individual database management system. The database designer thus needs to have intimate knowledge of the internal workings of database management systems in order to be able to predict performance, e.g., response times and operation cost, for a proposed database design alternative. In the sequel we shall therefore give a short overview of the structure of database management systems, including an introduction to the three most frequently used data models in commercially available database management systems: the hierarchical model, the network model and the relational data model.

Database performance may be predicted by leaning on experience, on mathematical analysis of the proposed design alternatives, or on simulation of the interaction between the proposed application system and the database management system. Analytical tools are available only for very simple situations. One may calculate performance properties of single data structures in the computer's main memory. One may also calculate performance properties of single files of data on secondary storage devices.

For systems of many interrelated files the complexity of the analytical models grows fast. Many parameters must be estimated in order to reflect the operational properties of the data management systems that are to be used. The cost of building an analytical model may become too large when compared to the precision to be expected in the performance predictions. Simulation tools may be made available by the vendors of database management systems, but are far too expensive to be built otherwise. That leaves us with the experience and knowledge of the human designer as our most valuable resource for database design. Therefore it should also be tacitly understood that the short introduction to databases that follows is far too limited to make a useful database designer.

Semantic data models try to capture the human's perception of the meaning of the data that are stored in a computer's database

One problem that is inherent in data modeling is the difference between the human's perception of an organization's information processing, and the computer's need to organize the data in a particular way for efficient storage and performance. This gives rise to three different modeling levels that reflect the needs for describing:

- The world as it exists, with or without computerized information processing,
- The information processing without consideration for performance or data storage issues, and
- The software and databases of computerized information systems without consideration for the effects that the operation of the software will have on the world that surrounds the software.

We see that the last two points are taken care of, as far as data modeling is concerned, by modeling on the logical level and on the physical level as described above. The first level of "world modeling" – or reality modeling as it is often called – is not well taken care of by any of the widely used data modeling methods.

The past decade has seen the emergence of numerous data models with the aims of providing increased expressiveness to the modeler and incorporating a richer set of "world modeling features" into the database. This collection of data models may be loosely categorized as "semantic" data models since their one unifying characteristic is that they attempt to provide more semantic content than the hierarchical data model, the network model, and the relational model.

By far the best known and most used of the "semantic" data models is the entity-relationship model, usually abbreviated to the ER-model. Although the ER-model is fairly poor in providing concepts for reality modeling, it is a very widely used model, probably because of its simplicity. We shall therefore review also the ER-model in the following.

4.1 Files and Databases

Data are stored on a wide variety of storage devices. The various device types have different performance characteristics in terms of access times, storage capacity and so on. They also have different costs. Database design is concerned with finding a good balance between performance and cost, provided that all of the other concerns are satisfied, such as data security, flexibility relative to future changes to the database, etc.

The computer's primary storage has the most desirable features in terms of access time. But primary storage – the computer's random access memory, the RAM – is also the most expensive of the storage alternatives. It is therefore always in short supply for data-intensive applications. So one has to rely on the secondary storage media that are outside of the computer. Examples are magnetic tapes, disc packs, and optical storage discs. More data can be stored in secondary storage devices than in RAM, for a considerably smaller storage cost.

One cannot achieve savings in storage cost without giving up something. Accessing data from secondary storage takes much more time than accessing it from RAM. The proportional relationship between the access times is 100 000 : 1 and up. So when deciding to store something in secondary storage rather than in RAM, one makes a potentially costly decision. On the other hand, there is often no choice. The volume of data that has to be stored is in many companies now so large that one of the important costs to consider when choosing storage technology was, until fairly recently, the cost of floor space necessary to accommodate the storage devices. Was a new building for the disc drives needed, or not?

The price differential between the various storage media are changing gradually

Present price/performance trends indicate that RAM storage and disc storage will cost approximately the same around 2010. This means that comparatively less data will have to be put out on secondary storage in the years to come. This change of price structure will have impact on the way that we think about data, the way that we make models of the data's storage. The data models that we are going to describe in this chapter are all rooted in the old situation where RAM storage was exceedingly costly compared to secondary storage. Because conceptual changes come slowly, those data models are going to be with us for a long time to come, even if their *raison d'être* has disappeared. From this point of view it may be worthwhile to investigate data modeling in terms of data storage.

Data that are placed off on secondary storage are collected into files

A file is usually defined to be a collection of related data, organized as a collection of similar records. A number of simple fundamental data structures such as linked lists and binary trees have been developed for handling data in primary storage. The study of file structures is essentially an application of data structure techniques to the special problems associated with storage and retrieval of data that are organized (in collections of similar records) on secondary storage devices. A file system is a software product that provides mechanisms for defining files and for storing, updating and accessing data records in the files. A file system views its files as being unrelated, and will handle only one file at a time.

A database is a collection of interrelated files. A database management system, usually abbreviated to DBMS, is a software product that provides a comprehensive mechanism for defining a database and for storing, updating and accessing data in it. Generalized database management systems came about because of apparent shortcomings in the use of file systems when the number of uses for the same data was growing.

What happened was that files that had been developed for one purpose, were "discovered" by others, and used for new purposes that more often than not were unrelated to the original purpose for which the files had been established in the first place. Of course the new users of the files needed to change the file content, the format and so on – just a little bit – to suit their needs better. This in turn brought about a need to change the original application programs, because of the tight binding between files and application programs. Chaos threatened unless one could achieve a higher degree of so-called data independence, so that one could change the physical storage of data, without needing massive changes in the application programs.

There were a number of associated problems that needed a solution. One of the problems had to do with keeping control of the increasing redundancy of stored data. For example, several applications that needed access to customers addresses would simply establish their own address files if they were not satisfied with what was available. The cost of maintaining a number of overlapping, independent files would become staggering after a while.

The evolution from file systems to DBMS also brought about a need for new data modeling concepts

In the good old days the terms used to describe data were quite simple. A punched card was divided into fields, the card itself was a record, and a number of records constituted a file. A field could be divided into subfields, e.g., a date field could be thought of as composed of the three subfields day, month and year, but that was also as far as the available data modeling concepts could take one. This modeling hierarchy was taken over to tape processing and later to disc processing. We got variable length records and repeating fields, but we could still use the same words to describe the data.

With the development of database technology, this simple vocabulary no longer seemed enough. Even the term "file" became questionable, since a great deal of the value of database techniques lies in taking the files from each application and integrating them in a database that all applications can use. Each new major DBMS undertaking introduced a new set of modeling concepts that were suited to the needs of just that DBMS. For example, IBM's Information Management System [IMS74] describes data as "data fields", which are combined into "segments", which are combined into "databases". The CODASYL DBTG proposal [DBTG74] describes data as "data items", which are combined into "data aggregates", which are combined into "records", the relationships between which are expressed as "sets". In relational databases [CODD70] "domains" are combined into "tuples", which are the elements of "tables".

The dust has still not settled after the major terminological upheavals of the 1970s, and a consensus has yet to emerge on one single set of preferred terms. The situation is even further complicated by the emerging object oriented database management systems [DITT91]. We therefore have to conform to the terminological requirements of the database management systems that are available for us to use, and learn how to live in a multilingual world of data description languages.

4.1.1 File Organization Techniques

A *file* consists of *records*. A *record* consists of *data elements* and/or *groups* of data elements. A *data element* is the smallest logically meaningful unit of data in a file. A *group of data* elements consists of data elements. A group may be a *repeating group*. The *structure of a record* is the relationship among its data elements. A *key* is an expression derived from one or more of the data elements within a record that can be used to locate that record within the file. A key that uniquely identifies each record within a file is called a *primary key*.

When we build file structures we are imposing order on the records of the file. The purpose of that is to make it possible to retrieve and change the data in the file at reasonable costs within a reasonable time. Factors which affect the selection of file organization are:

volatility – the number of additions, deletions and changes to the records in a file,
hit rate – the number of records processed during one "typical" time period, compared to the total number of records in a file,
size – record size, and number of records,
growth – the rate of growth of a file, and the anticipated future size of the file,
availability – commercial availability of suitable file system on the available computers, standardization policy in the organization, price.

There are three major modes of accessing files: sequentially, directly and through an index. There are a number of different file organization techniques available for the support of the various access needs. The file organizations differ in the way that they group records into storage blocks, in the way that the blocks are located on the storage devices, and in the way that the blocks are searched for and retrieved from the storage devices. Whereas the organization of files in data elements and records is a way of maintaining the logical organization of data within the file, the organization of records into storage blocks is done strictly as a performance measure. As such, the size of the block is usually related more to the physical properties of the storage devices than to the content of the data. A number of analytical models have been developed which permit the calculation of average access times, required storage space and so on, for the various file organization alternatives.

Sequential access to a file means reading the file from the beginning and continuing to read until the desired records have been found

Sequential access is applicable for large files which have low volatility and high hit rates in their processing runs. The physical organization of files for the support of sequential access may differ vastly with the storage medium. One may store the blocks of records such that the blocks have to be read one after the other (Fig. 4.2). The records of each block may then be sequentially accessed in primary memory. This is the way it would have to be done if the storage medium were magnetic tape, for example. If the storage device is a disc drive one could choose to distribute the blocks on several discs, so that several blocks could be transferred from secondary storage in parallel and made available for sequential search. This has the same effect as increasing the block transfer rate in the former example. There are a number of alternative ways of giving the files a physical organization. Mathematical models for the calculation of performance parameters are available.

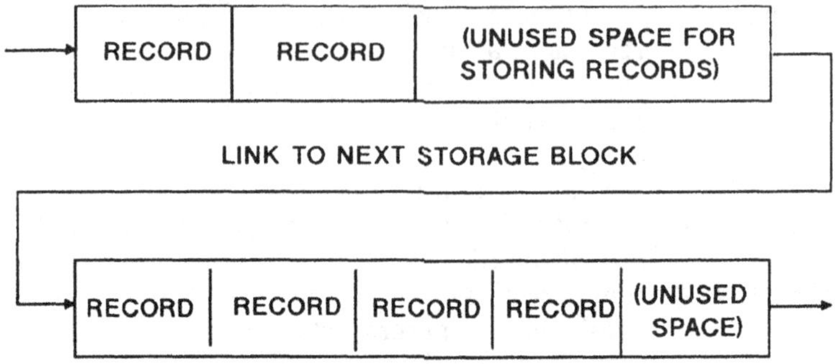

Fig. 4.2. Records are stored in blocks located in secondary storage

Direct access to a file means going directly to the storage location of any wanted record and reading it

Direct access is predicated on knowing where the beginning of the required record is in the secondary storage device. Direct access addressing techniques involve converting a key into a unique positional indicator that can be used to find the desired records without the need to perform multiple read operations. The most common techniques for doing this are either to use the primary key directly, e.g., customer number 743928 has the 743928th record in the file, or to use a hashing algorithm to transform a key into a unique (or almost unique) positional indicator. The first method implies that the needs of the file organization should determine the properties of the file's primary key, e.g., the structure of the customer number. It goes without saying that such practices should not be encouraged.

The hashing technique is by far the most effective approach to direct file access. Assume that the records of a file are organized in blocks, each block containing several records, and that the blocks are to be stored on a direct-addressable storage device like a disc. We would want the hashing algorithm to return the storage address of the block in which a record is stored, when the algorithm is applied to the key of that record. We could then read the block in one read operation and access the record directly.

With hashing, it is important that the records are distributed evenly over the available blocks. If not, some of the blocks would quickly become filled to capacity and beyond, while others would remain empty. When overflow occurs some means has to be found to deal with it. Overflow usually results in the transfer of records to an "overflow" area, and therefore implies several secondary storage accesses in order for a record to be retrieved. The various hashing algorithms deal with these and other relevant issues in different ways. Mathematical models are available for the calculation of performance parameters like average access length, required storage size, etc., for the different hashing algorithms.

Indexed access to a file means going to the storage location of any desired record via a separate address file

The address file is usually called an index file, an index or a secondary index. The last few pages of many books contain an index. Such an index contains a list of topics and the page numbers where the topics can be found. An index always works in conjunction with a "main" file which contains the records that we want to access. Each entry in the index file contains a key value and the addresses to the location of records having these specific values for the key elements. Thus, we confine most of our searching to a comparatively small file, the index file, instead of laboriously searching in larger files.

An index makes it possible to impose order on a file without actually rearranging the file. The order must, however, be maintained when the content of a file is changed. Whenever a new record is inserted into a file, the index files must be updated accordingly, in much the same way that the index of a book would have to be changed if additional pages were added to the book. The index files must also be updated when records are changed or removed from a file.

A file may have any number of index files associated with it. One may, for example, index a customer file on customer name, customer address, the last two digits in the customer's telephone number, or whatever else may be of interest for supporting the retrieval of data from the file. One should keep in mind, however, that every new index adds an extra overhead cost to the file maintenance, because of the required updating of the index files.

When files grow large, their indices may also become so large that not all of an index can be kept in primary store at the same time. The index files must also be placed in secondary storage and given a proper file organization, so that their contents may be quickly retrieved. Special file organization schemes have been developed for the storage and retrieval of index files, e.g., B-trees.

Also for indexed file accessing there exists a well developed mathematical apparatus for the calculation of performance parameters, like average access lengths, required storage space for a file of a given size and logical organization, and so on.

4.1.2 Database Management Systems

Simply stated, the purpose of a database management system (DBMS) is to make it possible for a large number of users to share a common repository of data without destroying for each other either the quality of the data or the availability of the data in the database. This requires that "firewalls" be constructed between the database and its users, so that the actions of one user will not infringe on the freedom and possibilities of any other user.

Database management systems employ a so-called three-level architecture of data [ANSI77]. The levels are called internal, external and conceptual. The internal level refers to the physical storage of data. The external level refers to the individual application's logical view of the data that it refers to in the database. The conceptual level refers to a "global" view of the relationship between the physical database (the internal level) and the data references from the users' applications (the external level).

The term "conceptual" has created a lot of confusion. It has been associated with "real world" modeling, because it has been required that the modeling tools on the conceptual level should be able to represent everything of interest in those parts of the "world" that are modelled by a database. To avoid unintended associations and misunderstandings, we shall use the term "global" instead of "conceptual". So, we shall hereafter use the terms internal, external and global views of data. We reserve the term "conceptual" to be used to denote some particular aspects of "real world" modeling which will be explained later.

To each of the data modeling levels is associated a data description language which is used to define so-called "schemas"

We have internal, external and global schemas. The internal schema describes the physical storage of the database, while the global schema describes the logical organization of all of the data in the database (Fig. 4.3).

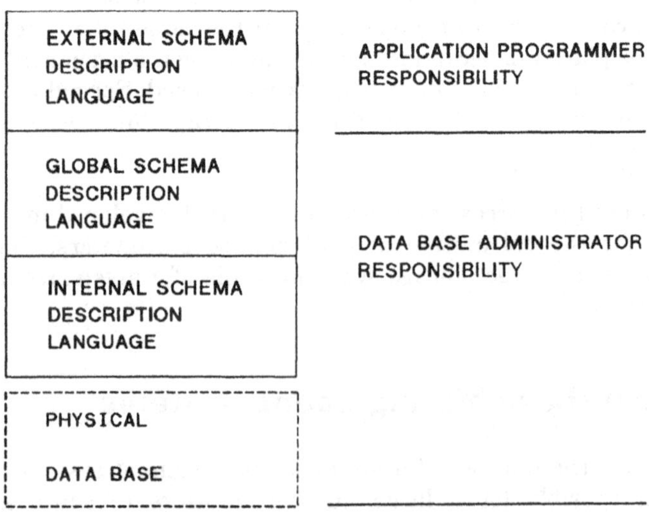

Fig. 4.3. Layers of DBMS data description languages

An external schema describes how an individual application views the logical organization of those data in the database that are of interest to just this application. Each external schema can therefore be viewed as a subschema of the global database schema. There are many external schemas, one for each computer program that uses data from the database.

The internal schema language is used to specify the physical layout and structure of data on the storage devices. It is specified how logical records are organized in storage blocks, and how buffering and overflow is controlled, and it is specified which addressing and searching techniques, such as indexing and hashing, can be used. The internal schema is concerned with all of the data in the database. The performance requirements of all of the database applications have to be considered when the physical structure of the database is designed. At some future time, systems may become clever enough to produce the physical layout of the database entirely automatically. However, the state of the art is far from that capability at present.

The application programmer's view is facilitated by the subschemas, and is often different from the view that is represented by the global logical database description. Furthermore, these views are different from the physical organization of data. The DBMS uses the data descriptions to derive the global logical records from the physical records, and to derive the records required by the application programs (described in the subschemas) from the global records.

A DBMS comprises a number of standard software components

The major software components of a DBMS are:

- Database Manager
- Data Definition Language (DDL)
- Data Manipulation Language (DML)
- Host Language Interface
- Query Language
- Report Writer
- Screen Formatter
- Data Dictionary
- Utility Programs

The Database Manager provides the ability to store, update and retrieve data. It works with the host operating system and the access methods that are supported by the computer environment.

The Data Definition Language (i.e., the schema language) could be:

- A data declaration facility in an application programming language, e.g., an extension of COBOL's Data Division;
- A facility provided by the DBMS, and that is independent of which host programming language is used;
- An independent data description language which could be employed by future database management systems.

The Data Manipulation Language provides us with syntax that can be used in standard programming languages to find the desired records in the database. In practice, different database management systems have widely differing facilities for describing schemas, and for specifying the manipulation of the data in the databases. However, the relational DBMSs have come a long way towards the use of standard data description languages and data manipulation languages. But even here the situation is far from ideal. There are still too many differences between the various contenders in the market to make it possible to change the DBMS without some reprogramming of the application programs.

The Host Language Interface enables programs written in standard programming languages such as COBOL to access and manipulate data in the database. This is very useful because programs that use the database can be written in the most appropriate language for the task at hand.

The Query Language is a natural language-like facility that enables people with little or no computer knowledge to access data and display it on a terminal quickly and easily. No programming in the traditional sense is required.

The Report Generators are similar to query languages in the sense that their programming languages have a strong natural language flavor. The Report Generator Languages vary in complexity from the simplest query-language-like facilities to almost full-function programming languages.

Screen Formatters help programmers to construct formats for display terminal screens, without having to go through the laborious process of specifying a multitude of complex control parameters required by the operating system of the computer.

The Data Dictionary is a very important feature of any DBMS. In its most basic form the data dictionary is merely a documentation tool, although a most useful one. It documents data definitions, relationships between records and between data elements, report formats, and other information that describes the database and the application systems. More sophisticated data dictionary systems are active, in the sense that they are used for the operation of the DBMS. They may generate the actual definitions needed by the DBMS to create and use the database. They may show the effect that proposed changes to the schema definitions will have on the various system components. They may even prevent the implementation of adverse changes.

The data dictionary is one of the prime tools of the Database Administrator. Because of the global nature of databases, there is a need for central coordination and control in order to protect them from misuse. The Database Administration function has been set up with this in mind. From starting up as a mostly technical function providing expertise for the physical database design, for database security, backup and recovery, the DBA function has evolved into a full-fledged data resource management function. The added responsibilities are that of corporate data modeling and that of maintaining a corporate data dictionary. Both of these are difficult and complex tasks, which are closely connected. Most organizations experience difficulties because things are defined differently across departments within the company, because things have the same name even if they are different, and so on. We shall return to these questions in later chapters.

4.1.3 Data Security

Data security issues may be divided into issues of

- *privacy*: protecting the database against unauthorized use,
- *integrity*: ensuring that the database is an accurate model of the part of the model it represents,
- *recovery*: protecting the database against errors which occur during system failure.

Database privacy concerns authorization to read and to modify the database

Databases usually contain some confidential data that must be protected from unauthorized prying. Authorization may be granted for reading, updating, inserting or deleting data, to individuals, to groups of individuals, or to special terminals that are physically protected and restricted. Authorization statements may take the form of statements such as: "A and B are the only employees who are allowed to see the contract proposals for new widget systems". Such statements have to be converted into constraints on the access paths in the access maps. For example: "Users A and B are the only ones who are allowed to follow the access paths into the PROPOSALS data store" (see Fig. 2.19).

Privacy requirements are identified fairly early in the logical design phase. The actual implementation of mechanisms for enforcing the constraints is, however, dependent on which facilities are offered by the DBMS that is to be used. The design of a privacy system involves decisions concerning how users/uses are to be identified, how access rights are to be stored and retrieved, and what actions to take when an attempt is made at unauthorized access to data. Unfortunately, many privacy schemes are implemented in practice in an ad-hoc manner [FROS84].

Privacy issues have previously been most relevant for large, multiuser systems, with shared databases. Because of today's communications environment, where thousands of computers are connected to each other through the public telecommunication network, the protection of the data of small and even single-user systems has become an issue. Unauthorized intrusion of, e.g., computer "viruses" proves that the protection of data against unauthorized prying has become a general issue facing every computer user.

The integrity of databases must be maintained by protecting them against invalid alteration or destruction

Assume that a customer places an order for a non-existent kind of widget. If such an order is accepted into the database, then the database is no longer an accurate model of the "world". Faulty data may occur for a number of reasons, ranging from incorrect observations of the world, data corruption through errors of coding or of transmission, data corruption through hardware faults, or through concurrent update, whereby two or more users are permitted to change the same piece of data at the same time. Many techniques have been developed for protecting the database against such errors. For example, parity checks are used to detect errors in transmission, and locking mechanisms are used to reduce errors due to concurrent update.

Other aspects are not so well developed. One of these is concerned with semantic integrity and its enforcement. By semantic integrity, we mean the compliance of the database with constraints which are derived from our knowledge about what is and what is not allowed in that part of the world that is represented by data in the database. An example has been given above, saying that one of the business rules is that we may not accept any orders for non-existent widgets. Other constraints may be that we cannot accept orders unless the person who places the order is already recorded in the database as one of our customers.

Semantic constraints can be of varying complexity. The subject is still not so well understood that general purpose integrity checking facilities have found a place in DBMSs. Consequently, ad-hoc procedures tend to be written and implemented in the application programs that handle the data input to a system. The maintenance of integrity is also a difficult and poorly understood subject. However, it is of crucial importance that the database is designed such that the required level of confidence in the data can be maintained. It is unrealistic, in most cases, to strive for a completely correct database. The cost for the associated integrity constraint checking may become staggeringly high. The designer's decision is to find the right compromise between database correctness and its operational cost.

The recovery of the database after a system failure is usually an overriding issue

The value of the data in the database is mostly so high that to lose it would be a disaster for the owner of the database, and might very well put him out of business. There is therefore usually no question about having recovery mechanisms or not, but rather of how speedy the recovery should be, and of how much it should cost. For some applications it is enough to have a copy of yesterday's database to start out from if today's database is lost, for others it is a catastrophe to lose even a single data element.

Recovery may be achieved in a variety of ways, by using locking strategies, backup files, recovery routines of different types, and even duplication of hardware. To a large extent, the recovery system is dependent on the physical storage structure, and the design of these two parts of the system should be carried out concurrently.

4.2 Data Model Alternatives

Although there are many competing DBMSs in the marketplace, they fall naturally into three categories based on the data model each one supports. The three data models are called *hierarchical*, *network*, and *relational*. A fourth model is the *entity-relational* model of data (the ER model), which is used mostly for logical database design. The ER model is not supported by any major DBMS, so ER specifications have to be translated to either a hierarchical, network or relational form prior to any DBMS implementations.

Before we embark on a discussion of the data models, we should bear in mind that few of the popular DBMSs available today are pure examples. Many commercial database products use their own terminology and support data models that are also not "pure". We should further keep in mind that it is difficult to make strong generalizations about the advantages and disadvantages of each model because so much depends on the specific implementation.

To facilitate the description of the first three models we shall use a simple database which consists of three simplified data stores from the Widget Inc. example company, for storing data about customers, widgets, and reservations of widgets that have been ordered by customers. Each customer may order many widgets of various types, and each widget type may be ordered by various customers. Each ordered widget is considered to be "on hold" until it has actually been shipped to the customer. Each ON_HOLD-record thus represents one customer's accumulated unshipped orders of one particular widget. The data store contents are

```
CUSTOMER  = CUSTNO, NAME, ADDRESS
WIDGET    = WIDNO, WIDNAME
ON_HOLD   = CUSTNO, WIDNO, QTY
```

Each ON_HOLD record is updated with the appropriate quantity whenever widgets are ordered or shipped. CUSTNO, WIDNO and (CUSTNO,WIDNO) are the primary keys of the three data stores.

The relationships among the data stores are depicted in Fig. 4.4. The double-headed arrows denote one-to-many relationships, i.e., one customer may appear in many ON_HOLD-records, but one ON_HOLD-record can only refer to one customer.

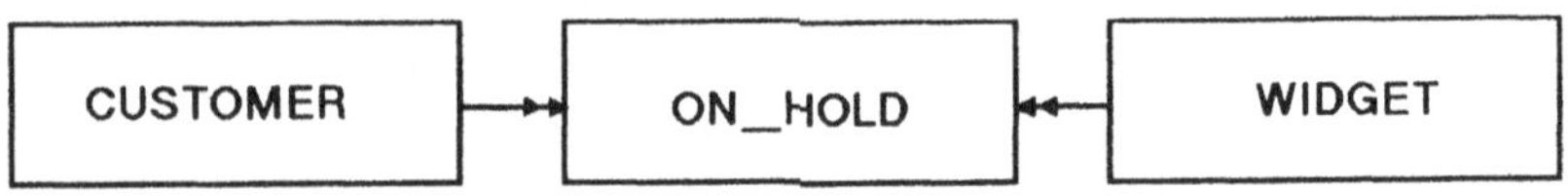

Fig. 4.4. The example database

4.2.1 The Hierarchical Data Model

The hierarchical data model has developed from the data storage and handling techniques commonly used for sequentially organized files. The hierarchical model requires, as its name suggests, that each record may be subordinated to one other record only. In the example above that means that an ON_HOLD record may be subordinated either to its CUSTOMER record, or to its WIDGET record, but not to both of them at the same time. The two possibilities are

```
#CUSTNO                       #WIDNO
  NAME,ADDRESS                  WIDNAME
  *ORDERED_WIDGETS              *CUST_RESERVATIONS
    #WIDNO                        #CUSTNO
      QTY                           QTY
```

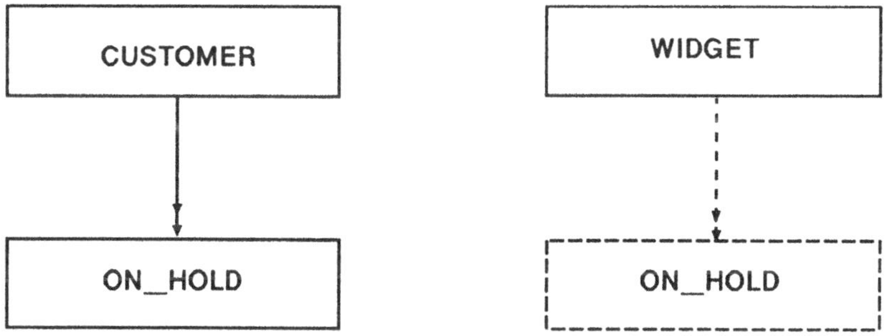

Fig. 4.5. Hierarchical schema for two alternatives of the customer-widget-on_hold database

We may choose either of these two organizations. But we may not choose both because that would leave us with redundant storage of the quantities on hold. This is further illustrated in Fig. 4.5, where it is assumed that the ON_HOLD records are made subordinate to the CUSTOMER records. This is indicated by showing the alternative (drawn in broken lines) where ON_HOLD records are subordinate to WIDGET records.

Note that a sequential file links ON_HOLD records to CUSTOMER records by physically storing the on_holds in the CUSTOMER record, while the hierarchical DBMS uses pointers to establish those linkages. A pointer is a part of a record that contains the address of the next logically related record. A record in a hierarchical database may therefore contain many pointers, depending on how many logically related records it has. Details of the widget-customer-on_hold example are shown in Fig. 4.6.

The ON_HOLD records are here said to belong to the CUSTOMER records, or to be members of the CUSTOMER records, while the CUSTOMER records are said to be owners of the ON_HOLD records. To be able to access an ON_HOLD record it is therefore necessary to specify the customer to which it belongs. This is all very practical if we want to know what some customer has reserved. But it is very impractical if we want to know which customers have reserved a particular widget. To answer that question we should have to search through all of the ON_HOLD records, as these are organized customer by customer. If, on the other hand, WIDGET records had been chosen to be superior to ON_HOLD records instead of CUSTOMER records, the opposite situation would apply for the two requests.

This asymmetry is at the root of the problem of using the hierarchical data model for the representation of data which do not posses a genuinely hierarchical structure.

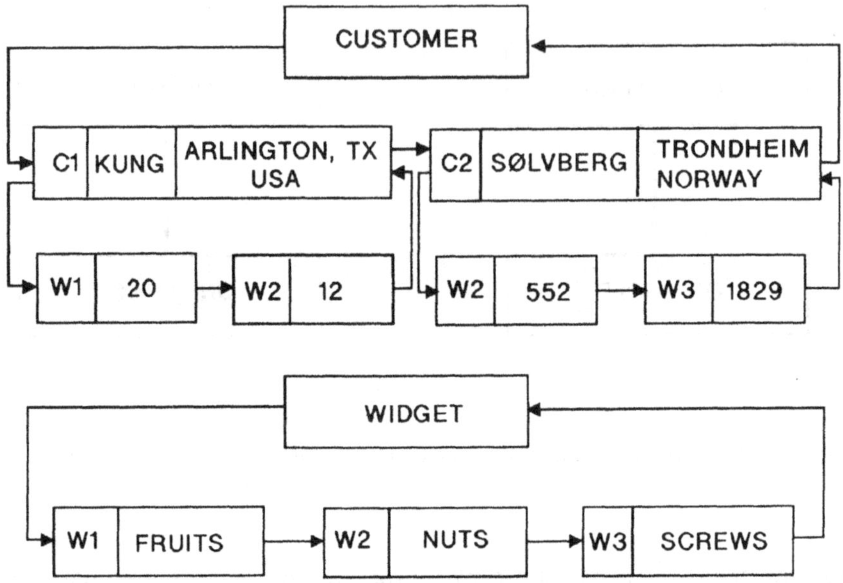

Fig. 4.6. Details of hierarchical database for widget reservations

4.2.2 The Network Model

The asymmetry of the hierarchical model is overcome in the network model. In the network model a record may have many owners. That means that the database schema of Fig. 4.5 can be modelled directly without further change. The network model links the member records of each owner in a ring structure connected back to the owner. In the customer-widget-on_hold example each reserved widget is a member of two rings (Fig. 4.7).

In the figure, the customer number and the widget number are shown to be included (in parentheses) in each of the ON_HOLD record occurrences, in order to make the figure a little more clear. Both are redundant data, as the proper key values can always be found by following the respective ring structures of pointers back to the owner records. Whether or not these data are included depends on the tradeoff between performance efficiency and data storage cost.

Entry to the database may be to a WIDGET record, to a CUSTOMER record, and/or direct to an ON_HOLD record if it has been specified that the address of this record can be generated from its key. We may now easily find the widgets reserved for one particular customer, as well as which customers have reserved which widgets. The price paid for the flexibility of the network model is the complexity of traversing the links in order to get to the data. Since the links may be represented by physical pointers in the

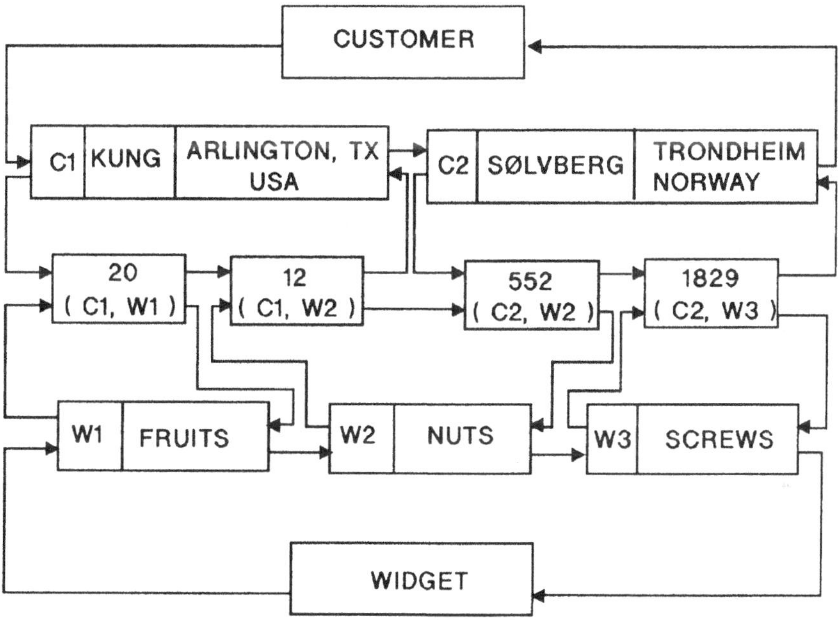

Fig. 4.7. The customer-widget-on_hold example represented in a network database

database, the network data model contains some of the major aspects of the physical data model. This transparency may enable the application programmer to anticipate what will happen at the storage level when issuing database commands.

In the example that we have used so far, we have only dealt with one-to-many relationships, e.g., one customer may have reservations for many different widgets, but one particular widget may only be on hold for one customer. The example has been set up in this way because the network model does not permit direct modeling of many-to-many relationships. In practical data modeling many-to-many relationships are abundant. In the example shown in Fig. 4.8, arcs with double-headed arrows in both directions indicate many-to-many relationships. The figure is to be read as follows: each customer may issue many orders, and each order may be for widgets of various types, but each order may only be issued by one customer. This is in good accordance with common business practice.

To model many-to-many relationships in a network database, it is necessary to convert them into one-to-many relationships. This is done by introducing special intersection records in order to relate each record of one record type to all of its related records of the other type, and vice versa. An intersection record exists uniquely within the confluence of two or more records connected by a many-to-many relationship. Each set of intersection records may be given a name which illustrates its function. The data that are

Fig. 4.8. An example database with many-to-many relationship

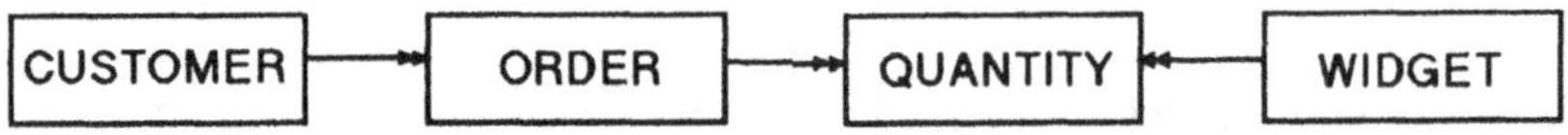

Fig. 4.9. Many-to-many relationship converted into one-to-many relationships

associated with the many-to-many relationship itself will become attached to the intersection records. In the example, one would expect that the quantities ordered of the various widgets would be attached to the intersection records. It would feel natural to call this record set QUANTITY (see Fig. 4.9).

We see that the QUANTITY records would contain those data that are normally found in the orderlines of customer orders, excluding those of the data elements that are specifically related to only the order per se, or to the widgets that appear on the orderlines.

4.2.3 The Relational Data Model

The relational data model originated from an attempt to define a fundamental theory for the structure of data that would enable data manipulations to be expressed with the same rigor and ease that is common in mathematics [CODD70]. General mathematical terms were borrowed and given a meaning restricted to their application for data modeling purposes. Basic terms are *relation*, *tuple*, *attribute*, *domain*. The terms will be explained below.

The data are organized in relational tables

A relational database consists of one or more two-dimensional flat files called *relations* or *tables*. We will use the latter term whenever possible, in order to avoid confusing the concept with the more general mathematical concept of relation.

Each *table* contains rows and columns roughly analogous to records and data elements in a non-relational setting. Each *row* in a table is called a tuple. Each *column* in a row must have a value. No two rows in a table may be identical. A relational table may be viewed as consisting of a set of nonhierarchical records of fixed length. We shall use the terms *record*, *row* and *tuple* interchangeably.

Each *column* represents the values for a specific attribute. A *domain* is the set of all possible values for an attribute. A *key* is an attribute or collection of attributes that always makes each tuple within a specific table *unique*. A key is sometimes called a *primary key* to distinguish it from a *foreign key*. An attribute that is primary key in one table may appear as an ordinary non-key attribute in some other table, in which case it is called a foreign key of that table.

A *valid relational table* cannot contain duplicate rows, that is, no two rows can have pairwise equal values in all of their columns. It follows that there will always be at least one collection of attributes that uniquely identifies each tuple. The collection may have to include all the attributes in a relation. Each such collection of attributes is called a *candidate key* for the relational table. One candidate key must be chosen as the primary key. It is obvious that all of the attributes of the primary key must hold a value for the key to be functioning properly.

A valid relational table can also not contain any repeating groups. The possible hierarchical organizations of the ON_HOLD records were previously proposed to be

```
#CUSTNO                      #WIDNO
 NAME,ADDRESS                 WIDNAME
 *ORDERED_WIDGETS             *CUST_RESERVATIONS
  #WIDNO                       #CUSTNO
   QTY                          QTY
```

or written a little more compactly as

```
    #CUSTNO, NAME, ADDRESS, *ORDERED_WIDGETS (#WIDNO, QTY)
and
    #WIDNO, WIDNAME, *CUST_RESERVATION (#CUSTNO, QTY)
```

where # indicates that the following term denotes a key term, and * indicates that the following term denotes a repeating group of attributes.

In order to represent the ON_HOLD data in a relational database, the repeating groups have to be removed, giving the resulting data structures

```
    #(CUSTNO,WIDNO), NAME, ADDRESS, QTY
and
    #(WIDNO,CUSTNO), WIDNAME, QTY
```

respectively. Please note that the primary keys are enclosed in parentheses preceded by "#".

The relational tables above may be decomposed further into the two pairs

#(CUSTNO,WIDNO), QTY
#CUSTNO, NAME, ADDRESS

and

#(CUSTNO,WIDNO), QTY
#WIDNO, WIDNAME

and we see that we have arrived at the logical data structure of a network model (see Fig. 4.8).

There is a one-to-one correspondence between the network model and the relational model

This is illustrated in Fig. 4.10, where the customer-widget-on_hold database is represented in the relational view. By comparing the relational database of the example system with its network representation (Fig. 4.7) we see that the primary key of the ON_HOLD "file", (CUSTNO,WIDNO), must be stored explicitly in the relational table, while they are implicit in the network solution because they are also contained in the links of the ring. In the figure that illustrates the network solution we have indicated a possible redundant storage of the primary key by enclosing the relevant key-values in parentheses in each ON_HOLD record. We can now see that if we remove the links between the ON_HOLD records in the network solution, there would be no redundant information, and with a little reorganization we end up with the relational database of Fig. 4.10.

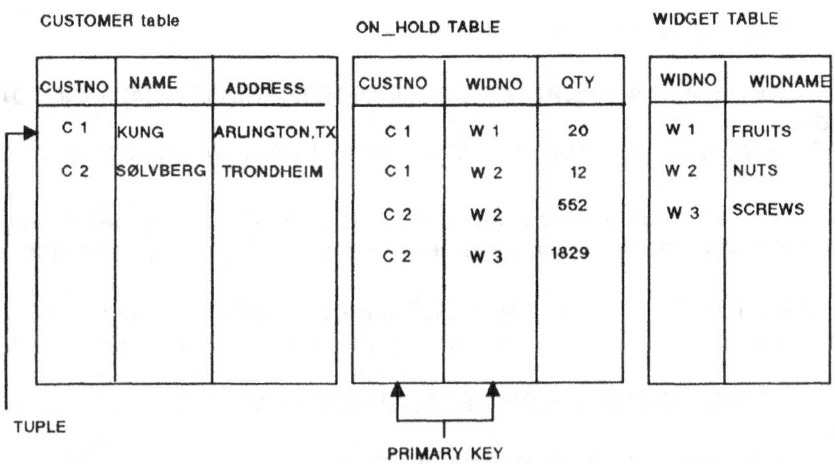

Fig. 4.10. Relational database for widget reservations

It is always a relatively simple matter to translate from the network model to the relational model and vice versa, provided that a unique key can be defined for each relational table. The possibility of translation should always be taken into consideration when choosing data modeling tools in a practical systems development situation.

The relational model is conceptually easier to master than the network model because it is so far removed from considerations about physical storage of data. It may therefore provide inexperienced users with workable applications in a very short time, in particular for single user applications with low data volumes, where performance, data integrity and other mundane operational problems are of no great importance.

The basic operations on relational data are *selection*, *projection*, and *join*

All of the data in a relational database are represented by means of relational tables. There are no pointers and no owner or member records. This means that only one type of data is dealt with. The fixed nature of the relational table means that an algebra of operations may be defined on the data in much the same way that mathematical operations are defined. The three basic operations are called *selection*, *projection* and *join*. Examples are provided in Figs. 4.11 and 4.12.

Selection provides a *horizontal* slice of data from a relational table, that is, those rows ("records") are selected that satisfy some selection criterion. For example, if the operation were to select all rows from the ON_HOLD table that have the value C1 in the CUSTNO column, all of Kung's widget reservations would be returned after the execution of the operation.

Projection provides a *vertical* slice of data. It enables us to examine specific columns from a table with redundant rows removed. For example, if the operation is to project the ON_HOLD table on its CUSTNO attribute, this is the same as asking which CUSTNO values are present in the relational table. The project operation would take out the CUSTNO column, remove the redundancies in the column and return the values C1 and C2.

Join operations combine data from two or more tables (Fig. 4.12). The relational tables to be joined must have a common domain, that is, they must have some common attributes that share some values. For example, the ON_HOLD table and the WIDGET table have the common attribute WIDNO. If the WIDNO value of one particular ON_HOLD record is found in one of the WIDGET records, we may form a new, composite record from the data elements in those two records. We have done this for all of the records in the two tables (Fig. 4.12).

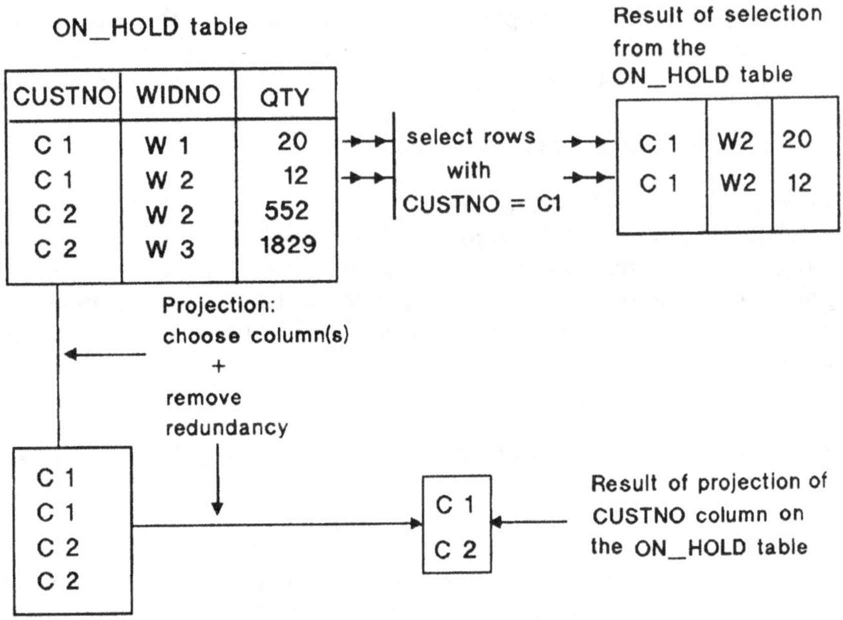

Fig. 4.11. Relational operators "select" and "project"

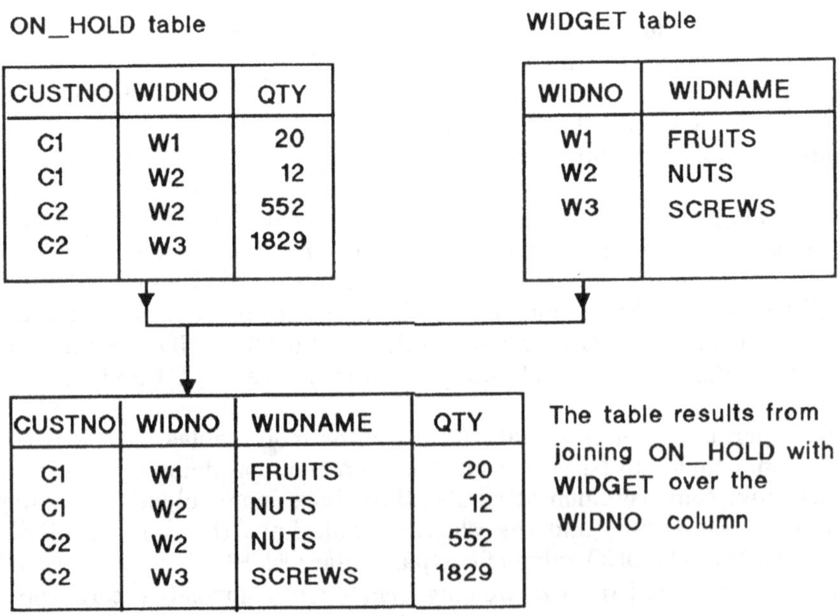

Fig. 4.12. The relational operator "join"

Relational databases may be decomposed into *normal forms*

The problem of defining the contents of data stores is to identify which data are of importance to the applications, and to arrange those data in an effective manner. We want a data arrangement that is performance effective as well as effective in reflecting the properties of the world outside of the database. In response to particular information requirements from some of the users, we may create data structures whose elements have little relationship to each other, while strongly correlated data are spread out in a number of different tables. Usually there is a need to restructure the first proposals.

The process of restructuring consists of a first task of decomposition, followed by a grouping of the smaller data structures that come out of the decomposition into more comprehensive data structures with more desirable properties than the original ones. This is done in one form or another regardless of which data model is used. In hierarchical databases and network databases the decomposition is informal. There is no way of evaluating the quality of the result except by employing human inspection. The relational data model has decomposition defined in formal terms. In the relational model, decomposition is called normalization.

Normalization converts a relational table into smaller tables until a level of decomposition is reached where little or no data redundancy exists. There are several so-called normal forms: the first normal form, the second normal form, and so on, abbreviated to 1NF, 2NF, 3NF, etc., respectively. Each new normal form is governed by progressively stricter rules. Originally there were only three normal forms, but these have later been extended by a 4NF and a 5NF.

A relational table is in 1NF (i.e., first normal form) if

- it is a valid table,
- a unique key has been defined for each row, and
- all attributes are functionally dependent on all or part of the key.

In a valid table, each column represents only one attribute, each row is unique, values for each key are present, there are no repeating groups, and the sequence of the rows in the table is unimportant.

Functional dependency exists when a unique value of one attribute can always be determined if we know the value of another. For example, a customer's name may always be determined if we know the customer's unique identifier, e.g., the customer number. On the other hand several customers may have the same name, so a customer number may usually not be inferred from the customer's name. We therefore have that the customer's name functionally depends on the customer's number, but not vice versa.

Any valid table is usually in 1NF, no matter how complex its key is or what interrelationships there may be among its data elements

A valid table is usually in 1NF, because it is most unusual to have columns in a table that have no relationship whatsoever to the key of the relational table. Relations in 1NF may suffer from two kinds of complexity:

- some of the non-key attributes may depend on only part of the key, not on the whole key,
- some of the non-key attributes may be interrelated.

This is easy to visualize. We shall use our previous example of customers having widgets on hold. In order to be able to use the example to explain all of the three first normal forms, we shall expand it a little bit by adding on the widget price and the money value of the widgets that are on hold for the various customers. After the removal of the repeating groups the data structures may be

#(CUSTNO,WIDNO), NAME, ADDRESS, QTY, VALUE

and

#(WIDNO,CUSTNO), WIDNAME, WIDPRICE, QTY

where VALUE = WIDPRICE * QTY, and where WIDPRICE functionally depends on WIDNO. The two tables are in 1NF but there are interrelationships among their attributes. The tables may be decomposed further to 2NF – the second normal form.

A relational table is in 2NF if every non-key attribute in the table is dependent on the entire key. All of the non-key attributes are then said to be fully functionally dependent on the primary key

By decomposing our example tables to 2NF we get

#CUSTNO, NAME, ADDRESS
#(CUSTNO,WIDNO), QTY, VALUE

and

#WIDNO, WIDNAME, WIDPRICE
#(WIDNO,CUSTNO), QTY

The four tables are all in 2NF. Because the last table #(WIDNO,CUSTNO), QTY is obviously included in the table #(CUSTNO,WIDNO), QTY, VALUE we may remove the former and end up with the three 2NF tables

#CUSTNO, NAME, ADDRESS
#WIDNO, WIDNAME, WIDPRICE
#(CUSTNO,WIDNO), QTY, VALUE

But within the relational tables there are still complexities that may be removed. We see that the VALUE of the widgets that are on hold cannot be found unless the quantity on hold (QTY) is known. We shall simplify the tables further by decomposing to 3NF – the third normal form.

A relational table is in 3NF if it is in 2NF and all of the transitive dependencies between the non-key attributes have been removed

We have a 3NF table if all of the non-key attributes are fully dependent on the primary key, and no non-key attribute is functionally dependent on any other non-key attribute. In our example we have to decompose the last table further, in order to remove the dependency between QTY and VALUE. We get the two tables

```
#(CUSTNO,WIDNO), QTY
#(CUSTNO,WIDNO), VALUE
```

both of which are in 3NF.

Although the third normal form provides us with relational tables that are so simple and "elementary" that one would consider them non-decomposable, it is possible to go further in the decomposition by using additional knowledge about how the various records relate to each other. Consider the situation that some of our customers ordered their best-selling widgets on Mondays and Thursdays only. That would mean that there would be a certain repetitive pattern in the appropriate relational table. Monday records could be repeated on Thursdays, for instance. This kind of information may be used to decompose the files into smaller ones by dividing the records into different groups. Higher-order normal forms deal with such situations by using this kind of world knowledge for decomposition purposes. Although of theoretical interest, the practical significance of 4NF and yet higher forms is questionable, so we will not discuss them further here, but leave them to specialist treatment.

The 3NF relations have desirable properties. They are a logical model of the data stores, as free as possible from consideration of physical implementation. Data can be structured in 3NF relations by knowing only their key and their functional dependencies. There is no reference to the way that the data is processed within the computer. It may therefore be assumed that the 3NF relations will change only because of changes in the methods of operating the business that is to be supported by the database. Thus 3NF data may be considered to be relatively invariant to the physical implementation of the database, and can therefore serve as the basis for implementation considerations.

By starting out with many data stores and decomposing all of them to 3NF, we may easily detect redundancies. If there is overlap between the 3NF components that come from the normalization of two (or more) different data structures, then we have detected redundancy between the original data structures. This may be most helpful when the database is so large that no single person has a sufficient overview of the database.

Normalization may lead to a large number of small relational tables. This may be undesirable both for performance reasons and also because it may be difficult for the poor system analysts and maintenance programmers to keep track of too many tables at the same time. It is therefore good that 3NF relational tables provide us with a well defined basis from which we can start to merge tables in order to decrease the number of logical files in the database. We may merge two or more 3NF tables into a new 3NF table, provided that we adhere to the 3NF rules that the tables share the same primary key, and that the new table does not contain any transitive dependencies between any of its attributes.

The normalization process is dependent on our ability to identify the primary keys of the relational tables, and on our ability to find the functional dependencies between the elements of the data structures

Both the keys and the dependencies are necessary in order to perform the normalization correctly. Keys and functional dependencies reflect properties of the world outside of the data structures rather than properties of the data elements themselves. So, when building a relational data model there is a tacit understanding that there is already a model of the world that is agreed upon, and that the data modeling task is to translate the relationships in the "world model" into relationships between the data elements, such as functional dependencies.

We shall illustrate through an example how the world modeling has to be taken into account. Assume that we have a data store consisting of customer orders of the following structure

```
ORDER
   ORDER_IDENTIFICATION
      ORDER_NUMBER
      ORDER_DATE
   CUSTOMER_DETAILS
      CUSTOMER_NUMBER
      CUSTOMER_NAME
      PHONE
      SHIPPING_ADDRESS
      BILLING_ADDRESS
   WIDGETS_ORDERED          .......... This group of data items to be
      WIDGET_NAME                      repeated one or more times
      WIDGET_NUMBER
      QUANTITY_ORDERED
```

Let us try to express the relationships between customers, widgets, orders and their respective attributes through functional dependencies between the data elements of the data structure above. The notation is simple: Functional dependency is expressed by a single-headed arrow going from the independent attribute to the dependent attribute, e.g., $A \rightarrow B$ is to be read " A determines B", and means that the value of B always can be determined if we know the value of A. The functional dependency $A,C \rightarrow D$ means that the D-value can be determined if we know values of A and C.

The functional dependencies of the data structure shown above may be expressed as follows:

```
ORDER_NUMBER → ORDER_DATE
ORDER_NUMBER → CUSTOMER_NUMBER
CUSTOMER_NUMBER → (CUSTOMER_NAME, PHONE,
                    SHIPPING_ADDRESS, BILLING_ADDRESS)
WIDGET_NUMBER → WIDGET_NAME
ORDER_NUMBER,WIDGET_NUMBER → QUANTITY_ORDERED
```

With this knowledge about the semantics of the data structure we may happily start to normalize and thereby contribute to the creation of a superb relational database. The problem is, however, that the functional dependencies do not become available unless someone first performs a thorough analysis of the underlying relationships between the entities that exist external to the database. That is, it is necessary to know the relationship between orders, customers and widgets before one can determine the functional dependencies among their attributes. In order to support such analysis, data models are needed which capture the underlying semantics of the data structures better than the relational data model.

4.2.4 The Entity-Relationship Model

The hierarchical and network models provide the users with means to reason about an organization on the data record level. Even if the relational model is further removed from the physical storage level, it permits only discussion of organizational issues in terms of data structured in relational tables. Many relationships in an organization are too complex to be mapped onto relational tables in a straightforward fashion.

Data models were needed which provided the user with more expressiveness than the relational model and which were further removed from the physical storage level than the hierarchical model and the network model. In other words, the new data models had to provide more semantic content than the relational model provides. Instead of focusing on the data, one had to focus on what the data represents. A database would be viewed as a representation of our perception of the part of the world that is of interest to us ("the universe of discourse").

The Entity-Relationship model [CHEN76] unifies features of the traditional models with enhanced semantic content. Perceptions of the world are expressed in terms of entities, relationships and attributes. There is a simple graphical notation for the ER-model which has contributed substantially to its popularity. The graphical notation is similar to the network model's graphical notation. At the representational level, the data structures for entities and relationships strongly resemble relational tables.

In the ER approach to data modeling the key idea is first to describe the entities and relationships of the enterprise (the part of the world that is of interest to us), without regard to data storage considerations, and later to translate this description into a user schema for the database system that is to be used. ER-modeling may therefore be viewed as an intermediate stage in logical database design.

An entity is a "thing" which can be distinctly identified in our minds and is of interest to the organization. An entity may be a person, a place, or anything else that an information system is to store and maintain data about. Entities can be classified into different entity types, such as PERSON and PLACE, and are represented in rectangular boxes in the ER-diagrams.

Relationships may exist between entities, e.g., customers may order widgets. Relationships can be classified into different relationship types. For instance, the relationship type ORDER_UNIT between the entity types CUSTOMER and WIDGET represents ordering associations between customers and widgets. Each customer may order several kinds of widget, and each kind of widget may be ordered by several customers, making this a many-to-many association. Relationship types are represented by either lines or diamond-shaped boxes in the ER-diagrams (Fig. 4.13).

Relationships may be either

- one-to-many (1:N),
- many-to-many (M:N), or
- one-to-one (1:1),

as is indicated in the figures.

Fig. 4.13. ER-diagram for customers ordering widgets

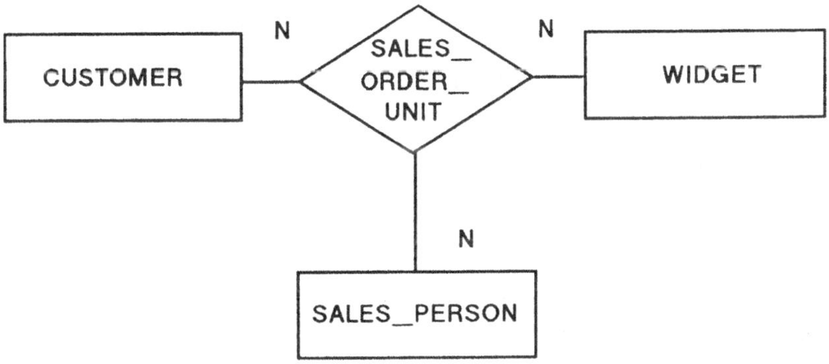

Fig. 4.14. ER-diagram for widgets being sold to customers by salespersons

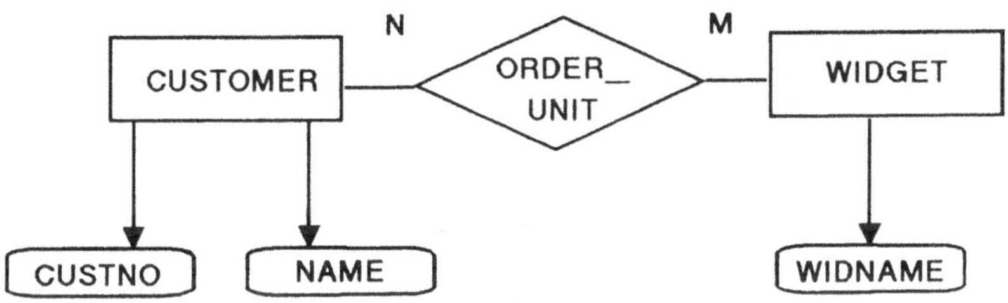

Fig. 4.15. ER-diagram for customers ordering widgets, with attributes for the entity types

It is possible to define a relationship type among more than two entity types. For example, let us define a SALES_ORDER_UNIT relationship which also involves salespersons in addition to customers and widgets (Fig. 4.14). Note that a three-way relationship cannot be replaced by three binary relationships. Usually higher-order relationship types are better represented by entity types, as explained in the next section.

Entities have properties which can be expressed in terms of attribute-value pairs. For example, in the statement "the NAME of CUSTOMER x is Vincent Lum", NAME is an attribute of customer x, and "Vincent Lum" is the value of the attribute NAME. Values can be classified into different value types, such as ADDRESS, QUANTITY, NAME. In the ER diagrammatic notation, a value type is usually represented by a circle, and an attribute is represented by an arrow directed from the entity type to the desired value type. In Fig. 4.15 we have chosen to replace circles by rounded rectangles to represent value types because it is difficult to place long names within circles.

The ER-model is extremely simple. The only modeling concepts that are used are entity types and relationship types for "real world" modeling, and attribute-value pairs for data modeling. There are no facilities for sub-classification in the ER-model, and there are no facilities for relating the database entities to the "worldly" entities except for the attributes that relate the value types to the entity types. These shortcomings have been solved in various ways, by adding new facilities to the model.

At this point it should be stated that there is nothing wrong in having few modeling concepts in a method. There is always a trade-off between simplicity and expressability in a modeling approach. An approach that employs only a few modeling concepts may be so much easier to learn and to use than the more comprehensive method that this far outweighs its lack of expressability for situations more complicated than the simplest ones. One most important feature of any modeling approach is that it contains the minimal number of modeling concepts that gives the maximal utility to the user of the approach, for the use that the user needs the model for. In this respect it seems that the ER approach has been successful, judging from its popularity, in spite of its obvious shortcomings.

Relationships may be viewed as entities

The ER-model supports classification into entity types, relationship types and value types only. So every phenomenon of interest must be classified into one of those types. One immediately runs into difficulties. Is an "order" a relationship between a customer and some widgets, or is it an entity in itself? Should an ORDER_UNIT type be classified as a relationship type or an entity type? Obviously an order may be viewed both as an entity having its own existence and as a relationship between two other entities. Which classification should be chosen? In the final analysis this boils down to a question of how important the order phenomenon is for the organization that is being analyzed. Important phenomena are elevated to the status of entity types, while not-so-important phenomena are given a relationship status or are represented through attribute-value pairs.

To support the treatment of phenomena which may be classified either way, the ER-model has modeling facilities that enable one to keep them both as entity types and as relationship types [CHEN85]. In the ER-diagrams such types are represented as diamond-shaped boxes circumscribed by rectangular boxes (Fig. 4.16).

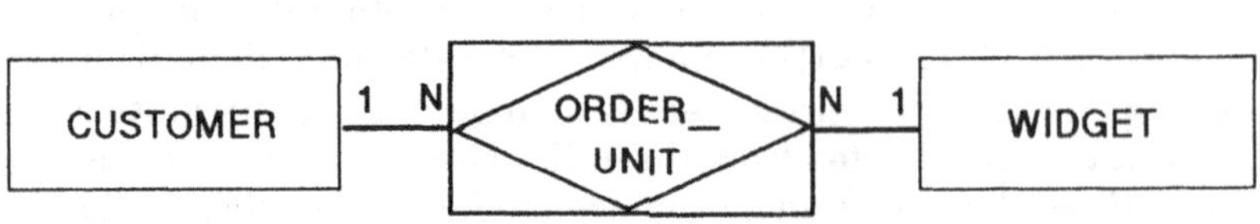

Fig. 4.16. ER-diagram for customers ordering widgets, with three entity types

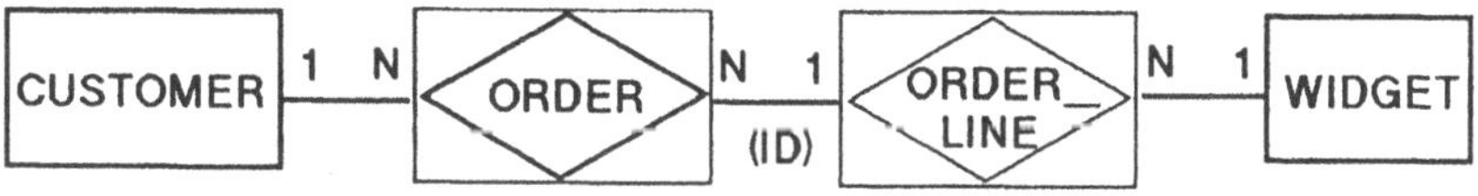

Fig. 4.17. ER-model for customers having (composite) orders for widgets

By introducing this "composite" entity type one makes it possible to enforce that relationship types cannot have attributes. If the "importance" of a relationship type increases, so that it becomes natural to attach attributes to the type, this may be achieved by reclassification, as shown above.

If we look more closely at this example, we find that our ER-model is still not a good representation of the real-world situation that we are trying to describe. Customers usually place their orders in groups. They do not order one kind of widget at a time. They collect many individual orders into larger orders. Each individual order-unit is called an orderline. In the ER-diagram of Fig. 4.16 the composite entity type ORDER_UNIT represents all of the orderlines in the system. We see that one orderline may only relate to one customer and to one widget, while one customer may relate to many orderlines. One kind of widget may appear on many orderlines as well.

To represent a picture of the world which is more consistent with the realities, we shall have to reformulate the ER-model of our example. We introduce a new composite entity type ORDER from the (1:N) relationship between CUSTOMER and ORDER_UNIT, and we rename ORDER_UNIT as ORDER_LINE. The new ER-model is depicted in Fig. 4.17.

Weak entities are uniquely identified only via their relationship to other entities

Entity types may relate to each other in various other ways which are difficult to capture with only three basic modeling constructs. One of these is called existence dependency. We have existence dependency when the existence of an entity depends on the existence of another entity. One frequently used example of an existence dependency is the dependency of the CHILDREN entity type on the EMPLOYEE entity type in a personnel database. In such a database, children are only of relevance as long as they are children of an employee. If an employee leaves the company, and the associated data is removed from the database, one would also want the data about his children to be automatically removed from the database. We see that existence dependencies are closely related to an assessment of the relative "importance" of the entity types being considered. Children do not cease to exist because one of their parents quit working for an organization, but they cease to be of interest for that organization. The CHILDREN entity type is in such a situation said to be a *weak* entity type, and is represented by a double rectangular box in ER-diagrams (Fig. 4.18).

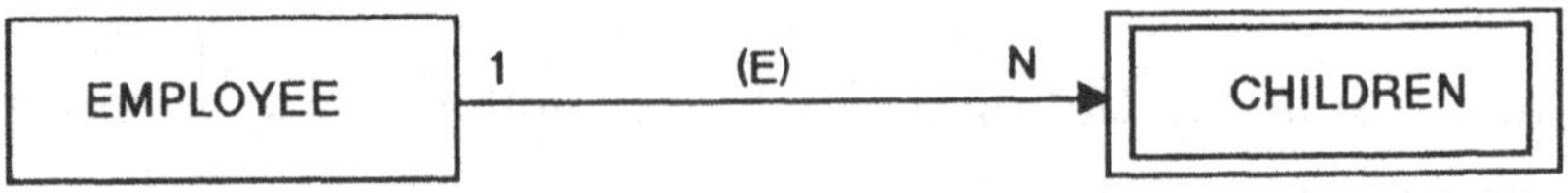

Fig. 4.18. An existence-dependent relationship type (E) and a "weak" entity type

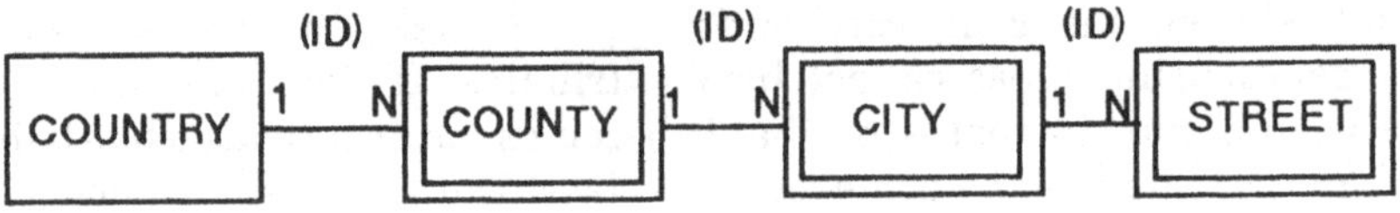

Fig. 4.19. Identifier-dependent relationship types (ID) and "weak" entity types

Existence dependency is closely related to identifier dependency. We have identifier dependency when an entity cannot be uniquely identified by its own attributes but has to be identified by its relationship with other entities. We may encounter this kind of situation whenever one entity type is considered to be so much "part of" another entity type that it does not pay to identify the latter on its own.

A typical example is the identification of streets in a city. In most cases it is not advisable to give streets unique identifiers above the city level, for example on the country level. Only for very special cases would this be meaningful, for example for streets like Champs Elysees, Pall Mall and La Rambla, but certainly not for 43rd street. For most streets the only sensible thing to do is to refer to the city where the street is located when identifying the street. When one entity type is identifier dependent on another entity type we let the former be a weak entity type (Fig. 4.19). Note, however, that existence dependency does not necessarily imply identifier dependency. The CHILDREN entities may perfectly well be identified through their own attributes independent of their parent's identifications.

Attributes and value types

We have previously discussed the difficulties one may encounter in the classification of phenomena into either relationship types or entity types. There is a similar uncertainty in the definition of attributes. Consider the previous example of representing a geographical location through interrelated entity types. If we are not particularly interested in geographical locations per se, as is most often the case, we would rather choose to view this as a value type which we would call ADDRESS, and later on define ADDRESS to be an attribute of the appropriate entity types.

Again we see that the classification of phenomena into entity types, relationship types or value types, depends on how explicitly we want the various phenomena to be treated, and how independently we want to view them in relation to each other. In the example above, a value type ADDRESS will always be used to describe a property of some other entity types. Only if the houses, the streets, the cities and so on are being classified into entity types, will the geographical location be granted an independent existence in the ER-model.

The classification of phenomena into the three categories of entities, relationships and values is highly subjective, and depends on the purpose of the classification. In practice, however, one may choose to reserve the entity type category for those phenomena that must have an identifier of their own in the system being analyzed, including identifier-dependent entity types. Properties of the entity types can be expressed in terms of attribute-value pairs, while relationships between the entity types must be expressed by relationship types.

In some cases an attribute may have more than one value for a given entity. An employee may, for example, have more than one child. In the graphical notation of the ER-model this is indicated by putting a “1:N” symbol in the attribute arrow to indicate that it is a multivalued attribute (Fig. 4.20). This is similar to the repeating group concept in the hierarchical data model. However, many attributes are single-valued. If there are no other indications in the ER-diagrams, the attributes are tacitly understood to be single-valued.

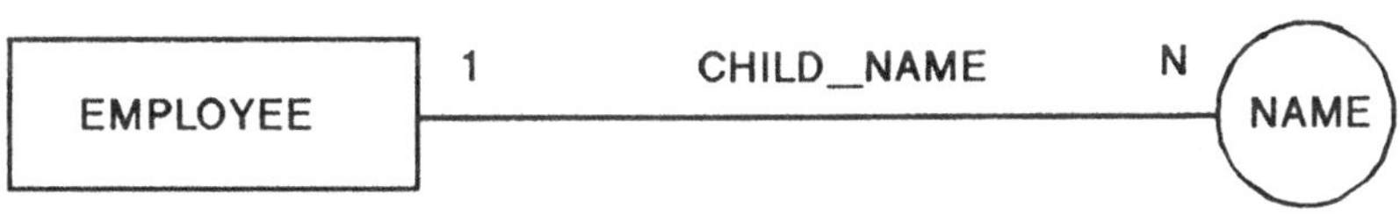

Fig. 4.20. An entity type and a multivalued attribute

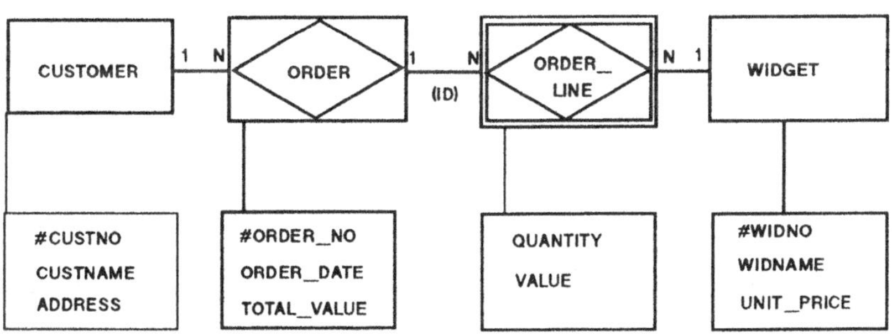

Fig. 4.21. ER-diagram for customers ordering widgets, with attributes represented in tables

The graphical representation for attribute-value pairs is too clumsy to be practically useful. It is better to use a combination of graphs and tables for representing the ER-model (Fig. 4.21). Please note that we have defined two composite entity types, ORDER and ORDER_LINE. The latter is a weak entity type, and is identifier-dependent on ORDER. This is so because it is impractical to let every orderline have a unique identifier within the system. We usually distinguish between orderlines by giving each of them a unique orderline number within the order to which they belong. Attributes which are unique identifiers are marked with "#" in the usual manner that we have used previously.

The value type definitions must contain a number of additional details, some of which are listed below:

- Name, synonyms and acronyms by which the data element is also known,
- Size, accuracy (number of decimals), dimension ($, %, meter, inch,....),
- Relationships to other value types, e.g., a DATE may consist of DAY, MONTH, YEAR,
- Value ranges.

The attribute definitions must contain details like

- Identification of key-attributes,
- Derivation rules,
- Editing/lay-out rules.

All of this has to do with collecting knowledge that is useful also for the creation of the physical database. Not all of these details are easy to visualize in a diagram. For example, derivation rules are more practically represented in formulae and tables, e.g., VALUE = PRICE*QUANTITY.

The ER-model is a data modeling tool

The ER modeling constructs are too weak to be able to capture a rich enough set of real-world features to make the model practically useful as a tool for modeling the "enterprise", for which it originally was intended. One example of a serious shortcoming is that the ER-model is strictly a one-level model. It is not possible to express subclassification as such, that is, that one entity type is a subtype of another entity type. This deficiency alone prevents the ER-model from effectively supporting multiple views of an organization and its database. This and other shortcomings contribute to the fact that the ER-model is more frequently used as data model during logical database design than as a tool for the modeling of the "enterprise".

Used as a data model during logical database design the ER-model has several strong properties compared to the network and the relational data models. The most important of these is that one is encouraged to reason about the organization in organizational terms, rather than in terms of the data chosen to represent the appropriate organizational properties. We are free to represent our model of entities and relationships in a network which is not restricted in its expressability because of physical implementation considerations, e.g., the ER-model permits many-to-many relationships, which are forbidden in the network model.

Unfortunately, the lack of expressibility in the ER-model leads its users to view the entity type structure that is developed for supporting the definition of the database as being itself an extended network model of the database. The ER-structures (with their many-to-many relationships) may be easily translated to the more restricted network model, with the addition of specific network model features. Alternatively the ER-structures may be translated into a relational model, each entity type becoming a relational table of the attributes as they are specified in the ER-model. It is therefore quite to be expected that the users of the ER-model end up by modeling data entities instead of organizational entities, and that they tend to perceive the entity types as being prototypical relational tables instead of being of the non-data variety.

4.3 Issues in Database Physical Design

The basic objective of the physical design phase is to provide an operational solution for the storage of data that satisfies the required performance and security requirements. Depending on how important those requirements are, it may or may not be advisable to use an available DBMS, or to procure one. The physical design task is to create a database schema that supports the applications' information processing needs, as stated for example in the access map, subject to existing constraints of hardware, database management systems, and other relevant software.

The result of the logical design serves as the starting point for the physical design. Even if there are various approaches to the logical design, the design results differ mostly only in form, not in content. The design results are the definition of the data elements, records and relationships between them that are needed to satisfy the functional or business requirements of the system. The logical design results serve as requirements that the physical design must satisfy. These requirements have to be translated into specifics, that is, into a form that conforms to the physical storage models, and they have to be analyzed with respect to storage and performance effects. Based on the results of the analysis, physical storage structures can be designed which satisfy the requirements in an optimal way, given that the conditions for optimal solutions can be expressed.

Most databases are implemented by the use of database management systems. It is therefore proper to first give a short overview of strong and not-so-strong properties of the three major kinds of DBMSs, before giving a short explanation of how the logical design is translated and analyzed prior to being forged into a physical data structure.

4.3.1 Properties of Database Management Systems

The major differences between DBMSs have to do with [GAYD88]

- how data are stored and retrieved,
- how data relationships are represented,
- which data security controls are available,
- ease of use of the system.

Hierarchical and network DBMSs have many similar characteristics, that are different from the properties of relational systems. There is, however, so much overlap between the properties of DBMSs of the three kinds, that few DBMS products can be considered "pure" with respect to the data model that they are claimed to support.

For physical storage, the hierarchical and network systems make extensive use of pointers embedded in each record, so that related records can be easily found. Relational systems do not use pointers, but use flat files in conjunction with secondary indices. Hierarchical and network systems also make use of indices, but they do not rely solely on those, as do relational systems.

Most systems, regardless of data model, also place frequently used records in physical proximity in order to enhance performance. The term "physical proximity" refers to the placement of records that are often used together, e.g., a NAME record and an ADDRESS record, in the same or in adjacent physical storage blocks in secondary storage, so that they may be retrieved in one retrieval operation rather than through a sequence of retrieval operations.

Access paths cannot be defined during run-time for hierarchical and network systems, but must be defined prior to execution of the application programs. In relational systems one may in principle access any piece of data, in conjunction with any other piece of data in the database, through the application of the relational operators *select, project* and *join*. This capability may, however, be very costly, and may also exist only if the database is skillfully designed.

Hierarchical DBMSs use tree structures for the representation of data, while network DBMSs use a variant of the tree structure called a *plex structure*, which permits relationships to be stated between any two nodes in a tree. Relational systems do not use any structural representation, but rely on two-dimensional tables.

Hierarchical and network DBMSs were developed in order to support massive, data-intensive, batch processing applications that handled millions of records. They therefore generally provide more facilities for ensuring data integrity and recovery than do relational systems. Most relational systems were designed primarily to support small interactive ad-hoc information systems for few users. Therefore database security facilities were not emphasized. This "defect" is being gradually removed in the new product generations that reach the market. So relational systems are gradually making inroads into the corporate data processing market, which previously showed a quite cold attitude to the relational products because of their weak data security facilities.

Relational systems are easier to use than hierarchical and network systems. On the other hand, hierarchical and network systems provide the database designers with more design support facilities than they have when designing relational databases. Hierarchical and network databases may be tuned in order to obtain increased performance much more effectively than is possible for relational designs. It is not entirely unexpected that this results in increased (perceived) complexity for the users of the DBMSs. It should be noted, however, that the tuning of the databases concerns the physical placement of the data, and does not encompass structural redesign. On the contrary, database redesigns that encompass structural modifications can be very difficult and painful in hierarchical and network systems. So one had better be right the first time!

Hierarchical and network systems are appropriate when

- most relationships are one-to-many,
- most data manipulation is routine and predefined,
- performance is important,
- the database structure is fairly static,
- data security is important.

Relational databases have not been suited for large-scale transaction processing. Hierarchical and network databases can be designed around the most heavily travelled access paths. Relational tables are more "neutral" and tend not to support any specific access strategy. This has the effect that hierarchical and network databases can be designed so that they support *critical* transactions, at the expense of giving mediocre performance for transactions that have less priority.

4.3.2 Translation and Analysis of the Logical Design

In order to be able to perform the physical design properly it is advisable that

- the logical data structures are translated to the data model of the DBMS that is going to be used for the implementation of the database,

- the data volumes are analyzed for the various record classes and their interrelationships, and

- the access paths are analyzed in order to establish priorities among conflicting requirements.

Logical data models have to be translated to a DBMS data definition language

No matter where we start, the translation will result in the identification of records, from which a database will be built by the use of DBMS software. If a relational database has been selected as the implementation vehicle, we shall have to translate the logical data structure into a relational data model. If it is a network DBMS we shall have to translate into a network data model.

During the systems analysis phase one usually does not want to be constrained by DBMS requirements. For example, one prefers to be allowed to formulate many-to-many relationships if this feels natural in the modeling situation, rather than to be constrained to using the more restricted modeling vocabulary of one-to-many relationships. ER-models are widely used during the first data modeling phase for just this reason. Translations from ER-models to either relational models or network models are therefore the most frequently occurring data model translations. They are both quite simple, and have been mentioned already in the section on the ER-model.

It may be worthwhile to add here a remark about so-called "iterative" relationships, that is, relationships between entities of the same entity set. Assume that widgets may have substitute widgets, e.g., if somebody orders fruits, and we are out of fruits, then we will offer the substitute widget, which may in this case be nuts. In the ER-model of data this situation could be modelled by a many-to-many relationship between WIDGET and itself (Fig. 4.22). "Iterative" structures are illegal in the network model, as are many-to-many relationships, and have to be translated to an appropriate network structure as shown in Fig. 4.23.

Each widget is seen to have potentially many substitutes, but each widget may also potentially be the substitute of many other widgets.

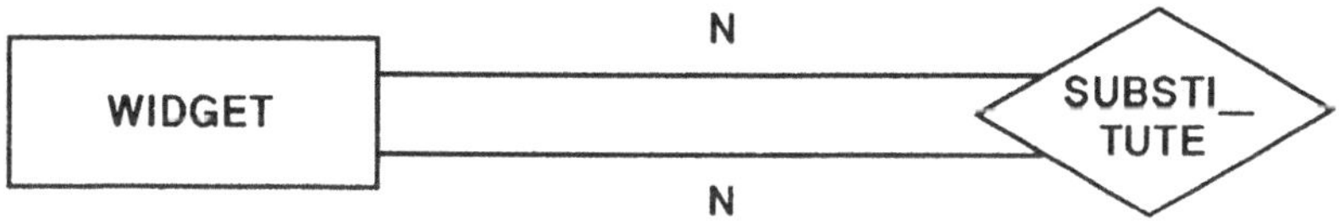

Fig. 4.22. "Iterative" ER-diagram for widgets and their substitutes

Fig. 4.23. Network diagram for widgets and their substitutes

Estimates of the data volumes are needed in order to determine the database storage requirements

Data volume estimates are needed because they have a big impact on database performance. The storage and accessing of half-a-million book records in a library system requires a far more thorough design than the accommodation of a few hundred employee records in a small firm.

Both the absolute volumes and the relative volumes are important factors in determining the storage and accessing arrangements. If there are ten thousand pending PROPOSALs and one million different kinds of WIDGETs, it is very different from having tens or hundreds of each. It is also very different, from a database design point of view, if there are one hundred different kinds of widgets involved in each proposal than if there are only five. In the latter case we would probably not care very much, because we would need a maximum of only five database accesses to retrieve the appropriate widget data for each proposal. In the former case, there is a large savings potential if we can come up with a good storage scheme. One hundred database accesses is probably in any case too much if we need to have a reasonably short response time for retrieving proposals, so we would have to come up with a good storage scheme anyway.

The volumes may be indicated in the diagrams as shown in Fig. 4.24. From the figure we see that there are 350 000 different kinds of widgets, that there are altogether 75 000 substitute relations, and that each widget that has substitutes has an average of 2.5 other kinds of widgets that can be used instead of it. Widgets that can be used as substitutes can substitute for an average of 1.5 other kinds of widgets.

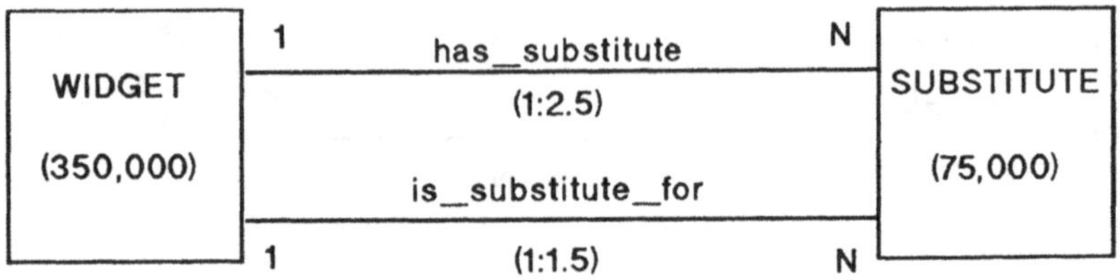

Fig. 4.24. Network diagram for widgets and their substitutes with indication of the relative volumes

When characterizing the volumes, it is advisable to look out for how the relationships are distributed. For example, it may make a big difference if only 10 % of the widgets are frequently used, and 90 % are almost never used, instead of an even usage of the different kinds of widgets in the proposals. Skew distributions of this type happen more often than not, so it is good advice to look out for them and to use them for one's advantage. In the example of Fig. 4.24 there are obviously skew distributions. The numbers indicate that only 30,000 (=75,000/2.5) kinds of widgets have substitutes, and that approximately 50,000 (=75,000/1.5) can serve as substitutes for others. Depending on the access requirements it may be decided to let widgets that can appear in substitute relationships get special treatment rather than to treat every widget as a potential participant in a substitute relationship.

Access path analysis provides the basis for database physical design

In previous sections we have seen how access needs have been transformed into an access map. The access needs must now be translated to fit the data model of the chosen DBMS, and further details have to be added in order to give a good basis for analysis. A thorough analysis is important for physical design in order to avoid slow and long access paths to frequently used data.

Access path analysis consists of

- detailed description of access needs,
- establishing access request priorities,
- determining the accessibility of desired data.

The access needs description including

- type of access (on-line query, update, batch, etc.),
- retrieval keys (Customer number, Widget number, etc.),
- data to be accessed (name, address, etc.),
- frequency of use (every second, monthly, etc.),
- natural language description of transaction.

Access request priorities may be established after an initial breakdown of the transactions into, e.g., query and update, high-volume and low-volume, online and batch processing transactions. For each transaction it should be noted if it is especially time-critical or not. The accessibility of the desired data must be ensured. If desired data cannot be accessed by the proposed design, one has to redesign in order to accommodate the request, provided the request is important enough to warrant the additional work.

4.3.3 Physical Design Approaches

The next design steps consist of *record design*, and *record set design*.

Record design consists of the following steps:

- group data elements into records,
- select keys and access methods.

For network systems one also has to

- decide on duplicate records, and
- assign each record to a physical storage area.

For relational systems those decisions are made by the DBMS, not by the database designer.

Record set design consists of deciding how the relationships between records in each set of records are to be handled and how sets are to be traversed. For relational systems those decisions are implicit in the DBMS, but for network systems it is possible for the designer to influence the physical structures by

- defining pointers and other navigation mechanisms,
- selecting record insertion techniques, and
- selecting owner-membership relationships.

The latter has to do with so-called referential integrity. For example, if we define customer records to be owners of order records, the network DBMS will not allow us to insert an order for a nonexistent customer without creating an appropriate customer record. The major consideration in selecting the owner-membership options is the degree of automation that one desires in maintaining database integrity.

It is beyond the scope of this text to go into details of physical database design. The reader is referred to the specialist literature, e.g., [GAYD88], [YAOS85], [ELMA90], [DATE77]. We shall, however, in the next section give an example of database design based on analysis of the transaction load on the database access map.

4.4 Database Design Constrained by Traffic Load Estimates

Methods for logical database design are mostly concerned with determining a database structure which satisfies the users' information needs and retrieval/update requirements. The physical database design methods aim at determining storage structures and physical location of data, such that performance requirements and users' requirements, as expressed in the logical database design, are satisfied. The major weakness of this phase distinction is that it is difficult to assess the quality of a logical database design, except by intuitive judgement.

In the previous chapter on software design a method was explained for calculating the usage of the data in the database relative to the users' load of external transactions. The objective of the method is to estimate the performance of logical design alternatives before making the detailed physical design decision. The method shows how users' transaction activities can be transformed to activity within the set of logical data stores. This provides aid for the subsequent schema design [OFTE81].

In the following, it is shown how to take advantage of the workload in the design of a DBTG database. We will not consider the earliest phases of the design process in detail. Instead, we will show informally how the results from these phases can be applied in a particular case. For comparison, we also suggest a DBTG schema which satisfies the users' information and functional requirements, but which is not based on performance considerations (Fig. 4.26).

4.4.1 The Example

A personnel department of a firm wishes to develop a computer-based information system to record some facts about its employees. The transactions are depicted in Figs. 4.25a-e (from [OFTE81]).

Some of the data stores need further explanation. An employee may work or may have worked in different departments, and many employees may work in one department. This indicates a many-to-many relationship between employees and departments. An employee may have many cars, but a car has only one owner. Thus, there is a one-to-many relationship between employees and cars.

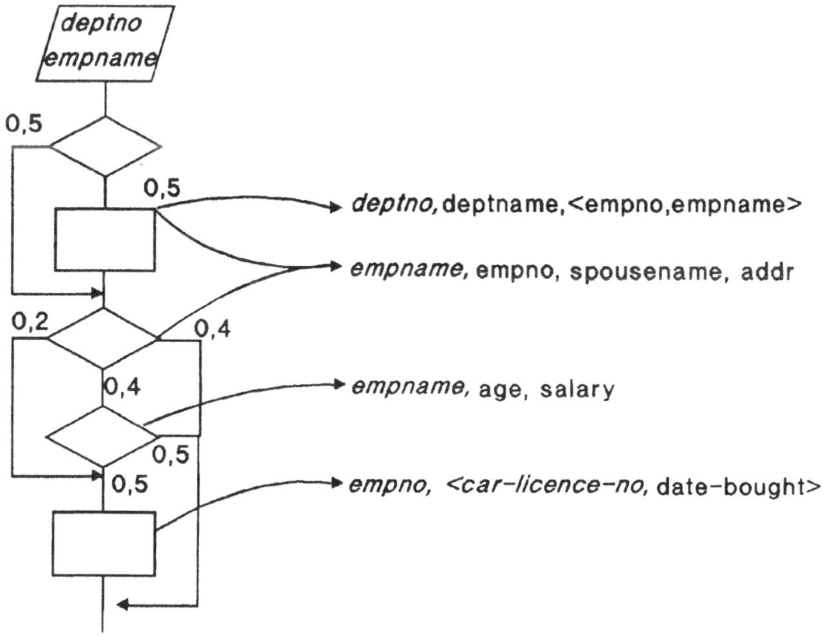

Fig. 4.25a. Transaction no. 1, requesting data based on knowledge of employee's name and department number

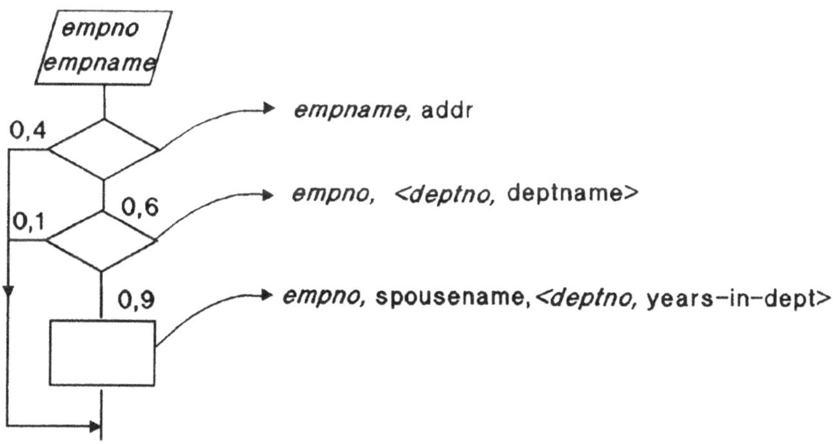

Fig. 4.25b. Transaction no. 2, requesting data based on knowledge of employee's name and number

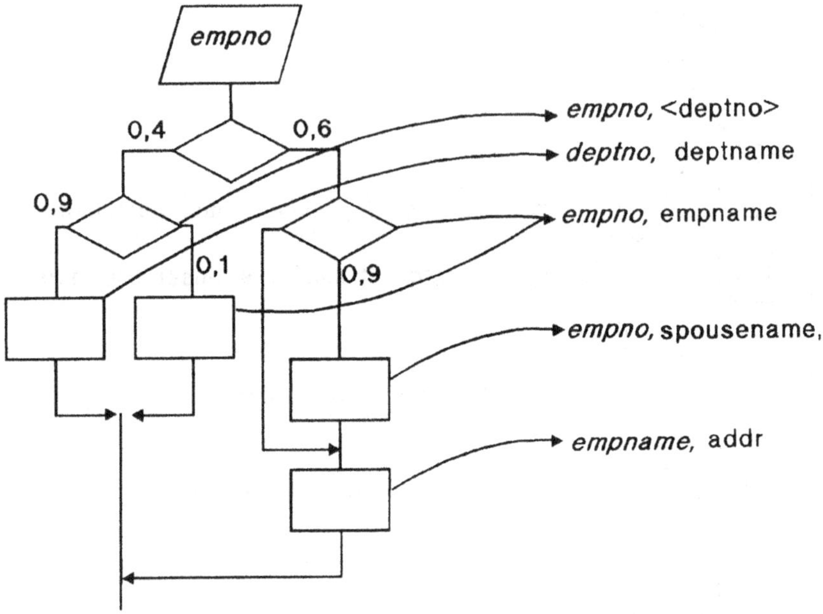

Fig. 4.25c. Transaction no. 3, requesting data based on knowledge of employee's number

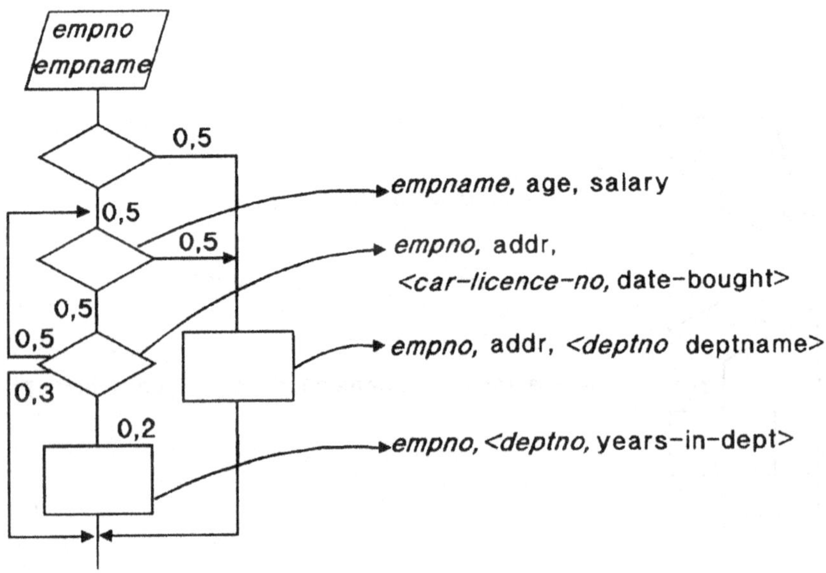

Fig. 4.25d. Transaction no. 4, requesting data based on knowledge of employee's number and name

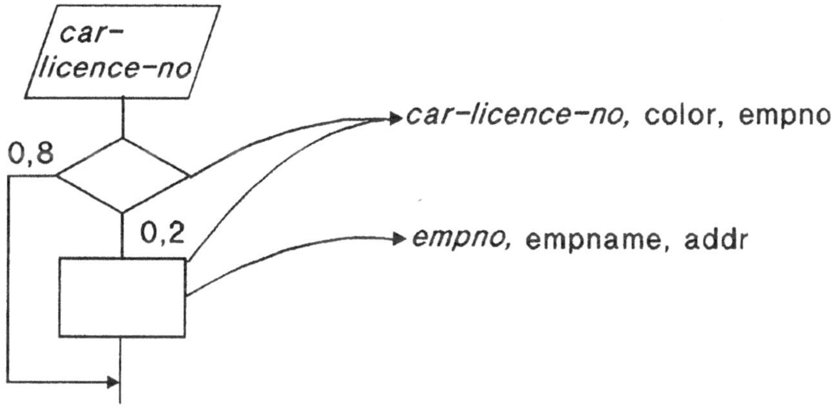

Fig. 4.25e. Transaction no. 5, requesting data based on knowledge of car's license number

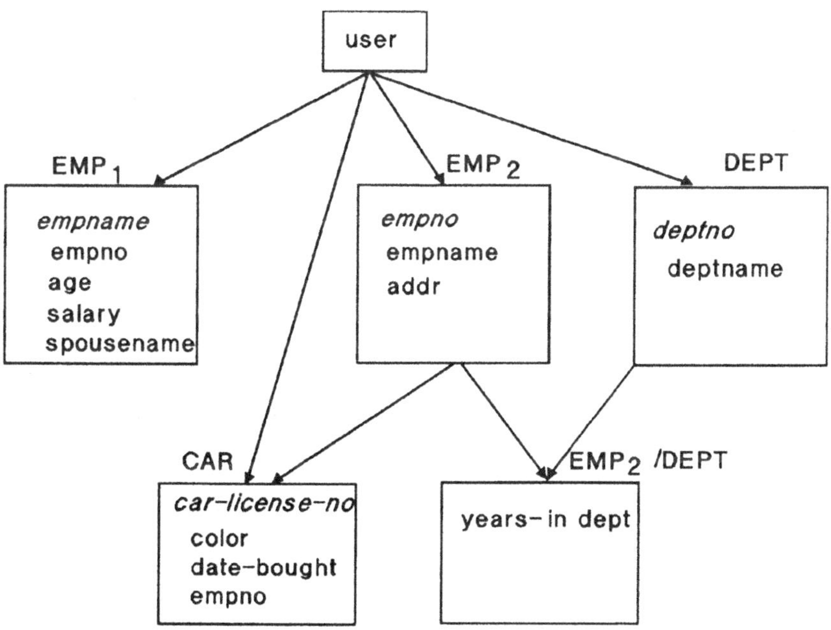

Fig. 4.26. Logical database structure satisfying information and functional requirements

A possible logical database structure is shown in Fig. 4.26. All functional requirements are satisfied by this solution.

So far, performance efficiency has not been considered. Therefore, some of the record classes have been defined quite arbitrarily. It is not obvious that age, salary and spousename should be accessed via empname, and that addr should be accessed via empno in order to arrive at a satisfactory solution. The redundancy of empno and empname in EMP_1 and EMP_2 is not desirable in a dynamic database.

Traffic load analysis will solve these and other design problems.

4.4.2 Traffic Load Estimation

The global information set becomes (keys written in italics):

a_1 *empname*, empno
a_2 *empname*, age, salary
a_3 *empname*, spousename
a_4 *empname*, addr
a_5 *empno*, addr
a_6 *empno*, empname
a_7 *empno*, spousename
a_8 *empno*, deptno
a_9 *empno*, *car-license-no*, date-bought
a_{10} *empno*, *deptno*, years-in-dept
a_{11} *deptno*, deptname
a_{12} *deptno*, empno
a_{13} *car-license-no*, color, empno

The traffic load on the global information structure is given by $\boldsymbol{E}'$, $\boldsymbol{G}'$ and $\boldsymbol{T}'$, as these are defined in Sect. 3.5. We recall that

$\boldsymbol{E}'=\{e'_i\}$ contains the number of entrances to each a_u, u = 1,2,...,13 in the global information set

$\boldsymbol{G}'=\{g'_u\}$ contains the number of times each a_u is active, and

$\boldsymbol{T}'=\{t'_{uv}\}$ contains the number of transitions from a_u to a_v, u,v = 1,2,...,13.

For the transaction frequencies f_1 = 50, f_2 = 1500, f_3 = 1600, f_4 = 100 and f_5 = 20 and the transition probabilities given in Figs. 4.25a–e, we get the **E**', **G**' and **T**' matrices as follows.

	1	2	3	4	5	6	7	8	9	10	11	12	13
E'=	50	50	50	1550	50	985	0	690	0	0	75	25	50

G'=	75	87	939	2535	121	1053	810	1623	53	817	1584	25	24

1	25	20	25	25	0	0	0	0	10	0	0	0	0
2	0	0	0	0	67	0	0	33	43	0	33	0	0
3	25	20	25	889	0	0	0	0	10	0	0	0	0
4	25	20	25	25	0	0	0	900	10	0	900	0	0
5	0	17	0	0	0	0	7	0	0	0	0	0	0
6	25	0	889	121	0	0	0	0	0	0	0	0	0
T'=7	0	0	0	0	0	0	0	0	0	0	0	0	0
8	0	0	0	0	0	64	810	0	0	810	576	0	0
9	0	17	0	0	0	0	0	0	0	7	0	0	0
10	0	0	0	0	0	0	0	0	0	0	0	0	0
11	25	0	25	25	0	0	810	0	0	810	0	0	0
12	25	0	25	25	0	0	0	0	0	0	0	0	0
13	0	0	0	0	4	4	0	0	0	0	0	0	4

4.4.3 Reasoning About the Consequences of the Transaction Traffic

Remember that ***E'***, ***G'*** and ***T'*** are not independent. In general, $g'_u > e'_u$ for all u, and the difference $g'_u - e'_u$ is the number of internal transitions to a_u.

We notice that g'_4 is much larger than g'_5. This indicates that there is a stronger relationship between empname and addr than between empno and addr. If we prefer to store addr in only one place it should be stored in a record class which has empname as key.

We also note that g'_6 is much larger than g'_1. This means that one asks for an employee's name, given his employee number, much more frequently than the other way round. Because there is a one-to-one relationship between empno and empname one of them is redundant. It is not necessary to store both empname and empno in the same record. A table which links empname and empno can be used instead. Preferably, this table should have empno as source and empname as target. If g'_1 had been of the same order of magnitude as g'_6, or if the smallest of them had been "large", then a second, inverse table might have been profitable. The table can be implemented using some hashing technique.

As opposed to Fig. 4.26, where data about the employees are stored in two record classes EMP_1 and EMP_2, we may choose to define only one record class. However, the traffic load estimates indicate that the employee data can be profitably separated into two records. We may choose to define only one record class. However, the traffic load estimates indicate that the employee data can be profitably separated into two records. We define EMP'_1 and EMP'_2 which have empname and empno as keys, respectively (Fig. 4.27). All of the record classes in the new schema of Fig. 4.27 are marked as EMP'_1, EMP'_2, CAR', DEPT'.

We now decide in which one of these records to store "spousename".

We had

a_3 *empname*, spousename $g'_3 = 939$
a_7 *empno*, spousename $g'_7 = 810.$

Since g'_3 and g'_7 do not differ significantly, our knowledge of the activity of a_3 and a_7 is not sufficient. But we notice that most of the transitions to a_3 comes from a_6, because $t_{63} = 889$ is much larger than any other t_{i3}. This means that spousename should be stored in EMP'_2.

Because of low traffic to a_2 from any other a_i, age and salary should be stored in EMP'_1.

The other record classes CAR' and DEPT' will be defined irrespective of traffic load estimates. Then we have the following record classes:

EMP'_1: *empname*, addr, age, salary
EMP'_2: *empno*, spousename
CAR' : *car-license-no*, color, date-bought
DEPT': *deptno*, deptname.

The hierarchies which are expressed by repeating groups are conveniently implemented as DBTG sets. If there are large numbers of references from a member record class to the owner record class the LINKED TO OWNER option in DBTG provides direct reference from the member to the owner. In particular, if the average set size is large one does not need to traverse a long set link to get access to the owner. The number of accesses also depends on other factors, e.g., record lengths and block factors. This will not be discussed here.

In the example there is only one one-to-many relationship between employees and cars. In this case it is not necessary to use the owner link because the number of transitions from a_{13} is very small. To store empno in the car records is not necessary either.

The many-to-many relationship between employees and departments will be implemented using an intersection record, as years-in-dept is the only attribute. Observing that the relationship between empno and deptno is stronger than the relationship between empname and deptno, EMP'_2 (not EMP'_1) should be used in this many-to-many relationship.

The traffic from EMP'_2 to DEPT' is large (g'_8, g'_{10}, $t'_{8,11}$) compared with the traffic in the opposite direction (g'_{12}). LINKED TO OWNER from the intersection record class to DEPT' and physical adjacency between the intersection record class and EMP'_2 will reduce the access cost. We also note that $e'_{11} + e'_{13}$ is rather small. This indicates that the access methods we choose for DEPT' and CAR' are less critical than those used for EMP'_1 and EMP'_2.

After this informal discussion we arrive at the database structure in Fig. 4.27. Structurally, this schema is fairly similar to the schema which was not based on traffic load estimates, Fig. 4.26. But some record classes are different. The traffic load estimates provide a means for deriving efficient database structures. The traffic load supports the definition of records, and it indicates where physical adjacency and owner links are especially useful. We have argued that the choice of access methods for the different record classes also should be based on the traffic load estimates. In our example, the final schema contains very little redundant data.

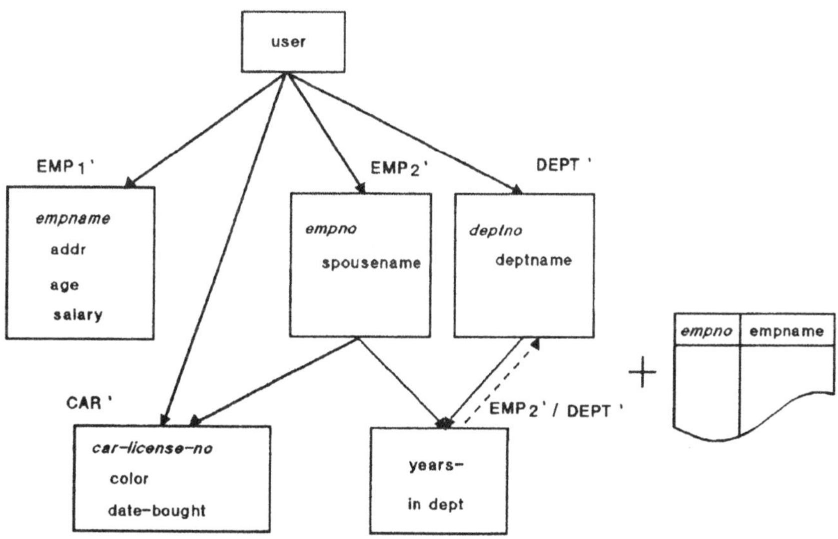

Fig. 4.27. Final database structure

4.4.4 Refining the Database Design

We recall that the traffic within the transaction structures is mapped into traffic within the global information structure using $\boldsymbol{Q}$. For a particular transaction k the number of deletions can be found if we redefine $\boldsymbol{Q}_k$, so that $q_{iu,k} = 1$ if the process $\alpha_{i,k}$ deletes or updates a_u (update can be viewed as delete followed by insert).

The traffic load also pinpoints which data items of a record are requested more frequently than others. March and Severance [MARC77] divide data items into two groups according to their frequency of use and store the groups in different places. The group which is requested most frequently should be easier to access than the other. We may introduce an additional record class to take advantage of this. For example, consider EMP'$_1$ which consists of a_2 (*empname*, age, salary) and a_4 (*empname*, addr).

Suppose that g'_2 is much larger than g'_4 (in contradiction to the previous figures!). This means that most references to EMP'$_1$ will not request the address of an employee, but that the requests are for his age and salary. But addr has then to be transferred every time age and salary are requested. In this case one should consider having a separate record class containing only addr; see Fig. 4.28.

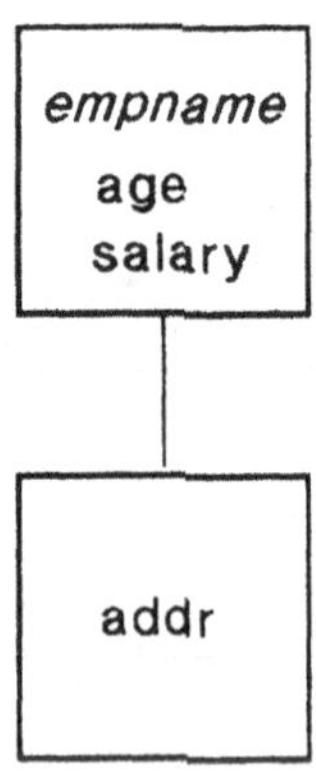

Fig. 4.28. Separation of data items according to frequency of use

4.4.5 Interpretation of the T-matrix

In order to arrive at a design which functionally supports the information requirements, $\boldsymbol{T'}$ should be interpreted as follows.

If $t'_{uv} = 0$,

then there is no traffic from a_u to a_v in the global information set. No access path in this direction has to be implemented. If $t'_{vu} = 0$ too, no access path in either direction is necessary. If $t'_{vu}>0$, access from a_v to a_u is required. The database designer may decide to implement the access path in different ways. This depends on the size of t'_{uv} and the capabilities of the database management system at hand (it can be implemented as a hierarchical father-son relationship, by secondary indices, etc.). Alternatively, a_u and a_v can be grouped into one record type, if this is feasible semantically.

Consider the global information set $\{a_1, a_2, a_3\}$.

a_3 conveys, for each car,

a_1: (empno, address)
↓
a_2: (car-license-no, type)
↓
a_3: (car-park, time-interval),

information about allowed car parks within a certain park area. Because $t'_{12} > 0$ and $t'_{21} = 0$, a hierarchical path should be implemented between a_1 and a_2. The same argument holds for a_2 and a_3. If the database management system is restricted to hierarchical database structures the path from a_3 to a_1 cannot be implemented implicitly. In order to answer questions like "which employees use car park...at...o'clock" a secondary index with car park as source and empno as target is required.

If $t'_{uv} > 0$ and $t'_{vu} > 0$,

then a bidirectional access path has to be implemented. Logical pointers, secondary indices or grouping of a_u and a_v into one record type may solve the access problem.

Consider the following example which forms a network:

a_1: (empno, address)
↑↓
a_3: (car-park, time-interval).

This structure cannot be implemented directly in a hierarchical database management system, but by introducing a logical pointer a bidirectional access path can be achieved. To obtain access efficiency the most frequent access direction should be reflected implicitly by the physical database structure, whereas the other direction should be implemented by using a logical pointer. If a set is accessed sequentially (e.g., to produce reports) it is cumbersome to follow logical pointers, so the opposite solution may be preferred.

As indicated in the example above, adjustments to the particular database management system at hand play an important part in the design. Physical adjacency in the storing of records which are requested in a predefined sequence is also an important design parameter, since it tends to reduce the number of block accesses. The retrieval/update ratio should also be considered. This ratio affects the amount of redundant storing of data one should allow for. The advantages of redundancy for retrieval operations have to be balanced against the disadvantages of redundancy for update operations.

Exercises

1. Give examples of semantic integrity.

2. Give example of business/industry applications which fit the hierarchical, network and relational data models. State your reasoning for each of the examples.

3. Assume that you want to store some information about your friends in a file named FRIENDS. Show the logical and physical record structures of the FRIENDS file. Do your own selections for the fields of the FRIENDS file.

4. What are the advantages and disadvantages of having a direct file "blocked"?

5. What are the factors affecting the block size?

6. Where do we need
   sequential files,
   direct files,
   indexed files
   in business applications? Give example applications.

7. Compare database management systems with conventional file organizations. State the advantages and disadvantages.

8. State how the ER model differs from other data models which are hierarchical, network and relational.

9. A database is to contain information concerning products, suppliers and customers. A product may be supplied by one or more supplier and may be demanded by one or more customers. A customer may demand one or more products.

   (a) Design a suitable relational data structure for this data,
   (b) Convert the data structure in part (a) into a network data model.

10. Explain the implications of weak entities in the physical design of databases.

Chapter 5

Rule Modeling

In the analysis and design of information systems, we are frequently required to specify information system functions that evaluate complex combinations of conditions and perform appropriate actions. The specification of these functions depends on the logic of the problem that is under consideration. Since the functions are implemented by processes, the task of specifying these functions is by some authors [GANE78] called the specification of process logic. The task requires special treatment because there are a variety of problems that need to be considered. Some application problems need to perform a large number of actions, while others may perform only a few. On the other hand, there are applications that require evaluating many combinations of conditions to determine which of the actions are to be performed, while only a few combinations are possible in other applications. Therefore, we need different tools to solve different problems. Some of the most widely used tools will be studied in this chapter. These tools are commonly called *structured tools*, i.e., structured English, decision trees, and decision tables.

The specification of the process logic must consider other issues. That is, the specification must be complete and consistent. Completeness means that every possible combination of conditions has been taken into account. Consistency means that each combination of conditions leads to only one set of actions; otherwise, the system would be in a conflicting situation, e.g., having to fulfill and to reject the same order. In formulating the process logic, we frequently encounter situations in which several combinations of conditions lead to the same set of actions. However, some of the conditions of the combination do not materially affect the selection of the actions. Therefore, we may remove these "redundant" conditions from the combination without changing the logic of the problem. The simplified formulation will be much more concise and easy to comprehend. In this chapter, we shall study methods for checking completeness and consistency and removing redundancy in a specification.

The specification of process logic is an important task between "programming in the large" and "programming in the small". According to a study on the psychology of programming by Sime, Green, and Guest [SIME73], two tasks, called *taxonomizing* and *sequencing*, must be performed to produce a program. The first task is concerned with identifying and specifying the conditions that lead to a particular set of actions. This is called the specification of the process logic. The second task is to convert the taxa and the actions into program code. An empirical investigation of the above mentioned tools has been conducted by Vessey and Weber to determine how they facilitate the taxonomizing and sequencing tasks [VESS86]. The basic conclusion of their experiments along with opinions of other authors will be presented at the end of this chapter.

5.1 Rule Formulation

Rules may be specified in a symbolic form which makes it possible to formally manipulate them, or they may be formulated as algorithms. We shall explain the difference through an example. Consider the task of finding the derivative of the function $f(x)$ for $x = x_0$. We may approach the task in two different ways. We may either solve the task by numerical evaluation or we may apply symbolic evaluation to first find the derivative $f'(x) = df(x)/dx$ and then calculate $f'(x = x_0)$.

If we apply numerical evaluation alone, we have to calculate an approximate solution $f'(x_0) \approx (f(x_0 + \Delta) - f(x_0 - \Delta))/2\Delta$, where Δ is a small number. If $f(x)$ is viewed as a rule, we see that we may either process the rule directly by using it as an algorithm, or we may manipulate it symbolically by deriving its derivative $f'(x)$ prior to a numerical evaluation to find $f'(x_0)$. This difference will be further discussed in Sect. 5.1.1. We will also discuss some difficulties in finding precise formulations of rules, because of ambiguities when using natural language, in Sects. 5.1.2 and 5.1.3.

5.1.1 Rule Processing Versus Rule Manipulation

The modeling of process logic is a particular case of rule modeling. The rules that can be specified by the tools that are studied in this chapter, decision tables and decision trees, are of the form

if (some combination of conditions is true) **then** (do some action)

The conditional part of the rules are conjunctions of individual logical conditions or their negations, of the form $C_1 \wedge C_2 \wedge .. \wedge$ **not** $C_i \wedge .... \wedge C_n$. In the example of vegetable preparation introduced in Chapter 2, the conditions C_i and actions A_j were

C_1: vegetables are crispy
C_2: vegetables are hard
C_3: vegetables are leafy
C_4: vegetables are green

and

A_1: fry vegetables
A_2: chop vegetables
A_3: boil vegetables
A_4: steam vegetables
A_5: grill vegetables

The conditional parts of the rules were

$C_1 \wedge C_3$ – vegetables are crispy and leafy
$C_1 \wedge$ **not** C_3 – vegetables are crispy but not leafy
not $C_1 \wedge C_2 \wedge C_4$ – vegetables are not crispy, but hard and green,

and so on, and the full rules became

R_1: **if** $C_1 \wedge C_2$ **then** A_1
R_2: **if** $C_1 \wedge$ **not** C_3 **then** A_3

and so on.

The action part of a rule will be performed if the condition part of the rule is satisfied. The action part will produce new data. This may lead to some other combination of conditions being satisfied. Some other processing rules may subsequently be fired.

Assume now that we have another set of processing rules that represent recipes for food. Assume that the food consists of meat and vegetables, that the meat is either chicken (C_5), pork (C_6), beef (C_7), or mutton (C_8), and that the different actions A_6 , A_7,.... are to make different courses according to the processing rules

if chicken and fried vegetables **then** A_6
if beef and boiled vegetables **then** A_7

and so on.

We can formalize this in the form of processing rules if we assume that

action A_1: fry vegetables, produces a post-condition C_{11}: vegetable is fried
action A_2: chop vegetables, produces a post-condition C_{12}: vegetable is chopped
action A_3: boil vegetables, produces a post-condition C_{13}: vegetable is boiled
action A_4: steam vegetables, produces a post-condition C_{14}: vegetable is steamed
action A_4: grill vegetables, produces a post-condition C_{15}: vegetable is grilled

In this case we may formulate processing rules

if $C_5 \wedge C_{11}$ **then** A_6 (if chicken and fried vegetables then make course no. 6)
if $C_7 \wedge C_{13}$ **then** A_7 (if beef and boiled vegetables then make course no. 7)

and so on for determining which courses to cook.

After replacing actions $A_1 - A_5$ by conditions $C_{11} - C_{15}$, the rules R_1 and R_2 above become

R_1: $C_1 \wedge C_3 \rightarrow C_{11}$
R_2: $C_1 \wedge$ **not** $C_3 \rightarrow C_{13}$

All of the other rules can be reformulated into this form, provided that all actions are substituted by their post-conditions. This reformulation makes it possible to view all rules as belonging to the same rule set, and to make conclusions from the total rule-set by automatic reasoning methods. It is evident that one may derive the rules

$C_1 \wedge C_3 \wedge C_5$ $\rightarrow$ course no. 6, resulting from action A_6
$C_1 \wedge$ **not** $C_3 \wedge C_7$ $\rightarrow$ course no. 7, resulting from action A_7

from the rules above. So we have arrived at the conclusion that if we have crispy and leafy vegetables, and chicken, we may make course no. 6. Furthermore, if we have crispy vegetables which are not leafy, and we have beef, we may make course no. 7.

We see that when we formulate the logic of the problem in terms of processing rules, we first have to determine how to process the vegetables, then we have to do the processing, and based on the result of the vegetable processing we may determine which courses to make.

The alternative way of treating the rules is to try to determine the consequences of the rules when having a particular initial condition, e.g., determine which courses it is possible to make given that we have chicken and mutton, and our vegetables are crisp and leafy. In other words, we try to draw conclusions from the collection of rules that represent our knowledge about the system. The techniques for doing such knowledge manipulation come from mathematical logic. They are basic to expert systems, for example, and make it possible to reason about a system even when the knowledge about the system is incomplete.

We have decided to concentrate on the convential structured tools for the formulation of processing rules in this text. This is partly because we want to concentrate on how to approach the development of systems of rules. That does not depend so much on the actual form of representing the rules. But it is also because we need to restrict the volume of the text.

The new rule-processing techniques have not yet been fully integrated into the conventional systems development methods. They will be integrated over the next few years. A better mathematical basis will therefore be needed by systems developers. We have indicated this by providing a small mathematical basis in the appendix. This is also for the benefit for those who read the last parts of this text, where we make use of some of the mathematical techniques.

Many rules are too complex to be formally manipulated

The rules that we have dealt with so far are very simple rules. For more complex rules there is a lack of adequate rule-processing languages. One should bear this in mind when evaluating the existing methods for rule processing.

A rule is a law or custom which guides the behavior of an organization. Rules may be vague or exact, they may be simple or complex. Some rules may even contradict each other. Rules are found on all levels of an organization, from the general policy level, e.g., “it is recommended not to buy gas guzzlers”, to very detailed rules that specify restrictions on individual data elements, e.g., “a social security number has 9 digits”.

A rule modeling language must be so expressive that the relevant rules can be captured and specified. Furthermore, the language must provide formality whenever the knowledge to be captured invites it, e.g., when a rule is to be part of a computer program. A rule language should also be easy to learn, to understand and to use.

Examples of rule modeling languages are natural language, logic, and programming languages. Each of these have their advantages and disadvantages. Natural language is very expressive, but inadequate when formality and precision are needed. The advantage of logic is its formality, and the possibilities for automatic deduction that this leads to. Its main disadvantage is its total lack of user friendliness, in that knowledge is represented in a way that is very different from that of natural language. Programming languages address themselves to specifying rules for the manipulation of data on a detailed level. They are also not appropriate on higher levels of abstraction, due to a lack of facilities for describing the vagueness inherent at these levels.

On the policy level in organizations the rules must necessarily involve the normative concepts of obligation, permission, prohibition and recommendation, e.g., "it is permitted to wear shorts if the weather is hot". Such concepts can usually not be expressed in the usual first-order mathematical logic which is the basis for most methods for automatic reasoning about rules. These concepts are expressible in deontic logic [FOLL70], which is one particular branch of logic that has emerged from the philosophical community. Other types of logic deal with temporal issues, e.g., "if a person is hired then at some time in the future his work performance shall be evaluated". There are also logics that deal with possibilities and necessities (modal logic), e.g., "some tax returns will necessarily be evaluated", "an individual tax return will possibly be evaluated". Common to all of these exotic logics is that they lend themselves to formal analysis only in the simplest of cases. They are still in the research stage, and have still not reached the field of practice.

5.1.2 Ambiguity in Range Specification

Using narrative English to specify process logic may create ambiguity unless special care is taken. Consider, for example, the following statement advertised by a travel agent:

> Children between 2 and 12 pay half price; children under 2 years old get 90 % discount.

Under 2	Between 2 and 12	Discount percentage	Comment
[0,2)	(2,12)	├─ 90% ─┤ ? ├─ 50% ─┤ 0 ─── 2 ─── 12	Discount for age 2 undefined
[0,2)	[2,12]	├─ 90% ─┼─ 50% ─┤ 0 ─── 2 ─── 12	
[0,2]	(2,12)	├─ 90% ─┼─ 50% ─┤ 0 ─── 2 ─── 12	
[0,2]	[2,12]	├─ 90% ─┤ ├─ 50% ─┤ 0 ─── 2 ─── 12	An inconsistency at age 2

Fig. 5.1. Four possible interpretations of a policy

The ambiguity of the statement arises when a discount is computed for a child exactly 2 years old. Should the child be given a 50 % or 90 % discount? This depends on how one interprets the prepositions "between" and "under". The phrase "under 2 years old" has two possible interpretations: one includes children exactly 2 years old; the other excludes this case. These two possibilities can be represented by two integer intervals, i.e., [0,2] and [0,2). Similarly, there are at least two possible interpretations of "between 2 and 12", i.e., [2,12] and (2,12). Therefore, the above example may result in four different discount policies as shown in Fig. 5.1.

In the second case, a child of age 2 can have only 50 % discount, while in the third case, the same child will have 90 % discount. In the first case, the discount percentage is undefined for children who are exactly 2 years old. That is, the policy is incomplete. In the last case, there are two discounts for a child of age 2. That is, we have a conflicting policy.

When formulating a decisive policy, we very often satisfy ourselves with the use of English phrases such as, "under", "between...and...", "up to", etc., to specify ranges of values. The above example indicates that narrative descriptions may lead to incompleteness and/or inconsistency. These deficiencies might not be detected until the system is implemented and tested.

5.1.3 Ambiguity in and/or Combinations

Ambiguity may also arise when formulating policy statements involving the "and" / "or" conjunctives. For example, a transport company may have the following discount policy:

> Passengers over 67 or students under 26 and traveling at least 150 km one way receive a 50 % discount.

Let us suppose that "over 67", does not include 67 and "under 26" does not include 26. We still have the problem of knowing whether the elderly passengers have to travel at least 150 km one way to receive the 50 % discount. In order to make the problem explicit we define the following propositional symbols:

A: The passenger is over 67.
B: The passenger is a student under 26.
C: The passenger travels at least 150 km one way.

The condition part of the discount policy can now be formalized by:

$$A \vee B \wedge C \qquad (5.1)$$

For this simple example we have two possible condition formulations:

$$(A \vee B) \wedge C \tag{5.2}$$
$$A \vee (B \wedge C) \tag{5.3}$$

Expression (5.2) requires that the elderly passengers must travel at least 150 km one way to receive the 50 % discount. In this case, the formulation of the policy should have been:

> Passengers who travel at least 150 km one way, and in addition, are either over 67 or students under 26, pay half price.

Expression (5.3) is more relaxed than (5.2). That is, passengers who are over 67 are entitled to obtain a 50 % discount. In this case, the formulation should have been:

> Students under 26 and traveling at least 150 km one way, or passengers over 67, pay half price.

5.2 Simple Rule Modeling Tools

We shall now present an overview of the structured tools through a simple example using the Teenage Travel Service (TTS). The TTS gives a 50 % discount for students who are less than 12 years old. Students who are 12 or more may obtain a 50 % discount if no fixed seats are required in advance. Passengers who are not students do not receive a discount. The *decision tree* for the discount policy example is depicted in Fig. 5.2.

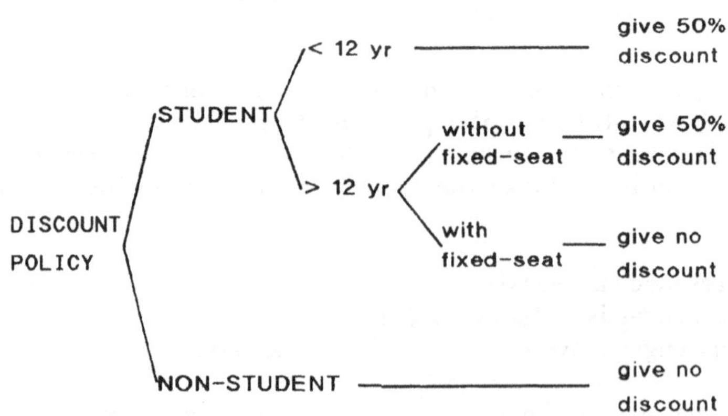

Fig. 5.2. Decision tree for the discount policy example

The root of the tree is labelled "DISCOUNT POLICY" which gives a name to the tree. The four nodes labelled by "give ... discount" are called terminal nodes and indicate the actions to be performed. A branch of the tree is a sequence of nodes leading from the root to a terminal node. The nodes of each branch, except the root, define a rule. For instance,the root of the last branch consists of the nodes labelled by "non-student" and "give no discount". This rule states that "for non-students, give no discount". The first branch states that "for students less than 12 years old, give a 50 % discount". Similarly, the second rule states that "for students who are 12 or more, and require no fixed seats, give 50 % discount".

A decision table displays the logic of a process in a tabular form. It contains four basic components: the condition stub, the action stub, the condition entries, and the action entries. These four components have been identified for the decision table shown in Fig. 5.3.

In this example the condition stub has three conditions, each represented by a predicate, e.g., "is a student". The condition entries opposite to the condition stub define the possible combinations of the conditions. In particular, each column of condition entries gives one combination. For example, the three Ys in the first column postulate that "the passenger is a student, AND is less than 12 years AND has a fixed seat". Since each of the conditions can take two values, there are three conditions in total, we therefore have eight possible combinations.

The action stub has two actions. Corresponding to each combination of conditions the appropriate action entries are marked by 'X'. This means that if that particular combination of conditions is true, then the corresponding actions will be taken. That is, a column of condition and action entries define a rule. In particular, the decision table in Fig. 5.3 has

Rule No. Condition stub	Condition entries							
	1	2	3	4	5	6	7	8
Is a student?	Y	Y	Y	Y	N	N	N	N
Is < 12 years?	Y	Y	N	N	Y	Y	N	N
With fixed seat?	Y	N	Y	N	Y	N	Y	N
give 50% discount	X	X		X				
give no discount			X		X	X	X	X
Action stub	Action entries							

Fig. 5.3. The decision table for the discount policy example

eight rules, numbered 1 through 8. The first rule says that "if the passenger is a student, and is less than 12 years and has a fixed seat, then give a 50 % discount". Rule 3 states that "if the passenger is a student, and is not less than 12 years and has a fixed seat, then give no discount".

One may wonder why the decision table in Fig. 5.3 has 8 rules whereas the decision tree in Fig. 5.2 has only 5 rules. The reason is that the decision table contains some redundancies so the decision tree in Fig. 5.2 has been simplified. Techniques for simplifying decision tables will be discussed in Sect. 5.4.

Another point to be made is that the condition entries are not limited to Y or N in the general case. They can be filled by other values, e.g., 100, "married", etc. The decision table in Fig. 5.3 is called a "limited entry decision table". Decision tables allowing condition entries to take values other than Y and N are called "extended entry decision tables", and will be studied in Sect. 5.4.3. Also note that in general, a rule of a decision table may have more than one action entry marked, although in our example each rule involves only one action.

Structured English expresses process logic by using the nested **"if-then-else-so"** pattern. Fig. 5.4 illustrates the discount policy example expressed in structured English.

```
if student
then
    if less than 12 years
    then give 50 % discount
    else { more than or equal to 12 years }
    so
      if not with fixed-seat
      then give 50 % discount
      else { with fixed-seat }
      so  give no discount
else { non-student }
so  give no discount.
```

Fig. 5.4. The discount policy in structured English

Note that the **"if-then-else-so"** pattern has been indented to indicate the nesting.

We have briefly introduced the three structured tools. Note that Figs. 5.2 – 5.4 give only one possible formulation of the discount policy example using the three tools. In fact there are many other equivalent formulations. A tool is a language; hence, anything can be expressed in many different ways.

5.3 Decision Trees

A decision tree displays the process logic in a horizontal tree structure. We will begin with another TTS example which states the following discount policy:

> TTS provides reasonable prices for group travel: 2–9 persons traveling together will receive 25 % discount; 10 persons or more traveling together receive a 40 % discount. We also take into account the occupation of the passenger: Students are eligible for a 50 % discount and military service personnel need pay only 25 % of the standard rate.

For this example we can identify three conditions:

(1) Travel category (TC), which has two values: individual travel (IT) or group travel (GT).

(2) Group size (GS), which has three values: 1, 2–9 or ≥10.

(3) Occupation (OCC), which has three values: student (S), military service (M) or other (O).

5.3.1 Standard Decision Tree Development

The standard decision tree for this example is depicted in Fig. 5.5. The root of the tree is labelled 'TTS group disc.'. The first level (i.e., the level next to the root level) of the tree is indicated by the name of the first condition, that is, TC for travel category. Since the first condition has two values, we draw two branches from the root and label them accordingly.

The second level of the tree is indicated by GS. Since the second condition has three values, the tree extends each of the existing branches by three new branches. Each of these new branches is labelled accordingly. We then extend each of the existing branches for the other conditions, and so on.

After all the conditions have been used to extend the tree, we begin to consider each of the branches and complete the action sequences. Clearly, if a problem involves n conditions, then the standard tree will have n+2 levels. One of the extra levels represents the root; the other extra level represents the actions. Suppose that the i-th condition has m_i possible values. The total number of possible branches will be $K = m_1 * ... m_n$. Our example has 2*3*3=18 branches representing 18 rules (see Fig. 5.5).

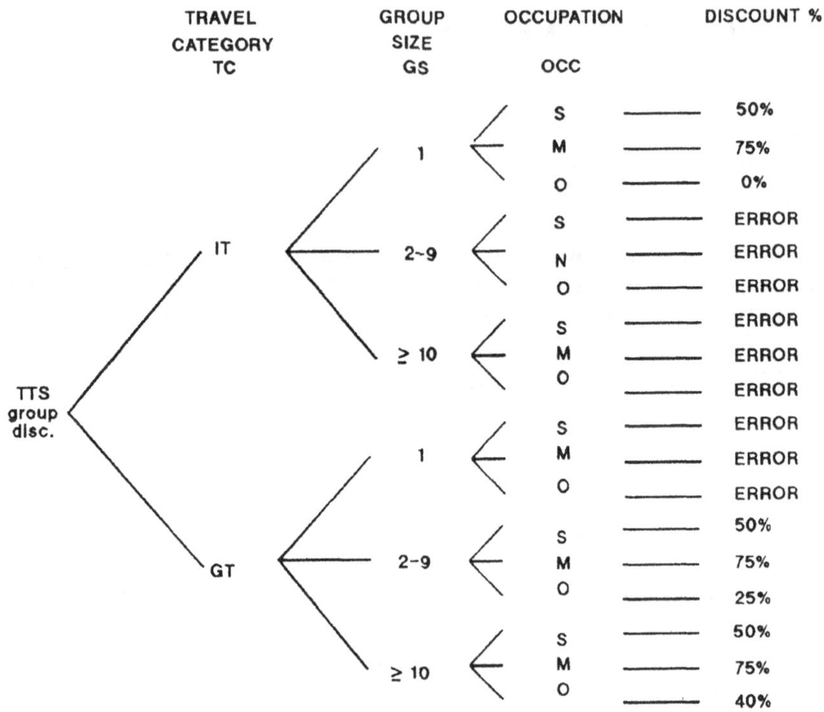

Fig. 5.5. A standard decision tree

5.3.2 Progressive Decision Tree Development

The advantage of the standard approach is that completeness is automatically ensured. This is particularly useful when the process logic is not clear enough. Since the standard approach requires that all the possible combinations be listed explicitly before entering the action sequences, we are forced to clarify the problem logic. If the problem logic is not clarified, then we will not be able to associate an action sequence with some of the combinations. However, the standard approach may introduce many redundancies. For instance, the branch containing IT, 1 and S states that "the case is an individual travel of one person who is a student". Clearly, "individual travel" and "one person" express the same thing. Therefore, this rule formulation is inefficient.

When the decision tree is developed by using the standard approach, it is called the "breadth-first" approach. All the branches of each level have to be extended before processing the next level. The opposite strategy to the breadth-first approach is the "depth-first" approach which always explores the uppermost branch of the tree until it reaches an action sequence. When it is clear that the combination of conditions that have been inspected so far can lead to action, we need not consider the other conditions. We can extend the branch directly to the appropriate action sequence and we may then consider the next branch in a similar manner. This approach is also called the "progressive approach". Its effectiveness depends on sufficient knowledge of the problem logic, so that redundancy is minimized.

The disadvantage of the progressive approach is that completeness checking must be conducted after the tree is built. We shall present a method for checking the completeness in Sect. 5.3.3.

The basic steps of the progressive approach can be laid down as follows:

(1) Create the root node and give it a meaningful name. This is the uppermost unfinished branch.

(2) If all the branches are finished, stop; otherwise, continue.

(3) If the uppermost unfinished branch can lead to action, then complete the branch by an arc leading to the appropriate action sequence and goto (2); otherwise, continue.

(4) If a condition has been entered at the next level, then extend the uppermost unfinished branch with a number of branches, each corresponding to a value of the condition at the next level, goto (3); otherwise, continue.

(5) If some conditions have not been entered, then enter a condition which is likely to lead to action in the next level and goto (3); otherwise, stop, the policy is incomplete.

Let us develop a decision tree for the ticket pricing example described in the previous subsection. We will discuss in detail how it can be developed by using the progressive approach.

Step (1) Create the root node and name it "TTS". This yields Fig. 5.6a. Since steps (2)–(4) cannot be applied, we continue with step (5). Suppose that the TC condition is chosen in step (5); therefore, we extend the root with two branches as shown in Fig. 5.6b.

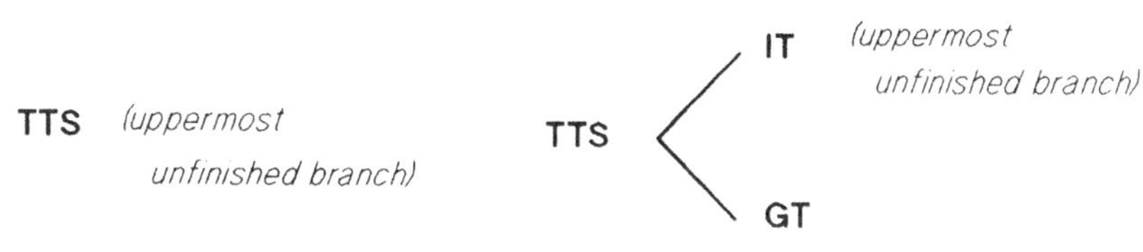

a) Create the root node b) Enter a condition

Fig. 5.6. Create the root and enter a condition

Now we continue with step (3). The uppermost unfinished branch cannot lead to action and there is no condition entered in the next level; therefore, we perform step (5) again. Suppose we consider the occupation condition next (Fig. 5.7a). At this stage, step (3) can be applied; therefore, we complete the uppermost branch and continue with step (2).

The test in step (2) failed since not all branches are finished. Therefore, we continue with step (3). The uppermost unfinished branch in Fig. 5.7b leads to 0 % discount (Fig. 5.8a); similarly, the third branch can lead to 75 % discount, as shown in Fig. 5.8b.

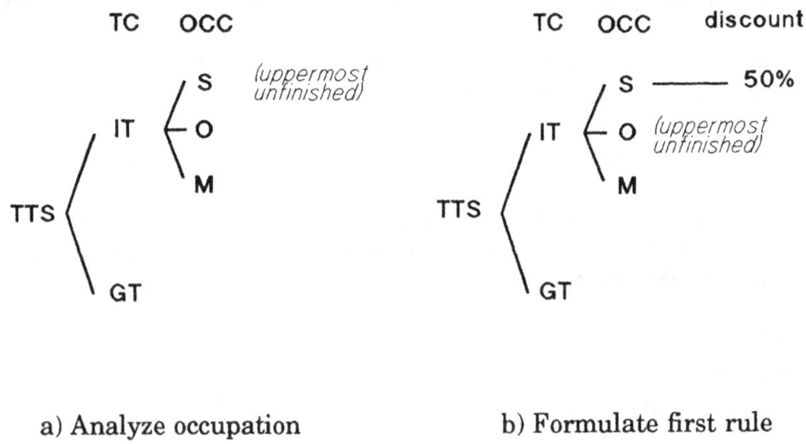

Fig. 5.7. Extending the current branch to an action

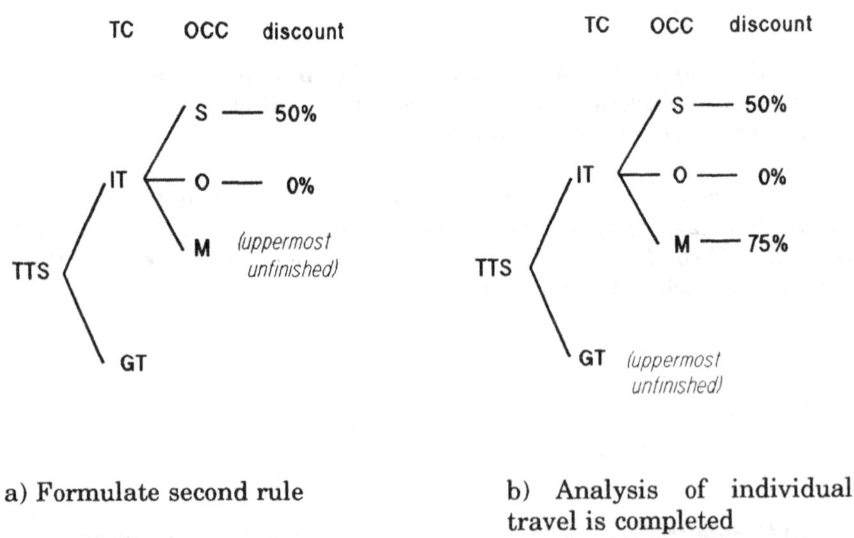

Fig. 5.8. Progressive development of three rules

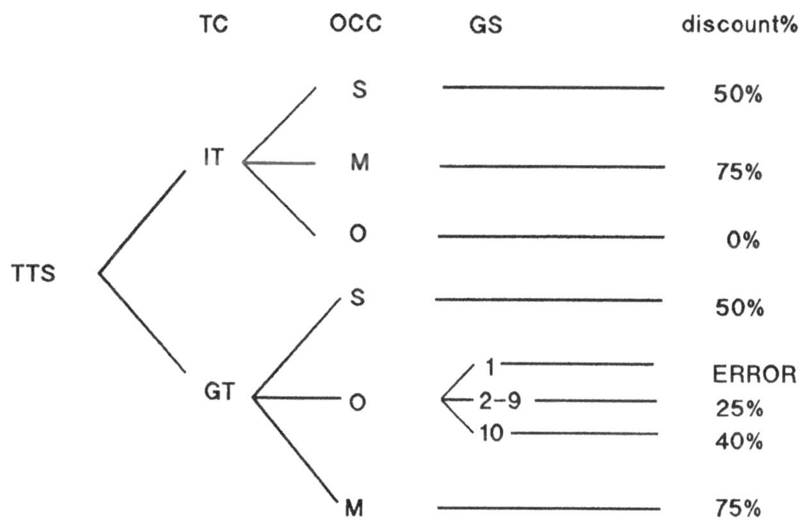

Fig. 5.9. Decision tree that is progressively constructed

Now the branch ended with GT becomes the uppermost unfinished branch. After performing step (5), the fourth branch is completed. The complete decision tree is shown in Fig. 5.9.

The decision tree in Fig. 5.9 contains fewer branches than the tree in Fig. 5.5. As a matter of fact, if we consider the occupation condition first, then the resultant decision tree is much simpler. Nevertheless, it is not always practical to discover which condition should be considered first when the problem logic involves many conditions, some of which have many values. We shall study a method that can be used to simplify a decision by using propositional reasoning. In the next subsection, we will discuss how to check the completeness of a decision tree.

5.3.3 Completeness Checking of Decision Trees

Our discussion so far reveals that there are two aspects relating to the completeness of a process logic specification. On the one hand, we have to consider the completeness of the condition values, which normally depends on how the application problem is stated. For instance, whether the value set {red, blue, yellow} is complete with respect to a condition "color" has to be determined by examining the problem in question. It might happen that the application only deals with three colors. In this case the value set will be complete. A solution to such a completeness problem is to include in the condition values a special value called "others". That is, when the color is not red, blue or yellow, then the "others" alternative is taken. The other aspect of completeness concerns the construction of the decision tree, that is, ensuring that all the combinations of the condition values have been considered. It is this kind of completeness we discuss in this subsection.

We know that for a problem with n conditions where the i-th condition has m_i values, there is a total of $N = m_1 * ... * m_n$ possible combinations including those that are inconsistent. We observe that each of the decision trees in Figs. 5.5 and 5.9 has less than N branches. Therefore, it is very natural to ask whether the decision tree has covered all the combinations. If we can check that the decision tree has covered all these N cases, then all the possible combinations must have been considered. Since a decision tree constructed by either the standard or the progressive approach has no more than N branches, each of the branches must represent at least one combination of the condition values. Our task is to calculate exactly how many cases are covered by each of the branches. Clearly, the sum of these numbers must equal N.

Let us use Fig. 5.9 for illustration. In order to check the completeness, we augment Fig. 5.9 with one additional row and one additional column (see Fig. 5.10). The additional row gives the number of values of each of the conditions. The column gives the number of combinations covered by each of the branches. This number is computed as follows:

Let $N = m_1 * ... * m_n$ as defined above.

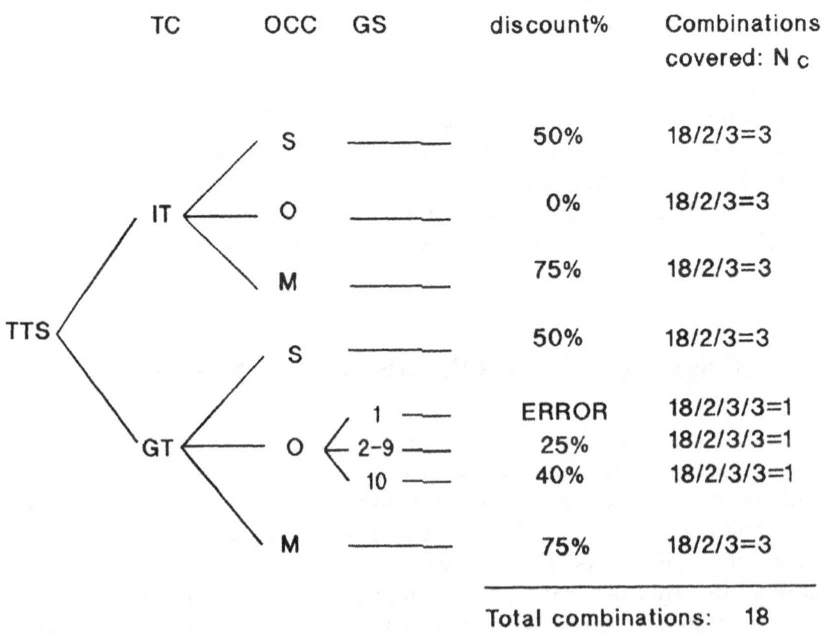

Fig. 5.10. Checking completeness of a decision tree

In our example,

$$N = 2*3*3 = 18.$$

Let n_{j1}, ..., n_{jk} be the nodes of a branch not including the root and the terminal nodes. Further, let n_{ji} be a node under the ji-th condition which has m_{ji} values.

The number of combinations covered by the branch is:

$$N_c = N/m_{j1}/.../m_{jk} \tag{5.4}$$

For example, the first branch contains {TTS, IT, S, 50 %}. However, TTS is the root node and 50 % is a terminal node. Therefore, we shall consider IT and S only. The number of combinations covered by this branch is N_c = 18/2/3 = 3. The sum of all the N_c is exactly 18; hence, we know that the decision tree is complete.

5.3.4 Syntactical Simplification

The two ways to simplify a decision tree are through syntactical and semantical simplifications. Syntactical simplification is in effect a logical simplification. That is, we use logical equivalences to simplify the rules expressed by a decision tree. Simplifications as such do not require knowledge of the application domain. On the other hand, semantical simplification requires knowledge about the application domain in order to eliminate redundancies among the rules. These two simplifications can be applied to both standard and progressively developed decision trees. In this subsection, we study syntactical simplification. Semantical simplification will be studied in Sect. 5.3.6.

As mentioned earlier, a branch of a decision tree expresses a rule. The conjunction of the non-terminal nodes not including the root is the conditional part of the rule. The terminal of the branch specifies the action(s) to be taken when the conditions are true. Therefore, the general form of a rule is

$$\textbf{if } C_1 \wedge ... \wedge C_p \textbf{ do } A_1, ..., A_q \tag{5.5}$$

where C_i, i=1,...,p, is a proposition expressing a condition; A_j, j=1,...,q, is the name of an action. Since the logical **and**($\wedge$) obeys the commutative law, the order of the conditions is insignificant. However, the order of the actions is important in many circumstances.

Consider, for example, the decision tree in Fig. 5.9. The 8 rules in Fig. 5.9 can be written as follows:

R1: **if** TC = 'IT' $\wedge$ OCC = 'S' **do** DISCOUNT 50 %.
R2: **if** TC = 'IT' $\wedge$ OCC = 'O' **do** DISCOUNT 0 %.
R3: **if** TC = 'IT' $\wedge$ OCC = 'M' **do** DISCOUNT 75 %.
R4: **if** TC = 'GT' $\wedge$ OCC = 'S' **do** DISCOUNT 50 %.
R5: **if** TC = 'GT' $\wedge$ OCC = 'O' $\wedge$ GS=1 **do** ERROR HANDLING.
R6: **if** TC = 'GT' $\wedge$ OCC = 'O' $\wedge$ $2 \leq GS \leq 9$ **do** DISCOUNT 25 %.
R7: **if** TC = 'GT' $\wedge$ OCC = 'O' $\wedge$ $GS \geq 10$ **do** DISCOUNT 40 %.
R8: **if** TC = 'GT' $\wedge$ OCC = 'M' **do** DISCOUNT 75 %.

Syntactical simplification is used to identify those rules having the same action sequence. These rules, in effect, express that if the conditional part of any one of the rules is true, then do the action sequence. This implies that we can replace these rules by a "compound" rule. The conditional part of the compound rule is the disjunction of the conditional parts of the rules having the same action sequence. We then try to simplify the conditional part of the compound rule by applying propositional equivalences, e.g., the absorption and the distribution laws. Simplifications as such can sometimes greatly reduce the number of rules of a decision tree without affecting the process logic. We illustrate this idea by analyzing R1 and R4, which have the same action sequence, i.e., DISCOUNT 50%. These two rules express that "passengers of category IT and occupation S" or "passengers of category GT and occupation S" get a 50 % discount. The compound rule that can replace R1 and R4 is:

R(1+4): **if** (TC = 'IT' $\wedge$ OCC = 'S') $\vee$ (TC = 'GT' $\wedge$ OCC = 'S') **do** DISCOUNT 50 %.

By the distribution law $(P \wedge R) \vee (Q \wedge R) \Leftrightarrow (P \vee Q) \wedge R$, the conditional part of R(1+4) can be rewritten as:

$$(TC = \text{'IT'} \vee TC = \text{'GT'}) \wedge OCC = \text{'S'} \tag{5.6}$$

The travel category condition (TC) has only two values: 'IT' and 'GT'. Therefore, the sub-expression (TC='IT' $\vee$ TC='GT') of (5.6) has exhausted all the possible values; it must always be true. By using $P \wedge$ **true** $\Leftrightarrow P$, equation (5.6) can be rewritten as:

$$OCC = \text{'S'} \tag{5.7}$$

Thus, the compound rule becomes:

R(1+4): **if** OCC = 'S' **do** DISCOUNT 50 %.

Similarly, we can replace R3 and R8 by:

R(3+8): **if** OCC = 'M' **do** DISCOUNT 75 %.

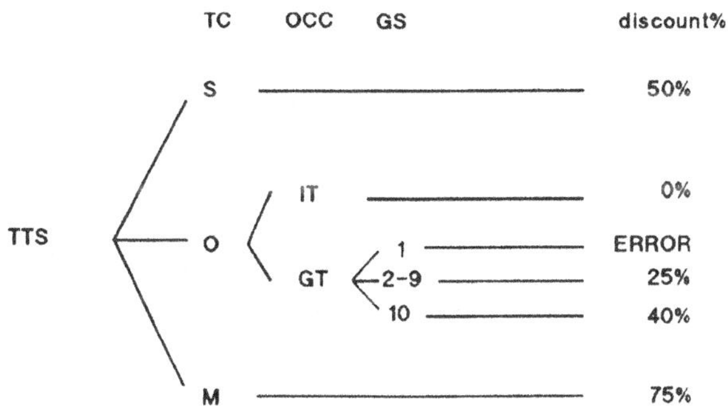

Fig. 5.11. Syntactical simplification of a decision tree

After simplification, we have only 6 rules. In particular, the conditional part of rule R(1+4) as well as R(3+8) consists of only one atomic proposition about the same condition, i.e., occupation. The conditional part of each of the other rules consists of more than one proposition about several conditions including the occupation condition. This suggests that we should consider the occupation condition before the other conditions. Therefore, we modify the decision tree in Fig. 5.9 and present it as shown in Fig. 5.11 below.

5.3.5 Syntactical Simplification Process

There are two methods for syntactical simplification. The one that we discussed in the previous subsection can be summarized by the following:

(1) List all the rules described by the decision tree.

(2) Identify a set of rules having the same action sequence. Form the disjunction of the conditional parts of these rules.

(3) Apply the distribution law to any part of the disjunction to obtain subexpressions of the form

$$D_1 \wedge \dots \wedge D_m \tag{5.8}$$

where each D_i, i=1,...,m, is a disjunction of atomic propositions on some conditions.

(4) Eliminate any conjunction clause D_i that exhausts all the possible cases of a condition.

(5) Repeat steps (3) and (4) until no reduction can be done.

(6) Repeat steps (2), (3), (4) until no reduction can be done.

(7) Reconstruct a decision tree for the simplified rules. This can be done by considering the rules containing the least conditions and then those containing more conditions, etc.

(8) Check the completeness of the new decision tree.

The second method for syntactical simplification does not need to list the rules of a decision tree explicitly. It works on the decision tree directly:

(1) If two or more branches of the decision tree differ only on one condition C and the branches cover all cases of C, then cut all but one of these branches; otherwise, go to step (3).

(2) Short cut the node corresponding to condition C in the remaining branch.
Go to step (1).

(3) Calculate the length of each of the remaining branches.

(4) Construct a new decision tree by considering the branches in ascending order of lengths.

We illustrate the process by simplifying the decision tree in Fig. 5.9. In step (1), we find that the first and the fourth branches differ only on condition TC and these two branches cover all the cases of the condition. Therefore, we cut the fourth branch and short cut the first branch. This is shown in Fig. 5.12. Similarly, we do the same for the third and the last branch. The lengths of the remaining branches are also shown in Fig. 5.12.

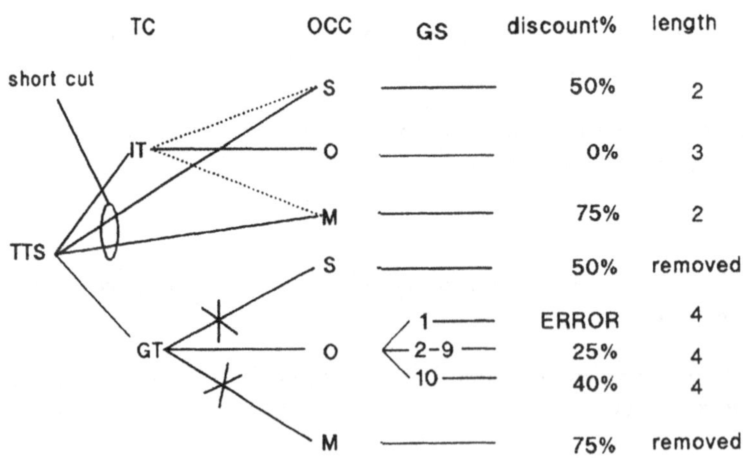

Fig. 5.12. Syntactical simplification

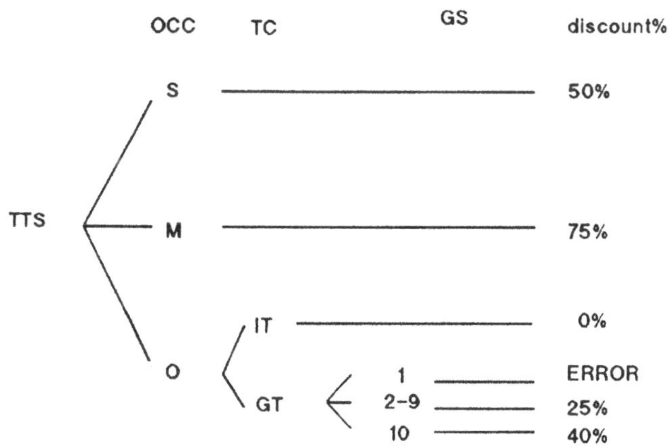

Fig. 5.13. Decision tree after direct simplification

A new decision tree is constructed. The first and the third branches have length 2, which is the smallest. Therefore, we consider these two branches first by producing them as shown in Fig. 5.13. We then consider the second branch of Fig. 5.12. It produces the third branch in Fig. 5.13. The other three branches with length 4 give rise to three other new branches.

5.3.6 Semantical Simplification

Semantical simplification requires that knowledge about the application domain is known. For instance, in the TTS example, we know that TC='ST' iff GS=1. Furthermore, TC='GT' if 2 ≤ GS ≤ 9 or GS ≥ 10. That is, propositions about travel category TC can be expressed in terms of propositions about group size GS, e.g., take the rules in Fig. 5.11:

R1: **if** OCC = 'S' **do** DISCOUNT 50 %.
R2: **if** OCC = 'O' ∧ TC = 'IT' **do** DISCOUNT 0 %.
R3: **if** OCC = 'O' ∧ TC = 'GT' ∧ GS=1 **do** ERROR HANDLING.
R4: **if** OCC = 'O' ∧ TC = 'GT' ∧ 2≤GS≤9 **do** DISCOUNT 25 %.
R5: **if** OCC = 'O' ∧ TC = 'GT' ∧ GS≥10 **do** DISCOUNT 40 %.
R6: **if** OCC = 'M' **do** DISCOUNT 75 %.

Substituting GS=1 for TC = 'IT' and (2≤GS≤9 ∨ GS≥10) for TC='GT' gives

R1: **if** OCC = 'S' **do** DISCOUNT 50 %.
*R2: **if** OCC = 'O' ∧ GS=1 **do** DISCOUNT 0 %.
*R3: **if** OCC = 'O' ∧ (2≤GS≤9 ∨ GS≥10) ∧ GS=1 **do** ERROR HANDLING.
*R4: **if** OCC = 'O' ∧ (2≤GS≤9 ∨ GS≥10) ∧ 2≤GS≤9 **do** DISCOUNT 25 %.
*R5: **if** OCC = 'O' ∧ (2≤GS≤9 ∨ GS≥10) ∧ GS≥10 **do** DISCOUNT 40 %.
R6: **if** OCC = 'M' **do** DISCOUNT 75 %.

The rules that are marked with an asterisk differ from their counterparts before the substitution. Therefore, we will only discuss these rules. By using the absorption law

$$P \wedge (P \vee Q) \Leftrightarrow P \tag{5.9}$$

we may rewrite R4 and R5 as:

R4: **if** OCC ='O' $\wedge$ 2≤GS≤9 **do** DISCOUNT 25 %.
R5: **if** OCC ='O' $\wedge$ GS≥10 **do** DISCOUNT 40 %.

The conditional part of R3 is always false because by the distribution law, we have:

$$\begin{aligned} & OCC = 'O' \wedge (2\le GS\le 9 \vee GS\ge 10) \wedge GS=1 \\ \Leftrightarrow\ & OCC = 'O' \wedge ((2\le GS\le 9 \wedge GS=1) \vee (GS\ge 10 \wedge GS=1)) \\ \Leftrightarrow\ & OCC = 'O' \wedge (\mathbf{false} \vee \mathbf{false}) \\ \Leftrightarrow\ & OCC = 'O' \wedge \mathbf{false} \\ \Leftrightarrow\ & \mathbf{false} \end{aligned}$$

This means that we can never apply R3. Therefore, deleting R3 will not affect the process logic that is being specified. Thus, we have five rules involving only two conditions, i.e., occupation and group size. The decision tree representing these five rules is shown in Fig. 5.14.

Semantical simplification can also be done directly. The following example, of semantically simplifying the decision tree in Fig. 5.13, shows how this is carried out. Assume that

$$\begin{aligned} TC = 'IT' &\Leftrightarrow GS = 1 \\ TC = 'GT' &\Leftrightarrow (GS = 2\text{–}9 \vee GS\ge 10). \end{aligned}$$

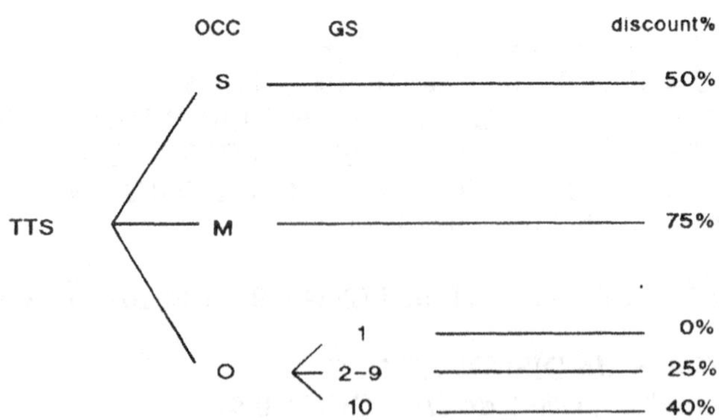

Fig. 5.14. The final decision tree of the TTS group discount

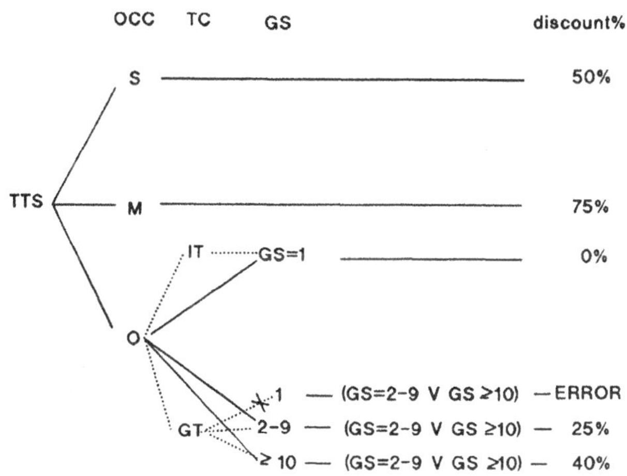

Fig. 5.15. Semantical simplification on decision tree

We insert the right-hand side of each of these equivalences in front of the terminal node of the corresponding branch. Further, the node corresponding to the left-hand side is short cut. The resultant picture is shown in Fig. 5.15.

The fourth branch becomes semantically inconsistent; hence, it is cut. The disjunction in the fifth branch can be eliminated since GS = 2–9 is semantically equivalent to 2–9 under condition GS. Similarly, the disjunction in the last branch can be eliminated.

The third branch needs some explanation. In this branch there is no node corresponding to GS in the decision tree of Fig. 5.13. This means that the rule is "indifferent" to GS. That is, the "invisible" node is in effect:

$$GS = 1 \vee GS = 2\text{–}9 \vee GS \geq 10$$

By the absorption law, we should have 1 under the GS condition in this branch. The simplified decision tree is identical to that in Fig. 5.14.

5.4 Decision Tables

Decision tables display the process logic in a tabular form. There are four basic components in a decision table: the condition stubs, the condition entries, the action stubs, and the action entries (see Fig. 5.16).

Like decision trees, there are two ways to construct a decision table: the standard approach and the progressive approach. The standard approach corresponds to a "breadth-first" evaluation of processing rules, while the progressive approach corresponds to a "depth-first" evaluation. Because both approaches are similar to corresponding approaches for decision tree development, we shall only explain the standard approach for decision table development.

5.4.1 Standard Decision Table Construction

In this section we will study the standard method for decision table construction and use the discount policy example of Sect. 5.2 to illustrate the method.

The standard decision table building process consists of the following steps:

(1) Compute the total number of rules by multiplying together the number of possible values for each of the conditions. That is, if the problem in question has n conditions and the i-th condition has m_i values, then the total number of rules is $K = m_1*...*m_n$. In our example, there are three conditions, each of which has two possible values, i.e.,

No.	Condition	Values	Number of Values
1	Is a student?	Y, N	2
2	Is less than 12 years?	Y, N	2
3	With fixed seat?	Y, N	2
Total number of rules K ≈ 2*2*2 = 8			

(2) Lay out a table and fill in the condition and action stubs; provide K columns for the rules (Fig. 5.16):

	1	2	3	4	5	6	7	8
Is a student?								
Is < 12 years?								
With fixed seat?								
give 50% discount								
give no discount								

Fig. 5.16. Layout of the table with 8 columns

By convention, the condition entries in each row will be filled from left to right in the following steps.

(3) If the i-th condition is being processed, then $K:=K/m_i$.

(4) Take each of its condition values in turn. Enter the condition value into K consecutive entries in the row opposite to the i-th condition. Repeat this step for each of the remaining values (of the i-th condition).

(5) Repeat the last step until the i-th row is completely filled.

(6) If some of the conditions remain to be considered, then i:=i+1 and go to step (3); otherwise, continue.

(7) Enter X into the appropriate action entries for each of the rules.

Figure 5.17 illustrates the result of steps (3) – (7).

	1	2	3	4	5	6	7	8
Is a student?	Y	Y	Y	Y	N	N	N	N
Is < 12 years?	Y	Y	N	N	Y	Y	N	N
With fixed seat?	Y	N	Y	N	Y	N	Y	N
give 50% discount	X	X		X				
give no discount			X		X	X	X	X

Fig. 5.17. Standard decision table construction

Another example is from the Galactic Banking Information System (GBIS), where the interest rate is determined by the following policy:

> The interest rate for minibank accounts is 0.5 % per month. The monthly rate for a normal account is 0.9 % if no withdrawal has been made in the last 12 months including the present month; otherwise, if only one withdrawal is made in the month, then the rate is 0.8 %, else it is 0.5 %.

The example has three conditions, each of them has two values:

C1: Is it a minibank? Y/N	2 values
C2: Withdrawal in 12 months? Y/N	2 values
C3: Only one withdrawal in the month? Y/N	<u>2 values</u>
	Total 8 rules

The decision table for the GBIS example is shown in Fig. 5.18.

	1	2	3	4	5	6	7	8
minibank?	Y	Y	Y	Y	N	N	N	N
withdrawal in 12 months?	Y	Y	N	N	Y	Y	N	N
≤ 1 withdrawal in the month?	Y	N	Y	N	Y	N	Y	N
monthly rate 5%	X	X	X	X		X		
monthly rate 8%					X			
monthly rate 9%							X	
error								X

Fig. 5.18 Standard decision table for the GBIS example

5.4.2 Extended-Entry Decision Tables

So far we have studied decision tables where the statement of each of the conditions is complete in the condition stub. The condition entries are limited to take only two values, e.g., N or Y. These decision tables are commonly called "limited-entry decision tables". When some of the conditions are not binary, i.e., have more than two possible values, then an "extended-entry" decision table can be used.

In an extended-entry decision table, the statement of a condition in the condition stub is not complete and has to be extended into the entry part. Consider the group size (GS) condition of the TTS group discount example. The condition has three values, i.e., 1, 2–9 and ≥10. Therefore, we may partially state the condition in the condition stub by using the expression "GS is". The entries opposite to this condition will be filled by 1, 2–9, or ≥10. Thus, a complete formulation of the condition consists of the expression in the condition stub and one of the entries, e.g., "GS is ≥10".

It is possible to use limited-entry instead of extended-entry conditions. For instance, the GS condition can be replaced by two subconditions "is GS =1" and "is GS = 2–9", each of which has values Y and N. When the values of "is GS = 1" and "is GS = 2–9" are both N, then "GS is ≥10" is assumed. The correspondence between these two formulations is shown in Fig. 5.19.

In general, an extended condition with n values can be replaced by k limited-entry subconditions, where $2^{n-1} \le k \le 2^n$. However, it is usually more convenient to display the process logic in terms of an extended-entry decision table; especially, when some of the conditions have more than two values. From the information modeling point of view, an extended-entry

is GS = 1?	is GS = 2–9?		
N	N		GS is ≥10
N	Y		GS is 2–9
Y	N		GS is 1
Y	Y		ERROR X

limited-entry formulation

extended-entry formulation

Fig. 5.19. Limited-entry via extended-entry formulations

	1	2	3	4	5	6	K:=6
within 6/22–8/12?	Y	Y	Y	N	N	N	K:=K/2=3
category is	HP	VCM	NN	HP	VCM	NN	K:=K/3=1
discount 0%			X	X	X	X	
discount 30%	X						
discount 50%		X					

Fig. 5.20. Extended-entry table for the Viking Hotels example

table often contains more information than a limited one. For instance, the travel category condition of the TTS example has only two values, i.e., 'IT' and 'GT'. When it is treated as an extended-entry condition, the table will display more precisely whether it is TC = 'IT' or TC = 'GT'. Treating it as a limited-entry condition will not have this advantage. In general, we should use an extended-entry formulation unless the condition is binary and its values can be clearly determined according to common sense.

An extended-entry decision table is shown in Fig. 5.20 for the following hotel discount policy:

> "The Viking Hotels offer summer discounts beginning from June 22 to and including August 12. To enjoy the special offer, the customer should possess a Hotel Pass (HP) or a Viking Club Membership certificate (VCM). The discount rate for HP is 30 %. The discount rate for VCM is 50 %. Only one of these can be applied. During the rest of the year, there is no discount."

In order to formalize the policy we may choose to have three binary conditions: "within June 22 to August 12" (within 6/22–8/12), "special rate?" (SR), and "possess an HP". Alternatively, we may choose to have a binary condition, "within 6/22–8/12", and a ternary condition, "category". The category condition has values HP, VCM, and NN (which stands for having no HP or VCM). In this case, an extended-entry table can be constructed. The total number of rules is 6. There are three actions: discount 0 %, 30 % and 50 %. By using the standard approach described in the previous subsection, we can construct the table shown in Fig. 5.20.

5.4.3 Indifference and Consolidation

A decision table may contain redundancies unless some special measure is taken. This is similar to decision trees. Consider, for example, rules 4–6 in Fig. 5.20. When the value of "within 6/22–8/12" is 'N', then the discount rate is 0 %, regardless of whether the customer possesses an HP or VCM. This suggests that we can consolidate the three rules by creating a "compound" rule which is indifferent to the category condition. The condition entry of the compound rule is filled by a special symbol '–', called the indifference (symbol). After consolidation we have only four rules as illustrated in Fig. 5.21, where a decision tree equivalence is also shown.

	1	2	3	4-6
within 6/22-8/12?	Y	Y	Y	N
category is	HP	VCM	NN	—
discount 0%			X	X
discount 30%	X			
discount 50%		X		

a) Decision table formulation

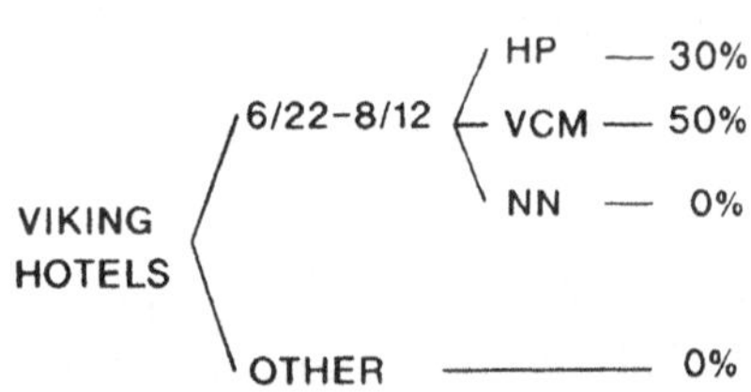

b) Decision tree equivalence

Fig. 5.21. Rule consolidation

Decision table consolidation corresponds to syntactical simplification of decision trees. Therefore, it does not require additional knowledge of the application other than what is contained in the decision table. The general procedure for decision table consolidation is as follows:

(1) Identify a set of rules having the same action sequence and whose condition values are not the same for only one condition.

(2) If one of the rules has the indifference symbol for the only condition, then delete the other rules belonging to the set; otherwise, go to step (3).

(3) If the condition values of that condition includes all the possible values, then replace the set of rules by a single rule using the indifference symbol for the only condition that is different.

(4) Repeat steps (1) – (3) until no rules can be consolidated.

The justification behind the procedure can be clarified as follows. Let the set of rules identified by step (1) be the following:

$$B \wedge C_1 \Rightarrow A$$
$$B \wedge C_2 \Rightarrow A$$
$$\vdots$$
$$B \wedge C_m \Rightarrow A$$

where B is a conjunction of atomic propositions representing the rest of the conditional part and C_i, i=1,...,m, is an atomic proposition on the condition that is different. By definition the rules of a decision table are conjoined together; therefore, by $(P \Rightarrow R) \wedge (Q \Rightarrow R) \Leftrightarrow (P \vee Q) \Rightarrow R$, the set of rules can be rewritten as:

$$(B \wedge C_1) \vee ... \vee (B \wedge C_m) \Rightarrow A \qquad (5.10)$$

By the distribution law of propositional logic, equation (5.10) can be rewritten as:

$$B \wedge (C_1 \vee ... \vee C_m) \Rightarrow A \qquad (5.11)$$

If C_1, ..., C_m exhaust all the possible values of the only condition in question, then $(C_1 \vee ... \vee C_m)$ is always true. Therefore, the truth value of the right-hand side of (5.11) is indifferent to the disjunction of the C_i; thus, we use an indifference symbol and rewrite (5.11) as:

$$B \wedge - \Rightarrow A \qquad (5.12)$$

This is realized by step (3) of the procedure.

The above discussion indicates that an indifference symbol stands for a disjunction of the form $(C_1 \vee ... \vee C_m)$, which exhausts all the possible values of a condition. If, for some i=1,...,m, C_i is an indifference symbol, then (5.11) can be rewritten as:

$$B \wedge (C_1 \vee ... \vee C_{i-1} \vee (C_1 \vee ... \vee C_m) \vee C_{i+1} \vee ... \vee C_m) \Rightarrow A \qquad (5.13)$$

	1	2	3	4	5	6	7	8
minibank?	Y	Y	Y	Y	N	N	N	N
withdrawal in 12 months?	Y	Y	N	N	Y	Y	N	N
≤ 1 withdrawal in the month?	Y	N	Y	N	Y	N	Y	N
monthly rate 5%	X	X	X	X		X		
monthly rate 8%					X			
monthly rate 9%							X	
error								X

Fig. 5.22. Two pairs of rules can be consolidated

	1+2	3+4	5	6	7	8
minibank?	Y	Y	N	N	N	N
withdrawal in 12 months?	Y	N	Y	Y	N	N
≤ 1 withdrawal in the month?	–	–	Y	N	Y	N
monthly rate 5%	X	X		X		
monthly rate 8%			X			
monthly rate 9%					X	
error						X

Fig. 5.23. Another pair of rules can be consolidated

RULE NUMBER	1–4	5	6	7	8
minibank?	Y	N	N	N	N
withdrawal in 12 months?	–	Y	Y	N	N
≤ 1 withdrawal in the month?	–	Y	N	Y	N
monthly rate 5%	X		X		
monthly rate 8%		X			
monthly rate 9%				X	
error					X

Fig. 5.24. The consolidated decision table

By the associative law $P \vee Q \Leftrightarrow Q \vee P$ and the equivalence $(P \vee P) \Leftrightarrow P$, equation 5.13 can be rewritten as:

$$B \wedge (C_1 \vee ... \vee C_m) \Rightarrow A \qquad (5.14)$$

However, since $(C_1 \vee ... \vee C_m)$ is the indifference symbol, (5.14) becomes:

$$B \wedge - \Rightarrow A \qquad (5.15)$$

This justifies step (2) of the procedure.

We shall apply the procedure to consolidate the decision table in Fig. 5.18. We see that rules 1, 2, 3, 4, and 6 have the same action sequence. Rules 1 and 2 differ only for "≤1 withdrawal in the month", as do rules 3 and 4 as well. Moreover, the condition values that are different cover all the possible values in each of these cases. Therefore, these two pairs of rules can be consolidated as shown in Fig. 5.23.

The two compound rules (1+2) and (3+4) in Fig. 5.23 can be consolidated again. The resultant decision table is depicted in Fig. 5.24.

We cannot consolidate the compound rule 1–4 and rule 6, since they have more than one condition that is different. The consolidation procedure can be used for limited-entry as well as extended-entry decision tables although we illustrate only a limited-entry example.

5.4.4 Completeness of Decision Tables

Completeness of a decision table can be checked by using the *rule count*. The *rule count* entries in decision tables are similar to the "Combinations covered" in decision trees (Fig. 5.10). The *rule count* entries of each rule can be calculated in two equivalent ways. If the rule does not contain the indifference symbol, then its *rule count* entry is 1. If the rule contains very few indifference symbols, then its *rule count* entry is the multiplication of the number of possible values of the indifferent conditions. If the rule contains many indifferent entries, then its *rule count* entry is computed in the same way as for decision trees. That is, if there are n conditions, each of which has m_i values, $i=1,...,n$, then the total number of rules is $K = m_1 * ... * m_n$. Suppose that a rule contains p entries which are not the indifference symbols. Let $i_1, ..., i_p$ be the corresponding conditions. Then the *rule count* entry of the rule is $K/m_{i1}/.../m_{ip}$. If the sum of all the *rule count* entries is K, then the decision table is complete; otherwise, some rules are missing.

We exemplify this by checking the completeness of the decision table in Fig. 5.25, which corresponds to the syntactical simplification of the decision tree of Fig. 5.5.

The total number of rules K = 2*3*3 = 18. The sum of the *rule count* entries is also 18; therefore, the decision table is complete. The decision table with *rule count* entries is shown in Fig. 5.26.

	1	2	3	4	5	6
TC	–	IT	–	GT	GT	GT
OCC	S	O	M	O	O	O
GS	–	–	–	1	2–9	≥10
0%		X				
25%					X	
40%						X
50%	X					
75%			X			
error				X		

Fig. 5.25. An extended entry decision table for the TTS group discount example

	1	2	3	4	5	6	
TC	–	IT	–	GT	GT	GT	
OCC	S	O	M	O	O	O	
GS	–	–	–	1	2–9	≥10	
rule count	6	3	6	1	1	1	total of 18 rules
0%		X					
25%					X		
40%						X	
50%	X						
75%			X				
error				X			

Fig. 5.26. Checking completeness of decision tables

5.4.5 Semantical Simplification of Decision Tables

As with decision trees, we can perform semantical simplification on a decision table. The objective is to identify and specify the semantical equivalences among the condition values. That is, to express each case of a condition in terms of other conditions. We illustrate this by simplifying the decision table in Fig. 5.25. As shown earlier, the TC condition can be expressed in terms of the GS condition. That is, we replace TC by GS. Further, each case of TC is replaced by its equivalent. An indifference symbol represents an assertion which is always true, i.e., a tautology. It will remain a tautology after the replacement; therefore, we do not have to do anything about the indifferent entries. The result of the replacement is illustrated in Fig. 5.27.

The conditional part of rule 4 is a contradiction; hence, the rule can never be invoked. That is, we can delete the rule without affecting the process logic. By the absorption law: $P \wedge (P \vee Q) \Leftrightarrow P$, the first entry in rule 5 can be absorbed by the third entry of the rule. Similarly, this can be done for rule 6. The third entry of rule 2 is a tautology. When it is conjoined with the first entry of the rule, the truth value is determined by the first entry. Therefore, the table becomes as shown in Fig. 5.28.

	1	2	3	4	5	6
GS	–	1	–	2–9∨≥10	2–9∨≥10	2–9∨≥10
OCC	S	O	M	O	O	O
GS	–	–	–	1	2–9	≥10
0%		X				
25%					X	
40%						X
50%	X					
75%			X			
error				X		

Fig. 5.27. Replacing a condition by its equivalences

	1	2	3	5	6
OCC	S	O	M	O	O
GS	–	1	–	2–9	≥10
0%		X			
25%				X	
40%					X
50%	X				
75%			X		
error					

Fig. 5.28. Semantical simplification of a decision table

5.5 Structured English

In practice, we frequently encounter program structures such as sequencing, selection, and repetition. The sequencing structure defines a total order over a group of instructions. That is, the sequence of instructions is to be executed in a strict order without branching and repetition.

The selection construct selects one alternative according to the outcome of a decision: for instance, the **if** ... **then** ... **else** statement and the **case** statement of a programming language.

The repetition construct allows one to program a process which repeatedly performs a set of instructions until a condition is fulfilled.

The **for**, **while** ... **do** and the **repeat** ... **until** statements are some examples of the repetition structure.

These programming structures can be used to specify process logic. So-called "structured English" is a language for specifying process logic in terms of English sentences with clearly defined conventions. The basic constructs are sequencing, selection and repetition as used in programming languages. The conventions are that the keywords must be capitalized and the structures should be indented to highlight the logical hierarchy.

The TTS discount policy in structured English is shown below:

```
if OCC is 'S'
then discount is 50 %
else
    if OCC is 'M'
    then discount is 75 %
    else
        if GS is 1
        then discount is 0 %
        else
            if GS is 2-9
            then discount is 25 %
            else {GS is ≥10}
            so discount is 40 %
```

Structured English provides a convenient way for specifying repetitions of a process. To illustrate this we shall specify the progressive approach for developing decision trees in structured English, as shown below.

The procedure can be described as follows:

```
create a root with a name.
while not all branches are finished
do
    while uppermost unfinished branch
          not lead to action
    do
        if a condition entered in next level
        then extend uppermost unfinished branch
        else
             if more conditions to consider
             then enter a condition
             else {policy incomplete}
             so  STOP with failure
    end-while
end-while
stop
```

5.6 Comparison of Decision Trees, Decision Tables and Structured English

We mentioned at the beginning of this chapter that different problems should be solved by using different tools, as in choosing whether to eat with a fork, a knife, or a spoon. Decision trees give a graphical representation of process logic. However, it is limited to decisive descriptions; although repetition or recursion can also be specified in a decision tree or a decision table [LOND72]. On the other hand, structured English does not have a graphical representation. Nonetheless, as shown in the previous example, repetition can be easily specified in structured English. Therefore, it is important to choose the right tool to solve the right problem.

As suggested by Gane and Sarson [GANE78], decision trees are best used for moderately complex decisions that result in up to 10–15 action sequences. When the number of different action sequences is large, then the decision tree description will contain too many branches even if it has been simplified. Decision tables are best used for problems involving complex combinations of up to 5–6 conditions. However, when the number of combinations is large, the decision table description will be difficult to comprehend.

A study by Vessey and Weber [VESS86] indicated that decision trees outperformed structured English and decision tables for the taxonomizing task. Traditionally, it was expected that decision tables would perform the taxonomizing task better than structured English, because a decision table was thought to be more understandable than a piece of structured English. Vessey and Weber's research revealed that this was not true. Their experiment with 124 information systems and computer science students showed that structured English was better than decision tables for the taxonomization. When performing the sequencing task structured English was proven more effective than decision tables. However, contrary to what was expected, decision trees evoked the same level of performance as structured English. According to Vessey and Weber, decision trees seem to combine the best features of both decision tables and structured English.

Completeness, consistency, and non-redundancy are desirable properties of a tool, and methods exist for ensuring these properties (see [LOND72]). However, before this text was written, we found no method that could be used to verify the completeness, consistency,and non-redundancy of a decision tree. The method described in Sect. 5.3.3 solves the completeness problem for decision trees. Redundancies of a decision tree can be removed by applying syntactical and semantical simplifications. Finally, the construction of a decision tree by either the standard or the progressive approach cannot produce inconsistency, because two identical combinations of conditions are always represented by one branch.

Figure 5.29 summarizes the weakness and strength of the tools with respect to a number of features. A brief description of each of these features is listed here:

- *User friendliness*: is it easy to learn the tool, easy to understand?
- *Expressiveness*: is it powerful enough for process logic specification?
- *Logical verification*: does it facilitate the verification of logical correctness of a specification?
- *User verification*: is it easy to validate the semantical correctness with the users?
- *Programmer friendliness*: how easy is the tool for a programmer to use for specifying a program?
- *Machine-readable*: is it possible to specify the process logic on a terminal and how easily can this be done?
- *Automatic checking*: how easy is it to automate the logical verification?
- *Changeability*: is it easy to change a process logic specification?

Feature	Decision Trees	Decision Tables	Structured English
1. User friendliness	Very good	Poor	Good
2. Expressiveness	Limited	Limited	Very good
3. Logical verification	Very good	Very good	Poor
4. User verification	Very good	Very good	Poor
5. Programmer friendliness	Very good	Poor	Good
6. Machine-readable	Very good	Very good	Very good
7. Automatic checking	Very good	Very good	Good
8. Changeability	Very good	Very good	Good

Fig. 5.29. Summary of tool features

5.7 Process Logic and Expert Systems

Process logic and the rule base of an expert system are two different forms of knowledge representation. Usually, knowledge that is represented by a decision tree, decision table or a piece of structured English will eventually be translated into a piece of code, e.g., nested if-then-else statements, in some traditional programming language. Such code can then be used to evaluate the condition combinations and perform appropriate actions. Thus, the structured tools can be regarded as tools for knowledge elicitation and acquisition. Knowledge manipulation, however, is done in a traditional way using nested if-then-else statements.

Knowledge representation and manipulation in rule-based expert systems are done in terms of rules. An expert system rule usually assumes the form of productions:

if <antecedent 1> **and**
  <antecedent 2> **and**
  :
  : **and**
  <antecedent m>
then <consequence 1> [with certainty v_1],
  <consequence 2> [with certainty v_2],
  :
  :
  <consequence n> [with certainty v_n]

where the certainty clauses are optional and a consequence may be either a conclusive assertion or an action. The rule base of an expert system consists of collections of rules in the above form. Knowledge manipulation is achieved by repeatedly applying the rules to the facts stored in the database. If all the antecedents of a rule are evaluated as true, then the conclusions are asserted or the actions are executed.

The basic components of a rule-based system are depicted in Fig. 5.30.

Knowledge representation and manipulation in the form of production rules has certain advantages. First, knowledge and control are kept strictly separate. Knowledge is stored in the rule base, while control is exercised by the inference engine. Thus the rule base can be updated easily. Moreover, knowledge about the rules stored can be formulated as rules, called metarules, which can be used to speed up the search. Third, rules are a uniform and natural representation. The rule base specifies *what* to do *when* something has to be done but *not how* to do it. Therefore, the rules can be more easily understood by other persons as well as by the system itself.

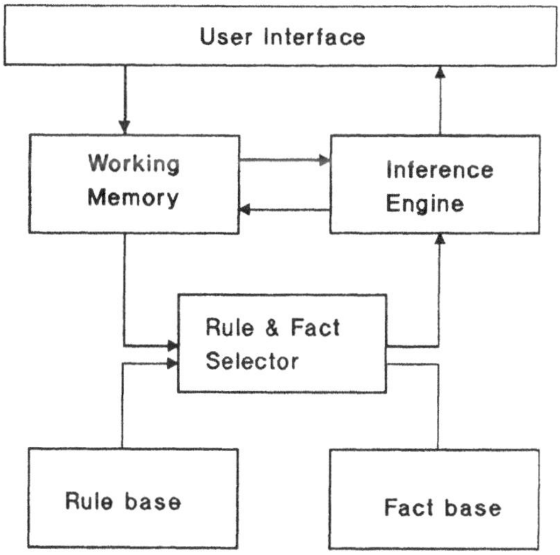

Fig. 5.30. Basic components of a rule-based system

An example of a rule base for determining the cohesion level of a subprogram

We have previously pointed out that functional cohesion is one of the desirable properties of subprogram design. One guideline for determining the cohesion level of a subprogram from its documentation is to write a single sentence starting with "The full purpose of this subprogram is to ...". The sentence can then be analyzed by using a set of heuristic rules.

Some heuristic rules are listed below:

R1: **if** the sentence contains the symbol "..." **or**
the sentence contains the word "etc."
then the subprogram is coincidentally cohesive
with certainty 0.9.

R2: **if** the sentence has a single verb **and**
the sentence contains the word "all"
followed by a plural object
then the subprogram is logically cohesive
with certainty 0.9.

R3: **if** the sentence consists of a single verb **and**
a non-plural object
then the subprogram is functionally cohesive
with certainty 0.9.

R4: **if** the sentence contains words relating to time
then the subprogram is temporally cohesive
with certainty 0.85.

R5: **if** the sentence contains words
relating to control sequence
then the subprogram is procedurally cohesive
with certainty 0.95.

R6: **if** the module is procedurally cohesive **and**
the sentence contains more than one verb
followed by one or more nouns
then the subprogram is communicationally cohesive
with certainty 0.85.

R7: **if** the sentence contains the word "initialize" **or**
the sentence contains the word "clean up"
then the subprogram is temporally cohesive
with certainty 0.85.

Note the certainty values are subjectively assigned. Therefore, the outcome is more of suggestive than conclusive in nature. Imagine a CASE tool that supports a data dictionary. Then the process description slot for all the processes can automatically begin with the phrase "The full purpose of this process is to" and let the analyst/designer complete the functionality description of the process. The key words of the description may be passed to an expert system. The expert system may use the heuristic rules to determine the cohesion level of the process being designed.

There are two inference strategies adopted by expert systems: the forward and backward chaining strategies

Consider a rule base containing the following rules:

R1: **if** X **and** Z **then** B
R2: **if** Y **and** Z **and** A **then** B
R3: **if** X **and** A **then** Z
R4: **if** B **then** C

These rules can be rewritten formally:

R1: $X \wedge Z \rightarrow B$
R2: $Y \wedge Z \wedge A \rightarrow B$
R3: $X \wedge A \rightarrow Z$
R4: $B \rightarrow C$

Suppose that the fact base initially contains the facts {A, X}. The fact base and the working memory together are called the database.

Forward chaining starts from the database

The rules in the rule base are scanned and evaluated with respect to the data base. Rules that can be fired are fired immediately. The firing of a rule produces intermediate results, which are stored in the working memory. The process is repeated until no new result can be produced.

For the above example, the first rule that can be fired is R3. The firing of R3 produces Z. So the database now contains A, X, Z. The next rule that can be fired is R1 and the firing of R1 produces B. So the database now contains A, B, X, Z. Now rule R4 can be fired. Firing R4 produces C. The database is now A, B, C, X, Z. Since no additional results can be produced, the inference process stops. The chain of inference is:

$$AX \rightarrow Z \rightarrow B \rightarrow C$$

where C can be regarded as the conclusion that can be drawn from the reasoning.

Some expert systems have the capability of providing explanations for how the conclusion was derived. For the above example, the explanation may look like:

- **since** A and X is true and by R3: **if** X **and** A **then** Z, one can derive Z.
- By R1: **if** X **and** Z **then** B, one can derive B.
- By R4: **if** B **then** C, one can derive C.

Backward chaining starts from the goal, or hypothesis, which the user wants to establish

Suppose the user wants to know if C can be derived from the fact base and the rules. In this case, the system will perform backward chaining. In backward chaining, if the goal is already in the fact base, then the process stops immediately since the goal is already true. Otherwise, the goal will be used to fire one of the rules, whose right-hand side matches the goal. The left-hand side of that rule becomes the subgoal that needs to be established. If no rule's right-hand side matches the goal, then the process also stops; the goal cannot be proved.

In backward chaining, the working memory stores the subgoals. The rules are scanned from the beginning of the rule base. A rule can be fired if its right-hand side matches one of subgoals stored in the working memory. The firing of a rule results in the removal of the matching subgoal from the working memory and the addition of the left-hand side of the rule to the set of subgoals. Subgoals that are already in the working memory or the fact base will not be added. The above process is repeated until no subgoal is in the working memory; in this case, the goal is proved.

Suppose the goal is C. Then the first rule that can be fired is R4. The firing of R4 replaces C by B in the working memory. Now rule R1 can be fired. The right-hand side of R1 is $X \wedge Z$, but since X is in the fact base, the subgoal is Z. Now R3 can be fired and firing R3 proves the goal, because both X and A are in the fact base.

We see that forward chaining is logical inference using the rules and the facts as the premises, while backward chaining is theorem-proving by refutation in mathematical logic. However, there are some technical differences. The essential difference is that logical inference in the framework of mathematical logic (some people call it automated theorem-proving) is basically a batch process, whereas expert systems allow information to be input interactively during the inference process. Thus, expert systems are capable of processing incomplete and uncertain information.

5.8 An Introduction to Logical Inference

The previous section ended by stating that logical inference through backward chaining is theorem-proving by refutation in mathematical logic. For those who have good knowledge of mathematical logic this is a straightforward statement that needs no further explanation. For those who do not have sufficient mathematical knowledge the statement is utterly incomprehensible. For the benefit of the latter group we shall give a short introduction to the subject of theorem-proving. We shall do so by explaining the simplest of the inference methods, the resolution method, as it is applied to the simplest logic, propositional logic.

The objective of an inference method is to prove whether a collection of logical formulas is in internal conflict or not. If an internal conflict is found, i.e., that not every formula in the collection can be simultaneously satisfied, the set of formulas is said to be inconsistent. For example, the two statements $x = 2$ and $x = 5$ can not both be true at same time.

5.8.1 Rewriting of Logical Formulas

Logical formulas may be complex. Inference methods require that the logical formulas have certain simple forms. So it becomes necessary to transform the original logical formulas to the required form. To this end rewriting techniques are used. We have shown examples of the rewriting of logical formulas in previous sections of this chapter. One example of rewriting is the removal of the logical implication sign. It can be proved that $P \rightarrow Q \Leftrightarrow \sim P \vee Q$, that is, the two expressions are logically equivalent. A logical formula may always be replaced by its logical equivalent. So the logical formula may change its form without changing its meaning.

Many commonly used rewriting rules can be proved by simple applications of truth-tables. The truth-tables for the basic logical connectives are

P	Q	~Q	$P \wedge Q$	$P \vee Q$	$P \rightarrow Q$
f	f	t	f	f	t
f	t	f	f	t	t
t	f	t	f	t	f
t	t	f	t	t	t

where t means true and f means false. We may construct truth-tables for new formulas, e.g., the truth-table for $\sim P \vee Q$ is constructed below, and we see that it is equivalent to the truth-table of the logical implication $P \rightarrow Q$.

P	Q	~P	$\sim P \vee Q$	$P \rightarrow Q$
f	f	t	t	t
f	t	t	t	t
t	f	f	f	f
t	t	f	t	t

So we have proved the logical equivalence $P \rightarrow Q \Leftrightarrow \sim P \vee Q$.

5.8.2 The Resolution Principle

There are several proof methods. The resolution method is most common for automatic theorem proving. The resolution method works on formulas that are in *clause* form. A formula is in clause form if it employs only the logical **or** ($\vee$) and **not** (~) connectives, e.g., $A \vee \sim B \vee C$. A collection of logical formulas is consistent if all of the formulas are true. This is expressed by connecting all of the formulas by **and** ($\wedge$) connectives. An example is

$$
\begin{aligned}
& (A \vee \sim B \vee C \vee \sim E \vee F) \\
\wedge\ & (A \vee B \vee \sim C \vee \sim G) \\
\wedge\ & (\sim A \vee \sim D \vee G) \\
\wedge\ & (A \vee \sim B \vee D \vee E \vee \sim F)
\end{aligned}
$$

The first step in the proof process is to compare the formulas, and to further simplify them in order to discover whether there are logical inconsistencies or not. The resolution method is based on the rewriting rule

$$(R \vee \sim Q) \wedge (S \vee Q) \Rightarrow R \vee S$$

This rule is simple to prove. Because each of the literals (non-logical symbols) can be only true or false, we may substitute the values **true** (t) and **false** (f) for Q, and we get

$$
\begin{aligned}
&(R \vee \sim Q) \wedge (S \vee Q)\\
\Rightarrow\ &((R \vee \sim t) \wedge (S \vee t)) \vee ((R \vee \sim f) \wedge (S \vee f))\\
\Leftrightarrow\ &((R \vee f) \wedge (S \vee t)) \vee ((R \vee t) \wedge (S \vee f))\\
\Leftrightarrow\ &(R \wedge t) \vee (t \wedge S)\\
\Leftrightarrow\ &(R \vee S)
\end{aligned}
$$

and we have proved the logical implication.

This rewriting rule implies that if we can find two formulas which are in clause form, where one formula contains the negation of a literal that appears in the other formula, then the two formulas can be simplified into *one* new formula which is also in clause form. The new formula may consequently be used for further simplifications using the same procedure as above.

So the first step of the proof procedure is to compare pairwise the **and**'ed formulas to see if one of them contain the negation of a literal that appears in the other one. If this is the case the two formulas are rewritten according to the rewriting rule above. This process goes on until further simplification is no longer possible or until an inconsistency has been discovered. The procedure is depicted below.

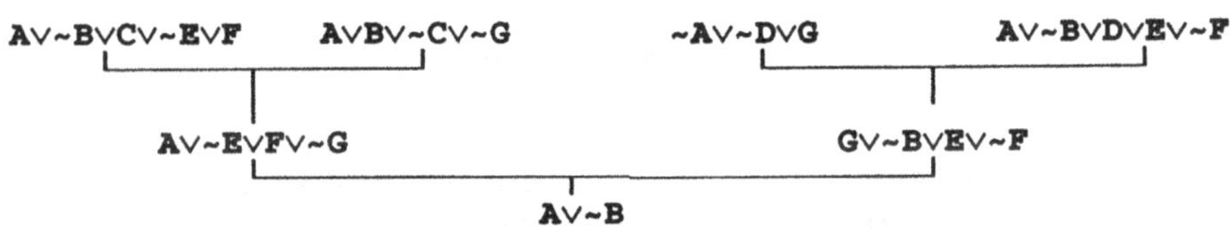

The original formulas become considerably simplified when using this approach. In the example we see that the original set of logical formulas is consistent if A is true or if B is false, because A∨~B is equivalent to the original formulas. But the set of formulas is inconsistent with the situation that A is false and B is true. We shall show this formally. Let us add the logical formulation of the latter situation (~A∧B, which means that A is false and B is true) to the original formulas, and continue the proof process, as shown below.

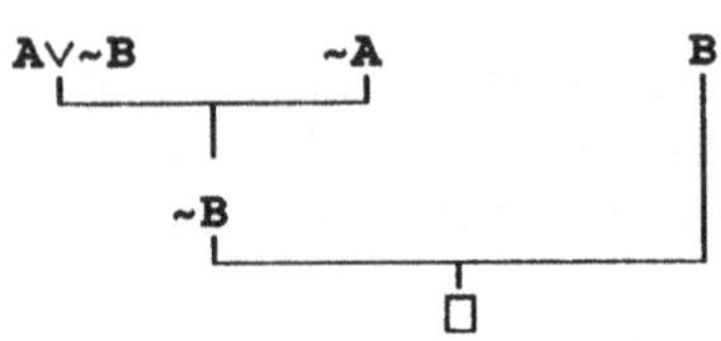

The empty square □ symbolizes inconsistency. We see that the set of formulas must be inconsistent because B and ~B can not both be true simultaneously.

5.8.3 Three Main Tasks for Applying Logical Proof

Mathematical logic may be viewed as a language for formalizing a story, e.g., the "story" of making Chinese food of the previous sections of this chapter. Theorem-proving is used for finding out whether the story that is told is true or not. There are three main tasks that must be performed:

- *the story must be reformulated into mathematical logic,*
- *the logical formulas must be rewritten so that they are suitable for the chosen proof method,*
- *the formal proof must be carried out efficiently.*

We shall show this through a simple example. Consider the following story:

If Tom brings Jane flowers then Jane feels very happy. However, Jane does not feel very happy if Mary does not write Tom a letter. In fact, Mary does not write Tom a letter. Nevertheless, if Tom talks to Jane then Mary writes Tom a letter or she has to leave Los Angeles. Anyway, Tom talks to Jane and even brings her flowers.

The first task is to reformulate the story in a way that is suitable for logical reasoning. That is, we must first formalize it in the chosen logical language. Let us assume that after analysis we have identified the following five *atomic propositions*:

A: Mary has to leave Los Angeles
B: Mary writes Tom a letter
C: Tom talks to Jane
D: Tom brings Jane flowers
E: Jane feels very happy

We may now reformulate the story by stating the following propositions:

$D \rightarrow E$	If Tom brings Jane flowers then Jane feels very happy
$\sim B \rightarrow \sim E$	Jane does not feel very happy if Mary does not write Tom a letter
$\sim B$	Mary does not write Tom a letter
$C \rightarrow (B \vee A)$	If Tom talks to Jane then Mary writes ....
$D \vee C$	Tom talks to Jane and even brings her flowers

The next step is to rewrite these five formulas so that they may be subjected to theorem proving. We want to bring them into clause form. In order to do this we need to have suitable rewriting rules. Let us rewrite the above formulas into clause form:

$D \rightarrow E \quad \Leftrightarrow \quad \sim D \vee E$
$\sim B \rightarrow \sim E \quad \Leftrightarrow \quad \sim\sim B \vee \sim E \Leftrightarrow B \vee \sim E$
$C \rightarrow (B \vee A) \quad \Leftrightarrow \quad \sim C \vee B \vee A$
$D \vee C$
$\sim B$

and we get the formula

$(\sim D \vee E) \wedge (B \vee \sim E) \wedge (\sim C \vee B \vee A) \wedge (D \vee C) \wedge \sim B$

which represents Mary's love story.

The third and last step is to apply the resolution method to find whether the formulas are logically consistent. The rewritten formulas for Mary's love story are shown below followed by the resolution proof.

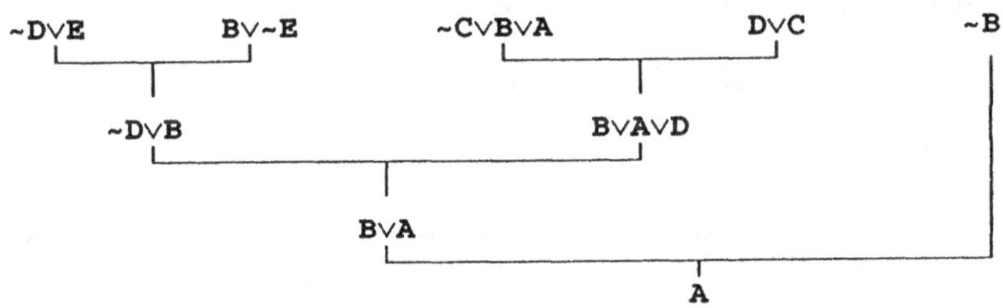

So we see that the only logical conclusion that can be drawn from this story is that Mary has to leave Los Angeles!

5.8.4 Some Properties of Proof Methods

It can be proved that the proof process always terminates for propositional logic. We only need a finite number of steps to prove whether a set of propositional logic formulas are consistent or not. The number of steps required may however be very large, and increases exponentially with the number of propositional symbols in the set of formulas. A propositional symbol is either a literal, e.g., A, B, or one of the logical connectives ~, ∧, ∨, →, or a parenthesis) or (.

Since the consistency or inconsistency of a set of propositional formulas can always be determined the propositional logic is said to be decidable. *Predicate logic*, the logic of general formulas about all or some of a given domain of objects, does not have this property. If a set of predicate logical formulas are inconsistent we can eventually detect the inconsistency. However, if the set is consistent the proof process will run forever. Predicate logic is consequently said to be semi-decidable.

The semi-decidability of predicate logic has led to an approach which is called proof by negation. The rationale behind this approach is as follows. Assume that we have a collection of predicate logical formulas that are consistent, and we add a new logical formula to the collection. If the new formula is consistent with the formulas that are already there, it becomes impossible to prove that directly, because of the semi-decidability of predicate logic. On the other hand, if we negate the new (consistent) formula, the negated formula must contradict the formulas that are already there. If we apply the resolution proof we would find inconsistency in the new collection. Consequently the new formula is consistent with the formulas that are already in the collection, because the negation of the new formula leads to inconsistency. Let us apply this to Mary's love story. We shall ask whether Mary has to leave Los Angeles, that is, whether A is true. Consequently we add the statement "not A" (~A) to the collection of logical statements, and we get

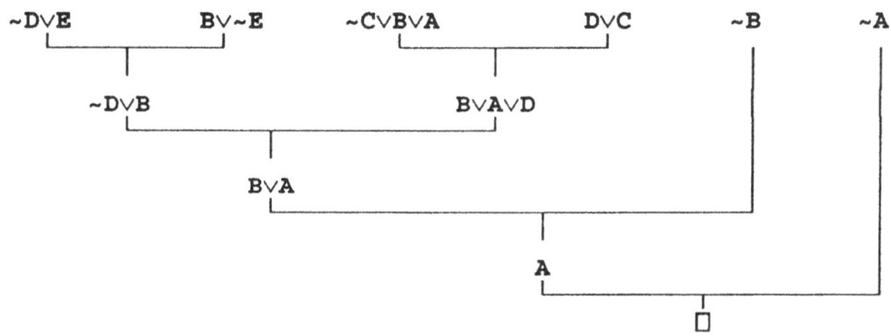

and we have proved that A is consistent with the original story.

It goes far beyond the scope of this text to delve further into the mysteries of mathematical logic. The reader is encouraged get acquainted with this branch of mathematics in order to be able to judge its strengths and limitations and apply it sensibly to practical problems.

Exercises

1. Show the ambiguities in the following set of formulas. If there exist no ambiguity, state the reason why.

   $A \vee B \vee C \vee D$
   $A \vee B \vee C \wedge D$
   $A \vee B \rightarrow C$
   $A \rightarrow B \vee C$

2. (a) Redraw the decision table progressively for the example given in Sect. 5.3 by using the conditions in the following order :

Root	:	TTS Group Discount
Condition 1	:	Occupation
Condition 2	:	Group Size
Condition 3	:	Occupation

   (b) Compare the resulting decision tree with the one given in Fig. 5.9. Which one has more branches?

   (c) Check the consistency of the decision tree you have obtained in part (a).

3. Is the order of the conditions important when developing the decision trees progressively?

4. Which of the following structured tools are completely sufficient to express the process logic? Why? Why not?
   (a) decision trees
   (b) decision tables
   (c) structured English

5. State the basic differences between limited-entry decision tables and extended-entry decision tables.

6. Given the following problem definition for a public telephone billing system

   " ... if the fixed-rate method is used for billing, then a minimum monthly charge will be assessed for consumption of less than 200 time units. For the consumptions above 200 time units, Type 1 billing method will apply. In the case of using a variable-rate method for billing, Type 1 method will be applied for the consumptions below 200 time units and additional consumption will be billed according to Type 2 Schedule. ... "

   (a) Prepare a decision table by applying progressive consolidation.
   (b) Check the completeness of the decision table.

7. Express the logic of the problem given in Exercise 6 by structured English.

8. Mary's love story was described in Sect. 5.8.3. Try to draw the following conclusions:

   (a) Tom does not take flowers to Jane.
   (b) Tom talks to Jane.
   (c) Jane does not feel very happy if Mary has to leave Los Angeles.
   (d) Jane feels very happy if Mary has to leave Los Angeles.
   (e) Mary does not have to leave Los Angeles, but the city is calm.

9. Prove the consistency/inconsistency of the following stories

   (a) If information systems are important, then we must build one. If we can live without an information system, then it is not important. If we must build an information system, then we can live without it. Therefore, information systems are important.

   (b) Information systems are important and expensive. If information systems are important, then either we do not build one or we cannot live without it. If information systems are expensive, then we must build one and we can live without it.

Chapter 6

Information Systems Evolution: The Software Aspect

Most information systems are in a continuous state of evolution. Contributing to this state of affairs is that systems objectives change over time, available implementation technologies change, the cost/performance ratios change for the available implementation technologies, the needs of the users of the systems change, the capabilities of the available human resources change, and so on. For computerized information systems, the consequence of system evolution is that of a pressure for software modifications, because many of the features that are being changed are usually implemented in software. Because software and persons often interact very closely, software changes frequently have associated with them modifications of the behavior of the persons who have to interact with the software. One should therefore always have the human component in mind when software changes are contemplated in an information system.

Information systems modifications come by as a complicated interplay between many factors. There are technological factors and there are human factors to take into consideration. Some of the well-known human factors are: human resistance to change, and human capacity for change. Other factors are: the cost of software maintenance, and the organization's administrative ability to stay in control of a process of change. Three aspects are of particular importance for the evolution of computerized information systems:

- *Software maintenance*: the cost of modifying software systems, the limits for changing the functions of software systems, the reliability of a software system that has undergone a series of modifications.

- *Human sides of organizational changes*: acceptance/rejection of software systems, implementation strategies, threats, opportunities, training, education.

- *Software systems quality*: technological level, user interfaces, systems integration features, decision support capabilities.

The human aspects are of particular importance. Most software in information systems is, after all, developed in order to serve human needs. Much of the software is intended to support human activity through close interaction with persons, for example, customers and employees interact with the software through terminals and workstations. The overall quality of an information system is critically dependent on both its computerized components and its human components and their interaction. If the human subsystem is of mediocre quality, then the whole information system will be of mediocre quality, even if the software components are of superb quality.

There have been particular problems in building information systems because the use of computers has been alien to so many

Ignorance about computers and their properties has been widespread. The first applications of computers in organizations brought about substantial changes in job content for many in the clerical staffs of the companies. This brought about changes in the "power structures" within the companies. Very often the anticipated changes brought resistance among those who felt threatened. Software systems that were successfully installed and used in one company could become complete disasters in other companies.

These initial human-factor difficulties now seem to be on the decrease. Many companies are installing their second, third, or fourth generation of information systems. Computer technology is not "new" any more. People are usually more interested in how their company's computerized systems can be improved to make the company more competitive than are they in fighting to have things the old way. Therefore, technical problems of how to achieve reliable modification of a company's software, and how to ensure that the technological level of one's company is advanced enough to keep the company competitive also in the future, are taking priority.

This is not to say that software can no longer fail to be successfully introduced in an organization. On the contrary, it is still possible to behave so stupidly that a software installation becomes a fiasco because of human factors. It is, however, easier than before to achieve a good result, because the "ignorance" factor is less serious today than it was in the first years. Some of the lessons learned then are still valuable today. We shall discuss some of these problems in the context of installing so-called "common" software systems.

Changes in an information systems environment bring about a need to modify the software of the system

Early information systems software was of the "custom-tailored" variety. When a company needed to beef up its payroll operation, its Data Processing department would be asked to build a payroll system that suited the needs of the company. The DP department would analyze the payroll needs of the company and build a software system for doing the task, with as high a quality as the DP department could manage.

Most companies soon found themselves in the intolerable situation of having to modify ever increasing volumes of information systems software, to keep track of changing requirements for the software's functionality. Every change in the employee-employer agreements on rules for calculating the wages had the potential of resulting in a software modification in the employer's payroll system. Even changes in the taxation system or government-induced changes in the rules for calculating social security taxes might require the companies to modify their internal software systems.

The response from the companies was partly a cry for increasing standardization, and partly an effort to improve their tools and methods for software production and maintenance. Because of an increasing number of less than successful experiences in the installation of interactive software systems, increasing emphasis was also put on the human-factor problems. This was manifested in increased user education and training. Much effort was also put into finding better ways for encouraging the future users of a system to participate more actively in its development and installation.

Software modification costs depend on the complexity of the software, as well as on the approaches that are used to cope with system complexity

We shall in the sequel view the problem of information systems evolution from three different perspectives that we feel are particularly well suited to cast light on the basic problems of the human factors, the technological factors and their interplay. They are the perspectives of:

- *The installation of software in an organization,*
- *The evolution of large software systems,*
- *The administration of change in software systems.*

The installation of software in an organization brings to the surface the problems of getting software accepted by the people who are going to work with the software on a daily basis, and whose efficiency (and job security) depends on the total quality of the information system of their organization.

Large software systems seem to take on a life of their own after they have been installed and put into operation. A characteristic feature is that the complexity of large systems seems to increase steadily with every new modification, unless special action is taken to remove complexity. So software systems seem to be "aging", becoming more and more fragile, as the process of change takes its toll.

Software "aging" may be counteracted by administrative means. Methods for supporting software evolution include configuration control methods, software documentation standards, software engineering data bases, and general project management methods.

6.1 The Role of Standard Software in Information Systems Evolution

The cost for developing a piece of software differs dramatically with its intended use. For software that is intended to be used only a few times by the software developer himself, one would expect the development costs to be substantially less than the development costs for the same piece of software when it is intended for sale to others, that is, when the software becomes a *product*. The increased reliability will carry a price tag, as will the different kinds of documentation that are needed when software is to be given a wider distribution. It is also to be expected that software which is intended to be part of a larger software system will be more expensive to develop than the same piece of software when it is intended to function in isolation. A piece of software that is to be part of a system has to be interfaced to the other components of the system. Each interface is associated with program development. This adds to the cost.

F.P. Brookes has suggested that if it takes one unit of effort to develop a *program*, then it will take three units of effort to develop the program into a *program product* [BROO75]. He has also suggested that it will cost three units of efforts to develop a program if it is going to be a part of a *program system*, and nine (= 3 x 3) units of effort to develop a program to become a *program system product*, that is, if the program is going to be part of a system that is going to be sold as a product.

6.1.1 Common, Standard, and Custom-Tailored Software

There is an additional dimension for software classification. This has to do with standardization of functional properties of software systems within organizations. We may classify software systems into:

- *Custom-tailored systems*, which are programmed to meet specific, local needs, and which are not intended to be used for any other purpose;
- *Standard systems*, which are generalized software packages that are intended to be installed in organizations with no changes made to their code and data definitions;
- *Common systems*, which consist of a standard "core" system, that has pre-defined interfaces for adding on locally defined custom-tailored systems.
- *Application platforms*, which are generalized common systems supplied by platform vendors, to be sold to many different companies having similar application problems, e.g., a bank platform is sold to many different banks in many different countries.

All of these four varieties of software are "program system products". They are usually quite large, so they have to be organized as systems. They are also used and modified by people other than their original developers, so they are "products". Custom-tailored systems are, as the name indicates, in-house developments. Standard systems are usually understood to be the kind of systems that are offered in the commercial marketplace. Common systems and application platforms lie somewhere between the two others. The major difference between the "core" of a common system and a standard system comes from a difference in the division of responsibilities for the development and maintenance of the two. Differences between the four kinds of software are depicted in Fig. 6.1.

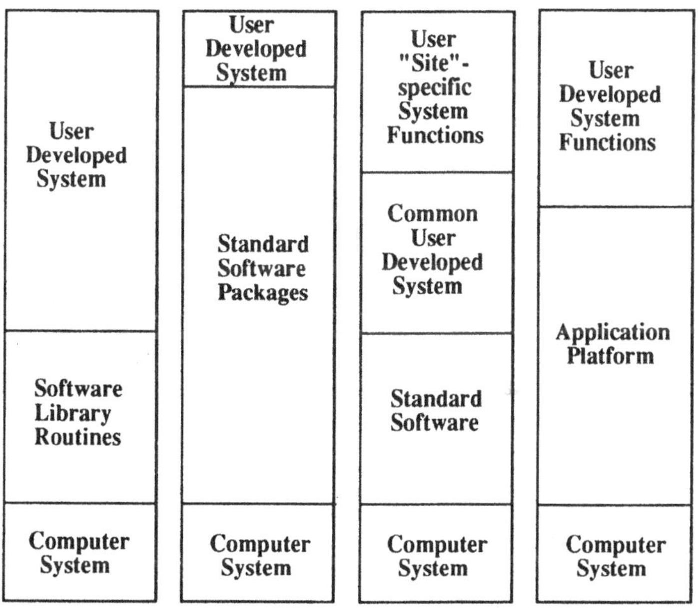

Fig. 6.1. Four types of relationships between software vendor and buyer

Standard systems are the responsibilities of their vendors. If the buyers are not satisfied with a vendor, they may choose to make their future business with other vendors. Common systems are, on the other hand, usually developed for internal use in large companies. The current trend is to organize a company in a number of autonomous decentralized units, as opposed to the centralized organizations of the past. Each organizational unit is equipped with its own software. Within one company this leads to a number of similar information systems, consisting of a functional "core" which is adapted to local needs in the different organizational units. If the current trend towards organizational decentralization continues it is a safe bet that this will lead to an increased need for developing common systems within companies.

It may also happen that several companies team up to share the costs of making and maintaining a common core-system. In either case, the company that uses the core also has some economic interest in the development and maintenance of the core. This situation can lead to all kinds of organizational in-fighting between the providers of the core and the users of the core. We shall see in the sequel how to avoid some of the more destructive situations.

Common software systems (and application platforms) rarely involve any innovative technology. However, they involve complex problems of design and management. To take additional chances on unproven technology is most often deemed to lead to unacceptably high risks for project failure. This is one of the main reasons why many highly successful software vendors are perceived as being technologically behind.

Of paramount importance is also the standardization of user interfaces for the various software components. In an information system which comprises non-integrated software systems, an individual employee may very well interact with 5–10 different software systems on a daily basis. Standardization of interaction styles and user interfaces is a key factor in bringing down operational costs.

The rationale for developing common software systems is partly one of economy of scale. But there is also usually involved an element of standardizing organizational practices. From a global point of view it may pay off to enforce standard cross-company ways of organizational behavior. What better way to enforce this than through a common software system? So one might often quite rightly ask if what is called a software installation should rather be called organizational development. It is often the case that the global benefits of installing a cross-company software system are not fairly appreciated from the local point of view. On the other hand, local problems of fitting the software to suit local needs are also often not appreciated and fully understood from the global point of view. Objectives and responsibilities should be agreed upon ahead of the installation, in order to avoid misunderstandings during the installation phase.

Hardly ever is it possible to install a software system in an organization without having to modify either the ways of the organization or the properties of the software, or both. This is of course more so for standard software and common software than for custom-tailored software. But even for custom-tailored software we cannot expect a perfect match between what the needs really are and what the software system offers. This is because it is so difficult to anticipate what is needed, ahead of the actual operation of the software. Modifications usually have to be made before, during, and after the installation phase, in order to get a better fit between what the software system is able to do and what one would like it to do.

6.1.2 Application Platforms, Common Software, and Information Systems Integration

Every organizational unit that uses a standard software system is using a copy of the same software to solve its problems. In the case that the functionality of the standard software does not fit completely with the problem at hand, it is very tempting to try to modify or extend parts of the code. The alternative is usually to modify the human procedures in the organizational unit, in order to fit the software at hand. There is a great temptation to choose to modify the software, because it somehow seems to be more "right" to change a machine's behavior than to change human behavior, if the machine lends itself to modification.

However, modification of standard software systems is frequently not possible, because of proprietary rights. Furthermore, it is almost never advisable to modify standard software even if this is permitted by the software vendor. This is because the software vendor will not accept responsibility for maintaining a software system that somebody else has meddled with. So the software buyer will have to assume the maintenance responsibility for the "standard" software system that has now evolved into a non-standard system.

It will also become increasingly difficult to take advantage of the vendor's new versions of the standard system, where old errors have been corrected, and new functionality has been added. One is faced either with introducing the local modifications, in full, into each new vendor version of the standard software, or with modifying the local version of the "standard" software with the vendor's modifications to his software – if one is able to find out all of the implementation details of the vendor's software modifications.

To put it in short, the software system that originally was a standard system will take on more and more custom-tailored features with each modification. In the end, each organizational unit will have to treat the previously "standard" system as a custom-tailored system. The economy-of-scale savings that were sought when buying standard software have evaporated completely. So one should never go beyond vendors' recommendations in modifying standard software.

In some instances common systems are designed to replace existing, incompatible systems. An important objective is to integrate the information processing of different parts of the organization into one comprehensive system. Standardization of data elements and compatible databases are key words for information systems integration. The situation is depicted in Fig. 6.2.

Integration of information systems applications is a formidable challenge. The databases of the many information systems that are in operation in a large company reflect various views of the company and its environment. The content and structure of each of the databases result from the particular abstraction of the world that was employed by the database designers. Systems integration requires that the databases are integrated. This requires that a common world-view is developed within the company.

Even if a company is completely free to determine the database schemas independently of commercially acquired application software, it is a formidable job to develop a common conceptual data model for the whole company. On of the reasons for this difficulty is that those who know enough to be able to form a consistent world-view of the company and its environment usually have been in management positions for a long time, and are not designing information systems any more. Software development is usually done by those who know much about computers, but who know too little about the overall operations of the company to be able to do a proper integration job. This problem is not going to become one bit easier to solve when application platforms carry their own vendor supplied world views into the companies!

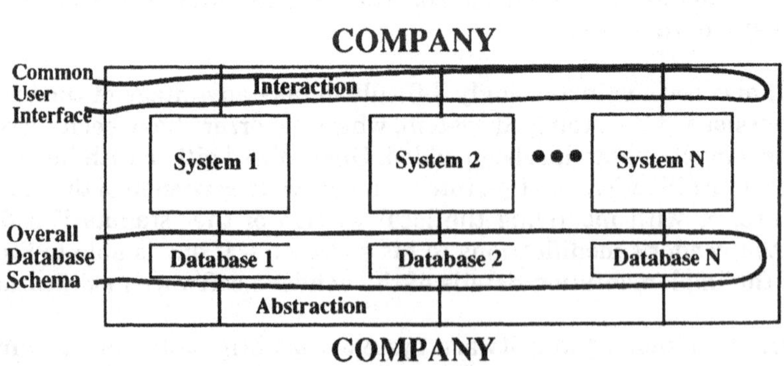

Fig. 6.2. System integration within a company

The tasks of the systems designers are different in a common system/application platform environment than they are in an environment of custom-tailored systems. The add-ons will have to be designed relative to the world-view of the common system. To be effective, the designers will have to be specialists of the inner workings of the appropriate application platform.

6.1.3 The Architecture of Common Systems

Common systems have an architecture that accommodates the need for modifying a standard solution of a general problem to fit the local problem at hand. A common system contains a *core* and local add-ons. The core is supposed not to be modified, and is therefore to be treated as a standard software system.

When common software systems are being designed, some of the major issues that have to be addressed are:

- The definition of the *core*: how large shall it be, if (and how) shall it be possible to modify it?
- Shall all of the users be required to use the same type of data processing equipment; e.g., if a DBMS is to be used, shall all common systems users be required to use the same DBMS?
- The maintenance responsibility; is it a local user responsibility, a central *core* responsibility, or a shared responsibility?

The key question when designing the core is that of which functions to standardize in a typical user environment, and to implement as parts of the (non-modifiable) core. One is thus forced to define what a "typical" user environment is, what is similar among the various environments, and what are the features which are different.

How large is the core of the common system?

The key question when investigating the possible use of a common system in an organization is which of the information processing functions to support by the standardized functions of the core. Is the core 80 %, 60 %, 40 %, or 20 % of a particular installation?

Opinions usually vary on this question. Those who develop the core tend to see it as a much larger part of the total system than do those who are going to fit it into the local environment, and to maintain the local add-ons. The core developers will usually try to have as much as possible of the (unmodified) core used in every installation. This may not be in the best interest of the local user, if there is a mismatch between his needs and what is offered by the common system. The local user may therefore resist using more than a restricted part of the core, and quite rightly so, from the local point of view.

There is often a tendency to "over-design" the core, that is, to make it larger and more complex than necessary. This is always to be expected when the central design team is young, vigorous, ambitious, and inexperienced. Is there a more worthy goal for the central design team than to make the world's best core system, so that the local users will have to design as few local add-ons as possible, preferably none at all? Here, as elsewhere, the road to disaster is paved with good intentions. The "world's best" core system may turn out to become a nightmare, if it is trying to cover every conceivable local situation. Over-design may be counteracted by providing better contact between those who design and those who are going to use and to maintain the common system.

Local needs may be in conflict with the adaption possibilities that are offered by the common system

So-called "technical problems" are more often than not related to differences between central and local expertise, and to subsequent misunderstandings between the two parties due to mutual lack of knowledge about the other's affairs.

When central design is to be supplemented with local maintenance, a minimum requirement is that the local teams are given complete and timely documentation of the core system and the options that its design offers. This is seldom the case. The cost of developing first-class documentation is usually very high, sometimes comparable to the cost of making the core system in the first place. So the system's documentation is quite often of low quality. The unavoidable result of low-quality (or lacking) documentation is increased maintenance costs.

A fair number of common systems are generalizations of successful local solutions to problems that are deemed to be of wider interest. What often happens is that a successful local installation is believed to have a potential market in similar businesses, after a suitable rewrite in order to make it more adaptable. The first new installations usually meet with less success than did the original installation, because the local needs differ from those of the original environment.

After a while it is decided to rewrite the original system, to make it more adaptable, and to improve cross-company standardization and integration. One decides to create a common system. Because one tries to take advantage of the knowledge already acquired, it is natural that one tries to use as much as possible of the previous design. Why make unnecessary changes to a design that has already proved itself?

The result may very well be that business practices from the original system are carried over to the new design, without anyone in the design team really knowing, because they are not aware of the lack of generality of the business practices of their original system. If this is coupled with low-quality documentation, the local installation and maintenance teams may encounter a very difficult situation indeed when trying to adapt common systems to their local situation.

Common systems may either be data base-centered or program-centered

In a database-centered view, the common system is viewed as a database on which programs are hung (Fig. 6.3.a). In a program-centered view a common system is a set of programs (Fig. 6.3.b). In the program-centered view, very little attention is paid to the consequences of changing the database. This view is therefore potentially very dangerous, because it may easily lead to incompatible databases at the different installations of the common system. The system integration goal of the common systems approach may thus become unattainable.

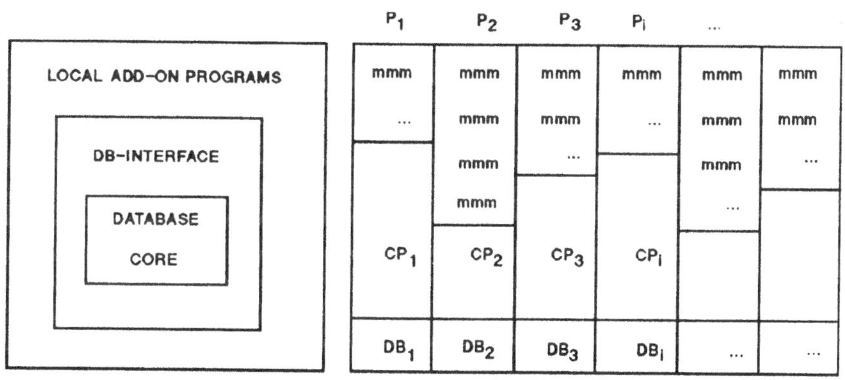

(a) Data base-centered view

(b) Program-centered view

P – Program
CP_i – Core Program
DB_j – A program's data base
m – Program modifications

Fig. 6.3. Common system, data base-centered and program-centered views

6.2 The Installation of Software Systems in Organizations

Installation of software in an organization is only partly a technical venture. This is easy to accept when one experiences severe discrepancies between the software's capabilities and the organization's needs for information processing. But even when there is a perfect match between software capabilities and organizational needs, it is possible to ruin an installation project. It is always possible to complicate things so much that chaos seems imminent and stress builds up.

The needs for changing an information system has quite often to do with the integrational aspects of an organization, and not with details at the individual level. Therefore, it is sometimes difficult for the individual employee to understand the motivation for installing the new software systems in the first place. The individual employee may have to pay a "price" for the modifications, e.g., in terms of changed working conditions, often for the worse during the installation period, without seeing any direct benefits from the changes. In case of severe mismatches between software capabilities and organizational needs, working conditions may deteriorate severely until the discrepancies have been removed. This may lead to negative attitudes towards the new software system.

Many difficulties with installing software in organizations can be traced back to inexperience on the part of those involved. Inexperienced software people are usually overly optimistic with regard to how many of the functions in an information system can be performed by a software system, and how much it will cost to develop and install it properly. When their first system version has to be modified, extra costs are incurred, both for the software modifications and for the organizational modifications. The latter are by far the most serious ones. If the discrepancy between the software functionality and the perceived needs of the people in an organization becomes too large, there may be a complete breakdown of trust between those who install the new software and those who are going to use it in the future. Disaster may then be the only outcome to be expected.

Factors that have to be considered when embarking on the installation of common software systems, include [KEEN82]:

- Installation approach (strategy, pace, focus)
- User participation (passive/active, hostage/co-worker)
- Education of users of all categories
- The roles of central and local groups
- Central authority versus local autonomy
- Technical issues (core size, maintenance responsibility)

6.2.1 Installation Approaches

Installation approaches may be characterized in terms of [KEEN82]:

- *Strategy* : parachuting or acculturation
- *Pace* : crash or filter
- *Focus* : technocentric or organizational

Parachuting stands for a philosophy of installing the software system in the organization with as few modifications as possible to the software. Standardization and integration are given higher priority than variation. The organization is expected to change its ways in order to interface to the software system, rather than the opposite. A consequence of this approach is that the system is usually felt to be enforced on the organization, from the outside, from "upwards" and "down".

Acculturation, on the other hand, is a strategy for making a software system into an integral part of the user organization from the start. It is a policy that software should be modified to accommodate local organizational variation. Local ownership of the (modified) software system is emphasized. Local autonomy is emphasized, as is education and user participation.

Crash pace during implementation gives priority to getting the software system up and running as quickly as possible. User education and other organizational issues are given priority only after the system is operational.

Filtering adjusts the installation pace to the organization's ability to assimilate the imposed changes. This approach delays the installation of the system, but is supposed to make it easier to institutionalize the changes.

Technocentric focusing sees installation as centering around technical development. The main responsibility for the installation project is usually given to the software people.

Organizational focusing places far more responsibility on the local users than technocentric focusing does. The installation project is headed by the future users of the system, rather than by the software people.

These three dichotomies are quite suggestive: The "hard" crash-pace, technocentric, parachuting style of inducing change in an organization is opposed to the "soft" organizationally focused filtering of changes into the organization in an acculturational style.

6.2.2 Who is the User?

The "user" has been at the center of the previous discussion. The various installation approaches differ mainly with regard to degree of "user involvement" in the installation project. So, who is the "user"? And what is "involvement"?

In general, a user is someone who employs or consumes somebody else's services or products. Therefore, the term user may denote many different groups of people in an organization. One cannot know who the user is until one knows who uses the word.

In small organizations it is easy. The users are those who interact with the software on a regular basis, and are served by the software in their regular work. The technical staff in the data processing department keep the software operational, and maintain the software systems, and they are definitely not the users.

When organizations grow it is not so simple any more. The various "user departments" develop their own software systems skills. They organize themselves internally into "users" and a technical staff whose task it is to give the local organization appropriate software services. The original DP-department has in the meantime changed tasks and responsibilities, and is now more and more involved in standardization and integration of data processing services across all of the user departments in the whole company. They encourage the installation of common systems, and other kinds of standardizations, and they regard the technical staff in the user departments as their primary users. The real users in the user departments are viewed as secondary users only, often called "end-users". So-called "naive users" are end-users who interact so seldom with the software system that they cannot be expected to remember the user interfaces and the interaction procedures every time. They must therefore have every step in the interaction procedure explained to them each time they interact with the software system.

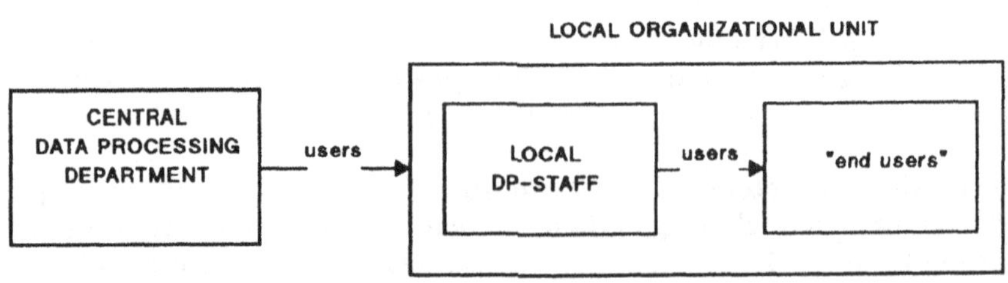

Fig. 6.4. "Users" and "users" of software systems

There is a conflict of interests in every organization where the use of computers play a substantial role, between the needs for standardization/ integration of software services, and the need for variation in local organization because of differences in local business needs and practices. This conflict of interest sometimes erupts in a fight over the control of the local DP-staff. "Down with the tyranny of the DP-departments" was a much-used slogan in the 1980s, when the end-users equipped themselves with personal computers and happily – without really knowing it – repeated most of the mistakes that previous generations of software people had already made many years ago when they first started to work with computers.

The same interest conflict can be found in the difference between the parachutists and the acculturists view on the "users" [KEEN82]. The (central DP-departments) parachutist point-of-view is that the users are those who are directly involved in developing and installing the software system, that is, the local DP-staffs (Fig. 6.4). The acculturist point of view is that the users are those who are affected by the operation of the software system, that is, the "end-users".

6.2.3 Installation Experiences

Keen et al. [KEEN82] have investigated the installation history of a common software system in a large U.S. bank which operates in almost 60 countries all over the world. The development of the common system started in the mid-1970s. We shall highlight the main observations of the study in the sequel.

Like all other banks, our bank also has to make its future profit from fee-based services, instead of from interest charged on loans. Customer service is the key to future success. The bank has therefore decided to implement a world-wide Information Processing System (IPS). The objective was ambitious for the mid-1970s: real-time processing of all bank transactions, standardized data storage at transaction level in all countries, and global customer integration.

The bank had a central development group in Milan, Italy. Each country had a local data center equipped with a standard mainframe computer. The largest countries had their own development staffs. The bank had a private data-net that operated in the U.S.A., Europe, and a few countries in the Far East and Latin America, but was also a heavy user of TELENET and SWIFT, an international banking network.

The IPS grew out of a system built in Milan in the early 1970s and was based on Italian banking practice. It was written in Cobol. There was no use of data base management systems. There were also no traces of modern programming disciplines such as structured programming. The system was, in short, technically quite old-fashioned even in the mid-1970s when the decision was made to implement it worldwide.

The system was organized as a common system with a core and provisions for defining local add-ons. It contained over a dozen functional models, one for each of the banks major products. After five years of world-wide implementation efforts, there were 24 different operational variants, with 15 more underway. The core system almost always needed local modification. This required close contact and cooperation between the local implementation group and the central "core" group in Milan.

Detailed studies were made of the implementation history of several countries. The results of the installations varied from the disastrous to the virtually problem-free. The results from four of the countries can be summed up as follows:

* In the first country the parachuting strategy was used. There was very little user involvement. There was strong control from the central group in Milan. The local DP-group was quite weak, and the local top management was "invisible". The installation project was plagued by poor quality of operations and severe lack of coordination. This resulted in implementation delays. A major reorganization was deemed likely at the time of the investigation.

* Parachuting and crash pace was used in the second country. The end-users were deliberately excluded from participation in the installation project to avoid potential resistance. There was a lack of commitment from local top-management. The crash pace implementation led to frequent operational breakdowns of the software system. There was no time available for modifying the core system to conform to local rules for accounting controls. After the system was installed there was widespread and continuing dissatisfaction. There were severe problems with data quality and control. More user training was obviously needed.

* Parachuting was also the strategy of the third country. Here the local DP-staff became heavily involved, but there was no involvement whatsoever of the end-users. Top management was of the end-user variety and did not get involved, but left the whole installation to a very strong operations manager. The local management was even strong enough to face up to the core-systems group in Milan. Changes were introduced to the original system that made the system incompatible with other countries' systems. Due to very hard work by the local DP-staff the system was installed in six months. Alas, it took several years of follow-up support to get the system stable. There is a continued cultural gap and conflict between the technical staff and the end-users. The installation has been characterized as a "psychological disaster".

* The fourth country experienced few problems, if any at all. They used an acculturation strategy. A strong education program was used to prepare and motivate the end-users for the changes that lay ahead. Formal contact groups consisting of end-users and technical staff were organized at a very early stage. The top management involved itself explicitly along with senior representatives of the end-users. There were no technical problems and no organizational problems. Some end-users complained that their involvement started too late.

One of the very distinct differences between parachuting and acculturation is that parachuting takes everybody by surprise, except those that become directly involved. Acculturation prepares everybody concerned to become involved. Unexpected problems may therefore be smoothly resolved before they lead to crisis.

In this particular example the core system was based on Italian banking practices. The systems documentation was partly in English and partly in Italian. Seemingly small differences in local banking practices could result in substantial modifications of the core system code. The bilingual documentation did not contribute to make the code modification any easier. The technical staff in Milan was not aware of the local differences until several installation disasters had taken place.

Even if the particular difficulties of multinational installations are removed, parachuting seems to lead to:

- Strong control from the core-system staff,
- Competence fights between core-system staff and the locals,
- Weak user participation,
- Passive local management,
- Global system incompatibilities because of local software modifications,
- Installation delays,
- Weak system stability,
- Conflicts between end-users and technical staff.

Acculturation also has its pitfalls, even if those are not mentioned in the study that has been cited above [KEEN82]. The major problem with acculturation is that it increases the possibilities for having delays of software systems installation because of unwarranted local resistance. Hidden agendas occur quite often in the "administrative waltz" of companies. Human resistance to change is a fact of life, also when new software systems are being installed. Changes that are needed because of long-term survival may hurt in the short run. Of course it is all for the better if those that may potentially be hurt by the changes, can be enlisted to participate in the change process in order to modify the changes to their own advantage.

6.2.4 Features of an Installation Strategy

There is an approach to the installation of common systems in organizations [KEEN81], [KEEN82]. The approach is accultural in style, and puts great emphasis on the need for preparing the organization properly ahead of the installation of a software system.

The six main stages of the approach are called

(1) Expectation setting
(2) Technology mobilization
(3) Data conversion
(4) Core installation
(5) Local adaption
(6) Evolution

The first stage is concerned with developing realistic expectations in the development group and among top management for what the future may bring. A common system is brought in from the outside. It is therefore a wise first move to get acquainted with experiences from other installations, in order to be able realistically to estimate the resources, time-frame and commitments required. Mechanisms for user involvement should be identified, and arrangements should be worked out for acquiring the necessary support and education.

The next stage, technology mobilization, is concerned with getting the wider organization ready for the venture. Most important is the education of the users, and the creation of a cooperative climate in the organization. These two ends are closely connected. It is difficult to be successful with one without being successful with the other. It is during this stage that:

- One organizes user groups, which are to become the internal change agents, coordinators and educators, and

- One uses education to inform, to publicize, to build skills, to provide a common set of concepts and words for user involvement, to demonstrate top-management commitment, and to create a forum for discussion.

It is of crucial importance to get high-quality people into the user groups. The best people are hard to get. Their managers will not give them up unless they are convinced of the importance of the project, and unless the project is given priority by top management.

The users should be given the responsibility for leading the education process, even if most of the teaching would have to be done by outside teachers from the central or local DP-staff. The education process takes time and substantial resources. On the other hand, if the users are not properly educated, the whole installation may be jeopardized.

After this mobilization, the necessary basis for a successful installation has most probably been established, and the technical work can begin with the activity in which the users play the main role, namely the data base creation or conversion. Getting the data right is a key challenge, and may require big efforts, depending on the quality of the existing data.

The core installation stage is followed by its adaptions to local requirements. It is of special importance to maintain the integrity of the common system during local adaptions if standardization and corporate integration requirements are to be met.

The final stage is to put the system into operation, to transfer it to whoever is to take the operational responsibility, be this a user department or a local operations group. More training will be needed for the users. Formal liaisons must be set up. Users' representatives may move into the local DP-department, or someone from the local DP-staff may move out into the user department, to act as a coordinator for the operations.

6.3 Evolutionary Behavior of Large Software Systems

One view of the software system evolution process is to see it as an interaction between the software system itself, and the human organization that implements the process. The function of the project organization is to make changes to the software system. During its lifetime all kinds of changes may become necessary. The computer may be changed or replaced. New devices may be added. New uses may add new requirements. In general, software modifications are desirable [BELA76] in order to remove

- *Faults,* which are system deficiencies that are related to the difference between anticipated and desired behavior of a system.

There are two types of "faults":

- *Defects,* which are faults that are related to changes in the software system's environment, and

- *Errors,* which are faults that are related to the difference between actual and anticipated behavior of a system.

When faults manifest themselves, the human project organization is required to take corrective action, that is, to "repair" the software system. So the evolutionary process of a software system is governed by the properties of both the software system itself and of the human "software maintenance" organization. The effects of these two entities on the evolution process are difficult to distinguish from each other. It is quite hard to ascribe observed evolutionary behavior either to the particularities of the maintenance organization or to the particularities of the software system itself.

Software modifications are made available to users in a sequence of versions, or releases. When a new release is exposed to an operational environment, its actual usage often differs from the usage that was anticipated by its developers. So the release of new code may most often result in the discovery of new and unexpected faults. It is therefore a good maintenance strategy to produce sufficiently many releases to prevent a build-up of too many undiscovered faults. On the other hand, too many releases in too short a time will create an unstable situation for the users of the system. They will not get used to the idiosyncrasies of one system version before they have to replace it with another. Sometimes faults will not manifest themselves until after the system has been used for a while. Excessively frequent releases may therefore lead to faults not being discovered sufficiently early, in which case they will accumulate.

Studies of software evolution are hard to find. For a study to be of value it has to be conducted over a time-span of several years. This is necessary in order to cover a large enough number of software releases that sufficient data will be available for analysis. Belady and Lehman have published a study of the evolution of IBM's operating system for their 360-series computers [BELA76]. Three types of model were developed to describe and explain the observed behavior of OS/360 over a period of twelve years, through more than 20 system releases. They are:

- A statistical model for describing the observations;
- An error propagation model, for explaining part of the observed evolutionary behavior;
- Resource allocation models, for describing the effects of management decisions about the allocation of project resources, on the evolution of a software system.

Furthermore, the observed data led the two investigators to propose three quite general principles associated with systems evolution. They presented the three propositions in the form of "laws" that are claimed to govern the behavior of software systems and their maintenance organizations. The "laws" are about:

Continuing change: a software system that is being used undergoes continuing change until it is judged more cost effective to freeze and recreate it,

Increasing unstructuredness: the unstructuredness of a system increases with time, unless specific work is executed to maintain or reduce it, and

Smooth system growth: growth-trend measures of global system attributes may appear to be stochastic locally in time and space, but statistically they are cyclically self-regulating, with well-defined long-range trends.

The three laws seem self-evident. They are also backed by substantial practical experience. They can be supported by simple arguments, e.g., all systems have to change unless they are to disappear by becoming obsolete and losing out to new and better competitors. When a software system is changed, many components usually have to be modified. As a result of the modifications, the system's structure inevitably degenerates. Further modifications become more expensive to perform when the system's complexity increases. But it is the task of the project organization to respond to external change (e.g., growth) as quickly as practically possible. Thus, if a proper response cannot be delivered at once, a well-run project organization will try to adapt during the next few releases. So the third law is a consequence of having a sensible development organization.

6.3.1 An Analysis of Observed Evolutionary Behavior

The development of the operating system for IBM's 360 series of computers started in 1963-64 when this first "family" of computers was conceived. It was the largest software development project that had ever been tried, until then. It seems clear, in retrospect, that the development organization was operating on its margins over some periods, and that OS/360 on several occasions was on the verge of collapsing.

The growth rate of OS/360 was phenomenal. It was, after all, an operating system for a whole family of computers. Each new "family member" added to the complexity of the internal family relationships, and subsequently to the size and the complexity of the operating system. The increase in systems size over the observed evolution period for OS/360 can be seen in Fig. 6.5. OS/360 grew to four times its initial size after approximately 15 versions had been released. This is comparable to another operating systems product from the same company (the DOS system), but very different from other observed evolution histories, of more "normal" software systems [LAWR82].

Even so, it is of great interest to analyze the evolutionary behavior of software systems like OS/360 (and DOS). This is so because of the great

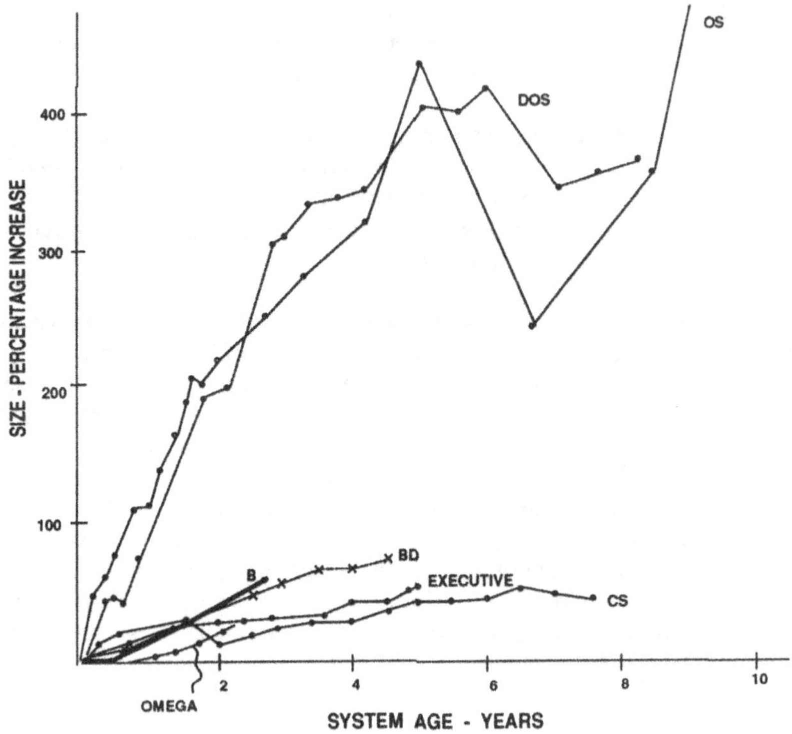

SYSTEM NAME	SYSTEM TYPE	First Yr of data	Last Yr of data	MODULES			No.of Users
				Initial size	Final size	No.of Releases	
OS/360	Operating System	3/1966	3/1975	1152	5300	21	V.Many
OMEGA	Operating System	2/1972	9/1974	335	388	9	Many
DOS	Operating System	*	8 years	438	2141	31	V.Many
B	Operating System	10/1975	7/1978	1728	2413	6	Many
EXECUTIVE	Systems Support	11/1973	11/1978	657	967	12	2 Sites
BD	Application	1/1973	12/1977	42	66	Not Released Based	Few
CS	Application	4/1972	12/1979	971	1483	58	6

* The first year of data collection on DOS is not known

Fig. 6.5. Size of various software systems, as a function of age (from [LAWR82])

demands that systems with such high growth rates put on their respective development organizations. Effects that might easily be suppressed and hidden in "normal" maintenance situations are bound to surface and become visible when organizations are put to their limits in coping with change and complexity.

One particular difficulty in making anything worthwhile out of an analysis of evolutionary behavior of software systems is that the appropriate data are usually not collected while the action is taking place. Nobody will collect data unless it is for some valid purpose. To prepare for analysis of system attributes ten years down the road does not seem very important for people who are working long days and nights to keep new software releases from collapsing, meeting deadlines which are always too short.

For the OS/360 analysis there were only three independent attributes that could be observed in retrospect. They were the system's age, its size, and the number of different components that had been investigated and/or modified ("handled") during the development of a new release. Other measures that may be derived from these three parameters are release interval, handling rate, and complexity.

The directly observable parameters are

- r – release sequence number
- D(r) – the system's age (in days) at release r
- M(r) – system's size measured in number of "modules"
- MH(r) – the number of system modules that have received attention (have been "handled") during the release interval

The derived parameters are

- $I(r) = D(r) - D(r-1)$ = the inter-release interval, that is, the interval (in days) between releases,
- $HR(r) = MH(r)/I(r)$ = the "handling rate", that is, the number of modules handled per day during the release interval
- $C(r) = MH(r)/M(r)$ = the system's "complexity", that is, the fraction of the modules that are being "handled" during a release interval

In general, there were large, apparently stochastic, variations in the individual observations from release to release. Nevertheless, all of the observations indicated general upward trends in the parameter values. The

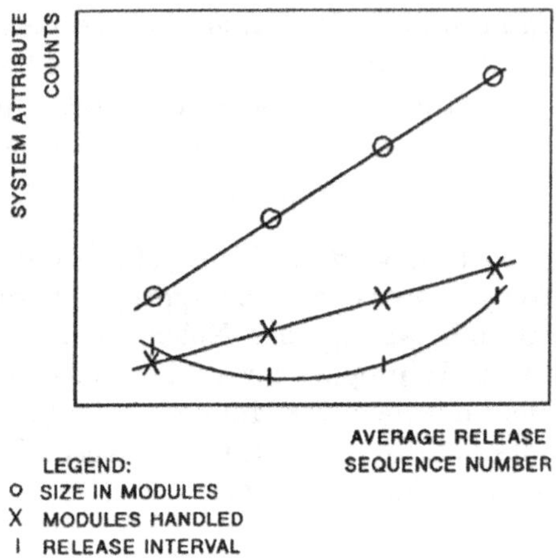

Fig. 6.6.
Average growth trends of system attributes
(from [BELA76])

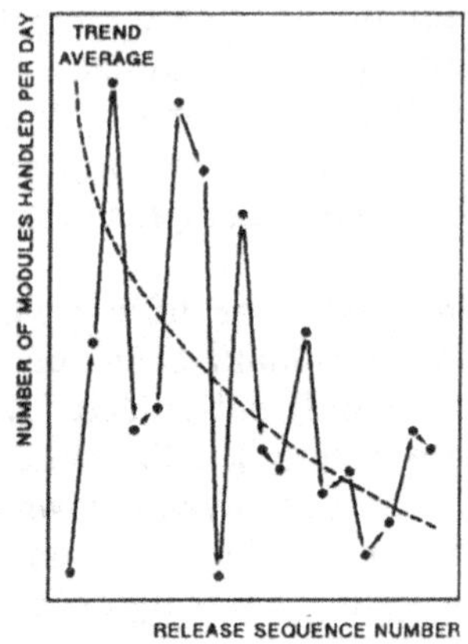

Fig. 6.7.
Serial and average growth trends of a particular attribute
(from [BELA76])

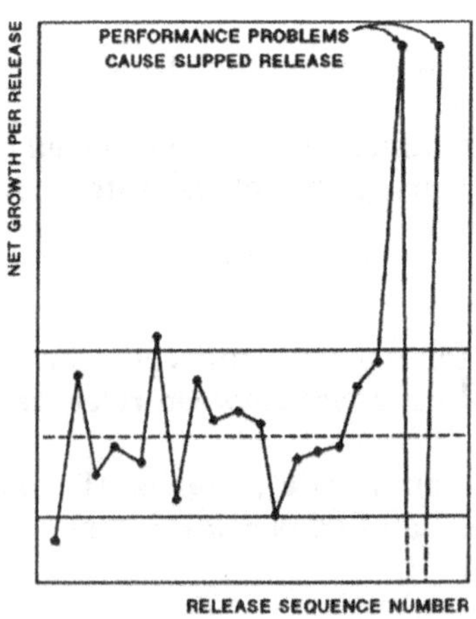

Fig. 6.8.
Net growth of OS/360 releases
(from [BELA76])

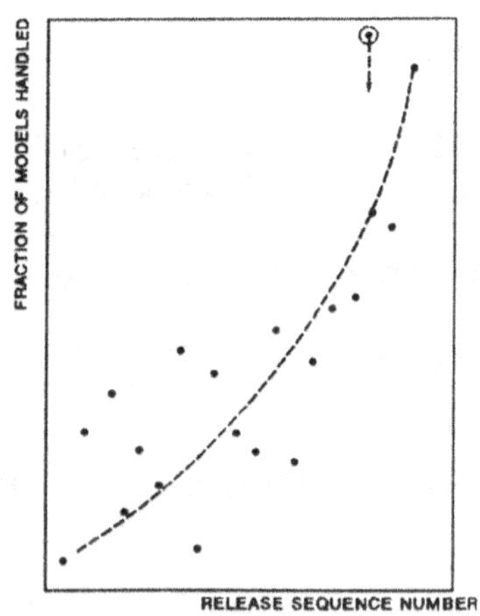

Fig. 6.9.
Complexity growth during the interval prior to each release
(from [BELA76])

trends were clearly exposed when observations were subjected to averaging (Fig. 6.6). Some of the observation series also indicated the possibility of cyclic patterns (Fig. 6.7).

This encouraged the investigators to try to describe the observations with mathematical equations which were fitted to the observations by standard curve-fitting methods. They chose to use equations that had a long-term trend component in the form of a polynomial, and a short-term cyclic component to represent cyclic variations like those exhibited in Fig. 6.7 and Fig. 6.8.

One example of how the mathematical apparatus is used can be found in a description of how the system's size grew from release to release. The net growth of OS/360 for each release is depicted in Fig. 6.6. The growth process is cyclic, but there also seems to be an upper bound as well as a lower bound to the net growth per release. In the two cases when the net growth significantly exceeded the upper bound, there were serious delivery delays.

Because the net growth seems to vary around an average value, as seen from Fig. 6.8, it is natural to guess that the observations may be described by an equation of the form

$$M(r) = a + b*r + S(r)$$

where S(r) is a cyclic component. A least-square fit to the available data returns the equation

$$M(r) = 760 + 200*r + S(r)$$

were the system's size M(r) is measured in "modules". The "module" concept was well defined within the OS/360 organization, so we do not have to bother about the exact meaning of the concept. The small number of available observations prevents a statistically significant determination of the cyclic component. Note that the equation directly reflects the first and third of the "laws" that were described in the introduction of this chapter.

The fractions of the modules that were handled during the interval prior to each release is depicted in Fig. 6.9. We see that this measure is steadily growing with each release, although the data points are fairly widely scattered for early releases. Belady and Lehman have proposed that this measure is used to indicate a system's complexity. The argument is that there is a wider "ripple" from a repair in a poorly designed system than in a well-designed system, that is, more system components would have to be modified in the poorly designed system than in a well-designed system. A poorly designed system is tacitly assumed to have a higher complexity than a well-designed system.

Even so, this complexity measure is clearly inadequate. In real situation there are various independent complexity factors involved. C(r) only measures complexity very indirectly. It does not distinguish between the complexity of the fault that is to be repaired in a system, and the complexity of the system itself. Assume a case where a simple fault is to be repaired in a complex system, and compare that to a case where a complicated fault is to be repaired in a simple system. The simple fault may require the handling of a large fraction of the system's modules because of the "unstructuredness" of the system. The complicated fault may require that a similar fraction of the modules be handled, but in this case because of the complexity of the fault. In both cases the complexity measure C(r) will be found to be of approximately the same size, even if the two systems are of very different complexity.

A large number of faults are usually repaired prior to each release. The faults may be evenly distributed over the various parts of the software system, or they may appear in "clusters" related to different subsystems, for different releases. Especially at a time when a number of new subsystems are put into operation, one would expect that faults appeared in "skew" distributions. New subsystems are usually not well enough tested at the time of a release. Therefore, it is to be expected that most of the faults that are discovered would relate to the new and untested subsystems than to the older parts of the system. When the system growth has decreased would one expect that the faults were more evenly distributed over the software system from one release to the next. This effect may account for the wider distribution of data points for early releases than for late releases in Fig. 6.9.

In spite of its weaknesses, C(r) is the best complexity measure that can be had with the available data. If we assume that the repair activities during different release intervals are of similar difficulty relative to systems of the same complexity, then the complexity measure C(r) would certainly be expected to classify the systems reasonably correctly relative to each other, with respect to their degree of "unstructuredness".

6.3.2 Basic Assumptions of Different Models of Evolutionary Behavior

The previous analysis indicates that software systems evolution is heavily influenced by fault repair activity. In order to be able to properly control the evolutionary process, it is necessary that the process itself be understood. Belady and Lehman have proposed several abstract models for explaining the observed evolutionary effects. The models were based on the following assumptions:

- System modification activities implies the making of errors,
- There is a delay from making an error to its discovery,
- There is a delay from discovering an error to its repair,
- Some errors are ordered in the way that one must be repaired before the other can be detected,
- Documentation is viewed as an integral part of the software system,
- Documentation is used for internal communication within the project development group,
- Project group members must be educated in the documentation,
- The documentation must be updated to reflect new system modifications,
- Documentation deficiencies decrease the effectiveness of the software modification process.

The first of the models emphasizes the internal distribution and propagation of errors in the software system. The role of the project group is simply to eliminate observed faults.

In the second model the project group is free to use their efforts on various tasks, e.g., error repair, documentation or other activities that are deemed useful. Different error generation rates would be expected from different distributions of available resources on project activities.

6.3.3 The Impact of Error Propagation on Structural Degeneration

Belady and Lehman's error propagation model gives a simplified view of the "structural aging" of software systems. The model is based on the fundamental assumption that new errors may be introduced when a software system is repaired. It is also assumed that not all of the faults have been discovered at the time of repair. Therefore, it is to be expected that some faults remain undiscovered after the known faults have been extracted. Such faults are called residual faults. Both of these assumptions are known to hold in all too many practical situations. A primitive model is depicted in the diagram of Fig. 6.10, which captures these two effects.

The distinction between residual and newly generated errors is fundamental to Belady and Lehman's error propagation model. The primitive change activity of Fig. 6.10 is the basis of the network of Fig. 6.11, which

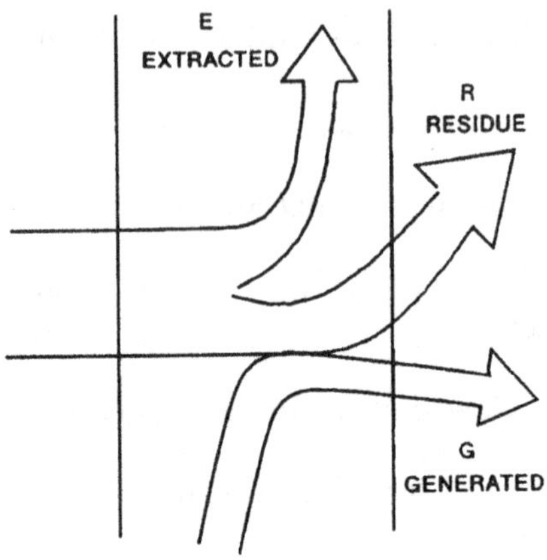

Legend:
E – extracted faults
R – residual faults
G – generated faults

Fig. 6.10. Primitive model of fault penetration (from [BELA76])

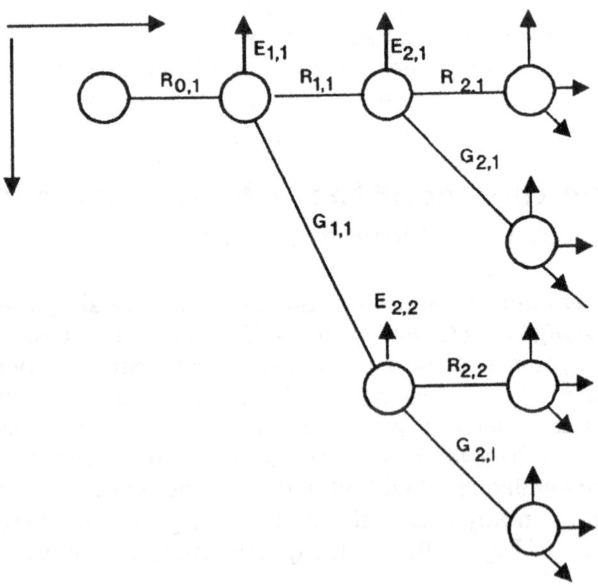

Fig. 6.11. Network showing faults extracted and faults generated (from [BELA76])

captures the essentials of the error propagation model. The variable i is the version number (release number) and is a measure of the age of a system. The variable j is used to introduce a tree structure that represents the various possible fault propagation paths of the system.

The model represents an increasingly large and complex network of fault trajectories. For each node in the network where new errors are generated, there is an addition of a new fault class. It is easy to see that the number of fault classes doubles for each new version, if new errors are generated for all fault classes of the previous version ($i - 1$). That means that the number of fault classes of version i of a system is 2^{i-1}, provided that no repair can be made within a fault class without making at least one new error. This is the extreme case, and gives a maximum number of fault classes. The number of fault classes in a realistic situation would therefore most probably be less than 2^{i-1}.

Belady and Lehman propose that the number of fault classes $C(i) < 2^{i-1}$ may be used as a measure of a system's complexity. Errors that are generated during fault repair are thus seen as the most important reason for structural degeneration of software systems.

In order to get a better understanding of the model, we shall look at the impact on C(i) resulting from different assumptions about the relationships between E, R, and G.

1) Assume that, for each node, E = G, that is, as many faults are extracted as are generated. Under this assumption the system appears to be in a steady state. The faults are "chased" around in the system, but the number of faults is the same, no matter how many are caught and removed. Even so, the number of fault classes $C(i) \sim 2^{i-1}$ would be expected to increase with each new version of the system.

2) If we further assume that, for each node, *a fraction* p *of the faults are extracted, and* G *is very small*, then the fault decay follows a geometric distribution giving the remaining number of residual faults of version i to be $R(i) = R(0)\,(1 - p)^i$ whereas altogether $R(0)(1 - (1 - p)^i)$ faults are extracted.

   If $G = 0$, the system approaches an error-free state asymptotically, and the number of fault classes C(i) remains constant.

   However, if $G > 0$, no matter how small it is, C(i) will increase, even if the increase would be considerably less than the most extreme situation of the doubling of C(i) for each new version.

6.3.4 The Impact of Resource Allocation on Structural Degeneration

Structural degeneration is deemed to be important because it is assumed that if there is a complicated interaction among the components of a system, then this will lead to complicated error propagation. Furthermore, if there is complicated error propagation, then error repair will lead to a gradually increasing complexity of the software structure. An increase in structural complexity is in itself regarded as a system fault, which needs repair. Periodical restructurings are therefore needed in a software system that is undergoing change. Resources have to be made available both for preventing structural degeneration from taking place too fast, and to rejuvenate the aging system through restructuring operations.

Based on the previous analysis, it seems reasonable to assume that the increase of $C(i)$ is less than a doubling for every new version. We assume that $C(i) \sim 2^{G(i)}$, where $G(i) < 2^{i-1}$ is a monotonically increasing function, such that $G(i+1) > G(i)$. A consequence of this assumption is increased structural deterioration. The project organization will have to prevent excessive increase in system complexity. This requires that the system documentation is of high enough quality, that the project group members possess the right knowledge about the system, and that the project organization functions effectively. This desirable state of affairs may effectively reduce the effect of growing complexity, and can be represented symbolically as

$$C(i) \sim 2^{G(i) - DAL(i)}$$

where DAL means "Documentation, Accessibility, and Learning".

According to this very qualitative model, complexity increases in a software system may be counteracted by investing in the increased competence of the project development group, and in improving the quality of the documentation of the system. The critical question for a development project's management to answer is, how much of the total resources shall be used for such "unproductive" endeavors.

Let us assume that there is an available budget B which bounds the total activity of the project development group. Let us further assume that the budget B can be divided into the three parts

P – for planned fault extractions,
A – for administration, learning, documentation, etc., and
C – for complexity control, including unplanned "fire-fighting",

such that $B = P + A + C$.

We see that A is associated with DAL(i), while P and C are associated with G(i) in the previously introduced complexity model. Neglect of A-type activities will therefore result in increasing the complexity of the system. More resources have to be made available to keep an increasingly complex system under control. That is, the C-part of the budget has to increase at the expense of the A-part and the P-part. Note, however, that the C-part of the budget cannot be measured explicitly and apart from the P-budget. Whenever the C-part increases because of increased system complexity, this is seen as a necessary increase in the P-budget in order to repair the same number of faults as before. Each fault repair becomes more costly because of complexity "ripple".

A-activities and P-activities can be planned for ahead of time. C-activities are not plannable in the same way. They just have to be done in order to keep the system operational. They are of a "fire-fighting" nature. Increasing complexity of a software system can only be indirectly observed. Unit costs for the repairing of faults are proportional to (P+C), and will thus be seen to be on the increase. System "crashes" will be experienced more frequently, so that emergency actions will have to be undertaken more often than before. Increased complexity is experienced in that less useful work seems to come out of the P-resources as the system comes of age.

We shall build a simple model in order to highlight some of the consequences of making the wrong allocation of project resources. We assume that $a(t)$, $p(t)$ and $c(t)$ are the levels of efforts that are associated with A-type, P-type and C-type work at a time t after the evolution process started. Let us further assume that for every level of complexity in a software system there is an "optimal" allocation of resources between A-activities and P-activities

$$a_{opt} = kp$$

so that the system's complexity is kept at a constant level, and neither increases nor decreases.

Assume further that there is a "planning factor" m, which reflects the real allocation of resources

$$a = mkp$$

such that A-type work is less than optimal when $m < 1$, and optimal for $m = 1$. If management is "starving" the A-activities with budgets that are too low, then the outcome is an unavoidable structural "aging" of the system. The economic consequences for this structural deterioration have to be absorbed later, in the form of an increase in maintenance costs. On the other hand, if the A-activities are given more money than the "optimal" amount, that is $m > 1$, then the complexity of the system will be made to decrease, and the unit cost for fault repair will therefore also decrease.

Let us now assume that the change (per time unit) of the complexity work, that is, in the C-part of the budget, is proportional to how far away the A-budget is from the optimal value associated with that system complexity. For the time t we get the differential equation

$$\frac{dc(t)}{dt} = l(a_{opt} - a)$$

where l is a proportionality constant. Substituting $a_{opt} = kp$ and $a = mkp$ gives the equation

$$\frac{dc(t)}{dt} = l(1 - m)kp(t)$$

But $B = a + p + c = (mk + 1)p(t) + c(t)$, and therefore $p(t) = (B - c(t))/(1 + mk)$. When substituting for $p(t)$ we get the differential equation

$$\frac{dc(t)}{dt} = \frac{lk(1 - m)}{1 - mk}(B - c(t))$$

which has the solution

$c(t) = B - (B - c(0))e^{-qt}$ where $q = lk(1 - m)/(1 + mk)$

We see that the complexity increases exponentially the further we get from the optimality point, and the situation becomes gradually worse (or better) as time evolves (depending on the value of m). For $m < 1$ the evolution pattern is sketched in Fig. 6.12.

A simple model like this can only give a limited, qualitative picture of the relationship between resource allocation and the evolution of system complexity. Management has to decide when it does not pay off to improve the quality of a system further. Management also has to decide when more resources have to be used for preventive activities in order to avoid further structural deterioration of a system.

If the real level of A-activities is less than optimal, the following effects can be expected in the course of time, during fault repair:

- more new errors are introduced during fault repair, which leads to
- increasing system complexity and increasing $c(t)$, which leads to
- decreased resource allocations to A- and P-activities because "fire-fighting" will always be given priority, which leads to
- signs of stress and chaos in the project organization, which leads to
- externally required fault-repair rate cannot be met.

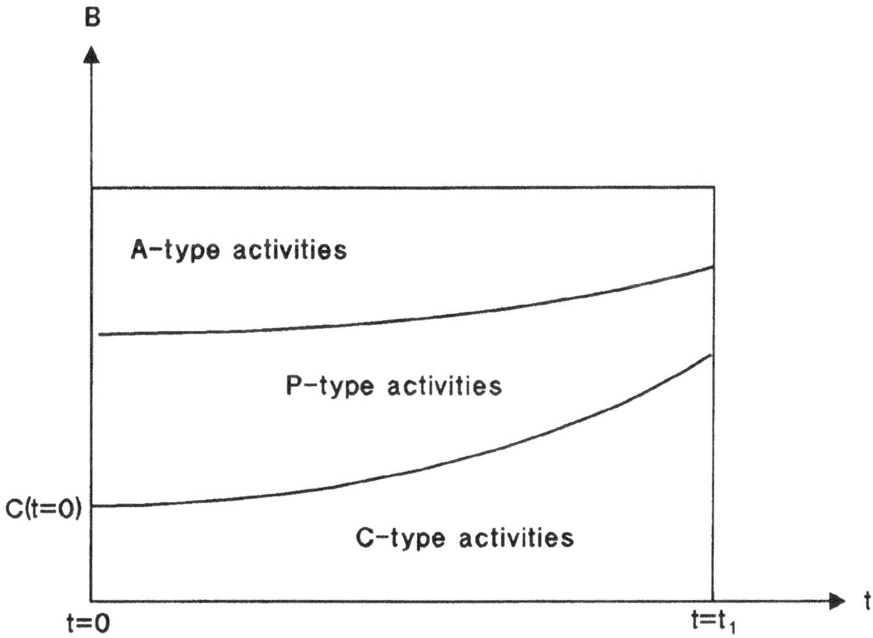

Fig. 6.12. Evolution in resource allocation when systems complexity increases

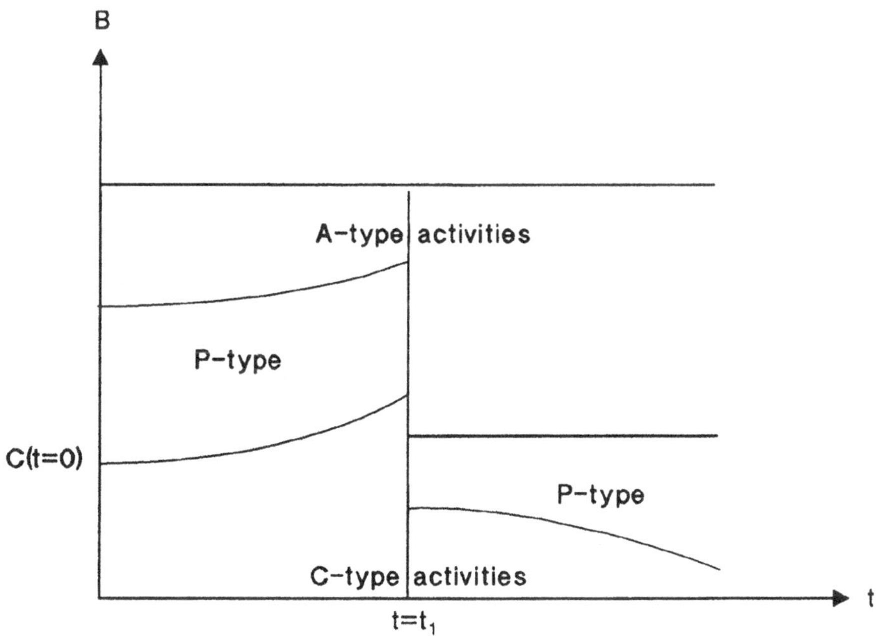

Fig. 6.13. Management decreases complexity and increases A-type work at $t=t_1$

A possible course of action is depicted in Fig. 6.13. After an initial increase of system complexity is experienced, management decides to take action. Two decisions are made. First, the system complexity is decreased in a one-shot operation, e.g., a restructuring of the system. The high system complexity is treated as a fault, it is repaired, and brought down to a more acceptable level. The second decision is to increase the relative portion of the A-budget, in order to avoid future degradations of the system due to complexity increases. In the figure it is indicated that management has given so many resources to A-type activities that complexity is steadily decreasing.

A simple model like this has, of course, very serious weaknesses. We are not able, for example, to estimate the optimality factor k. We also cannot estimate the initial complexity value $c(0)$. We are not able to distinguish P-type work from C-type work. Another example of model weakness is that we have not taken into consideration that it takes more and more resources to search for and repair the "last" errors than it takes to repair an error when errors are in abundance. That is, we have not taken into consideration that the more "perfect" a system becomes, the harder it becomes to improve it further. To represent this effect would increase the mathematical complexity of the model beyond a reasonable level, for the purposes of this modelling exercise. Alas, there is no way to give explicit guidance in these matters unless one can measure complexity directly. And that is far beyond the current state of the art.

Exercises

1. State the differences between an information system and a software system.
2. What are the advantages and disadvantages of standard software systems?
3. Find some examples of common systems in your environment. Discuss what features they have that make them common.
4. Discuss the advantages and disadvantages of software installation approaches.
5. What kind of installation strategy should be followed for custom tailored systems?
6. Explain the relationships between "system aging" and the system complexity.

Chapter 7

Managing Information Systems Development Projects

Information systems have to be treated like other organizational resources. The systems have to be developed in an orderly manner, and they have to be maintained and kept operational just like any other company resource. Over the years the information systems resources of organizations have increased in economic importance. To use 3–6 % of the turnover for information processing is quite normal in modern companies. This means that development, maintenance and operation of the information systems must be properly planned and managed.

Because of the complexity of computerized information systems, project failures are quite often reported. Metzger [METZ73] gives a list of the most frequent reasons for project failures:

Poor planning, Ill-defined contract, Unstable problem definition, Poor planning, Inexperienced project management, Political pressures, Poor planning, Ineffective change control, Unrealistic deadlines, ...

The list could be several pages long. The *poor planning* item would always be present as the most important reason for project failures. Metzger claims that the term *technical difficulty* would never appear on the list. That is not because projects are never technically tough. It is rather because if somebody agrees to do a job that is technically beyond the state of the art, that is indeed poor planning (research projects are obvious exceptions).

Poor planning sometimes means failure to consider a job from all points of view, sometimes there is no plan at all, and sometimes it means that the plan is a fairy tale. In order to ensure that plans are developed and that deviations from the plans are discovered and reported, most organizations rely on standard, mandatory planning procedures and planning formats. The planning standards are usually found in project development handbooks in the companies. They imply that there are standard systems development methods agreed upon in the organization.

The objective of a project development standard is to provide the procedures to be followed from initiation to completion of a project. The form of a development standard is usually a checklist which covers *what* should be done at each point in the life cycle. The question of *how* each task should be performed is not addressed. Checklist methods can therefore be said to be "result-oriented". They may be quite detailed, and prescribe content and form of the various project documents. Checklist methods are frequently offered as commercial products.

The details of project planning may differ from standard to standard, but essentially they are the same. Information-system projects may be viewed as proceeding through the stages of definition, design, programming, implementation and operation. The work in the various project stages must be managed and controlled. To facilitate easier management and control, each stage may be further divided into one or more phases, starting with a project proposal, and ending with a post-implementation evaluation.

Such a series of project phases is called a "system life cycle". Each phase is supposed to produce one or more "milestone" documents, which contain the results that have been produced in that phase. These reports are the basis for management's control of the information-system project. They are usually subjected to formal review. They are the basis for assessing a system's quality.

Project control is usually implemented through formal review meetings. There are technical reviews, budget reviews, and reviews of the overall quality of a project. The actual division of a project in phases is dependent on the management and control needs of the project development organization. An organization which mostly develops very large systems will have different needs from one which only develops small systems.

Organizations that develop their systems as cooperative efforts between their user-departments and their data-processing department have specific management and control needs in order to take care of the inter-departmental interactions. These may be quite different from the management and control needs of software-centered organizations which construct systems to be delivered to customers according to pre-defined specifications, but where the customer is supposed not to play any active part in the project after the initial specification has been agreed upon.

A typical "life cycle" for a software-centered organization may consist of phases for definition, design, programming, system test, acceptance, installation and operation [METZ73]. The world is seen from the software developer's point of view. The user is regarded as "the customer". He (or she) is supposed to participate in the initial definition of the information system, and is given a role in the acceptance phase, when the customer may reject the software if it falls short of expectations.

Information system engineering is concerned with the quality of the whole information system, not only the software system. The management and control structure of information-system development organizations reflects this concern by putting more emphasis on the first stages of the development projects. The life cycle of individual projects is usually seen as comprising the following phases:

Phase 1 :	Pre-project study	(Feasibility study)
Phase 2 :	Requirement specification	(Systems analysis)
Phase 3 :	System modeling and evaluation	(Systems analysis)
Phase 4 :	Functional specification	(Logical design)
Phase 5 :	Data processing system architecture	(Physical design)
Phase 6 :	Programming	
Phase 7 :	System installation	
Phase 8 :	Project evaluation	

The terms *systems analysis*, *logical design* and *physical design* appear frequently in the literature. We do not want to use those terms here, because they may bring with them associations which do not reflect fairly upon the content of the various life cycle phases. However, we have indicated their approximate meaning by enclosing them in parentheses and positioning them by the appropriate phases.

The eight phases of the individual projects' life cycles are usually seen to be preceded by a phase of project selection and masterplan formulation. This preceding phase is sometimes called the phase of "strategic planning". The candidate systems are identified and analyzed, priorities among projects are decided, and a total plan for the organization's information system development is formulated.

Following the eight development phases is a phase of operation and modification. This is the maintenance phase, which may also be called the phase of information system resource management. The database resources and the software resources of the organization have to be properly managed, so that they do not fall prey to frailty because of aging and structural degradation.

7.1 Project Selection: The Master Plan

The objective of this "strategic" planning phase is to select the "right" information systems to be implemented, and to develop a master plan for the organization's systems development and systems management efforts during the next planning period. The "right" information systems are those that are most important for the organization in achieving its objectives. The formulation of a master plan is therefore one of the most important decisions to be taken by an organization with respect to its data processing policy.

In order to establish a master plan for an organization's information systems development, it is necessary to first:

- Identify candidate systems, and then
- Provide justification for each candidate project, before
- Establish a priority list, and finally
- Form the development projects of the master plan.

This should also be followed by a planning step to

- Determine the corporate data base structure

in order to provide for future systems integration.

Identification of Candidate Systems

The objective is to identify those organization areas that are of key importance, and which can be improved by better information systems support. The activities are [BROO82]:

(i) Analyze the key organizational areas with respect to those aspects which have the greatest effect on the organization's performance (e.g., decision support improvements),

(ii) Determine those constraints that are most important in limiting the performance in the key organizational areas (e.g., poor customer service, outdated information, long response times),

(iii) Determine how improved information processing can remove the constraints,

(iv) Make a tentative cost/benefit analysis,

(v) Decide if a possible project will be consistent with the organizations general policy guidelines for the development of information processing systems.

Project Justification

The objective is to provide economic justification for giving priorities to the candidate projects. The steps are:

(i) Determine cost items,
e.g., analysis, design, implementation, testing, training, new hardware/software, operations costs, cut-over costs, etc.,

(ii) Determine benefits,
e.g., reductions in staff, working capital, operational errors, improved productivity, service, working conditions, and improved management decision quality,

(iii) Make financial evaluation to find when benefits outweigh costs: Standard economic evaluation techniques may be used, e.g., payback analysis, break-even analysis, but care must be taken to include the intangibles in the analysis.

Establishing the Master Plan

The objective is to formulate the master plan for the organization's information-system development. This requires the matching of

- the priorities of the user departments, who are going to pay for the new developments, and
- the priorities of the information system department, who has an organization-wide responsibility for information processing resources and for information integration, and
- the available budget limits and the general company policy.

The plan formulation includes the following steps [BROO82]:

(i) Identify systems that should be developed together, because they share common subsystems, databases, etc.

(ii) Determine the development sequence,
e.g., some systems will provide data that are used by other systems. Form tentative groups of such "strings" of systems that might be candidates for being developed together. Form a tentative development plan from these groups.

(iii) Evaluate benefits that cannot be measured (intangibles) for the various candidate systems groups.

(iv) Determine the requirements that the tentative plan put on computer workload, need for new software, staff competence, staff workload.

(v) Rework the cost/benefit evaluations for each system group.

(vi) Formulate a long-term development plan (e.g., 2–4 years) that maximizes the expected returns. Both tangibles and intangibles must be recognized.

Determining the Corporate Database

The objective is to establish a logical data model for the whole organization in order to ensure that the different subsystems can be properly integrated and interfaced. The steps are:

(i) Determine the data content of the most important data stores,

(ii) Perform a tentative database design.

Specification of the corporate database is a vital part of the project planning process. This is because of the quality of the specification affects the development cost and the future operational costs of the organization's information systems in the most profound way. Therefore, the specification of the logical data model for the whole organization should be reviewed at the same time as the master plan is considered.

7.2 The Project Life Cycle

A standard approach to information-system design is usually adopted by an organization to facilitate common ways of handling project management. Standardization also assists in the training, interteam communication, staff interchange and documentation aspects of systems development work. Although the need for standards differ with the size of the information systems undertakings, some standardization of reporting procedures and forms is necessary for all but the simplest tasks.

Even if there are standard "checklist methods" available in the market place, many organizations have produced their own standards. The following overview of the information systems development life-cycle is based on the premise that there are project control reviews at the end of each project phase. The description is therefore oriented more towards describing the content of the phase-reports than in proposing how the work should actually be carried out. The description is based on life-cycle descriptions in [BROO82], [METZ74], and [MIDA88]. We have used the phase arrangement of [MIDA88], which is an internal company standard.

Those readers who are already familiar with other life-cycle phase arrangements will find that there are more distinct phases in the first stages of the project in our description than what is usually found in the literature. A trend to this effect has become quite usual in practice over the last decade. It reflects an increasing tendency to put more work into problem definition and planning, instead of getting involved with software design and programming as soon as possible, which was a much more common work style in the early years of the computer.

7.2.1 Phase 1: Pre-project Study

The objective of the pre-project study is to develop a sufficient basis for deciding if a project shall be carried out or not. This phase is called the pre-project phase, because the work in the phase is done prior to the start of the project.

An effective approach is to let user representatives and information-system engineers work together in order to:

- describe problems and goals in the intended project area,
- delimit the project's level of ambition,
- perform a preliminary cost/benefit analysis,
- submit a preliminary project plan.

The project may already appear in the company's master plan, if such a plan has been developed. If there is a master plan, then some of the work of the pre-project study may already have been done.

The phase report should contain:

(i) The scope of the project, the problems to be solved and the goals to be met. Present systems deficiencies should be described, and new possibilities should be explored.

(ii) The size of the project, in terms of functions and data, people affected, other systems affected, organizational effects.

(iii) Time and cost estimates, for development, operation and maintenance.

(iv) Cost/benefit analysis.

(v) Analysis of possible constraints, e.g., the future operational environment of the system to be developed.

The first edition of a Project Manual will be developed in the pre-project study. The Project Manual will follow the project (if it is decided to launch it) for the project's life-time. The Project Manual contains descriptions, plans, and standards that are associated with the project.

The pre-project study report will have to be reviewed by the user organization, that is, by the "customer" who is going to pay for the project. The future system users must evaluate the report in order to assure both themselves and the other involved parties that they agree to the problem description as well as to its proposed solution.

7.2.2 Phase 2: Requirement Specification

The objective of the requirement specification is to define the users' requirements for a system that can solve the problems and satisfy the goals that were stated in the pre-project report, and accepted by the appropriate review committee.

The most effective approach is also in this phase to form groups of information-system engineers and user representatives, who are charged with collaborating in developing the "right" systems requirements. It is effective to let the users' representatives be responsible for the content, that is, for the completeness and the correctness of the requirements, and let the information-system engineers be responsible for the report writing and its presentation.

The phase report should contain:

(i) A complete description of problems and goals;

(ii) Analysis of the current system functional processing, and an evaluation of their strong and weak properties;

(iii) Requirements for a new system, including functional and data requirements, and operational properties;

(iv) Constraints for the further development imposed by policy and standards, e.g., mandatory use of particular hardware and software;

(v) Revised cost/benefit analysis;

(vi) Allocation of responsibilities for the system's development, installation, operation and maintenance.

The functional requirements may be expressed by high-level data flow diagrams, in order to show which of the business functions that are involved in the project. An Entity-Relationship type data model may be appropriate to give a high-level view of the data that are related to the problem area.

The requirement specifications must be read and understood in detail by everybody involved, both on the user side and among the information-system engineers. This document is the basis for the work in the further phases, and will serve as the baseline for measuring later project results.

7.2.3 Phase 3: System Modeling and Evaluation

The objective of the work in the system modeling and evaluation phase is to develop a logical model of functions and data that is sufficiently detailed to give users, management and the project group a realistic understanding of the properties of the system and of its implications. Several alternatives should be worked out to a level of detail sufficient for them to be evaluated.

A recommended approach is to develop several alternative proposals which would satisfy the requirements of Phase 2 and to analyze and evaluate them in order to develop further details of the system's requirements specification. The chosen alternative should be worked out in some more detail.

The phase report should contain:

(i) The criteria that have been used in the evaluation of the various systems alternatives (technical, functional and economic criteria),

(ii) The evaluation of the various alternatives,

(iii) Requirement specification, final version,

(iv) The model of the preferred alternative (functional model, data model, subsystem hierarchy, e.g., in the form of data flow diagrams or entity relationship diagrams),

(v) Implementation evaluation (technical alternatives, organizational consequences),

(vi) Plan for next phase, revised cost/benefit analysis.

7.2.4 Phase 4: Functional Specification

The objective of the work in the functional specification phase is to decide in detail *what* the information system is going to do.

A recommended approach is to specify in detail all of the automated and manual functions in the information system, and the associated data. The detailed specifications should be read and understood by the user representatives in order to enhance their understanding of the system, so that they may require changes while the system's properties still can be changed at a reasonably small cost.

The phase report should contain:

(i) A function model, e.g., detailed data flow diagrams;

(ii) A detailed functional specification, that is, input/output specifications (content, layout), detailed process logic, validation rules, error handling;

(iii) Man/machine dialogues, that is, how the various functions are related through interactions between the software system and its environment, content and layout of dialogue interfaces;

(iv) Data archive descriptions, that is, formal definitions of all of the data elements in the data base, and a logical data model, e.g., in the 3NF relational data model;

(v) Manual routines;

(vi) Software interfaces to other systems;

(vii) Verification that the requirements specification is satisfied by the planned systems properties;

(viii) Preliminary acceptance test plan (procedure, responsibilities);

(ix) Plan for next phase, revised cost/benefit analysis.

7.2.5 Phase 5: Data Processing System Architecture

The objective of the work in the data processing system architecture phase is to design the automated part of the information system. In the previous phases the concentration has been on describing *what* the future system shall do. In the DP system architectural phase the concentration is on *how* to build a system that will satisfy the stated requirements.

A recommended approach is to form a design group of the best software engineers that are available. Some of them should preferably have been involved in the previous project phases, in key positions, and some of them should continue as leaders of the next phases.

The quality of the systems architecture is critical for the final success of the project. Design errors that are made during the architectural phase are usually very expensive to correct, sometimes impossible.

The phase report should contain:

(i) The overall design concept,

(ii) Standards and conventions, for names of programs and data, interface formats, message formats, error handling procedures, coding standards, version control, etc.,

(iii) Physical data base design,

(iv) Software structure, interconnections and interfaces, definition of basic software modules, definition of major programs modules,

(v) Final input/output specification,

(vi) Description of the operational environment, including analysis of operational costs,

(vii) Evaluation of how well security and auditing requirements are satisfied,

(viii) Test plans,

(ix) The implementation plan,

(x) Plan for next phase, revised cost/benefit analysis.

The architecture of the data processing system is, of course, of primary importance for those who are going to do the further development of the system, the programming, the testing, and so on. Most of the work in this phase has to do with the internals of the software system, and is therefore of a very technical nature. The users are only involved in parts of the work, most notably in the work that has to do with input/output specification, security requirements, installation plans and operational costs.

7.2.6 Phase 6: Programming

The objective of the work in the programming phase is to develop the software according to specifications. The approach to this phase is far more straightforward than for the previous phases. The design specifications are usually fairly detailed, if the previous phases have been performed properly. The activities of the programming phase may, nevertheless, be both long-lived and complex. The various software components must be coded, bought in the market place or acquired in other ways, and then properly tested, documented and integrated into a reliable, complete system.

The most important phase results are:

(i) The software and its documentation, source code, cross-references, test documentation;

(ii) Test results (functional tests, performance tests);

(iii) The system documentation, including operations procedures, maintenance documentation, and user manuals;

(iv) The system installation plan.

7.2.7 Phase 7: System Installation

The objective of the work in the system installation phase is obviously to install the system in its operational environment. A recommended approach is to invest plenty of resources in user training. Even the world's best system may come to nothing if nobody is able to use it properly. It is also recommended that one delay the installation if the results are not as planned. The effects of decisions made in earlier phases are felt during this phase, and "quick fixes" are usually not possible.

The results of the installation phase are of another character than the results of the previous phases. The installation phase result has more the character of a confirmation of the soundness of the previous work, at least, if everything goes well.

The major installation phase result is the system's operation, hopefully without any serious problems. The documentation of the result of the installation phase includes:

(i) Reports of the installation, and of the conversion from the old system to the new system when appropriate,

(ii) Reports of the user training program,

(iii) Acceptance test results,

(iv) Formal acceptance documents.

7.2.8 Phase 8: Project Evaluation

The objective of the project evaluation phase is partly to close the cycle back to the user, that is, to review the actual results of all the work, relative to the user's expectations. But the work is also aimed at a systematic collection of project experiences in order to help future project participants to avoid pitfalls that are obvious only to those who have experienced them.

The project evaluation report should contain:

(i) A review of technical and cost aspects, e.g., timeliness of system output, system controls, accuracy of data, response times, development time and cost, operational costs;

(ii) A review of benefit claims that were used to justify the project, e.g., staff reductions, improved decision quality;

(iii) A review of user satisfaction with the system, e.g., ease of use, quality of manuals and training, impact on job satisfaction;

(iv) Recommendations for changes to the system, and/or to project development methods and procedures.

7.3 Project Evaluation and Control

The objective of project control methods are to ensure that the project produces the expected results within time and cost budgets, while at the same time the organization's guidelines and standards are conformed with. The critical aspects of project control are therefore:

- a well defined set of reporting points,
- a project plan which defines anticipated rates of progress and resource utilization,
- reports from the project at the appropriate reporting points,
- reviews of the performance of the project team against the project plan.

Probably the most important technique available for achieving overall effectiveness in project development is the checkpoint review. Major quality reviews are conducted at the end of each project phase, and are based on the written report for that phase. Additional, less extensive technical reviews may be conducted within each phase according to a pre-planned schedule (Fig. 7.1).

The Quality Review meeting is the forum for formal acceptance of the results of a project phase. The quality review is done by the project's Quality Control Committee. The recommendations of the Quality Control Committee are the basis for deciding whether the project should proceed or be stopped, how much money should be allocated for the next project phase, and so on. Quality Review meetings may be arranged after every phase, or only for a few of the major phases, depending on the size and economic importance of the project. The Quality Review meetings are formal meetings, with attendance from all of the interested parties, that is, from the user departments, the information systems department, internal auditing, and so on.

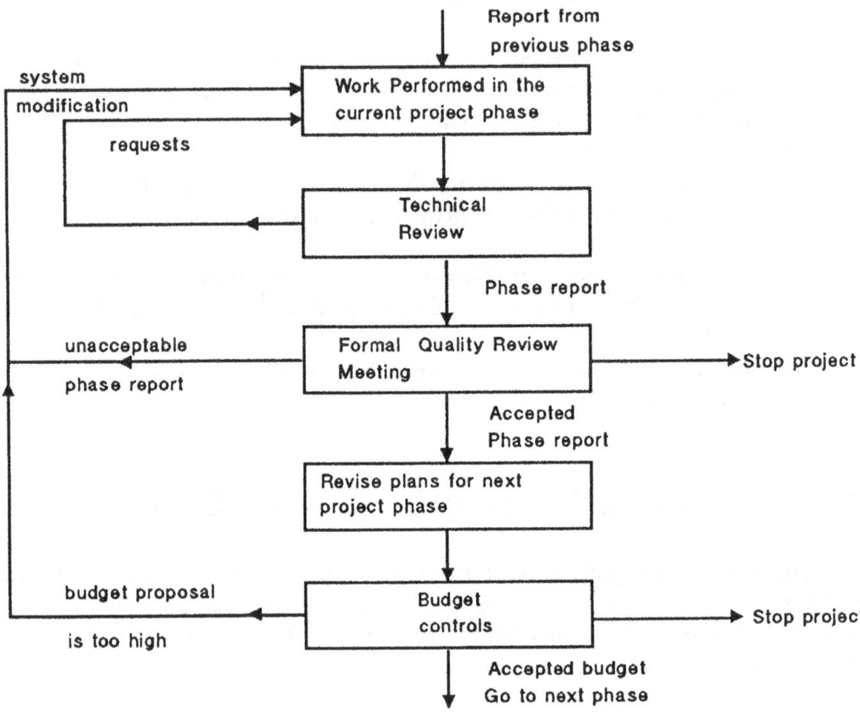

Fig. 7.1. Project control procedure for a phase (adapted from [MIDA88])

Technical reviews are needed in all project phases. The objective of the technical review is to have somebody take as much time as necessary to form an opinion on the work that has been done so far in the project. Experience shows that it is very difficult to find people who are willing to do this, unless it is defined to be part of the job. It is hard work evaluating other people's work. But it is very important to have independent expert opinions as the project work proceeds. It is better to locate errors prior to the Quality Review meeting than to have the phase report rejected!

Technical reviews are more informal than the Quality Review meetings. A most important issue is to choose the right types of expert to participate in the technical review. Some persons are very good and constructive in their critique, some are only destructive. The latter type should be avoided like the plague.

7.4 The Information System Development Organization

There are various ways of organizing information system development projects. The organizational structure of the information system department of a company has a significant influence on the management style applied to development projects.

7.4.1 The Information System Department

The information system department of an organization has the overall responsibility for information system operation and integration. It is also given the responsibility for ensuring that needed competence resources can be found when needed. Information system departments can vary in size from the very small operations group, installing and maintaining commercially acquired software systems, to the very large operations and development organization, with hundreds of employees.

Information system departments are usually organized in groups with different technical skills (Fig. 7.2). There are groups for data communication, database management systems, operating systems, data entry equipment, and so on. In small organizations the skill-groups may be so small that they are one-person groups, or sometimes one person will be the local "technical oracle" for the whole company. Information system departments may also be organized according to the different market areas (usually called the "application areas") of the information system department. These usually coincide with the business areas of the company. The various business area groups may share common technical skill-groups, or they may be self-supported with, e.g., operating system skills, if they are large enough to carry the associated cost.

The "internal" technical projects which are associated with technical infrastructure, for example data communication and computer operation, are usually initiated by the information system department, and performed within its technical groups. The "external" information-system development projects may be initiated from the user divisions, or from top management. They will be performed by the appropriate application area group in the information system department, in cooperation with personnel from the user divisions, and supported by personnel from the technical groups of the information system department.

When development groups are needed for projects that do not conform to the organizational structure of the information system department, they will usually be formed on an ad hoc basis, and populated by personnel from

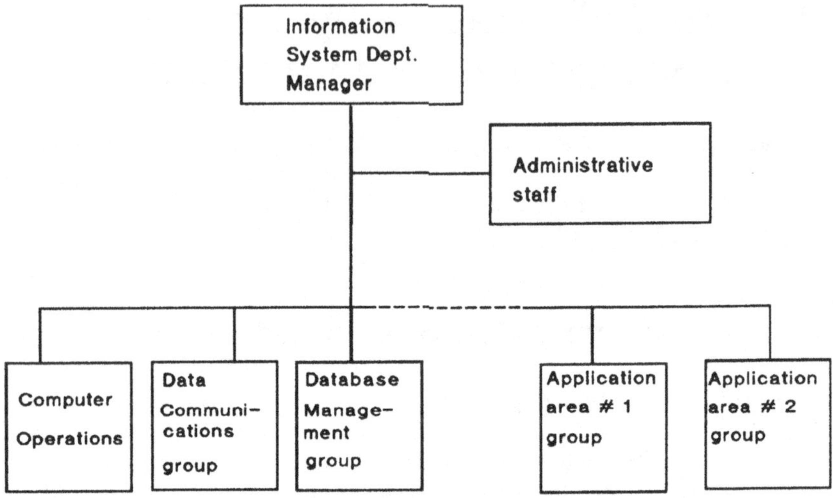

Fig. 7.2. Information systems department organizational structure

the permanent groups of the information system department. The project personnel will return to their permanent groups when the development project is dissolved. Only for very large projects that last for extended time periods will people be directly hired. This may sometimes give problems with the loyalty of the project's staff, if there are conflicts between the needs of the project and the needs of their permanent organizational group.

The two basic extremes in organizing development projects are

- The functional structure, and
- The project team structure.

The functional structure was originally developed for use in large information system development projects, that is, projects with more than 20–30 people working at a project at the same time. The project team structure is intended for organizations with several smaller projects.

7.4.2 The Functional "Large-Project" Structure

A "large-project" structure is usually organized according to the life-cycle phases of the projects. There is a systems analysis and design group, there is one (or more) programming group(s), there is a test group, and there may be a documentation and training group to take special care of user-liaisons. Unless the project is very large, it will be based on the shared technical resources of the information system department. For very large projects it may sometimes be effective to organize local technical skills groups for the sole purpose of supporting the project (Fig. 7.3).

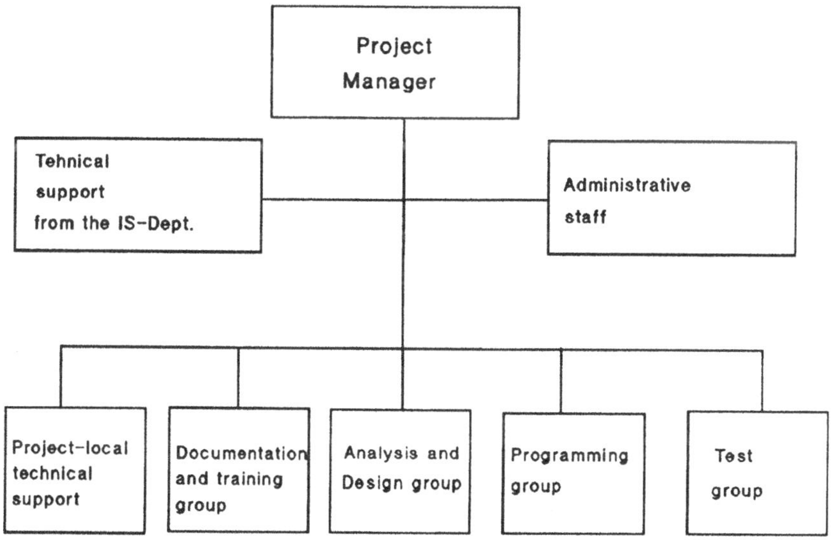

Fig. 7.3. Functional "large-project" organizational structure

The functional organization is very effective for establishing large, short-lived projects, where people can be "laid off" and sent back to their permanent groups in the IS department as soon as they have finished their job in the project. However, it is not a good structure for a permanent organization.

In the early days of computing many information systems departments (computer centers, data processing departments) were set up with a functional structure, when the scope of their application area was narrow. It made sense to specialize in skill groups of systems analysts, designers and programmers for data processing departments that had only one dominant application area, such as banking systems.

When additional application areas matured, they brought with them new requirements for technical knowledge. The analysts and designers of the functionally organized information system departments did not, in many cases, possess the required competence. The information system departments were, nevertheless, expected to play the lead role in developing the new, unfamiliar kinds of systems. The new user groups were also unfamiliar with the high complexity of large software systems, and the associated high development costs.

Unexpected high costs and long delays for developing new information systems proved fatal to many information system departments who had chosen this "large-project" organization structure for their overall internal organization. It was quite common to experience strong pressure from user departments to break the "tyranny" of the central data processing departments of organizations. In some cases this resulted in the establishing of divisional, local data processing departments which specialized in problems of that particular business area. Many central information processing departments became severely weakened by these struggles. On the other hand, this development also led to unnecessary duplication of technical skills, that could easily have been shared in a more integrated organizational structure.

There is an underlying assumption in the functional organizational structure, when it is used for the internal organization of an information system department, that a steady stream of jobs will arrive in the department, to be analyzed by a permanent group of systems analysts, who send their analysis to the systems designers, who develop designs which they give to the programmers, who finally turn them into an operational hardware/software system.

The functional project structure may be used for the internal organization of the information system department only if it can be assumed that the analysis/design group has sufficient expertise on the whole spectrum of jobs that arrive in the information system department. It is also a good structure if one can assume that the analysis and design group will always possess high competence on programming techniques and software products over the whole spectrum of jobs, so that they can be expected to develop design specifications of high enough standard for the (expert) programmers.

7.4.3 The Project Team Structure

All experience shows that the functional organization becomes too rigid when the variety of problems to be solved by the information system department increases. A project-team oriented organizational structure (Fig. 7.4) provides much greater flexibility than the functional organization.

Essential features of the project team structure are [BROO82]:

- Project managers are appointed for each project and are supposed to follow the project through until it is over.
- Team members are drawn from a pool of information systems engineers and programmers, and are supposed to stay on until the project is over.
- There is often little difference between the work of an analyst/designer and a programmer. Each person does the job at hand.
- Technical support is drawn from the technical groups of the information systems department.

The greatest difference between the project team structure and the functional structure is that all of the participants in a project team are supposed to follow the project through to the end, while in the functional organization each project participant is only supposed to do work within one or two project phases.

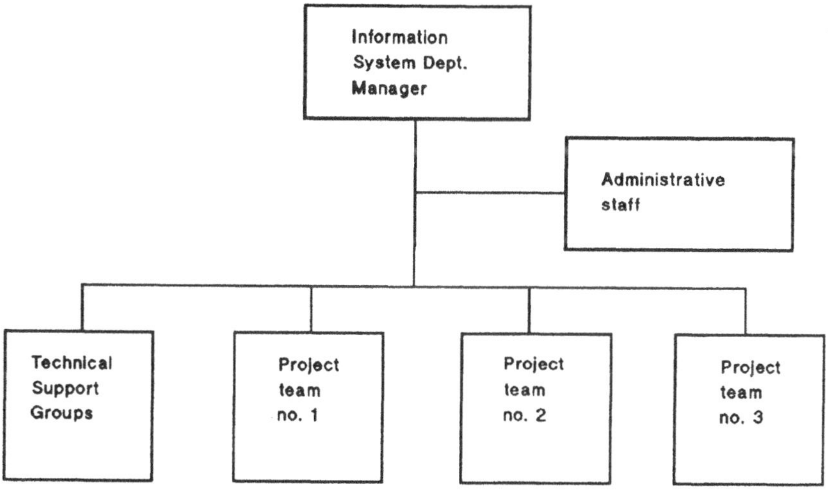

Fig. 7.4. The "project team" organizational structure

It is essential that each project team has the responsibility for a complete and independent project, so that the project team will have to proceed through all of the life cycle phases in order to finish their project. It is not recommended to organize project teams to work on subsystems of a larger system unless subgoals have been clearly defined for the subsystems that support the common goals for the system. One may easily create situations where nobody has a responsibility for the entire project, or those who feel the responsibility will lack the means and the authority to force the subprojects on the course toward common goals, instead of pursuing their local subgoals.

Exercises

1. Discuss the importance of project evaluation in the information systems life cycle.

2. Prepare relevant checklists to be used in each of the eight phases of information systems development described in this chapter.

3. Describe the technical review activities to be held during the life cycle. Identify technical review activity points in the life cycle.

4. Give some examples of tools/methods that can be used to describe the data processing system architecture.

5. Give some examples of automated project control/management tools.

Chapter 8

Information System Support for Information System Development

Information systems used to be built from scratch, because there were few, if any, components available to base the new systems upon. This is not the case any more. Information systems are increasingly built through the re-use, modification, integration and interfacing of commercially available software components. Programming, in the original meaning of writing commands for a computer to follow, is not as dominant an activity as it used to be only a few years ago.

A large part of the effort of a development project has been shifted from the actual writing of the software to finding out what software to write. This change has come about because too many projects failed, and is intended to secure that the software is useful for some worthwhile purpose when it is finally written.

Furthermore, because the information system development process is team oriented, rather than individual oriented, much effort is needed to ensure effective communication within teams and between teams. Much effort is also needed to ensure that team members know enough of what is going on in their projects, so that they do not make wrong design decisions out of ignorance.

Important trends in this change of the information system development process, from relying on programming skills alone to its emergence as an engineering design discipline, are:

- Increased need for practical ways of managing the many persons and tasks that make up the cooperative process of designing and building an information system.

- Increased need to move from ad hoc development strategies to a strategy based on engineering principles, e.g., the development of standards for programming and for specifications.

- Increased availability of a larger variety of computerized tools for the support of the various systems development tasks, in addition to the well-known tools for the support of the programming task, such as tools for diagramming, code generation, testing, verification, and so on.

The many-persons/many-tasks problems are enhanced when the information systems engineering department is given a geographically decentralized organizational structure. The tendency to decentralize operations has increased in most enterprises over the last couple of decades. The information system engineering departments are no exceptions to this trend. Decentralized system development environments must therefore be supported. Particular emphasis must be given to improved communication, in order to enhance the individual information system engineer's level of understanding and knowledge of the system to be built.

In spite of the recognition of the need for standardization, there is a plethora of available development methods, specification techniques, and programming languages. There is so far no indication that the industry, as a whole, will settle for a broad agreement on standards for information systems development in the foreseeable future. Therefore, support systems must be able to incorporate various techniques and tools as they become available and are offered in the marketplace.

During the 1980s a large number of so-called CASE tools were developed and marketed. CASE is an acronym for Computer Assisted Software Engineering. Most of the tools are based on system development techniques that were originally intended for manual use by humans, e.g., data flow diagramming. Few CASE tools have so far taken advantage of the potential of doing more sophisticated analysis of the specifications, provided that the specifications can be more formally expressed. Many of the CASE tools provide graphical support for drawing diagrams only.

Given these trends, we propose some requirements which a support system for information systems development should be expected to satisfy:

- The scope must be so wide as to support system development and maintenance through all phases of a system's life cycle. Various system development methods, project sizes, and system types should be supported equally well.

- The project-management aspects of system development must be supported in a flexible manner, so that the support system can be adapted to whatever management strategy a company may choose to implement.

- Relevant project information must be supported in such a way that it can be made available to all project participants, as needed. Project information is to be understood as software, systems documentation, company standards, company addresses, tool indices, and the like.

- Easy communication among project participants must be facilitated. This comprises end-users communicating with the information system engineers, designers communicating their perception of end-user requirements to the implementers, communication among designers and among implementers, and finally, the transfer of knowledge of the whole program system, on every abstraction level, to the staff responsible for the future maintenance of the information system.

- The re-usability of software components and the associated documentation must be supported. The development of modularised software systems must be supported and stimulated. This means in particular that solutions to the problems of interfacing software components must be provided for.

- Tools must be provided for effective software configuration management, and for version control of associated specification documents. This is of particular importance in a decentralized engineering environment, where the need to keep control of versions of design components, their interrelationships, and the status of each version is pressing.

Several of the preceding requirements go well beyond what can reasonably be expected in the current state of the art (as of 1992). A support system for information system design should therefore be designed in an open-ended manner, so that new techniques may be applied as they appear in the market place. Many of the relevant aspects have been discussed elsewhere in this text. In this chapter we shall take a closer look at the functional properties of such support systems, and at the content of the information-system engineering database, and we shall discuss the configuration management issue.

8.1 Contemporary Environments for Supporting System Development

Contemporary support environments consist of:

- tools for analysis, design, and code generation,
- specification database(s),
- specification database manager.

The various environments differ in how integrated and comprehensive they are with respect to the system development process. A comprehensive specification database may consist of four (overlapping) clusters of specifications (Fig. 8.1):

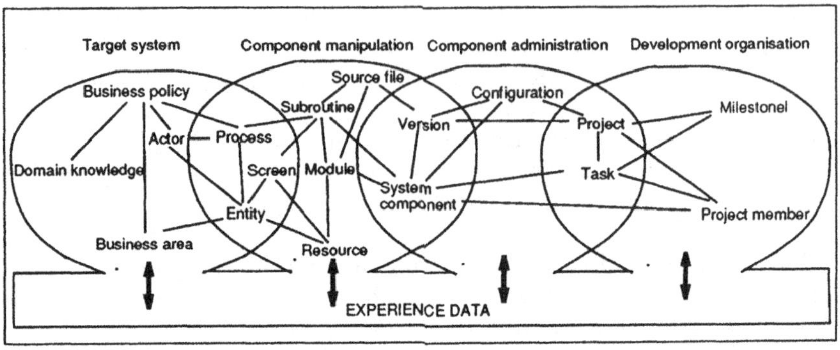

Fig. 8.1. Repository structure information systems development (from [ANDE90])

- A "world model" of the target system and its environment,
- A model of the software and data of the application system,
- A model of the various configurations of the application,
- A model of the development organization.

There are several ways of organizing the tools into tool systems. Individual tools for analysis, design, and code generation carry their own specification databases which are based on different views of information systems. These tools are mostly incompatible. The output of an analysis tool cannot usually be used as input for a design tool without being transformed and augmented by a human designer.

Analyst/designer workbenches (Fig. 8.2) organize tools for target domain modeling, requirements capture, software and database design within a common user interface. Most workbenches support several overlapping, incompatible tools. The specifications captured by the tools are stored in the workbench data dictionary.

An integrated CASE environment (ICASE) is a workbench consisting of compatible tools for the various life-cycle phases, so that the output from one tool can serve as the input to the next tool, from requirements capture to code generation. The granularity of the ICASE repositories is on the level of modeling constructs. This means that it is possible to provide version control on chunks of specifications which correspond to the modeling construct level.

An IPSE is an Integrated Project Support Environment. The term used in the U.S.A. is "software engineering environment". An IPSE provides tools for all phases of the life-cycle. The granularity of the design objects tend to be that of a document rather than a more fine-grained approach. The focus is on managing the systems development and the associated design products. First-generation IPSEs mostly have their project management data-

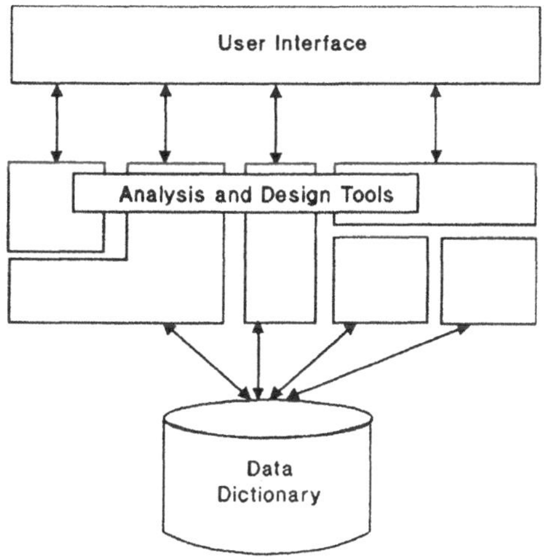

Fig. 8.2. Analyst/designer workbench

base separated from the design product database. Next-generation IPSEs are expected to integrate the two. The database manager provides version and configuration control facilities for the IPSE (Fig. 8.3).

The IPSE functions may be grouped into
- software engineering Kernel
- basic tool services
- software development and process management
- specific tools

The tools in the four groups are

1) SEE-Kernel
   - Process handling
   - I/O facilities
   - Database management
   - Distribution support
   - User interface support

2) Basic tool services
   - Version control
   - Configuration management
   - Automatic system building
   - Document preparation
   - Document management
   - Basic data structure manipulation
   - Security
   - Host-target communication

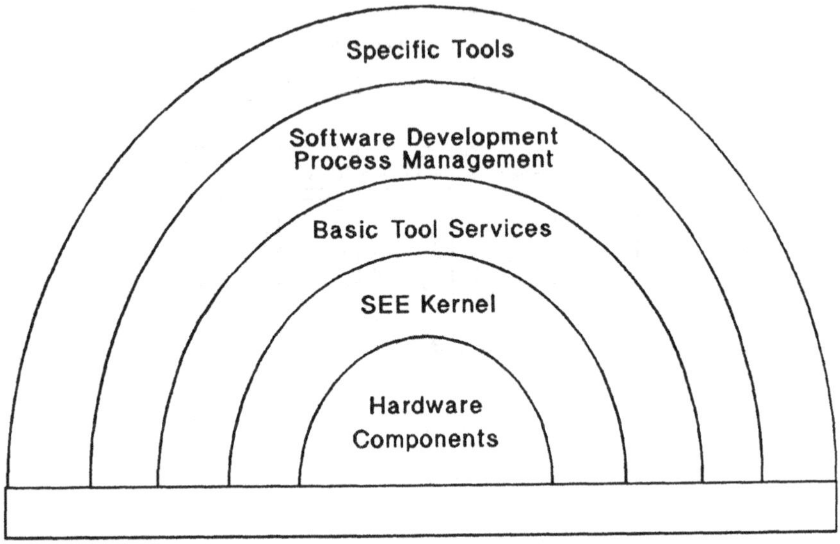

Fig. 8.3. IPSE architecture (from [SFIN87])

3) Software development and process management
   - Project management
   - Change request management

4) Specific tools
   - Requirements analysis
   - Architectural design
   - Detailed design
   - Coding and unit testing
   - System integration
   - Maintenance

IBM's AD/Cycle and Repository Manager (Fig. 8.4) is so far the most ambitious framework that has been proposed by a vendor. AD/Cycle has the structure of an application platform for application system development. The repository manager is at the heart of the system. A repository is a database of specifications. CASE tools are viewed as add-ons in the platform architecture, and may be acquired from third-party suppliers. The weak point in today's situation is found in the lack of a comprehensive system specification model. The consequence of this deficiency is that the CASE tools will remain incompatible for the foreseeable future. Independently of the quality of the available CASE tools, it seems that the overall quality of the new system development environments is now so high that they will replace the older, conventional programming environments during the next 10–15 years.

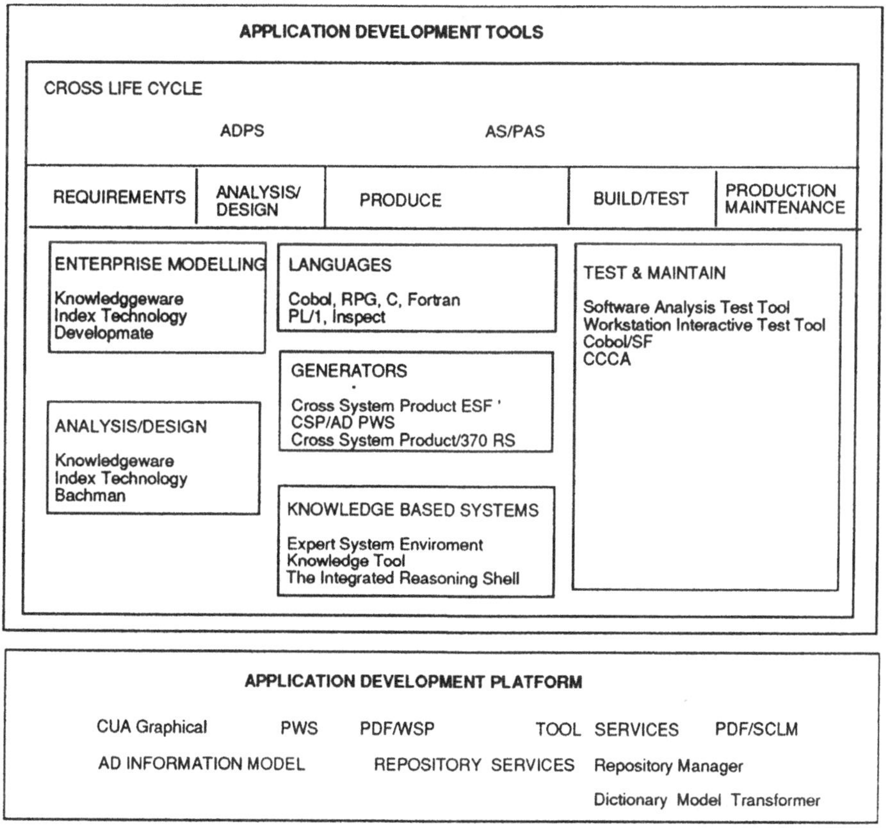

Fig. 8.4. AD/Cycle framework (from [IBM89])

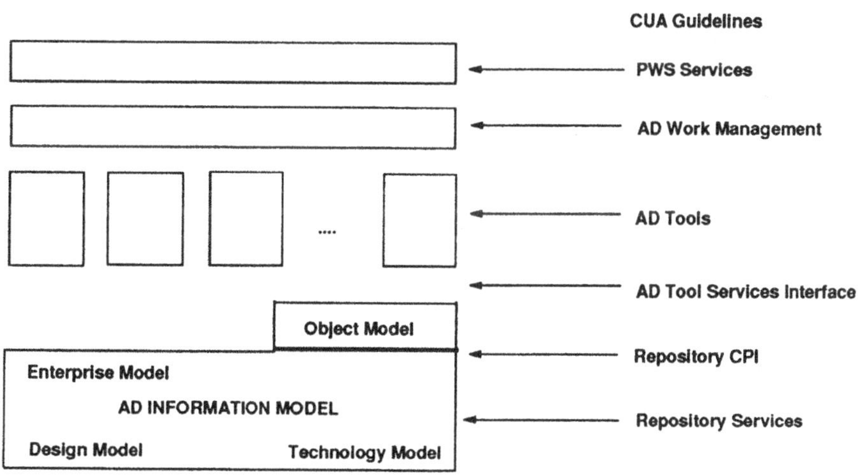

Fig. 8.5. AD/Cycle platform architecture (from [IBM89])

8.2 The Functional Properties of Support Systems for Information Systems Engineering (ISE-systems)

In order to build a support system for information system developers, their needs for information in their job-situation should be analyzed. A data flow diagram which shows various information processing tasks of the information systems engineer is depicted in Fig. 8.6. Processes, data stores and "external" entities are depicted on a very high level of abstraction, giving only few details, but concentrating on features that are felt to be central to the work of the information system engineer.

In the diagram there are four data stores, nine processes and four external entities. The external entities are project entities other than the information system engineers.

The data stores are:

D1: LIBRARY,
containing all types of commonly available information, e.g., company standards, tool descriptions, project status and overviews, specific results from previous projects, and general project experiences.

D2: SYSTEMS ARCHIVE,
containing all documents that describe the system that is being developed, its software, its design specifications and requirements specifications, relevant change proposals and error reports.

D3: PROJECT JOURNAL,
containing project plans, agendas and minutes of project meetings, reports on the project's economy, and on result production relative to the plans.

D4: TOOLS,
containing all of the computer-tools of the project, including CASE tools, data dictionaries, software development tools.

The external entities are:

PROJECT_MANAGER,
who is responsible for the project's results,

LIBRARIAN,
who is responsible for the selection and timeliness of all of the common, general project information, e.g., company standards,

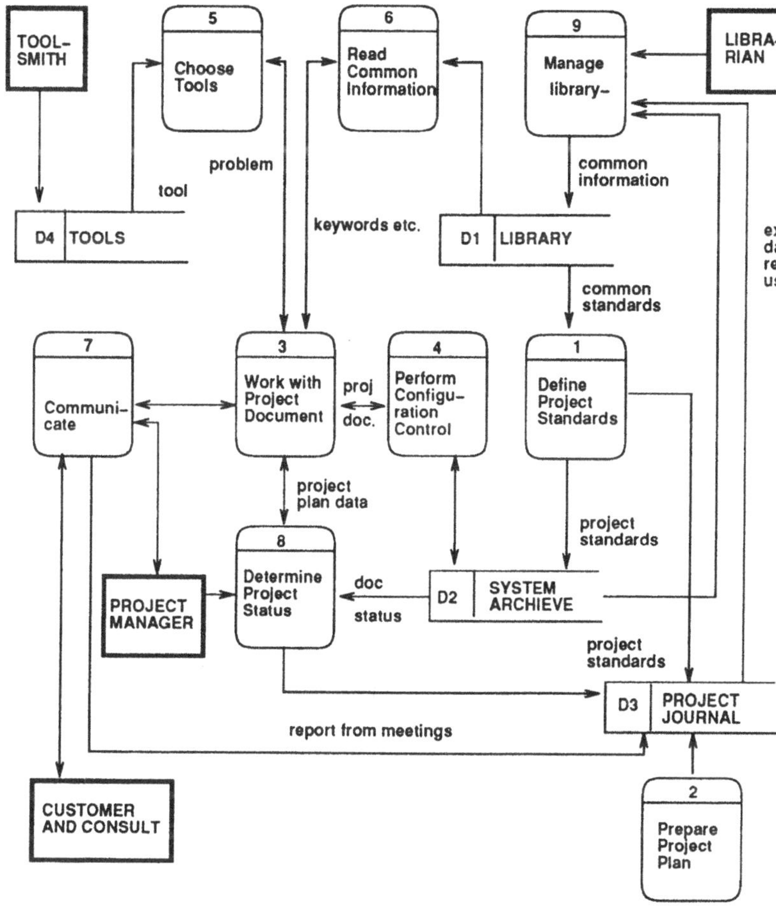

Fig. 8.6. The information processing tasks of an information system engineer

TOOL_SMITH,
: who is responsible for the system development tools,

CUSTOMERS and CONSULTANTS,
: who are persons not considered to be members of the project group, but who participate in meetings with the information system engineers, when their expertise is needed.

The data flow diagram reflects the various tasks that an information system engineer gets involved in during a development project. The various tasks have been related to each other in terms of data flows between the tasks that have to be there, regardless of what method is used for analysis, design, and implementation.

The DFD processes are:

P1: DEFINE_PROJECT_STANDARDS,
where systems development methods are evaluated and selected, where project management and documentation standards are selected and updated as the project proceeds.

P2: PREPARE_PROJECT_PLAN,
where the project plan is made.

P3: WORK_WITH_PROJECT-DOCUMENT,
where all of the documents produced in a system engineering project are written, read, and edited. Documents in this context are textual documents as well as modules and programs. This is the central process in the system, and it is continually repeated during a project. The P3-process interfaces with several other processes. While working on a project document one may need to select a tool, retrieve common information, record status information, perform version control, or communicate with the project management, a consultant or a customer.

P4: PERFORM_CONFIGURATION_CONTROL,
where it is controlled that the rules are obeyed that restrict the modification of project documents, in order to ensure internal consistency of the project documentation.

P5: CHOOSE_TOOLS,
where tools are made available for the information system engineers, as requested.

P6: READ_COMMON_INFORMATION,
where common information is retrieved from the project library.

P7: COMMUNICATE,
where meetings and other types of communication among project participants, and others, are supported.

P8: DETERMINE_PROJECT_STATUS,
where information about the project's performance, its use of resources and its result production, is collected and distributed.

P9: MANAGE_LIBRARY,
where selection and updating of library information is taken care of.

Based on a detailed analysis of the information processing needs of the information system engineer, the database and the functional structure of a support system for information system engineering may be developed.

The DFDs that describe the information engineer's tasks serve as a requirement specification to be satisfied by the support system.

In order to support the information system engineer in his various tasks, an ISE-system must contain subsystems for:

- DATABASE_ADMINISTRATION
- PROJECT_COMMUNICATION
- LIBRARY_ADMINISTRATION
- PROJECT_ADMINISTRATION
- REPORT_DEVELOPMENT
- SYSTEMS_DEVELOPMENT

The first three subsystems contain company-level functions, and serve all of the information system development projects of the company. The last three subsystems contain project-level functions.

The DATABASE_ADMINISTRATION subsystem
: contains functions for recording and maintaining user privileges, project definitions, staff allocations, company standards, tool descriptions, etc. The purpose of this subsystem is to prevent vital company data from being lost or changed by unauthorized personnel.

The PROJECT_COMMUNICATION subsystem
: contains the functions of an electronic mail system, with additional functions for personal archiving and planning calendars.

The LIBRARY_ADMINISTRATION subsystem
: makes available (for reading) company standards, tool descriptions, project experiences, and product libraries for general access.

The PROJECT_ADMINISTRATION subsystem
: consists of retrieval and maintenance functions for various project data, e.g., project team, project status, project description, project results, project plan, and progress reports.

The REPORT_DEVELOPMENT subsystem
: provides functions for developing and keeping access control of project reports.

The SYSTEMS_DEVELOPMENT subsystem
: contains functions for configuration management, specification development, software development, testing and implementation.

8.3 A Database for Supporting Information Systems Engineering

The particular example of an ISE-database that is shown in Fig. 8.7 is centered around the project concept. The project is the most important organizational entity for the individual team-member, on a day-to-day basis. It is therefore to be expected that the data be organized with respect to projects.

The database contains the data of the data stores of the previous section, but organized otherwise, in order to support many system developers instead of only one, as is the assumption of the analysis of the previous section. The complete ISE-database consists of ten major parts, called sub-databases. They are:

DB1: PERSONAL_DATA
: contains one entry for each project participant, for storage of electronic mail, access rights, project participation, and other personal data, e.g., personal notes, name, address, etc.

DB2: ADDRESSES
: contains common address lists of interest to all project participants.

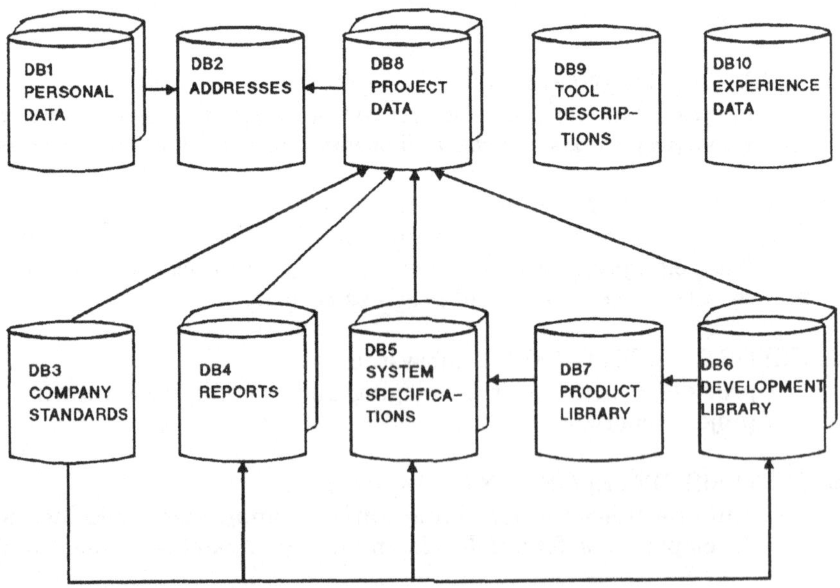

Fig. 8.7. An information systems engineering database

DB3: COMPANY_STANDARDS
contains documentation standards, programming standards, etc.

DB4: REPORTS
contains non-standard documents that are not related to particular parts of the system documentation, e.g., ad hoc reports for evaluation purposes, trouble-shooting analysis.

DB5: SYSTEM_SPECIFICATIONS
contains all of the standard project documents, for all project phases, including the software modules of each active project.

DB6: DEVELOPMENT_LIBRARY
contains document components and software components that are intended to be offered for use in other system than the one in which they are developed. Each project may have its own development library, or several projects may share one.

DB7: PRODUCT_LIBRARY
contains software and documents that are generally available for use by every project in the company.

DB8: PROJECT_DATA
contains project descriptions, project plans, and a project journal for each project.

DB9: TOOL_DESCRIPTIONS
contains descriptions of available tools.

DB10:EXPERIENCE-DATA
contains experiences from previous projects.

In addition, available tools (ref. data store D4 TOOLS of Fig. 8.6) may be stored in a separate sub-database, or they may be made available through the operating system of the projects' development computers.

We see that the sub-databases differ in complexity from simple files (addresses and tool indices), to complex networks of data (the system database). The structure of a SYSTEM_SPECIFICATION database is shown in Fig. 8.8. This structure is a simplified one. The database consists of documents, subprograms, and related information.

The database structure has to reflect the modeling concepts that are used for the representation of the software structure and of other specification objects, e.g., data flow diagrams. In this example of a SYSTEM_SPECIFICATION database we assume that all other specifications than the software itself are represented in (textual) documents. Subprograms are supposed to consist of data descriptions and (object) code, with interface, code shells and

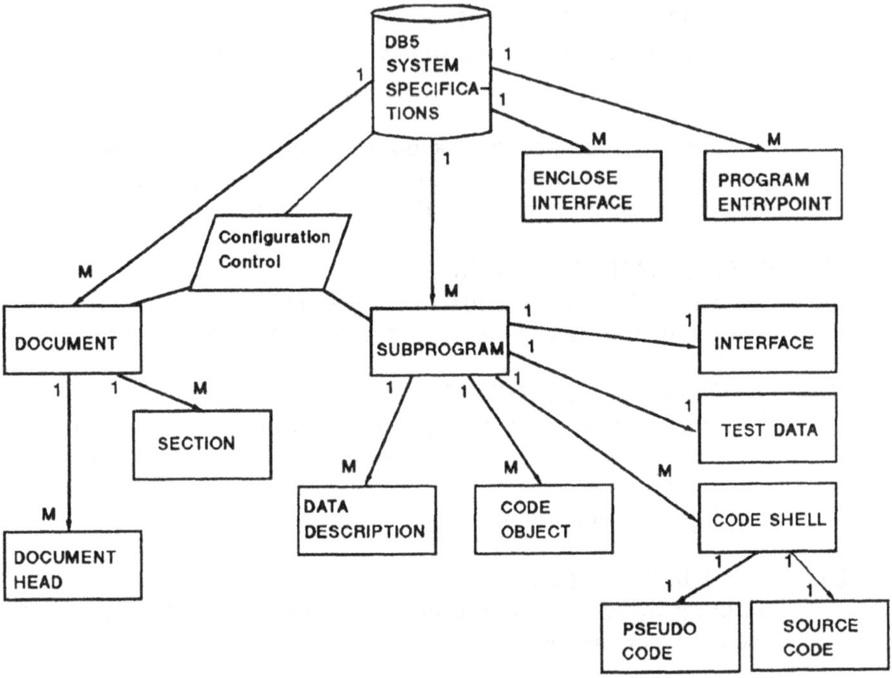

Fig. 8.8. Simplified structure of a system specification database

test data. The actual structure of the SYSTEM_SPECIFICATION database has to be determined with respect to the individual company's software model and specification models.

In Fig. 8.8 it is indicated that documents and software of an information system have to be subjected to configuration control. This is further elaborated in Fig. 8.9. In the figure it is assumed that a system may exist in several versions, as may subprograms and associated documents. Each version is associated with changes relative to some previous version. All of the changes to systems, subprograms and documents are stored in the database. Each system change is related to the appropriate changes of subprograms and documents. Each system version is related to the appropriate versions of subprograms and documents. Each subprogram version as related to appropriate versions of their associated documents.

As an information system evolves over time, a large number of versions may be created. It is no easy task to keep in control of all of the changes so that chaos is avoided. That calls for systematic approaches to the management of systems and their components, be they software components or documents. This is the theme of the next section.

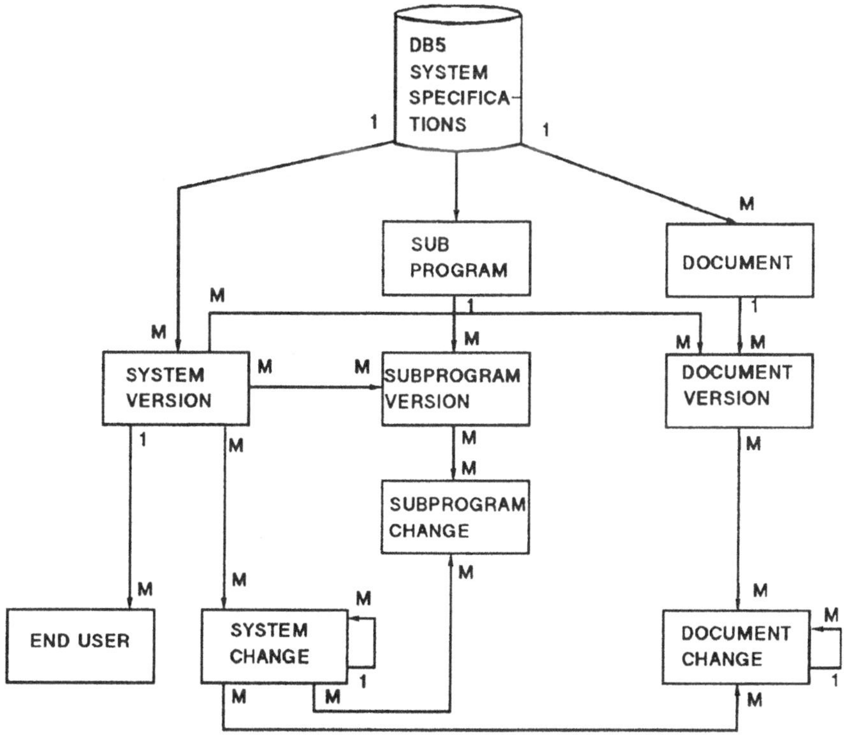

Fig. 8.9. Configuration management in a SYSTEM_SPECIFICATION database

8.4 Information Systems Configuration Management

Configuration management is the discipline of controlling the evolution of complex systems. A system's evolution is manifested in the creation of new versions of the system, each version being somewhat different from the others. Sometimes many versions are required to co-exist, sometimes only the last created version is valid, and the older versions are destroyed. Information systems evolve over time in two different ways:

- System variants are created, because different uses of a system may have different requirements for, say, functions and/or operational properties, and

- System revisions are made, because of new users' needs, and because of the removal of errors, so that the new version of the system will replace the older one.

System variants are intended to be operated in parallel, in order to serve different markets (users), while system revisions are intended to replace earlier versions of the system. In spite of that we may often find that different users employ different revisions of a system. The reason may simply be that some users do not need the (new) features of a system revision, and therefore do not want to install it. So even revisions of a system may often have to be operated in parallel.

System variants are sometimes called system families. Common systems are examples of system families. A system family must share a large number of components in order to make the family economically viable. The differences between variants may be small, e.g., error messages may be given in different languages for two software system variants (in English and in Norwegian), they may be larger, e.g., two software variants may have been developed for different operating systems, or the differences between variants may so large as to provide quite different functional properties.

The larger the difference between the system variants, the more each variant will tend to live a life of its own. The users of the various variants will have different priorities for the alternative ways of changing the system. The users' requirements will sometimes be in conflict. The various markets of a software product may have different functional requirements, and may therefore drive the evolution of the system variants such that their differences become larger and larger.

Just as systems come in variants and revisions, so do the systems' components: the software modules and the specification documents. Every part of a system may potentially be modified. There is a delicate balance between the benefits of re-use and the dangers of complexity. The more re-use that can be made of a component in different system variants the better, if the complexity of the system family does not increase too much. As the difference between the variants increases, and as more and more revisions have to be kept simultaneously operational, so the danger increases of losing control of the evolutionary process and ending in complete chaos.

Software configuration management is a specialization of configuration management. The major difference between the two comes from the fact that the software components may often be derivable from already existing ones, by automatic means. Consider the difference between the modification of a computer program and the modification of a car. A new version of a computer program may be produced (by a computer) from the older version together with a specification of the modifications to be made. Not so with a car. A 1992-model car is certainly not automatically produced from a car of the previous year.

Because of the ease of producing new software versions from old ones, once the modifications have been specified, it becomes necessary to keep track of derivability dependencies among software components. Consider the relation between a piece of source code and object code that may derived by the use of an appropriate compiler. The same piece of source code may become the basis for several different object codes, produced by different compilers, for different computers. Whenever the source code is modified, we would most certainly like new object code to be derived and properly inserted in new versions of the systems of which they are components. Software derivability dependencies may become much more complicated than this. Consider the added complexity of wide use of *include modules*, as an example.

The major problems to be solved by an information system configuration management may be summarized as:

- Version management, selecting the right versions of system components, and monitoring their development status,
- Change management, monitoring the changes made to components by several persons, supplying tools for merging competing changes later, and keeping record of what has been done to each component, and by whom,
- Efficient storage of revisions, avoiding redundant storage of data that are shared between several revisions of a system component,
- Software manufacture, putting together a software systems configuration by the use of tools such as compilers, linkers, pre- and post-processors, and so on.

8.4.1 Versions, Revisions and Variants

We have previously introduced the notions of versions, revisions and variants. (A version may be either a revision or a variant of some other version.) Systems may exist in several versions, as may system components. A new version of a system has been created if a new version of (at least) one of its components has replaced the component-version previously used in that system. New system versions may also be created by adding or removing components in the system structure. Note that system components are all kinds of specifications that belong to the system, for example software modules and design specification, regardless of whether the specification is formal or informal, for example textual.

The naming system for versions is an important part of the basis for managing information systems evolution. A two-level numbering system that identifies revisions of baseline versions is commonly used for single components, and for identifying system versions for their users.

The form is

<baseline version> . <revision>

where version 1.0 means baseline version no. 1, no revision,
version 1.5 means baseline version no. 1, revision no. 5,
version 2.7 means baseline version no. 2, revision no. 7.

This simple form does not capture the variant relationships, so we add one more term, getting the form

<baseline version> . <revision> (<variant>)

where version 1.0(Norwegian) means baseline 1.0, Norwegian variant,
and version 2.7(German) means revision 2.7, German variant.

This form captures the needs for identifying the various releases of systems and their components. A version is said to be released when it is made available outside of its development group. This form lacks possibilities, however, to capture temporary local revisions made in the development group when working on a new release. To capture also this need, we add on a <local revision> term, giving the form

<baseline version> . <revision> (<variant>) <local revision>

so that version 2.7(German)5

means revision 2.7, German variant, local revision no. 5.

Variant relationships and revision relationships between versions may be depicted in a variant graph [GRØN88], as shown in Fig. 8.10. The variant graph in the example shows that a baseline version 1.0 is developed into a 1.1 revision with three variants, for the English market (E), for the German market (G), and for the Norwegian market (N).

The German variant is locally revised twice, and then developed into a new revision 1.2 for the English variant as well as for the German variant. The 1.2 revision is not given a Norwegian variant. Next, a revision 1.3 is prepared for all three markets. A new variant of revision 1.3 for the Swedish (S) market is developed based on the Norwegian variant. Finally a new baseline 2.0 is developed for all four variants.

Every node in the graph represents a version of the system (component). The leaf nodes are the most recent versions delivered. There are three types of relationships that are represented in the graph: the *revision relationships, the variant equivalence relationships*, and the *variant splits*. The vertical arcs in the graph are revision relationships. The use of the two revision specifiers may be tailored to the needs of the individual companies.

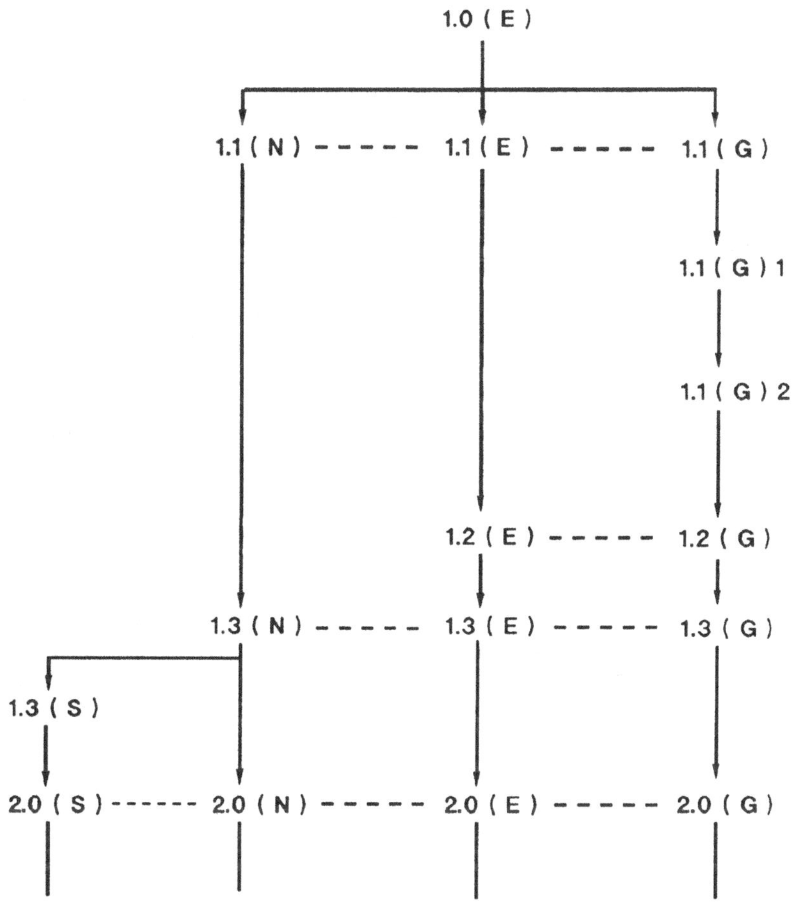

Fig. 8.10. Variant graph

One may want to reserve the first specifier for inter-variant modification, so that all "equivalent" variants have the same first revision specifier. It may be natural to reserve the first revision specifier for more profound modifications, and what is more profound than changes that are implemented in several versions? The horizontal arcs, the variant equivalences, signify that a change has been applied to several variants.

The variant graph captures all versions of a design object that have ever been produced, and thus represents the development history of the design object. Each system has its own version graph, as has every system component. Each system version consists of one version of each of its components. A detailed development history for a system may therefore be reconstructed by looking at the version graphs of the system's components.

8.4.2 Change Management

We have so far been looking only at the relationships between versions that are "finished", in the sense that the new components have been developed and made ready for inclusion in a system. However, the picture is more complicated than that.

New system revisions come about after proposals for change have been put forward and analyzed, and consequences in terms of component modifications have been analyzed and implemented. Each revision of a system has a life-cycle of its own. Requirements specifications must be worked out, the system and (some of) its components must be redesigned and implemented, pending suitable testing and approval.

The usual kind of documents and specifications must be developed in order to support the "re-design" process: requirements and design specifications, software modifications, implementation documents, test documents, and a user manual. The actual document structure depends on the requirements of the project development organization.

The development of a system, as well as of each of the system's components, proceeds through stages, starting with an initiation stage and ending with a delivery stage. In order to monitor the progress of a system and its components, one may introduce state values for characterizing their development state.

Suitable state values [ANDE85] may reflect life-cycle phases and important events for a system and its components:

- For determining a document's development state:

  Initiated, Ready for Approval, Conditionally Approved, Approved, Delivered, Retired, Scrapped.

- For determining a subprograms development state:

  Initiated, Designed, Implemented, Unit Tested, Integration Tested, Approved, Delivered, Retired, Scrapped.

- For determining a system's development state:

  Initiated, Specified, Designed, Implemented, Module Tested, System Tested, Approved, Delivered, Retired, Scrapped.

The development state of a system is related to the development states of its components, for example, a system cannot be said to be implemented unless all of its components have been delivered. The general rule is that a system cannot be in a more "advanced" state than any of its components, for example, for a system to have the state value 'System Tested', all of its modules have to have a state value higher than or equal to 'Integration Tested', and the integration test documents have to be at least 'Conditionally Approved'.

If status values of a system and its components are maintained to this extent, they may be used for estimating the progress of the development project. Taken together with accumulated experiences from previous projects, one may compute a "completion factor", that may support the project managements possibility to estimate the cost and time for the remainder of the project.

Let us turn our attention to the relationship between the system components and the persons that develop the components. In order to keep administrative control of the development, the ideal situation is to let each component be worked on by one person only, who is also the responsible person for the creation (or revision) of this component. This is the simplest way to ensure consistency and a minimum number of development errors. However, this may not always be desirable for other reasons.

It often happens that developers have tasks that require simultaneous modifications of the same documents or programs. This may originate when the development organization is geographically decentralized, or from an attempt to speed up a project by adding more people. If simultaneous modifications are permitted, one might end in a chaotic nightmare situation, unless proper administrative tools are provided.

One possible remedy is to apply the variant graphs of the previous section, by letting the various developers create their own temporary variants of the system components that they want to modify, but which they are not primarily responsible for. An example is shown in Fig. 8.11.

Assume that there are three system components A, B, and C, which are the responsibilities of persons #1, #2, and #3, respectively. We will let each of the three persons have the possibility of developing a temporary variant of each of the three components, and develop as many local revisions of their variants as they may wish, before a consolidation is required, and new (common) component revision are established.

The figure shows that version 1 of module A, which is the responsibility of person #1, that is, version A1(1), is further developed by person #1, who creates A1(1)1, i.e., component A, version 1.0, developer #1, local revision 1. But in order to develop a new revision of component A, he needs to modify B1, which is the responsibility of person #2, according to the variant identification B1(2). Person #1 develops a local revision of a new B1-variant, B1(1)1, in order to solve his problem. Furthermore, in order to develop his B-variant B1(1)1, person #1 needs to modify the current C-version, C1(3) which is the responsibility of person #3. He therefore creates

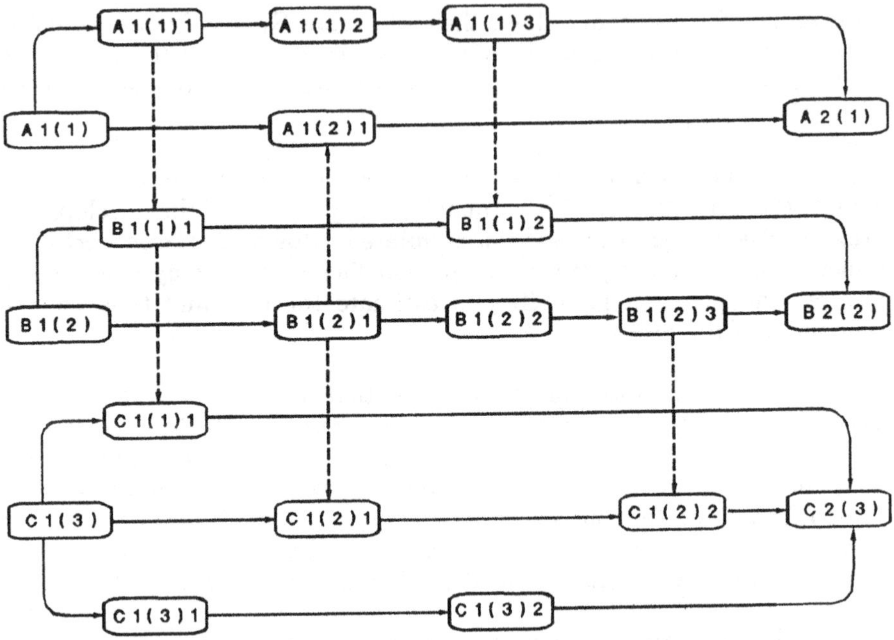

Fig. 8.11. Variant graphs applied to simultaneous modifications of system components

a temporary variant C1(1)1 of C. Person #1 may now go on to develop a further local revision A1(1)2 of module A, and subsequently even another local revision A1(1)3, which requires an additional modification of person #1's B1-variant, from B1(1)1 to B1(1)2.

The dotted vertical lines in the graph signify that the modifications in one component version are dependent on the modification of another component.

The other two persons proceed in a similar manner. They create their own temporary versions of system components as they are needed. And so the work continues until one of the developers, say #1, wants to create a new (permanent) revision A2(1) of the component for which he is responsible. The development graph shows that there is another temporary local revision A1(2)1, created by developer #2, that must be taken into consideration, and possibly accommodated, prior to the creation of a new component version.

There is no way to derive automatically whether there exists inconsistency or consistency between A1(1)3 and A1(2)1 without turning to symbolic evaluation of the code. Even that will most often prove to be futile. Therefore a "consensus meeting" is forced in order to negotiate the final contents of the next (public) version of A. The interested parties must meet and develop an agreement on the contents of A2(1).

By using techniques of this kind, one may introduce sufficient control so that several developers may be allowed to simultaneously work on the same system components. But great care must be taken to keep administrative control of what is going on, to prevent the project ending in chaos. Computerized support of the systems development project is a must if simultaneous modification of components by different persons is permitted.

8.4.3 Efficient Storage of Components

There are several ways of storing the new versions that are created during systems evolution. Each new version may be stored in its entirety. This would result in much redundant storage of data, because the differences between two neighboring versions are usually not very large. Most of their contents are shared. The newer version is most often a fairly small modification of the previous version, and complete storage of each new version would therefore quickly become quite expensive in terms of storage space.

Another alternative is to keep one base version stored in its entirety, and keep a log of all modifications done to that base version when new versions are being developed. The newer versions may then be generated from the base version whenever needed by doing a re-processing of all of the relevant modifications. This saves storage space, but may be somewhat impractical, because it will require re-processing also of all of the mistakes that were made during the preparation of the new version, for example recreating all of the internal revisions of a component.

A third alternative is to create so-called deltas, that consists of a script of commands that will generate one version from another version. A delta can be created by comparing the two versions. In its simplest form a delta consists of commands for deleting lines of text from, say, a document, and inserting lines of text into a document. The delta can therefore be automatically constructed by a suitable file comparator, when two versions are treated as line-organized texts. This technique can be applied to program texts as well as to other (natural language) texts. A delta thus represents the difference from one version to the next one, less all of the intermediate steps taken during the development of the newer version.

Delta storage is the most commonly used technique for conserving storage space [TICH85]. It makes the luxury of saving multiple versions of system components affordable. A sequence of versions may be represented by a base version and a sequence of deltas, where each delta represents the changes from one version to its neighbor in the revision sequence.

The deltas may be applied in a forward direction or in a reverse direction. The forward direction implies that newer versions are generated by applying the deltas to an older base version that is stored in its entirety. This means that the most recent version must always be generated by (a sequence of) delta-applications prior to its use. The reverse direction is generally preferred, because this implies that the most recent version is kept stored in its entirety, while the previous versions may be generated whenever required. Since the most recent version of a component is probably the one that is most frequently used, the reverse method is usually more effective than the forward storage method.

8.4.4 Software Manufacture

Software manufacture is the process of deriving executable software from source code. The handling of derived versions is simpler than the handling of source versions. Derived versions are automatically computed from their source versions. No human actions need to be supported and managed. Nevertheless, in order to manage the software manufacture process, it is necessary to keep track of the dependencies among the various systems components.

New versions of software systems are different from their predecessors because a number of systems components have been modified. The components that have been changed must be recompiled before a new version of the software system can be generated. But the large number of software components that are not influenced by the changes do not need to be recompiled. It is therefore important that the dependencies between the software components are preserved.

Dependency graphs [FELD79] represent the dependencies which exist between the various system components and their respective derived components. Dependencies may exist between source components, and they of course capture the dependency relationship between a source component and its derived component (Fig. 8.12).

A commonly used method for representing the dependencies is based on the application of time stamps [FELD79]. A time stamp represents the time of the latest generation of a new version of a software component. It is implicit in the method that if a new version of a (derived) software component has been created, then new versions of all of the software components that depend on the first component must also be generated. A software component B that is dependent on software component A is regenerated if the time stamp of component A is more recent than that of component B. In this way we need not regenerate more than a minimal number of components in order to have an updated operational software system.

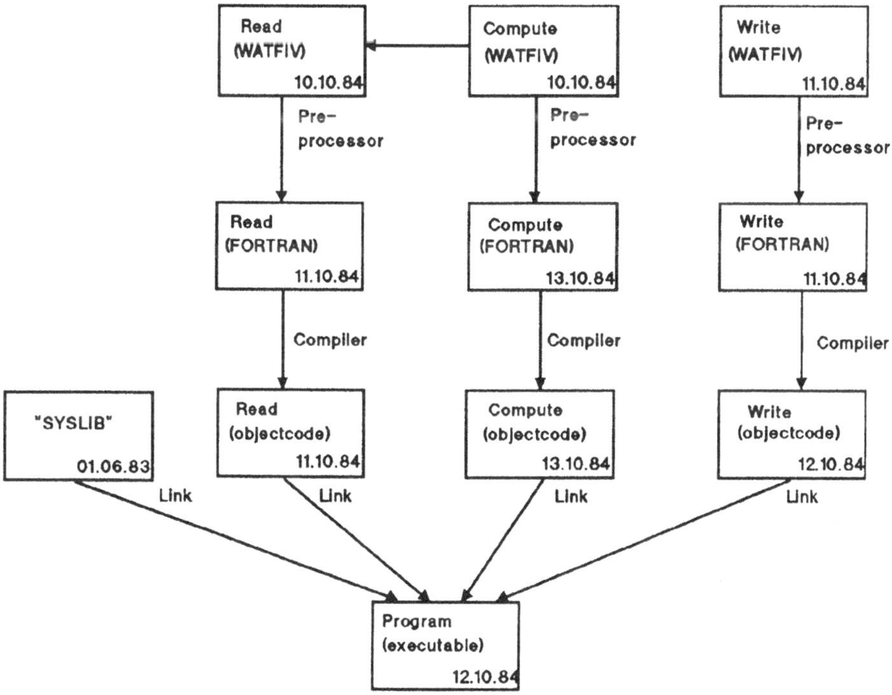

Fig. 8.12. A dependency graph with time stamps

Many of the dependencies of a software system may be automatically generated by analyzing the components' source code, for example hierarchical dependencies may be inferred from *include* and *call* statements in the program system. Other dependencies must be entered manually [WALD84].

The time-stamp mechanism is used in several widely used tools, for example MAKE [FELD79] of Unix. In spite of its popularity, time-stamping has a number of serious shortcomings when used for large-scale software configuration management [TICH88]. It is inappropriate for determining whether a derived component can be re-used. It is also insufficient for deciding from which variant a particular derived component has been generated.

There are a number of software configuration tools available in the market place. Research in the subject area is producing new methods and tools. However, it is beyond the scope of this text to go deeply into this subject, so we leave it for the specialist literature, e.g., [WAYN86], [TICH88], [TICH85].

Exercises

1. Define the following terms briefly :
   a) CASE
   b) CASE tools
   c) ICASE
   d) IPSE

2. State the characteristic features of CASE tools briefly.

3. Explain the role of the repository in ICASE.

4. Research the CASE literature and prepare a paper including a comprehensive classification of CASE tools.

5. What are the differences between information systems configuration management and software configuration management?

6. Discuss the importance of "versioning" of software.

Chapter 9

Engineering Design Principles for Unsurveyable Systems

The real challenge of systems engineering design is to provide operational solutions for systems that are so large and so complicated that no-one can survey every detail of the system and its behavior. This is a general problem in every engineering discipline. Engineers have learned to cope with the situation. They apply some common principles in their approach to design. These principles may be formulated differently in various branches of engineering. The first section of the chapter will explain the generally agreed approach to the engineering design process.

The design of large systems implies that different groups of specialists must be put into concerted action in developing a system that none of them can survey in all of its details. The second section of this chapter provides an overview of the properties of such unsurveyable systems. Principles for approaching the design task are explained. The notion of *constructivity* is discussed. It is argued that constructivity is a fundamental property that a system must possess in order to permit an orderly, controlled design process.

Not all system problems have the constructivity property. How to deal with such systems is discussed in the third section of the chapter. The presentation is based on Rittel's notions of "tame" and "wicked" problems [RITT72]. A mathematical problem is a typical "tame" problem. It can be extensively formulated, and a correct solution may be found. For "wicked" problems, the solutions may only reasonably satisfy a set of conflicting goals. Planning problems are typical examples.

9.1 The Engineering Design Process

The systems engineer is concerned with the design of systems consisting of logical, human, mechanical, electrical, and other physical components, such as computers, motors, and instruments, in a way that assures that these components will work together.

The theoretical aspect of systems engineering has to do with predicting the static and dynamic properties of the proposed system, prior to its realization. To reach this goal, mathematical tools such as differential equations are used. The aim is to calculate how the system's properties are influenced by the properties of the components and their interconnections. Sophisticated modeling tools have been developed over the years for various types of systems, such as process control systems, airplanes, bridges, roads, and so on.

The practical aspect of systems engineering has to do with the problem of constructing the different components separately, by different companies for instance, and then putting them together as a system. Early experiences show that often this approach did not work even for mechanical systems. The separate components would not work together, or would not even enable their proper inter-connection. System incompatibility became the thorn in the side of practical systems engineering.

Theoretical development takes time. Practical system-building must go on regardless of whether there is a satisfactory theory for the system in question. There is a general procedure used by engineers to obtain solutions to problems. This procedure, or problem-solving process, is common to all branches of engineering. It can be used even if there is no satisfactory systems theory on which to base design.

The engineering approach is characterized by a number of stages. Jensen and Tonies [JENS80] suggest six engineering design stages (Fig. 9.1):

- *Problem formulation* – a definition or description of the problem in broad terms without detail is proposed.
- *Problem analysis* – the problem definition is refined to supply essential detail.
- *Solution generation* – a set of potential solutions to the problem is developed.
- *Solution selection* – each of the solution alternatives is evaluated and compared to the others, and the best solution is chosen.

- *Design specification* – the chosen solution is described in detail.
- *Implementation* – the finished product is constructed from the design, tested, and installed.

To these six stages one might add a seventh stage:

- *Modification* – the system is changed as time passes to satisfy new requirements.

It is debatable whether modification should be regarded as a separate design stage because modification encompasses all of the six previous stages. Each time a design is modified one has to proceed from problem definition to implementation, in order to treat modification proposals in an orderly way.

The names and numbers of stages that are found in different textbooks for various engineering disciplines might differ, but essentially they are the same. The details of performing each stage might nevertheless vary considerably in different engineering disciplines, depending on the theoretical maturity and other particular features of each engineering branch. So the adaption of the general engineering design approach is different in software engineering than in, say, operations research, mathematical optimization, or electrical engineering. The design process encompasses activities from the recognition of the problem through the implementation of an economical, functional solution to that problem. It is this approach engineers use to apply their knowledge, skills, and creative ability in the development of new devices and processes.

The design process appears less orderly and less visible during the stages of problem formulation and problem analysis than in the later stages of design specification, implementation, and modification. The work in the first stages is to some degree concerned with developing objectives for the project, and with finding explanations that increase the analyst's perception of the problem at hand. In the later stages one is more concerned with finding detailed solutions for the problems that have already been defined in the previous phases of the project.

The first stages are mostly of a general, fuzzy nature. The problems are not clearcut technical, economical, administrative, or political ones, but a mixture of all of these. Consequently, the later stages of engineering design appear better controlled and more orderly – and easier to teach – than the first stages of design conception. Most of the remaining portion of this section will be devoted to discussing various aspects of the first stages of engineering design.

It is important to realize that the engineering design process is not conducted in a vacuum, i.e., with no outside communication. Each of the

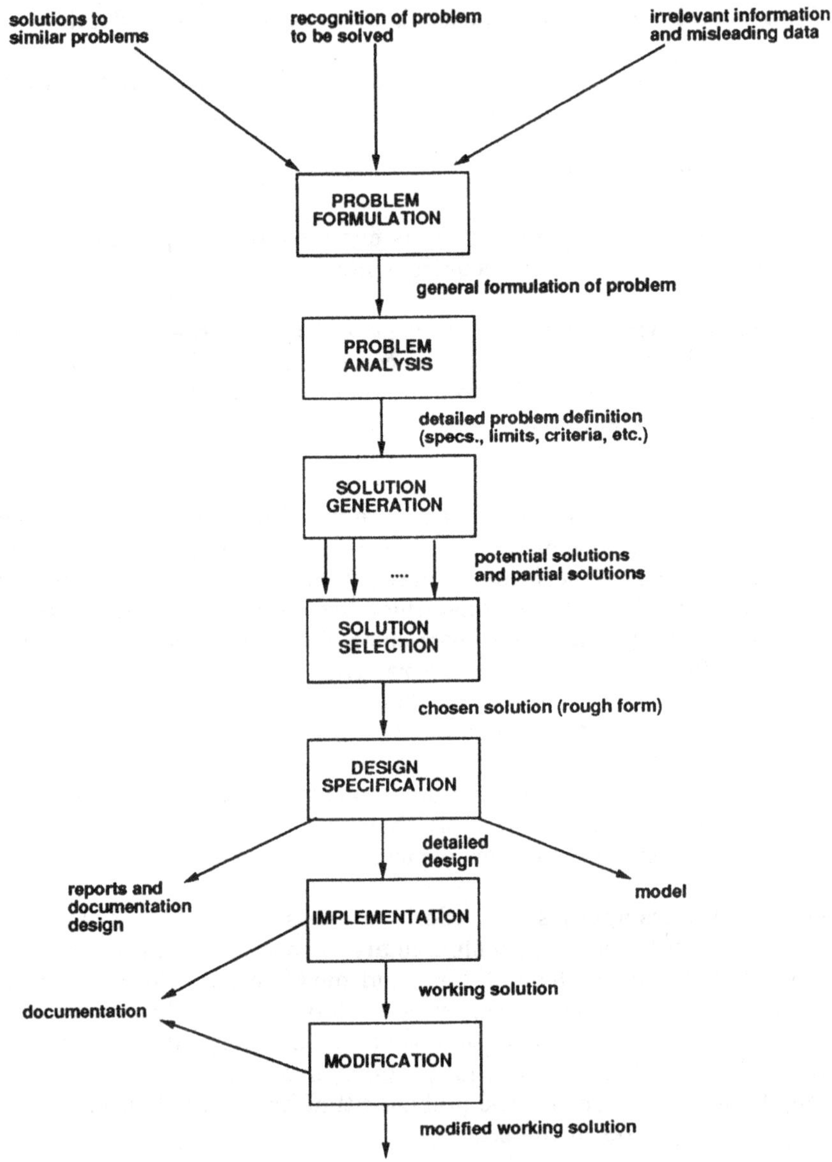

Fig. 9.1. The engineering design process (from [JENS80])

phases involves much communication with managers, users, customers, and colleagues encompassing both oral communication and written communication. The effectiveness of engineers and engineering projects is dependent not only on their technical skills, but to a large extent on their ability to communicate with others.

9.1.1 Problem Formulation

Before one can solve a problem, one has to define what the problem really is, then decide how to carry out a solution. Both of these needs are frequently short-changed. Often we form some fuzzy picture of the problem confronting us, then hack away at a solution without much forethought. Job promotion is rare for engineers who engage in such practices.

The problem formulation is primarily a point of view, i.e., a manner in which the engineer surveys the problem. One should not assume that the problem is obvious and that everyone knows what it is. Misunderstandings are frequent. Communication is very important in this phase so that everybody involved can contribute to developing a communal view of the problem.

Only in rare instances is the true problem presented to the systems engineer. The original problem statement is usually obscured by considerable irrelevant information, and distorted by misleading opinions, current solutions, and standard ways of viewing the problem.

It is tempting to formulate a problem by trying to adapt it to an existing solution. This is effective if the solution fits the problem. There is, however, a possibility of distorting the understanding of the real problem, because it is being analyzed in the perspective of a known solution. The approach may be described by an analogy: People whose only tool is a hammer will survey most problems as being nails. One quite often ends up adapting a current solution to a poorly understood problem. Therefore, a problem is solved, but it is frequently not the right problem to solve.

There is no single verbal or graphical method for formulating a problem. Nevertheless, there are rules of thumb and recommended approaches. One effective way of studying a problem is through systematic partitioning of the problem into subproblems. Our current discussion of the engineering design process is an example of such a partitioning. Figure 9.1 provides an example of visualizing problem partitioning, by graphical means. By using this approach we can gather information about subproblem inputs and outputs without engaging in a discussion about the details of the processes which represent the subproblem solutions.

Assume that we are going to design a radio. In the first design stage (of problem formulation) we may propose that the radio consist of a signal selection system (a tuner), an amplifier, a loudspeaker (all sequentially connected), and a chassis, to which the three other components are physically connected (Fig. 9.2). To put the essence of this approach in other words, we define the system to consist of a set of interconnected subsystems.

9.1.2 Problem Analysis

In the second stage of the traditional engineering approach the objective is to obtain detailed qualitative and quantitative characteristics of the system components and of their interactions. The characteristics cannot usually be given in absolute terms at this stage, but are treated as variables subject to constraints.

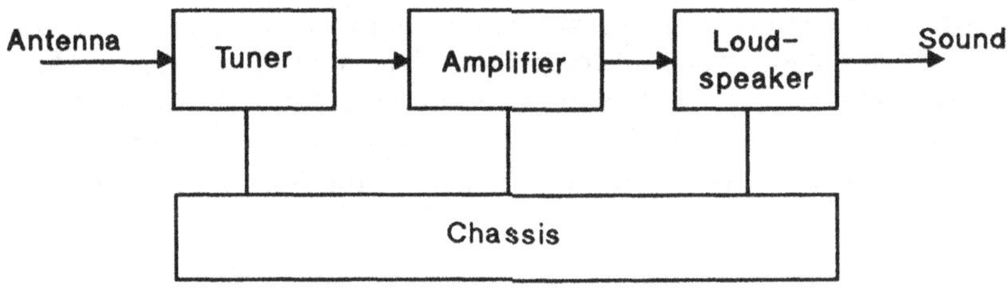

Fig. 9.2. Components of a radio

In the design of the radio example of the previous section, one would in the problem analysis stage constrain the weight of the radio (1 kg), the output effect of the loudspeaker (20 mW), the distortion of the output signal, the maximum price limit, etc. One would further describe the restrictions on the connection between the amplifier and the loudspeaker, stating the conditions for the output effect of the amplifier to match the input effect of the loudspeaker. Otherwise, if the amplifier output is permitted to become too large compared to the input capacity of the loudspeaker then in the final implementation we would have white smoke rather than sound coming out from the loudspeaker when the effect is turned on. (This reminds us that we have forgotten to specify that the radio must also be equipped with a power supply).

The outcome of the problem analysis is a set of solution variables (weight, output effect, input effect) and an accompanying set of restrictions on the solution (e.g., weight < 1 kg, output/input matching, etc.). The restrictions reflect laws of nature as well as customer wishes and management policy.

Next, we must resolve situations where there is conflict among the different components' properties. If it is apparent that the various imposed restrictions are incompatible, one or more of the restrictions must be relaxed if a feasible solution is to be found. The solution variables include the possible ways in which solutions to a problem can differ. For example, if the limits for signal distortion, input/output effect, weight, and price are given for the amplifier, there might be a number of different alternatives that would satisfy the constraints on these solution variables.

During problem analysis one should also develop criteria for selecting the best solution. The criteria usually include parameters of cost, reliability, ease of maintenance, accuracy, and efficiency. The relative weights of the individual parameters determine the order of importance of these criteria in any project.

The most effective way to analyze a problem of any significant size is through a process of decomposition. This is a process of *stepwise refinement* whereby we successively increase the level of detail by analyzing subsystem by subsystem.

9.1.3 Solution Generation

The development of alternate solutions to the problem follows problem analysis. Generally, solutions appear during the analysis as *byproducts* of the analysis. One will usually stop further decomposition when recognizing existing solutions and/or similar solutions to those needed for the problem at hand. These possible solutions should be collected and saved for the solution generation step during which an active search for alternate problem solutions is conducted.

Putting a strong emphasis on solution generation may be especially useful if the system components' properties can be selected from a wide range of possible values. This is often the situation in engineering design projects. If, instead, one chooses to concentrate on one particular solution found during the analysis phase, the engineer may severely restrict his perspective on design possibilities. Therefore, broadening one's mind to take other solution alternatives into consideration would be very helpful. The objective of solution generation is to maximize the number and variety of solutions from which a final selection can be made.

Alas, there is very little one can say about how to actually perform the search for solution alternatives. One seems to be heavily dependent on having access to experienced and creative people, as well as to effective solution generation techniques. It is quite expensive to develop solution alternatives. Even though, ideally, one should explore every conceivable option, this is usually not economically feasible.

It is interesting to see that engineering organizations that know their own strengths seem to rely on a limited number of solution alternatives which they try to adapt to the problems (or vice versa). Nonetheless, providing the broadest solution space possible from which to select a preferred solution is an advisable approach.

9.1.4 Solution Selection

To be useful, a system must do what it is intended to do with acceptable continuity and reasonable efficiency at affordable costs. We assume that we have a number of potential solutions to choose from. To guide the selection procedure it is usually recommended that:

1) Selection criteria are worked out and that the relative importance of the criteria are assigned.

2) The performance of the alternate solutions with respect to these criteria are predicted as accurately as possible.

3) The performance of the alternate solutions are compared on the basis of their predicted performance.

4) The preferred solution is selected.

Selection criteria may be based upon many factors including cost of production, functional performance, efficiency, reliability, maintenance, ease of use, etc., as well as political and human factors which are not readily measurable. If we deal with constructive subsystem structures with respect to these criteria, we can, in principle, proceed through the four steps of solution selection in an orderly manner. Otherwise we have to resort to more subjective methods. The judgements involved in such situations are value judgements based on vague, limited, or non-existent data. Some elements of the selection criteria may also be non-quantifiable. Remaining totally objective under these circumstances is impossible.

There is an apparent contradiction between the desire to have an objective selection process, and the fact that we sometimes have to base the selection on subjective value judgement of the components' properties. Therefore, one should limit the objective evaluation to measurable system properties and let the final choice be made subjectively on aesthetic value.

9.1.5 Design Specification

This design phase refines the rough solution to a level from which the end product can be built. The characteristics of the solution must be specified in sufficient detail so as to permit systematic review, analysis, verification, and validation. The level of detail must also be adequate to allow the manufacture of the product. Because the design may be implemented, operated, and modified by another person, it is essential that the design be carefully and effectively documented.

The actual design techniques used in this phase depend to a large degree on the particularities of the engineering domain. Techniques for designing a bridge are different from techniques for designing a computer. The specification languages, e.g., graphical language in technical drawings, differ in the various engineering disciplines. Nevertheless, the output of the design specification phase usually consists of two items: a set of detailed specifications and a model of the end product.

The detailed specifications include a report that completely describes the solution, both in words and in diagrams. The report also describes the performance of the design and presents a thorough evaluation of it. It is primarily through this report that the solution is transmitted to the outside world. Therefore the report must communicate the solution effectively. The second item, the model of the end product, may appear as a scale model of a bridge, a working electronic prototype, or a computer program that simulates some aspects of the end-product behavior. The model provides an effective means of illustrating and describing the final solution.

The detailed design specifications can be safely worked out only if implementation, manufacturing, and maintenance aspects are thoroughly analyzed. Technical reviews are conducted as the design evolves toward detailed engineering specifications. Off-the-shelf components are acquired and custom components are built. The prototype performance is evaluated, then manufacturing specifications are derived. The emphasis shifts from function and performance to ease of manufacturing [PRES82].

9.1.6 Implementation

The responsibility of the engineer seldom ends with the specification of the solution, but extends into the manufacture of the product, gaining acceptance of the design, training the user in the operation of the product, and installing the product in its operational environment. Before manufacture begins, quality-assurance methods must be established and a product distribution mechanism defined. A field service organization must be established for product maintenance and repair.

The implementation considerations are of course highly dependent on the type of product (one-of-a-kind, mass-product), the technical field (e.g., civil engineering, electronics) the system's complexity, etc. However, one common feature is that increasing emphasis is put on organizing orderly product test procedures along with the different subphases of implementation. Presently, products are rarely distributed that contain too many intrinsic errors if one wishes to survive the fierce competition in the market place.

9.1.7 Modification

After installation, the operational phase starts and the system is put into use in its intended environment. At this stage "hidden" errors may appear, alterations to some of the functional properties may be required, and enhancements may be suggested. At the same time, competition from other vendors in the marketplace may add to the pressure for modifying the system product. Furthermore, changes in the system's environment may suggest corresponding changes in the system.

There are three major classes of modifications:

- *Corrective modifications*, for repairing the system, to change unacceptable systems behavior (error-repair),
- *Adaptive modifications*, to change system functionality because of environmental changes, customer wishes, competitors' pressure, etc.,
- *Perfective modifications*, to change the system quality with respect to operational cost, modifiability, etc.

For a system product to be accepted as an operational system it must be both reasonably reliable and cost-effective. Corrective modifications and perfective modifications contribute to keeping the system product acceptable in the market place.

The adaptive modifications contribute to increasing the lifetime of the product in the market place by keeping it up-to-date in accordance with customers' needs. Adaptive modifications are done following the principles for system development described in this section, that is, through stages of problem formulation, problem analysis, solution generation and selection, design specification, and implementation.

9.1.8 The Engineering Design Process: Ideals and Reality

We have described the engineering design process as a sequence of stages indicating that each stage can be more or less finished before moving on to the next stage. This would be a very attractive situation from the point of view of managing the engineering design project. We could then make a final evaluation of the results of each development stage prior to moving on to the next project stage; and we would never have to look back.

Unfortunately the world is not that simple. At each design stage, new insights into earlier decisions become available. These insights may modify or invalidate the decisions that have led us to our present understanding of the problem, and force modifications to previous project stages. Each of these iterations cost both time and resources. It is therefore recommended that the work of each project stage be done as thoroughly as possible. The sequence of design stages is sometimes called "the life-cycle" of the system. It is also called "the waterfall" model of system development, implying that the results from one development stage "falls" into the next stage and feeds the corresponding development activity, the results of which fall into the next stage, and so on.

Viewing system development in such a way implies that there is a clear distinction between *what* to do and *how* to do it. In other words, it must be possible to separate the problem and its solution to such an extent that a complete problem formulation can be worked out prior to the solution of the problem. We shall see in the next section that this cannot always be done, and applies only to restricted classes of problems.

9.2 Properties of Unsurveyable Systems

The development of small, surveyable systems can usually be handled satisfactorily by groups of a few competent persons. The difficulties arise when the systems become so large that it is impossible for human beings to have enough expertise in a sufficient number of relevant component fields. A necessary prerequisite for success in the design of large systems is thus the enforcement of a balanced cooperation among groups of people with different expert competences.

A basic problem of system development is to find good ways of subdividing work among different groups of specialists. An important approach to solve this problem is to distinguish the systems problems, component problems, and the problems of interaction among components. Then a system designer does not need to be an expert in everything, but can concentrate on the genuine system concepts on his level of work. Component problems can be delegated to the proper specialists. The system designer must of course have sufficient knowledge of the "external" properties of the different system components in order to handle the interaction problems satisfactorily.

Børje Langefors formulated in the early 1960s a general approach to systems analysis. His approach is based on experience of engineering large technical systems like electrical networks and elastic structures. This chapter is based on Langefors' ideas, as described in his book *Theoretical Analysis of Information Systems* [LANG66].

9.2.1 Problems of the Whole and Problems of Components

The behavior of systems that have no human components is bounded by the laws of nature. If one knew the properties of a system's components and their interactions, one should in principle be able to infer the overall properties of the system through deduction and calculation according to the laws of nature. This can indeed be done for many technical systems.

Electrical networks and elastic structures are examples of such systems. They are big and complicated. The derivation of their properties has to rely on numeric computation of quantitatively specified properties. Let us use a simple example of analysis of electrical networks to highlight the notions of systems, components and interaction among components.

From college physics, we know that an electrical network can be analyzed by introducing branch currents and mesh currents (Fig. 9.3). Associated with each branch in the network is a branch current i_j, and with each mesh in the network a mesh current I_k is associated. The branch currents and the mesh currents are related to each other by equations. From Fig. 9.3 we read that $i_1 = I_1 - I_2$, and $i_2 = -I_2$.

The mesh currents **I** can be expressed as combinations of the branch currents **i** as

$$\mathbf{I} = \mathbf{C}\,\mathbf{i}$$

where **C** is the connection matrix of the network (Fig. 9.4). The first two rows in the **C**-matrix read $I_1 = i_1 - i_2$ and $I_2 = -i_2$, which is exactly the result of our above analysis, of the first two branches of the network.

We see that it is the interconnection among the system's components that define the mathematical relations among the mesh currents ("the system entities") and the branch currents ("the component entities").

Furthermore, it can be shown that if we know the impedance z_{ii} of each of the branches of the network, then we can calculate the system impedance matrix **Z** associated with the meshes by the formula

$$\mathbf{Z} = \mathbf{C}(z_{ii})\mathbf{C}^T$$

where $\mathbf{C}^T$ is the transposed connection matrix, and (z_{ii}) is a diagonal matrix of branch impedances.

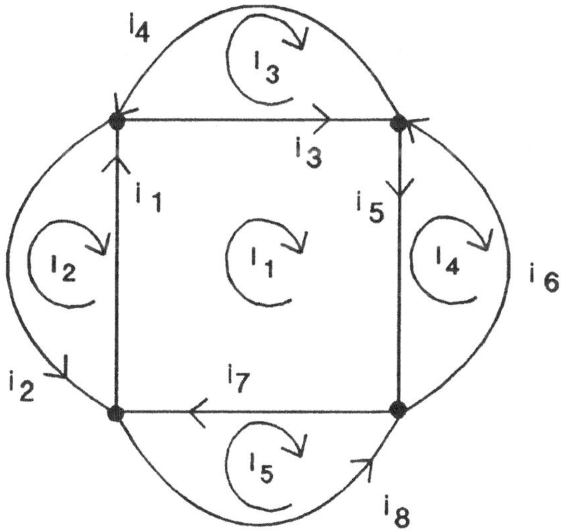

Fig. 9.3. An electrical network with branch currents $i_1, i_2, \ldots, i_8$ and mesh currents $I_1, I_2, \ldots, I_5$

	i_1	i_2	i_3	i_4	i_5	i_6	i_7	i_8
I_1			1		1		1	
I_2	- 1	- 1						
I_3			- 1	- 1				
I_4					- 1	- 1		
I_5							- 1	- 1

Fig. 9.4. The connection matrix **C** which relates mesh currents **I** and branch currents **i** of the network in Fig. 9.3

The important system-theoretical insight gained from this example is that it is possible to calculate a system property (the system impedance matrix **Z**) if one knows how the system components are interrelated (the connection matrix **C**), and if one knows the relevant properties of each component (the branch impedances z_{ii}).

For the class of systems for which such relationships among system and components are valid, we have a noticeable guideline for partitioning the system development work. First, we decompose the system and establish the interconnections among the components. Then, each component is analyzed to find the relevant component properties. Finally, the system properties can be calculated in a third, separate phase by combining knowledge about components with knowledge about component connections.

Another enlightening example from electrical engineering is the method of designing so-called electrical equivalences. Any linear electrical circuit (Fig. 9.5) which consists only of capacitances, inductances, resistances, and voltage generators can be shown to have an equivalent electrical structure that externally displays the same electrical properties as the original circuit, provided there is only one input signal and one output signal.

There are several methods for designing electrical equivalents. One of the older methods is based on the so-called four-pole theory. The two most frequently used electrical equivalences in the four-pole method are called the T-equivalent and the π-equivalent, because of their structural forms (Fig. 9.6). These electrical equivalences are called four-poles, because they have four external connection points, two for input and two for output connections. The basic property of four-poles is that it is always possible to assign values to the three impedances Z_1, Z_2, and Z_3 of the electrical equivalent so that for any input signal the four-pole will produce an output signal which is exactly equal to the output signal that the original electrical circuit would produce. An important deficiency of the four-pole theory is that it can only form equivalents of passive electrical components (e.g., resistors, capacitances), but not active electrical components (e.g., electric motors).

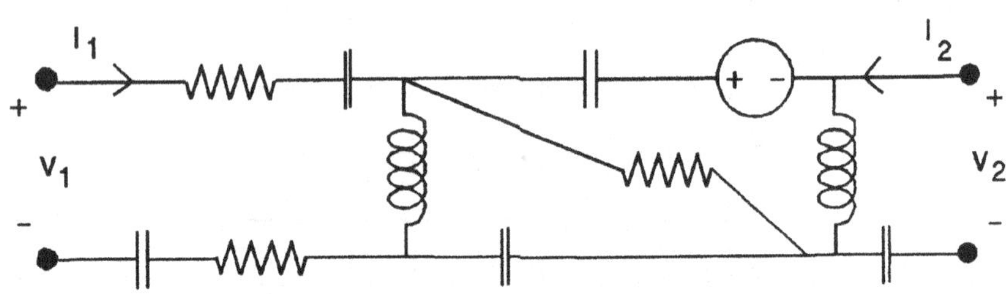

Fig. 9.5. An electrical circuit

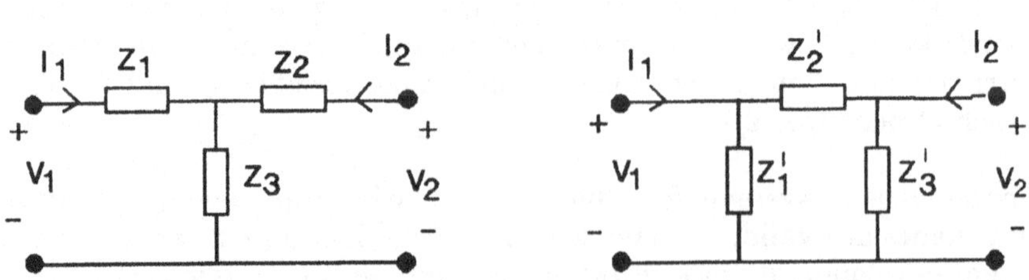

Fig. 9.6. Electrical T- and π-equivalents. Four-poles

The more general two-port theory superseded the four-pole theory in the 1950s. In the two-port theory the input signal, described by voltage V_1 and current I_1, is related to the output signal (V_2, I_2) by a set of four two-port parameters (Fig. 9.7). There are six possible choices of parameter sets, two of which are shown in Fig. 9.7. The basic property of two-port equivalents is that it is always possible to assign values to the two-port parameters of the chosen parameter set so that, for any input signal, the output signal calculated by the use of the parameters will be exactly the same as the output signal produced by the original circuit.

More complicated circuits can be considered as a system of connected two-ports (Fig. 9.8). We see that when we work on the two-port level, we in fact work with networks of interconnected two-port components. To each of the branches of the network is associated a four-parameter set which represents the relevant electrical properties of the branch. By combining the knowledge about how the two-ports are connected with knowledge about the local electrical properties of each two-port component, we can calculate the electrical properties of the whole system, by a formal procedure. We see that the class of systems covered by the two-port theory lends itself to systematic development methods, where component analysis can be separated from structural connection analysis, and the properties of the whole system can be formally derived from knowledge about connections and components.

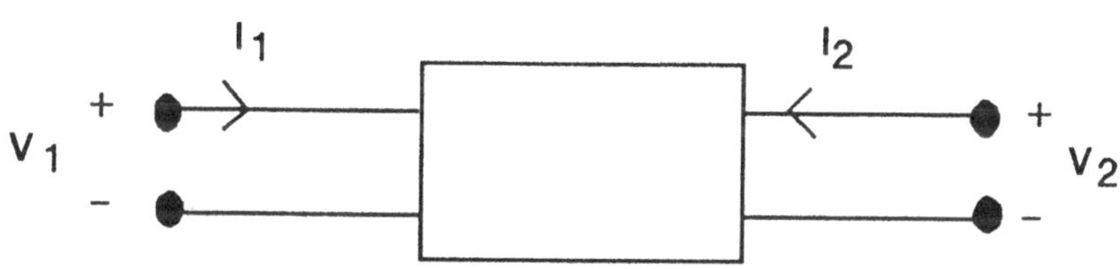

$$\begin{bmatrix} I_1 \\ I_2 \end{bmatrix} = \begin{bmatrix} Y_{11} & Y_{12} \\ Y_{21} & Y_{22} \end{bmatrix} \begin{bmatrix} V_1 \\ V_2 \end{bmatrix} \qquad \begin{bmatrix} V_1 \\ I_1 \end{bmatrix} = \begin{bmatrix} a_{11} & a_{12} \\ a_{21} & a_{22} \end{bmatrix} \begin{bmatrix} V_2 \\ -I_2 \end{bmatrix}$$

Parameter set no.1 Parameter set no.2

Fig. 9.7. Electrical two-port with alternative parameter sets

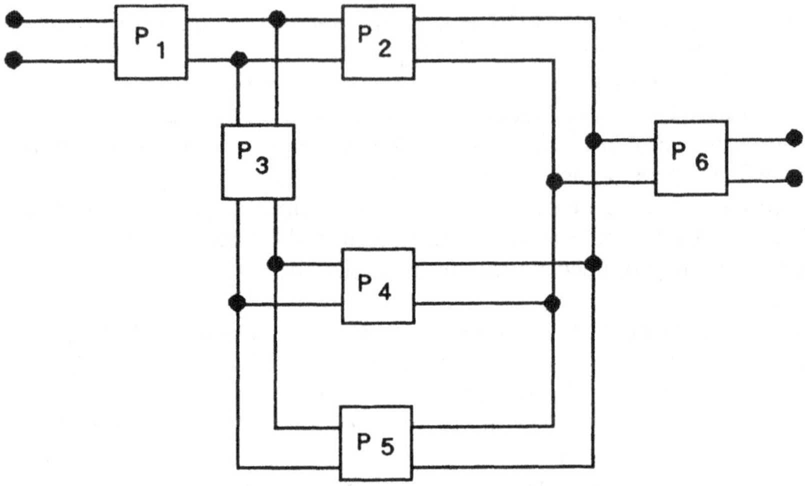

Fig. 9.8. Structure of two-ports

When faced with the task of analyzing new system types, we should always try to find a formulation of the system's properties that makes it possible to separate the component features and the connection features in ways similar to that explained above. This has proved to be difficult for systems which do not have to obey simple laws of nature. Examples of such systems are purely artificial systems like computer programs and systems that contain human components. Reactions and responses by human beings do not conform to easily expressible "laws of nature". Nonetheless, the insight gained from analyzing principles used for the design of physical systems, might be useful for approaching system problems of a more complicated and "lawless" kind. After all, the success of applying engineering design principles to very complicated system problems has been amazing. In constructing information systems we could also benefit from the successful experiences of applying the design principles of neighboring disciplines.

9.2.2 The System Concept

Webster's Dictionary for Everyday Use defines a "system" to be an

> "assemblage of objects arranged after some distinct method, usually logic or scientific; whole scheme of created things regarded as forming one complete whole; universe; organization; classification; set of doctrines or principles; the body as a functional unit".

So "system" is a general concept of order and relationships, as opposed to chaos and isolation. The concept of system can be applied everywhere, in every science and every craft, wherever human beings try to understand why things are as they are. So we have physical systems, social systems, biological systems, theological systems, dialectical systems, computer systems, management systems, technical/administrative systems, and so on and so forth.

Each of the sciences has usually developed its own "systems science", which is particularly well-suited for the problems at hand. So the concept of system is indeed one with several faces. We might, nevertheless, try to learn from the theories of physical systems. It might be practical to define the term as follows:

> *A system is a collection of entities, called systems components, which are correlated in some way.*

> *An entity is any physical thing or abstract notion that can be named and talked about.*

In physical systems it is common to have the correlation among the systems entities in the form of physical connections. In a mathematical model the correlations will appear as mathematical relations. In social systems the correlations might be of other types, for example that some person does not like another person.

It follows from the definition of an entity that a connection is also an entity, because we can name it and talk about it. So entities may be correlated through entities, according to the system definition above.

Langefors states:

> *Every system which is subject to influence from its environment is a subsystem of some larger system and every system component is potentially a system.* [LANG66].

The recursiveness in this statement follows from the previous system definition. The influence from a system's environment is equivalent to a correlation with other systems, so that both the system and its environment can be regarded as belonging to the same "supersystem".

We can now define the concept of structure [LANG66]:

> *A subsystem structure given to a system is a partition of the system into a set of subsystems together with a set of correlations among the subsystems.*

An example is shown in Fig. 9.9. The graph consisting of lines and nodes which are correlated is partitioned into one set of subsystems (the lines and the nodes), and one set of correlations. So the subsystem structure of the system has been specified.

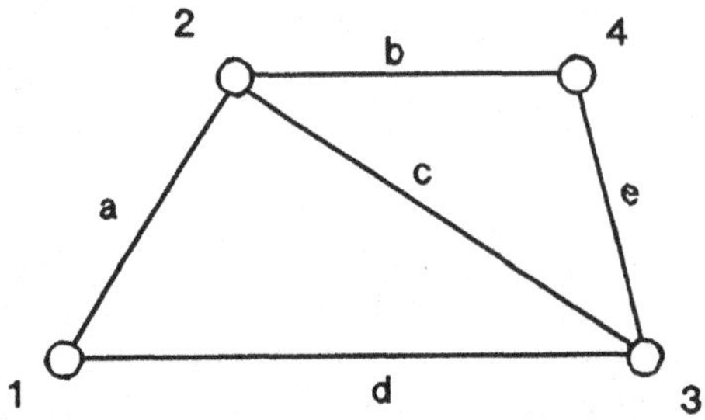

Entities:
Lines : a, b, c, d, e
nodes: 1, 2, 3, 4

Correlations:

a	with	1, 2
b		2, 4
c		2, 3
d		1, 3
e		3, 4

Fig. 9.9. A system consisting of lines and nodes

We shall further introduce the notions of *constructivity*, *applicability* and *implementability*, which are adaptions of definitions originally put forward by Langefors [LANG66]:

- *A constructive subsystem structure of a system is such that the properties of the system can be derived ("constructed") when the relevant properties of the system's components are known.*

- *An applicable subsystem structure of a system is such that the properties of the system components together with the correlations among the components result in the properties that have been specified for the system as a whole.*

- *An implementable subsystem structure of a system is such that every system component and correlation can be implemented.*

The electrical circuitry of the previous section is an example of constructive subsystem structures.

The last two definitions tell us the obvious: a design is applicable only if it satisfies the requirements that the system must satisfy. A design can be implemented only if all of its components and their connections can be implemented.

We can express this more imperatively:

> *The only possibility of designing a system to have specified properties is to design an applicable and implementable subsystem structure for that system.*

But we also find that:

Only constructive subsystem structures can be checked for applicability.

Therefore,

A subsystem structure can be said to be applicable if, and only if, it is constructive and its constructed properties satisfy the system specifications.

The feature of constructivity is of basic importance; this alone makes it possible to design systems so complex that no one can survey them in detail.

Most often it is not the case that the subsystem structures that we deal with are constructive with respect to every relevant property. In many decision situations non-technical factors such as the political environment must be considered. It seems impossible to have constructive subsystem structures which incorporate properties of the (business) political environment. If this was possible it would mean that we should be able to calculate the political acceptability of a system provided that we knew the political acceptability of each system component.

9.2.3 Dealing with Unsurveyable Systems

An unsurveyable system has such a large number of components and correlations that all of its structure cannot be safely observed and surveyed at the same time. Usually, we mean unsurveyable systems when we say "system". (Note that Langefors used the term imperceivable instead of unsurveyable in his original text.)

Systems of low complexity are usually surveyable systems, and can be developed by single persons, or a few persons working closely together, provided that the persons are sufficiently clever.

Systems of moderate complexity can be made surveyable, provided that the systems developers are clever enough and use enough time to understand all of the relevant systems features.

Systems of high complexity are inherently unsurveyable because of their large number of components and correlations. Clearly, such systems can only be developed if the work can be subdivided among different specialist groups who are not required to have total system knowledge.

Curiously, it is often systems of a little more than moderate complexity that are candidates for disasters. The reason for this is that the persons involved in the development are not aware that they are working with an unsurveyable system until it is too late.

Experience seems to indicate that:

> *People tend to neglect the importance or the existence of things they are not able to see or survey.*

Furthermore,

> *Every detail in a system, and in the work associated with describing the system, has a non-negative cost associated with it.*

Now, the definition of an unsurveyable system says that the numbers of its components and correlations are too large to be observed and/or surveyed at the same time. Therefore, it seems fairly logical to conclude that:

> *People tend to underestimate the complexity of an unsurveyable system, i.e., its number of components and correlations.*
>
> *Underestimation of systems cost and development costs is likely to occur when the system is unsurveyable.*

There is extensive empirical evidence to support these two conclusions. A rule of thumb seems to be that projects attempting to develop genuinely innovative results normally cost twice as much as the initial estimates. Experience also appears to indicate that most people tend to neglect the importance or the difficulty of other people's fields. It is clear that such underestimation might have very serious consequences for the successful development of unsurveyable systems. It can be combatted only by putting more effort into the early phases of system design.

There are some recommendations for how to develop unsurveyable systems that are generally accepted both in the literature and in practice. Examples of such recommendations are:

- System development should be done in stages starting with a rough survey stage (feasibility analysis) and then making each successive stage more detailed than the preceding one.
- System development should employ the principle of hierarchical decomposition, that is, base itself on "top-down" analysis, whereby each new subsystem is designed to satisfy some requirement on the level above.
- System development should strive to make use of available systems components, that is, employ a "bottom-up" technique by which available systems components are put together in structures that roughly satisfy the systems requirement.
- System development should be organized to support a "separation of concerns", such that separate stages are defined to take care of different

concerns, e.g., one organizational analysis stage, one man-machine interface stage, one database design stage, etc.

The recommendations are usually phrased in a rule-of-thumb manner. Even if each of the recommendations is supported by practical experience, they are not supported by convincing theoretical principles. Also, the detailed rules for guiding the development work usually lack operational precision. In some cases a particular recommendation is given very high priority at the expense of competing recommendations, possibly leading to disastrous results. A stereotypical example is when the recommendation to engage in hierarchical decomposition is given such a high priority that the system development team becomes occupied with generating systems specifications until they reach the algorithmic details of the system before they try to find already available solutions.

Another common error occurs when the recommendation to use available systems components is given priority at the expense of the development of satisfactory requirements specifications. The result might be that one develops a system for which there are no customers, because properties of the available components were permitted to unduly influence the system's properties. Therefore applying too much or too little of any of the various recommendations might be harmful to reaching the desired result.

9.2.4 Langefors' Fundamental Principle for System Development Work

The top-down decomposition approach and the bottom-up synthesis approach can be combined into one fundamental principle for system development work, as first formulated by Langefors [LANG66].

Langefors' fundamental principle for systems work is to partition the systems work into the four separate tasks:

a) *Definition of the system as a set of parts*
The task is to list the system's components.

b) *Definition of the system structure*
Define all relevant interconnections among the system's components.

c) *Definition of the system components*
For each component, define those of its properties which are relevant.

d) *Determination of the properties of the system*
Based on component properties and component interconnections, determine the system properties. Compare with the required system properties, and repeat the tasks a) – d) until satisfied.

This principle enables a natural division of the systems work into tasks, each task corresponding to a few components and/or interconnections. This makes it possible to let each individual task be treated separately by different independent groups of specialists.

We see that points a), b) and d) of Langefors' fundamental principle are concerned with the system as a whole, while point c) is concerned with each individual subsystem. This means that a central design team can operate efficiently without needing very extensive experience with each separate part and its function. Further, the specialized work on each system component can be performed efficiently even if the component specialists do not possess complete and detailed knowledge of the system as a whole. By adhering to Langefors' principle, one can naturally obtain a subdivision of labor and a specialization of skills.

The most efficient way in which Langefors' fundamental principle can be applied is when the work can be supported by mathematically formalized tools. This is often the case in engineering design. A prerequisite for this is that we deal with constructive subsystem structures.

If we can state the system's properties in a constructive form, then we can also support task d) in Langefors' principle by formal tools. That is, we can calculate the properties of a system if we know the properties and interconnections of the system's components.

9.2.5 A Guideline for System Development Work

We are led to formulating a guideline for system development that recommends "downward specification" and "upward construction". The guideline recommends that system development, on each level, is done by performing seven different work steps (Fig. 9.10).

First, develop an initial specification of the system's properties. Then, specify properties and interactions of a subsystem structure making the subsystem structure a constructive one if possible. Then, derive the system's properties from the subsystem structure and study the consequences of the proposed subsystem structure. Next, modify the proposal and iterate through the previous design steps when necessary. Finally, determine that all subsystems and their interconnections can be implemented. If unsatisfied at this stage, one has to return to the previous design step and modify the proposal. The detailed explanation of the guideline follows in the sequel.

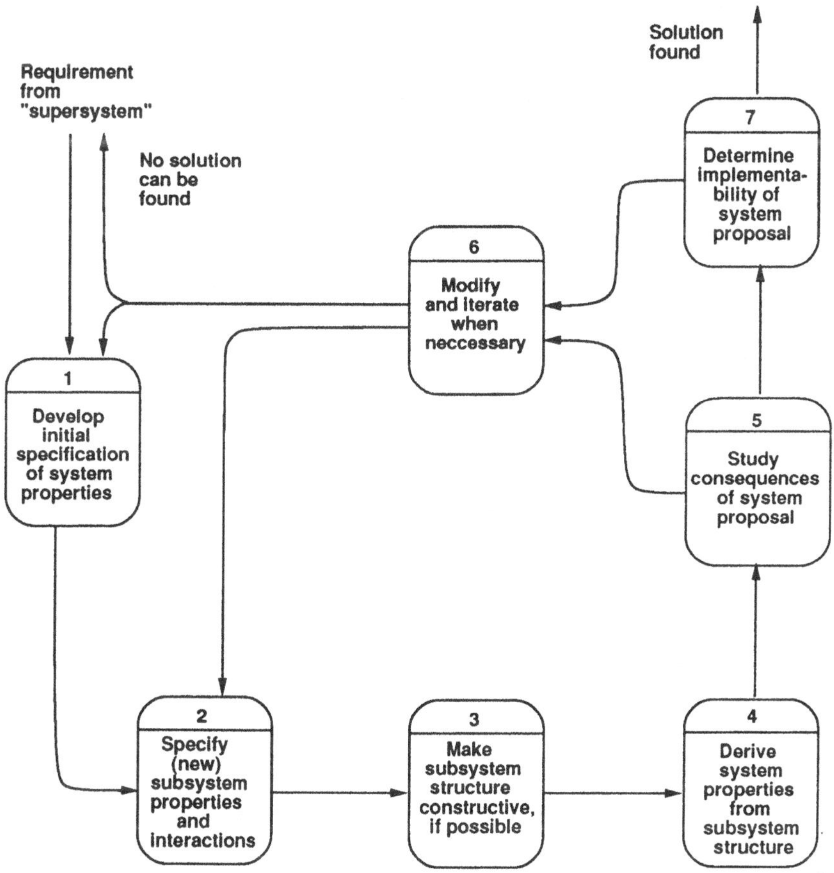

Fig. 9.10. Langefors' guideline for system design

Step 1: Develop an initial specification of the system's properties.
The result of this step is a statement of desired system properties, i.e., a *requirement specification*. The initial requirement specification usually reflects the needs of the "supersystem" of which our system is a component.

Step 2: Specify (new) subsystem properties and interactions.
The first three points of Langefors' fundamental principle are adhered to. The system's components (i.e., subsystems), their properties and their interactions are proposed. The results of this highly creative phase depends on the knowledge or the expertise on the subject matter, on the organization of the innovation process, and on the quality of the available design support tools.

Step 3: Make the subsystem structure a constructive one, if possible.
Because we know that only constructive subsystem structures can be formally checked for applicability, we concern ourselves in this step with achieving a constructive formulation of the problem at hand. This is a prerequisite for future use of formal structure analysis tools. But constructivity can only be achieved if a formal theory is available for the system type at hand. If not, we have to skip this step and resort to more intuitive methods for performing the remaining steps.

Step 4: Derive the system's properties from the subsystem structure.
The last point of Langefors' fundamental principle, the determination of system properties, is performed. This can be done formally if we have a constructive subsystem structure (e.g., an electrical network). Alternatively, we have to resort to intuitive methods. Quite often we have to rely on a combination of the two. In many cases some of the technical features of a system can be given a constructive formulation, while features concerning human interaction cannot be constructively formulated. Nonetheless, we are required to determine the system's properties, one way or another. We construct "upwards" based on system components' properties and interactions. The system's properties are derived either through formal or informal reasoning.

Step 5: Study consequences of the proposed subsystem structure.
This is the evaluation step. The objective is to compare the derived system properties with those stated in the requirement specification. If the derived properties satisfy the requirements, everything is fine and we can proceed to step 7 to determine whether the proposed structure is implementable. If it is implementable we declare our efforts a success and make ourselves available for working on other parts of the total system. If the derived properties do not satisfy the requirements, we proceed to step 6.

Step 6: Modify and iterate when necessary.
We have two alternative directions to follow when we want to modify the design proposal. Either we come up with a new proposal of a subsystem structure, or we attempt to change the requirements specifications. We will embark on the latter alternative only if we have exhausted every reasonable possibility for modifying the existing design proposal.

If we modify the previously proposed subsystem structure, we have to perform the steps 2–6 as many times as is needed to arrive at an acceptable result. If we choose to change the requirement specification, we have to determine the consequences of this change on the "supersystem" of which our system is a part. That means that we will get into a design loop similar to that in Fig. 9.10, applied to the supersystem.

As seen from the supersystem level we have determined that the subsystem on which we are working cannot be implemented (supersystem's design step 7). Therefore we have to embark on a new design cycle of the supersystem starting with step 2, and specifying some new subsystem properties and interconnections that may possibly render an implementable subsystem structure for the supersystem.

Step 7: Determine that all subsystems and their interconnections can be implemented.
Evaluations of implementability are performed several times during system design, not only at the end of the design cycle as indicated by appointing this task as the last step. When performing step 2, one would certainly not propose a subsystem structure which one did not believe it was possible to implement.

We meet three different situations when evaluating the implementability of a proposed design. For each system component and system connection we have one of the following situations:

a) There exists some available system that satisfies the desired properties of a component (or connection) in the system proposal.

b) There exists some available system that has properties sufficiently similar to the desired properties, so that it is deemed possible to reach a solution through modification of the available system.

c) There exists no available system that satisfies the desired properties of a component and/or connection, or is sufficiently similar to such a system.

In the first case, implementability is demonstrated, and we have completed our job. In the second and third cases we have to rely on performing a feasibility study, to establish a measure of confidence in the implementability of the component/connection in question.

This seven-step procedure which constitutes the guideline brings together the top-down and bottom-up styles of specification into one coherent framework

There is a "downward" specification and an "upward" construction. We have to iterate, modifying the design specification each time, until the derived properties satisfy the required system properties.

Following the guideline, we are led to a hierarchical system of design processes. In each of these design processes, every subsystem that has been specified in a previous process is designed by specifying an applicable subsystem structure for it. A system that is designed in this way will necessarily have a hierarchical structure.

9.2.6 The Feasibility Study

The purpose of a feasibility study is to determine whether it is possible to design a system with a prespecified set of properties within given time and resource limits. The task of the feasibility study is to prove/disprove the feasibility of a proposed system, at some level of confidence. Because we have limited time and resources for a feasibility study, we must accept that we cannot understand every detailed consequence of the proposed system. We have to reach our conclusions based on incomplete knowledge about the system to be constructed. The major problem in planning the feasibility study is to decide how far to go in the analysis given a rough, subjective estimate of the acceptable risks for drawing wrong conclusions.

Assuming that we employ the Langefors' seven-step guideline for systems design this can be easily answered: We may stop when all of the subsystems that we have reached in the decomposition process are sufficiently similar to existing systems, so that the surveyed probability of their implementability matches the accepted risk for reaching a wrong conclusion.

The chances of finding suitable available subsystems depends on how many similar systems have already been designed. But it also depends on whether scientific research has already established a set of facts and theorems that correspond to a large class of relevant subsystem designs. If one can determine that one's requirements can be satisfied by systems for which there is a known theory, this is equivalent to proving the existence of a solution for the problem at hand. In this way mathematical research has often produced results useful for applications without intending to.

There is of course a temptation involved in this. If a slight change of the requirements leads to a design proposal that can be proven implementable because some critical system part can be proven to conform to established mathematical theory, it is very tempting to make such changes in the systems requirements. By proposing a solution that is not completely satisfying but can be proven implementable, instead of proposing a fully satisfying solution that may not be implementable, one decreases the risk of system development failure. Therefore it is expected that systems proposals that can be given a constructive formulation are preferred because they contribute to minimizing risks in developing a new system or making a major change to an existing system.

One often hears the complaint that "hard technical" system properties are given preference to "soft human" properties when choosing priorities. Such behavior among system developers can be easily explained by referring to the arguments just mentioned. It is natural for people in an uncertain situation to try to minimize the risks. Those properties that are felt to be controllable will naturally be given more emphasis than other properties. Therefore the greater emphasis on "hard" versus "soft" properties is demonstrated.

9.3 Development of Non-constructive Systems

The traditional systems-development approach, of which the engineering design approach is an illustrious example, has achieved a large number of spectacular successes. In spite of this it is also evident that the traditional approach is not generally applicable so as to guarantee successful results for every type of system development problem. System development failures are often encountered, characterized by late delivery, high development cost, lower performance than expected, and so on.

Rittel has proposed that one major reason for system development failures is certain deficiencies in the premises for the traditional systems-development approach. He introduces the notions of "tame problems" and "wicked problems" [RITT72].

Examples of tame problems are mathematical problems and physics problems: that is, problems whose formulation rests on axioms and on laws of nature. Examples of wicked problems are planning problems, where the plans have to reasonably satisfy a number of conflicting goals.

Almost every system-development project will have a wicked nature in its early phases. As the development work goes on both goals and scope of the system will be refined and better understood. In the system-development work we should aim at taming as much of the system as we possibly can. For a tame system we can use the traditional systems approach with great confidence. The trick lies in the taming.

The notions of tame and wicked problems relate to Langefors' notions of constructive and non-constructive subsystem structures. A tame problem is a problem which can be formulated by using constructive subsystem structures. A wicked problem is not associated with a constructive subsystem structure.

Following Rittel's presentation we first describe the contrasting properties of tame and wicked problems, and then discuss how system-development techniques should be changed to better cope with the design of wicked systems.

9.3.1 Properties of Wicked Problems and Tame Problems Contrasted

Most research on creativity and problem-solving is about "tame" problems because they are so easy to manipulate and control. Yet all essentially systems-analysis problems are "wicked" because they involve the discovery and sorting out of conflicting goals. Unfortunately, little is known about the treatment of wicked problems or of the people dealing with them, because wicked problems cannot be simulated in a laboratory setting.

In the following we will list some of the most important properties which distinguish wicked problems from tame problems, and discuss each property briefly.

Wicked problems have no definitive formulation

A tame problem can be exhaustively formulated so that it can be written down on a piece of paper and handed to a knowledgeable person who will eventually solve the problem without needing any additional information. Examples are the solution of mathematical problems. Langefors' guideline for system-development work shows this very clearly. For a constructive subsystem structure we can formally derive the system's properties and compare them to the corresponding requirements on the system. In case of a discrepancy, we change the subsystem structure and iterate until the system's requirements are satisfied.

For a wicked problem it is not possible to give a complete statement of a system's requirements prior to the solution of the system problem. Assume that the problem is to develop a system for lending books from a library. It will not be long before the person who gets such a task will come back and ask for more information. How many lending stations should there be, where should they be located, how should the borrowers be identified, how should one keep track of unreturned books, and so on. If the task is to develop an administrative system for an organization we find a similar situation. The system developers will soon come back and ask how decisions should be made and what information should be used, and so on.

One might assume that answers to these questions should have been prepared ahead of time and given to the system developers as they got their initial task. However, the irritating thing is that one cannot know what the next question is because the next question is unique and dependent on the state of solution that the system developers have already reached. Assume that one decides to have fifty lending stations in the library system. A consequence of this is that one has to solve the problem of transporting documents from storage to the correct lending station. This problem would never arise if one chose to have only one lending station.

This discussion leads us to the second property of wicked systems.

Every formulation of a wicked problem corresponds to a statement of its solution and vice versa

The previous argument shows that it is not possible to understand a problem without solving it first, and that solving a problem is the same as understanding it. For a tame problem which has a corresponding constructive subsystem structure, the concept of abstraction is well-defined. System properties can be abstracted and formally derived from properties of the system's components. The derived properties can be compared with the stated properties, which constitute the problem formulation. We see that for tame problems, the problem formulation is one thing, and the problem solution is another.

For wicked problems there is no "formula" that relates system properties to component properties. The system properties can only be exactly described by explicitly making a list of the component properties. So the problem solution is the problem formulation, and the problem formulation is the problem solution.

Wicked problems can always be solved better

The solution of a tame problem is either correct or false. The solution can be tested and be proven to be right or wrong as with mathematical problems. Since solutions can be tested, it is possible to know when a solution has been found so one can stop working out further solution proposals.

This is not so for wicked problems. It is not possible to determine if a plan for a city is correct or false. We can only say that it is good or bad. And what is good for one person may be bad for someone else. There is no criterion-system or rule which can determine what is correct or false. Consequently, a solution to a wicked problem might always be improved. There is no natural stopping point to a wicked problem as there is for a tame problem. The solution process of a wicked problem stops when there are no more resources available to continue the solution. However, a lack of resources has nothing to do with the nature of the problem; one can always try to do better.

The difficulty of testing a solution proposal has also to do with the fact that there is no limit to what can be done to solve a wicked problem. For tame problems, there is an exhaustive list of permissible problem-solving operations. For the solution of mathematical problems there are rules that say which mathematical operators can be used. It is not permitted to invent new operators unless they are proven to be consistent with mathematical theory. This is different with wicked problems. Everything is possible and solutions are influenced by principle and imagination, hopefully constrained by law and ethics. An obvious consequence is that for a wicked problem there is neither an immediate nor an ultimate test with respect to the correctness of its solution.

Every wicked problem can be considered a symptom of another problem

Every tame problem has a natural form. If the problem is to make a computer program which can sort up to one million numbers, there is no reason to argue about whether it is the right problem to solve. It is a tame problem and it has a natural form. However, if the problem is that there are frequent out-of-stock situations in a company's raw material inventory, or frequent production delays, this may be surveyed as a symptom of another problem. The real cause of the problem might be the company's personnel policy, ineffective inventory models, or poor organization of the buying department.

A problem can be stated as a discrepancy when something is compared with how it ought to be. Finding the solution to the problem is the same as finding the cause of the discrepancy. The trouble is that in wicked problems there are many explanations for the same discrepancy and there is no way to test which explanation is the best. The reason for manufacturing delays in a factory might be that the manufacturing equipment is worn out, but it might also be that the workers are incompetent. The solution will follow a different path depending on the explanation one chooses. Choosing the explanation is the most decisive step in dealing with a wicked problem. Therefore one should not draw hasty conclusions when solving wicked problems.

Every wicked problem is essentially unique

Tame problems can be solved over and over again. If one has written a computer program for sorting one million numbers, one can write another program to do the same task. If one has solved this problem once, then one has solved the problem of sorting any combination of one million numbers. The trick of solving the sorting problem for one set of one million numbers is the trick of solving the sorting problem for any set of one million numbers. Tame problems can be partitioned into classes, and there are prototypical solutions for all classes of tame problems.

The solution of wicked problems can not be repeated because the process of solving changes the formulation of the problem. One cannot set up a factory, see how it works, demolish it and rebuild it over again until one is finally satisfied with its operation. One cannot undo what one has done in the first trial. Each trial is important. Each trial has consequences in terms of resources consumed, and credibility lost (or gained) by the system designers. Therefore, there is no trial and error. There is no experimentation in dealing with wicked problems. The wicked-problem solver has no room for error. He is solely responsible for what he is doing.

It is very irritating that every trial for finding a solution changes the problem. This means that one cannot easily carry over successful strategies from the past into the future since one never knows if the next problem contains characteristics that are sufficiently similar to the previous problems. In the treatment of wicked problems one should never decide too early what the nature of the solution should be and whether an old solution can be used again in a new context.

9.3.2 Principles for the Solution of Wicked Problems

There are some characteristic features of the process of solving wicked problems [RITT72].

For wicked problems there are no specialists

The expertise needed in dealing with a wicked problem is usually distributed among many people. Those people who are the experts (with the best knowledge) are usually those who are likely to be affected by the solution. However, expertise and ignorance are distributed among all participants in a wicked problem. Nobody is better qualified than another by virtue of degrees or status. If there are experts, they are only experts in guiding the process, but not with the subject matter of the problem.

Nobody wants to be planned at

The consequence of this is to try to make the people affected by the solution participants in the solution process. Recommendations for "participative design" of computer-based information systems have been frequently seen over the years [MUMF78]. Instead of introducing changes from the top down into the organization, the principle is to try to have those influenced by the changes maximally involved in the design of the new system.

The solution process must be "transparent"

When solving a wicked problem, a number of judgements are made for guiding the solution process. The judgements are most often based on opinions about how the solution *ought to be*. The chosen solution is influenced by the political, moral, and ethical attitude of the problem solvers. When looking at the solution, one cannot reconstruct its premises. One cannot construct the "ought-to-be" statements that entered into the argument leading to the solution. This is one more reason to involve many persons to participate in the solution process. Chances are then improved for having the premises for judgements and decisions out in the open, so that they can be debated and scrutinized. There is a need for methods that are "transparent" in the meaning that they help us in formulating problems in such a way that they can be both understood and communicated. This should help us in understanding why one person judges a solution differently from another person.

The problem-solver of wicked problems is a "midwife" of problems rather than an offerer of therapies

He helps to bring about a definition of problems rather than offer solutions. He is a teacher more than a doctor. It is a modest rather than heroic role such a system designer plays.

The problem solver makes careful, seasoned respectlessness a virtue

To find good solutions, it is often necessary to look at problems from new points of view and question the old ways of doing things. Moderate activism and optimism are important parts of the system designer's attitude.

It is important to find accomplices

Every treatment of a wicked problem is a venture into the unknown. It is impossible to evaluate every conceivable consequence of a solution-proposal. Therefore every trial to solve a wicked problem has uncertain outcomes. It may be too risky for one person, but if many join forces to share the blame and share the victory, then the risks may be tolerable and one would dare to embark upon the venture.

The solution of a wicked problem is a political, argumentative process

It is political because judgements and decisions are based on opinions about how the situations *ought to be*. It is argumentative because the decisions are always results of arguments built for and against the different positions. Every question to be decided must be combined with arguments. We always do this. Our solution methods should make the argumentation more explicit so that the problem solving process can be better understood and thereby better controlled.

These are the main principles of the process of solving wicked problems.

Exercises

1. What are the similarities and difference between the engineering design process and the information systems design process?

2. Define "surveyable" and "non-surveyable" systems. Give examples.

3. Give some examples of constructive and non-constructive systems.

4. Discuss the importance of the modification phase in the engineering design process.

5. Explain what is meant by a *feasibility study of a system*. What content should a feasibility report have?

6. Give some examples of wicked problems and tame problems.

Chapter 10

Information and Information Systems

The purpose of an information system is to collect, store, process and distribute information. The concept of information is itself not well understood. It is in some sense a relative concept, rather than an absolute one. That is, a data object is assumed to contain information if the receiver of the data interprets it in such a way that the received data adds to the knowledge of the receiver. Even if the concepts of information, data, and knowledge are interrelated in some sense, the relationships are not fully understood. There is also, so far, no complete theory of information.

As computers have become less expensive and more powerful, they have been increasingly used as tools for storing, processing and transmitting information. Computers deal with data. We therefore prefer to talk about data processing systems, when we discuss computer behavior. But humans deal in information and knowledge. Information systems comprise both humans and computers, and their interplay. So it becomes necessary to deal with information and knowledge, as well as data.

There is a difference between the concept of data and the concept of information. In normal everyday language, the term *data* denotes symbols which are represented in machine-readable form. There is a distinction between analogue representation and digital representation of data. When music is recorded on a conventional gramophone record, we use analogue representation when recording and playing the music. When music is recorded on a compact disc, the music is represented by digital codes, and recording and playing are performed by coding/decoding between the digital representation and the analogue representation, which is fed into the loudspeaker or received from the microphone. The term data is normally used in a restricted sense, to denote digital representation of symbols.

The term information is used in a much wider context than the term data. The term information implies that data is interpreted by human beings. In everyday language we talk about some data containing much or little information. That is, we imply some process of understanding the data by the person who receives the data. Since there is no generally accepted definition of the concept of information, in our context, we shall use information to denote data together with the human interpretation of that data. By our use of the word information, we approach what is known in the theory of languages as pragmatics: the relation between symbols and the effect that the symbols have on those who read them.

We make a similar distinction between the concepts of data system and information system. A system is generally understood to be a collection of partially connected parts, where each part and each connection is a system in its own right. The parts are often called subsystems. We usually view a data system as consisting of interrelated subsystems (parts) for collecting, storing, processing and distributing data. It is implicitly assumed that the data is digital data, and that the data system's operations are supported by computers. Another implicit assumption is that data systems are deterministic systems. We expect the data system to follow some formal rule for the manipulation of data, in every conceivable system state. It is implicitly assumed that the rules can be expressed in some programming language, and thus be embedded in computer software.

An information system is similarly viewed as consisting of interrelated subsystems for collecting, storing, processing and distributing information. The operation of an information system is expected to be supported both by machines and by humans. Information systems are non-deterministic systems where human decision making plays a central role. Aspects like quality of data with respect to quality of decision making are much more important in information systems design than in data systems design. Information systems are often viewed as having two subsystems, one informal human subsystem and one formal computer subsystem. Data systems are generally conceived as being support systems for information systems. Therefore, the design objectives of a data system must be determined in the context of the information system that will be supported by the data system.

10.1 Relationships Between Knowledge and Information

The confusion about the relationship between the concepts of information, data and knowledge is evidenced by a large number of proposals for a definition of the concept of information.

Here is a sample:

- Information is knowledge, intelligence, news (Webster's Dictionary)
- Information is the communication or reception of knowledge or intelligence (Webster's Dictionary)
- Information is knowledge in the form of facts (Longman Dictionary)
- Information is a process which leads to changing somebody's knowledge [RITT72]
- Information is any kind of knowledge or message that can be used to improve, or make possible a decision [LANG66]
- Information is the meaning that can be expressed by, or extracted from data by representation conventions [IFIP66]
- A datum is a representation of one quantity of information in digital form [LANG66]
- ...unorganized facts, or raw data, are transformed by some process into an arranged, ordered, and usable form known as information [FRAT80]
- "936392 is data, 93-63-92 is information" (Unknown amateur philosopher)

All of these definitions are highly informal. Most of them refer to the concept of knowledge. The definitions reflect normal, everyday, imprecise usage of the words information and data. In the literature one can find other definitions suffering from similar lack of precision.

One may, for example, say that information is data plus context, and knowledge is information plus an interpretation, without getting much closer to a formal definition of the concepts. Nevertheless, these "definitions" provide us with an intuitive feeling for the difference among these three basic concepts.

For our purpose, information is, in a general sense, some piece of knowledge that is represented by data, and which can be used to change somebody else's knowledge. This explanation is about as good as everybody else's. To improve our intuitive understanding of the concept of information, we shall discuss some aspects of knowledge and learning, where learning is the process of changing one's knowledge.

10.1.1 Types of Knowledge

There are a number of ways of classifying knowledge into different types. From the computer programming point of view it is quite tempting to classify knowledge into *declarative knowledge* and *procedural knowledge*. This classification reflects the distinction between a database and a computer program. The "declarations" of classes of objects and the relations among them are similar to the specification of the structure of data in the database and the associated constraints that are imposed on the data. The knowledge about how objects change their properties and form new relationships as time passes are usually embedded in computer programs and are therefore regarded as procedural knowledge.

From psychology we take the classification into *explicit knowledge* and *implicit knowledge*. Explicit knowledge can always be verbalized. It can always be expressed and explained to others. Implicit knowledge is often related to skills and cannot be easily explained to others. Everybody who knows how to ride a bicycle has the implicit knowledge of riding a bicycle.

Surface knowledge and *deep knowledge* distinguish the realms of empirical knowledge and knowledge about principles. Examples of surface knowledge are rules of thumb which relate characteristics of the problem at hand, with typical solutions for that problem. Deep knowledge consists of principles, axioms, and laws. Deep knowledge is concerned with abstraction, analogy and reasoning about causality. The possession of deep knowledge enables us to reason from first principles in new, unfamiliar situations. In our routine day-to-day work, we have to rely on surface knowledge. To reason from case to case, based on deep knowledge, would lead to an intolerable lack of efficiency. It would be like having to reinvent the wheel every time one planned to ride a bicycle.

One also sometimes distinguishes between *knowledge* and *meta-knowledge*, where meta-knowledge is knowledge about knowledge: how it is structured, how it should be interpreted, and so on. Meta-knowledge is often called conceptual knowledge.

Rittel classifies knowledge into five different types: *factual*, *deontic*, *explanatory*, *instrumental*, and *conceptual knowledge* [RITT72].

Factual knowledge is recognized by sentences of the form: x is, was, or will be the case. Whether, in a concrete case, the sentence is true or false, precise or fuzzy is another question.

Deontic knowledge has the standard form: x should be or should become the case. It reflects our convictions about what is or should be the case. It expresses our goals in a particular situation.

Explanatory knowledge has the general form: x is the case, because y ... This type of knowledge tells us why things are as they are or as they should be. We always use this type of knowledge when we search for solutions to our problems. In the very instant that we explain why things are not as they should be, we have laid down the direction in which to look for the solution. But explanatory knowledge is not enough. We also need some knowledge to provide us with the instruments to solve the problem once we have diagnosed it, and determined its cause.

Instrumental knowledge has the form: when x is done, y is the consequence. This type of knowledge provides us with the ways in which we can change something. When coffee powder is mixed with sufficiently hot water, then coffee is the result.

Conceptual knowledge is knowledge about the meaning of words and other vehicles of communication which we have to use to make ourselves understood. Conceptual knowledge and meta-knowledge are the same. As everyone knows from experience, most discussions arise from the question "what do you really mean?", the outcome of which is an attempt to provide "conceptual knowledge". The key to having meaningful communication between human beings is that the parties share the relevant conceptual knowledge. The alternatives are misunderstanding and the breakdown of communication. Similarly, in data communications, the sender and receiver have to share a set of rules that prescribe how the sequences of bits which are transmitted shall be interpreted. These rules constitute what is known as a communication protocol.

10.1.2 Knowledge, Information, and Information Processes

An information process leads to changing somebody's knowledge. We say that an information process has taken place when, for instance, an individual who knows something at time t_1 – that all bodies are heavy and fall to the center of the earth – knows something else at time $t_2 > t_1$.

The concept of information reflects a relationship between a knowledge provider and a knowledge receiver. A piece of knowledge contains information only with respect to the receiver of that knowledge. So the information content of a piece of knowledge may differ vastly depending on who evaluates the information content.

It is the receiver of knowledge who labels it "information" or not, depending on his previous knowledge. We shall use the word information to denote such knowledge that has the potential to change somebody's knowledge. We intend to concentrate on computer based information processing. We shall, therefore, talk about information in a restricted sense usually meaning explicit, surface knowledge, in terms of the classifications put forward in the previous section.

In the context of factual knowledge, the notion of "information content" can be easily explained. Assume that initially one knows that a person *X* exists. The statement "*X* is more than 60 years old" contains less information than the statement "*X* is exactly 76 years old", provided that the receiver of the statements has no previous knowledge about *X*'s age. This is so, because the last statement removes more uncertainty about *X*'s age than does the first statement.

Whenever we deal with factual, precise knowledge we can usually find some measure of information content with respect to the receiver's knowledge base. But real situations are much more complex. Only a small part of the knowledge that is relevant in day-to-day human discourse is based on unambiguous observations. The knowledge we communicate usually relates some degree of belief. Some things we are more certain about than others. Some knowledge we can express more precisely than we can express other knowledge. Sometimes we are forced to be imprecise, because we lack the words to express what we want. To measure the information content of imprecise knowledge with respect to an ambiguous knowledge base seems to be beyond the current state of the art.

10.1.3 Some Important Properties of Information

In normal everyday discourse, we use natural language to convey information. The vocabulary of any language community reflects the needs for particular concepts in that community. For example, in languages of the Arctic region, one may find almost 30 different words for describing different combinations of snow and wind. This reflects the needs of those communities to communicate precise descriptions of potentially hazardous weather conditions.

When we work with new problem areas, trying to design new information system support, we typically have to manage with relatively few words. The main reason for this is that in the early stages of analysis we do not understand the problem area well enough to define concepts precisely. So we are not always able to express knowledge, and information, in precise terms. From this perspective, information may be classified as being *precise*, *vague*, *uncertain*, or *incomplete*.

Precise information is the most common category in computer based information processing. Precision and accuracy of information have usually been regarded favorably. Precise information can, however, only convey a small part of the knowledge that we encounter daily. Precise information is typically characterized by "black or white" thinking. Anything which is not black must be white. First-order mathematical logic mimics such thinking.

In first-order logic, a statement which is not true must be false. Boolean algebra, which is the mathematical foundation of electronic computers, is based on such thinking as well. As the computerization of society expands, we must be able to represent information that is no longer precise. For instance, fatness cannot be precisely defined by weight alone. One cannot say that a person who weighs 80 kg is fat, whereas a person who weighs 79 kg is not fat. Thus, the precision of computers is a merit as well as a limitation.

Vague, fuzzy information is common in our daily life. When we say "the weather is fine" and "the city is big", we give qualitative judgements on weather and city properties. We do not intend to give precise measures for the beauty of the weather, or for the size of the city. We don't even say if we mean that the city is big in terms of population or in terms of area, or in terms of both.

Natural language is well suited for supporting vague, fuzzy statements. Our needs in everyday discourse are to be able to express qualitative statements. Our languages have developed to fulfill these needs. Most often, people do not have the time and interest to go into the finer details of a situation. Who cares if the city has a population of 2,433,775 or 2,433,776? In most situations one would be perfectly satisfied by knowing that the city is big.

Uncertain information is vague. However, it is vague mainly because the information supplier is not able to provide precise information, though precise information can be provided. A typical statement is "She is very rich, but I am not sure how rich she is." Somebody else may give a precise measure of her wealthiness, but not the one who expresses the statement above. A special kind of uncertain information is called indefinite information. For instance, the statement "John has moved to Los Angeles, or San Francisco," indicates that the speaker does not know to which city John has moved.

Incomplete information reflects the fact that the observer does not have complete knowledge. The difference between incomplete information and uncertain information is that incomplete information may be certain even if it is not complete. For instance, someone may know that Oslo is the capital of Norway but not know that Oslo is located in the southern part of the country.

Information processing systems have, so far, concentrated mostly on storing and processing precise information; although, in knowledge-based systems such as expert systems, the processing of uncertain, vague, and incomplete information is addressed. Generally, any piece of information will have some of these attributes:

- *Relevance* is the most important attribute of any piece of information. The evaluation of other attributes of information depends on the relevancy of the information in question, with respect to the goal of the organization or the individual who uses the information.
- *Correctness* of information refers to the degree to which the information reflects what is actually happening.
- *Accuracy* refers to the degree of vagueness in the information. Not every piece of information has accuracy as a property. For instance, the statement "Cats are animals" is correct but it is meaningless to evaluate the accuracy of the sentence.
- *Content* concerns the amount of knowledge which the information conveys.
- *Originality* measures how new or how different the information is with respect to existing knowledge.
- *Obsoleteness* refers to the usefulness of the information with respect to the time of observation. It might be that the system which is observed changes so fast that information loses its value over time.

10.2 Ways of Obtaining Knowledge

Knowledge is obtained by various methods. These include recording, and learning through acquisition, tutoring, or teaching.

Recording is an effective method for obtaining factual knowledge. Recording is either done by observation and measurement, or by transforming a set of facts from one form into another form. An example of the latter is transformation of a recording of speed versus time, to a measure of acceleration using the definition of acceleration as the rate of change of speed. Contemporary computing systems mostly handle recorded, factual knowledge.

The other types of knowledge such as deontic knowledge and instrumental knowledge are usually embedded in the computer's software in contemporary software systems. Deontic knowledge is used to formulate the preferred state of a system. Discrepancies between deontic knowledge and factual knowledge are used for directing the system toward the preferred state. Deontic, explanatory, instrumental, and conceptual knowledge must be obtained through learning. There are various types of learning, ranging from the most primitive brain-washing to acquisition of knowledge through sophisticated experimentation or through deep personal reflection.

Conditioning is a quite primitive form of learning. People are shown something until they can do it. *Conditional* conditioning amplifies the conditioning process by passing rewards and punishment according to how well the student responds to the teacher's wishes. *Persuasion* means that something is explained or stated until it is believed. This is the normal advertising approach. It is also the normal approach for spreading ideology, be it political or religious. These steps of learning are characterized by the learning process taking place gradually, eliciting responses that are increasingly in accordance with those expected by the teacher.

Planned conviction means that on the basis of an argument there is suddenly an "aha!" effect when something is understood differently [RITT72]. The so-called Socratic principle of teaching says that the teacher should ask the students such questions in such a manner that leads them to discover the answers themselves. This principle leads us to the notion of learning by planned conviction.

The difference between conditioning and planned conviction is similar to the difference between instruction and education. Instruction is mostly concerned with passing instrumental knowledge of the form: when x is done, y is the consequence. What is mostly taught are techniques and methods that have proven successful in the past. It is generally not required that either the teacher or the student understand why the techniques and methods have been successful. In education one strives to understand *why* some methods are successful while other methods fail. One is concerned with developing explanatory knowledge of the form: x is the case because of y. In engineering methods development, one is concerned with developing instrumental knowledge based on sufficient explanatory knowledge so that one can educate in the style of planned conviction rather than instruct in the style of conditioning and persuasion.

There are various teaching techniques that can be used both in a conditioning situation and a planned conviction situation. The most important ones are learning through understanding of theories, learning by analogy, learning by examples, and learning by observation. Both conditioning and planned conviction can be used as basic principles for the teacher-student relationship.

In most of our adult life we have no teachers; we must acquire knowledge on our own. We have to actively engage in knowledge development and actively try to learn from others. Learning through planned conviction is also a very important principle in the adult situation. This is especially the case when developing deontic knowledge of the form: x should become the case. For any human endeavor to be successful, there must be an agreement on the objectives of the project; on where one should go and what goals should be pursued. Using the learning principle of planned conviction implies that there is a process leading to consensus through persuasion.

If there are no teachers available, we have to learn by actively trying to tap the knowledge of those who know

The process is called *knowledge acquisition*, which means collecting unstructured knowledge from experts or other sources and arranging the knowledge in such a manner that it can be used by the non-expert. Typical techniques for knowledge acquisition are *interview*, *protocol analysis*, and *rule induction*. The first two of these are concerned with acquiring knowledge from a human expert. The rule induction technique addresses the problem of formulating rules based on an analysis of previously collected knowledge. A primitive form of rule induction is fitting a smooth curve to a set of observations and pretending that we have formulated a rule which can explain the experiment.

Interview is the most widely used method for collecting knowledge from human beings. The knowledge collector, i.e., the student or the systems analyst, engages in discourse with the human expert aiming to elicit as much information as possible from the hidden expertise. The method is most effective if the expert knowledge is explicit rather than implicit. In practice we may find that the knowledge that we try to grasp and explicate is implicit knowledge. Therefore, the results of interviews are quite often not up to expectations. The interviewed expert is simply not able to express his knowledge in an understandable manner.

Protocol analysis aims at explaining what the expert is doing and thinking while performing a typical task. Based on recordings of task-performing sessions, the analyst and the expert work out a so-called protocol which contains the rules that the expert seems to follow. One also in this case tries to explicate implicit knowledge.

How reliable and certain is the knowledge that we collect and represent in computer-processable form?

The previous discussion shows that only a small part of the knowledge that we have to handle in computer based information systems is completely reliable and precise. Let us introduce a degree of certainty c_1 for an element of knowledge K_1. Some things we know with more certainty than others. Let us imagine that this can be made measurable in some way, so that not only do we have some knowledge K_1, we also have knowledge about the certainty c_1 of the knowledge K_1. We know that we can change the knowledge K_1 by receiving relevant information, but we must then also be able to change our degree of trust in K_1; that is, the certainty c_1 of K_1 based on the information that we receive [RITT72].

By receiving information that reconfirms what we already know, we increase our certainty about our knowledge. By receiving information that is inconsistent with what we already know, doubt is cast on the correctness of our knowledge, and subsequently the certainty about our knowledge decreases. A third case exists when something that was known changes radically into its opposite. The old knowledge is substituted by the new knowledge. For example: from high-school chemistry, we know that noble gases do not combine. Because everybody knows this, there is no reason to doubt it, or to test it further. Hypothetically, let us suppose that some disrespectful student unknowingly conducts the experiment that shows that noble gases indeed do combine. In rare cases, Nobel prizes are awarded for spectacular new insights that are gained from such coincidences.

In its most extreme form, large volumes of knowledge have to be scrapped completely because of mounting counterevidence. Whole systems of knowledge can fall apart without a substitute because new information is inconsistent with the old knowledge. This is called a "paradigmic revolution". There are many examples of paradigmic revolutions in the history of mankind. The scrapping of the idea of the ether is one example, the breakdown of the geocentric world view another.

Information that casts doubt upon what we believe to be true is increasingly resisted the further up the scale we go. We prefer to absorb information that confirms what we already know. To unlearn something is much more difficult than to continue to believe what we already believe.

10.3 Formal and Informal Information

Only a small fraction of all of the information that is created and used in organizations can ever be formalized. By formalized information, we mean information which is represented in such a way that its semantics can be interpreted by formal rules. This means that photographs are informal information, as are natural language texts. Numbers represent formalized information because numbers can be combined by formal rules so that new information is created. It should be pointed out, however, that current progress in research might very well lead to computers that can understand natural language and pictures. These intelligent machines might be available in the not-too-distant future.

There is a large spread in the degree of possible formalization of information and information systems. Human judgement, human observation and human interaction are usually difficult to formalize. Technical systems where interaction, observation, and judgement follow the laws of nature can usually be formalized. Cross-breeds between technical systems and human organizations give rise to difficulties, as do systems that try to mimic human behavior. In information systems for the support of human decision making, one must be able to handle informal information as well as formal information. A substantial part of the relevant informal information does not even exist in a machine-readable form. Examples of such information are rules of thumb, opinions, gossip, and ideas. Whenever such information exists in machine-readable form, it is usually encoded in text or diagrams, rather than in a mathematical form. Investigations of office environments indicate that approximately 40 % – 50 % of an office's information processing applies to information that cannot be formalized.

There is a tendency to overemphasize what is formalizable at the expense of what cannot be expressed in quantitative terms. Most human beings prefer to deal with informal information rather than formal information. If one does not recognize this need, one might unknowingly fall into the trap of letting the information system provide quantifiable, formalized information when the real need is for communicating various informal views on the decision situation at hand. If the information system's reporting is not relevant to the real needs of the human decision maker, then the person will be uncomfortable with the system, deeming it less than useful, and subsequently cease to use it.

Modern information technology has the potential to manage informal information. We have workstations that can handle voice, picture and text. We have communication facilities for distributing and collecting informal information. This causes the border between formal and informal information to become somewhat blurred. Because computers support the processing and storage of both kinds of information it becomes increasingly difficult to distinguish between the two. However it is necessary to be aware of the informal nature of some information and to be able to incorporate it effectively into the information system.

Formal systems, informal systems, communication networks, and computers all operate upon information, but the information always has to be carried by something. Whatever it is – gesture, word, number, statement, or mathematical formula – it is convenient to refer to it as composed of signs, or "signals" if they are transitory. The word "sign" is used here in an all-embracing manner to include the numerical and alphabetic characters, words, sentences, messages of any length, as well as all the actions which through custom or convention have acquired some recognizable interpretation. Signs are used to get things accomplished. But what signs? How do they inform? How much information do they carry? What carries them? How efficiently?

Semiotics – the theory of signs – tries to answer these questions. The word "semiotics" comes from the Greek word for "symptom". The problems of semiotics can be discussed from different points of view: logical, linguistic, behavioral, organizational and psychological, to mention the most important.

The problems of semiotics are usually classified under three major headings: pragmatics, semantics, and syntactics. Stamper [STAM75] adds a fourth subdiscipline that he calls empirics which deals with the engineering problems concerned with signal transmission in telecommunication networks.

Syntactics is concerned with the description of formal rules for relating signs to each other. The syntactics of a language describes the rules for relating the language symbols, for example the words, to each other. Semantics is concerned with the meaning of signs; that is, the relationship between signs and what the signs are supposed to represent. Semantics has to do with the definition of concepts, and with attaching names to the concepts. Pragmatics is concerned with the relationship between signs and the effect that the signs have on people.

Our knowledge about the subdisciplines differ vastly. We know most about the engineering problems, a little less about syntactics, even less about semantics, and very little about pragmatics. The four aspects of information are closely related to the problems of communication among human beings.

Organization is only possible because we can employ signs to coordinate the work of many hands and eyes. There might be anything from ten to a million human components in an organization. It is quite beyond the capacity of men to understand the functioning of an organization in the same way as a team of engineers can comprehend the functioning of a machine. Each human component in an organization may respond to signs in his or her own way; we cannot describe how a single person behaves except in performing some simple, repetitive tasks. If we could describe an organization in detail, our knowledge would soon be outdated because people in the organization will change their goals or their perceptions of their tasks.

Each of the four mentioned approaches to studying the nature of information has a different role in an information system design project.

Empirics is concerned with the computer-resource level. How much additional computing power is needed if the transaction load of a particular software package increases by 20 %? How much communication capacity is needed? Which type of computing power and communication capacity should be made available; when and for what price? Questions of distribution of workload between human operators and machines, distribution of workload among machines, centralization or distribution of data processing are representative of the empirics level.

Syntactics is the computer-programming level. Information is represented by structured data in databases. Formal rules are defined for the interpretation of data, and these formal rules are embedded in computer programs. The behavior of these formal systems depend only upon a knowledge of the structural properties of systems of signs. Examples are algorithms for scheduling of production, or for the storage and retrieval of information.

Semantics is the systems-analyst level. The analyst's task is to establish the connection between signs and what they signify in the real world. Examples include the description of a market, a university administration system, or a bank. The term "conceptual modeling" is often used to denote system modeling activities on the semantic level. The tasks are to define the concepts of interest in the universe of discourse, to determine which signs should signify which concepts, and to find and describe how signs are created and used.

Pragmatics is the operational level. Everyone in an organization responds to signs. One must know how to do this intuitively, or the organization would cease functioning. Such knowledge cannot always be made explicit, and some people might not wish to regard it as a part of the knowledge of an information system; nonetheless, in practical affairs, it is the most important part.

Limitations of time and resources are enough to force a manager, confronted by a complex and unique problem, to rely upon his pragmatic knowledge. This should not excuse the failure to use critical semantic analysis, or general syntactic solutions, where circumstances permit. Our knowledge resides in the signs we use and how we use them. Therefore, it would be surprising if the study of information did not raise serious philosophical questions. However, it is beyond the scope of this text to do an in-depth analysis of these interesting and important problems. Thus, we leave the matter here.

10.4 The Information System and Its Environment

There are as many proposed definitions for the concept of an information system as there are definition proposals for the concept of information. Here are a few:

- An information system is an institution which is intended to serve the improvement and support of outside information [RITT72].

- Information systems are the organization's instrumentation. They inform decision makers on all levels about those variables which represent the state of the organization (e.g., inventory holding, staff numbers) and about those which represent changes, or rates of change, in variables affecting the organization (e.g., cash flow, production rates) [BROO82].

- A system of information sets needed for decision and signalling in a larger system (of which it is a subsystem) containing subsystems for collecting, storing, processing, distributing of information sets [LANG66].

Although these (and other) definitions are both broad and general, there seems to be general agreement that information systems deal in information and that they are designed to give service to some other systems which are not part of the information systems. The needs of "the other system" are therefore central for determining the functions and priorities of the information system. The "other system" is usually viewed as being the environment of the information system, when the focus of interest is on the information system's design and subsequent implementation.

The other system may well be seen as the primary system, rather than being reduced to a second-order status by the word "environment". Various approaches to information systems requirements analysis put different emphasis on the analysis of the information system's environment. Some approaches are mostly information systems oriented, and relegate the other system, the primary system, to a second-order environmental status, e.g., the structured analysis method. Other approaches do not distinguish clearly between the information system and its environment, in terms of the modeling concepts that are applied, e.g., object oriented simulation methods.

To decide what is inside and what is outside of the information system may sometimes be difficult because information processing is going on in the information system's environment as well as in the information system. So it is a decision of the system designer to determine what shall be inside the information system and what shall be outside the system.

An additional complication is that information is both a commodity and a basis for decisions about commodities. There is no longer a clearcut distinction between the information system and the "larger system" which it supports. The larger system may itself be an information system. Information systems that support information systems are becoming common. For an information system to support its environment efficiently, it must be based upon a model of its environment, and it must itself contain a model of its environment. But information systems which support information systems often introduce changes in the systems that they support. In a way, the information systems become parts of their environments. Consequently, they have to contain a model of themselves; they have to be self-referential.

Many "environmental" systems may be viewed as production organizations (Fig. 10.1). A production department (factory) transforms raw materials into finished products. The raw materials are acquired from vendors. The finished products are shipped to customers. There are inventories for raw materials and products. There are departments for selling products, for purchasing raw materials, and for keeping the accounts. The model is very general indeed, and a number of different situations fit in nicely, ranging from systems for the production of research results to systems for the production and distribution of widgets.

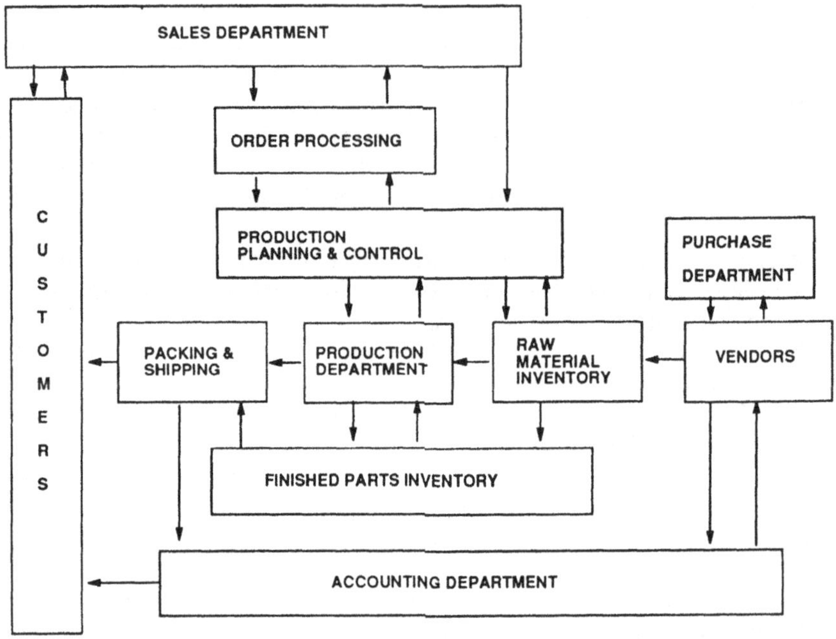

Fig. 10.1. Generalized structure of a production organization

A thorough understanding of the environment system, e.g., the production organization, is mandatory for the creation of useful requirements specifications for its supporting information systems. For example, one would hardly expect anybody to be able to create a workable information system for supporting the sales department of the production organization without having a thorough understanding of the functions of the sales department and of the operations of the department. Even if this is a pretty obvious statement, one should be aware of that a large number of information systems design failures have been due to lack of understanding of the other system on the part of the information system designers.

Information system design should always be preceded by a thorough analysis of the other system, e.g., the production organization. Issues of importance in a production organization are [HART69]:

- Timeliness of production: throughput, delays, transportation, queues.
- Quality of products, shipping, production, and planning.
- Usefulness of transportations, inventories, equipment, products.
- Necessity of production and planning processes: is the same work done several times?

- Completeness of production resources: timely availability of raw materials and components, of information, of floorspace.
- Cost: are product inspections made during the production process, or only at the end stage?
- Decision efficiency: for example, are customer orders produced optimally by grouping, batching, etc.?

It is unavoidable that proposals for changing the functions and operations of the primary system will be developed and put forward during the analysis phase. This phase is therefore crucial for the quality of the information system to be designed later in the project. The work must be handled with the utmost seriousness, and involve people who have both the necessary knowledge of the primary system, and who will shoulder responsibility in the new system.

10.5 Information Systems Viewed as Production Organizations

The systems we are mainly interested in are goal oriented systems which produce goods and/or services. Examples of such systems are industrial systems that produce gasoline from oil, electronic systems that transmit information from one location to another, legal systems that pass judgement, cybernetic systems that control robotics behavior, electric power systems that produce and distribute electricity, planning systems that achieve economical resource utilization, and so on.

A production system can be regarded as performing a sequence of transformations of entities, starting from an initial state and proceeding through intermediate states until some final state is reached. The *raw material – intermediate product – end product* sequence, which is found in conventional industrial production, exemplifies this view. Systems may be characterized in various ways, reflecting the peculiarities of each system, its purpose, and its environment. A production system must contain production activities and product repositories. Control activities must be performed to ensure that production resources are utilized in an effective manner. Production activities, as well as control activities, rely heavily on having access to information that reflect the state of the production activities and the production goals. The four subsystems production activity, product repository, control activity, and information repository and their interrelationships are depicted in Fig. 10.2.

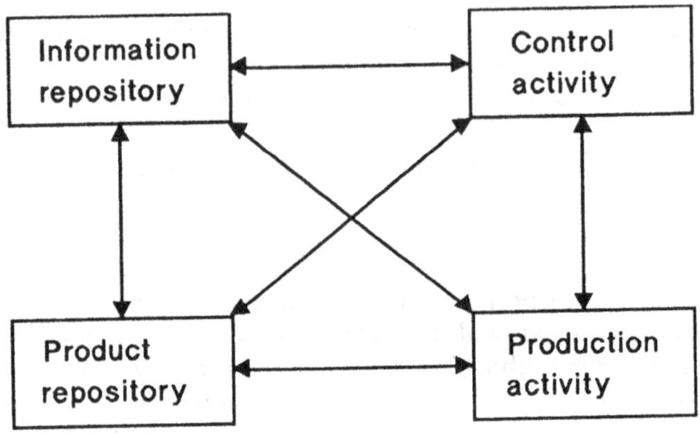

Fig. 10.2. An information system view of production

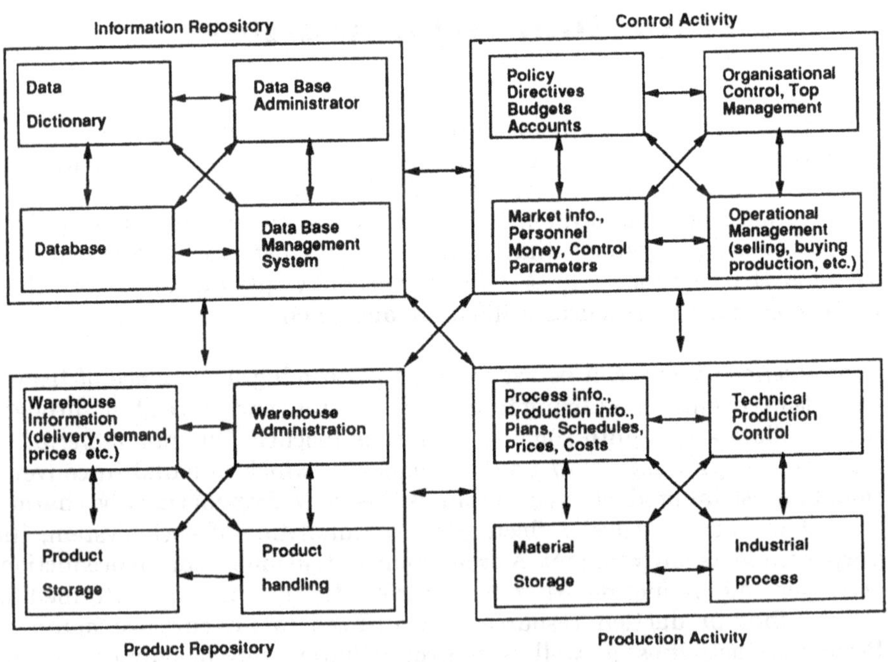

Fig. 10.3. The information system view applied to each of the subsystems of a production system

This model is not very precise, of course. It gives one of many ways of viewing a system. Different models describe systems from various points of view. Some models employ an economic point of view. Other models are based upon ecological viewpoints, or political, technical, sociological, or psychological premises which highlight corresponding system features. Our model describes the world as seen from an information processing point of view.

Each of the subsystems of Fig. 10.2 can be decomposed in a recursive manner so that each subsystem consists of production and control activities and the respective repositories. The result of the second-level decomposition is depicted in Fig. 10.3.

The Production Activity of Fig. 10.2 is perceived as representing industrial production as shown in Fig. 10.3. Material is transformed from raw material form into more refined forms through some industrial process. The latter plays the role of local production activity. The role of local product repository is played by the material storage, and the control activity is called technical production control. This subsystem has the task of coordinating the interaction among the components of the industrial process and the interaction between the industrial process and the material storage. Information about plans, schedules, prices, production costs, processing alternatives, etc., is necessary so that the production, control, and material storage system functions properly and fulfills the role of local information repository.

The Control Activity is perceived as a system for administrative control in the world of which the industrial technical production is a part. But a system for administrative control is also a production system, as seen from the information system point of view. The local production activity is called operational management. Examples of products resulting from operational management are control parameters that influence decisions about production scheduling. The products of operational management are mostly informational, as is also the case with the production basis which consists of market information, money, information about personnel skills, etc. The control activity role in a system for administrative control is concerned with distributing resources to different operational functions, such as functions for evaluating plans established by operational management, and so forth. Company policy, budgets, and reports on overall company behavior are typical elements of the information repository.

The product repository represents a warehouse in which the raw material and end-products of the industrial process are stored and managed. The production activity of the warehouse is concerned with the handling of goods by workers or machines. The warehouse must have an administrative branch, which coordinates warehouse activities, decides when and what to ship, and determines when to replenish product supply. Warehouse administration is dependant upon having access to information about buyers and vendors, demands and prices, etc.

The information repository is organized like a computerized database. The data in the database play the role of products in a product repository, and the database management system provides production activity. Data production is controlled by the database administrator, who relies on the information repository – the data dictionary – to control and schedule jobs effectively. The production resources are computers, data, database management software, and so on. The production resources should be utilized in a secure and economically satisfactory way.

The system model we have used here has apparent limitations. However, it shows that we can apply the same point of view of systems decomposition to several levels of system coordination. The model indicates that the features of information systems are fairly similar, independent of whether the systems are technical, administrative, or data processing systems. While products in a technical system are mostly associated with material entities, the products of an administrative system are mostly documents and reports. From the control activity point of view, the systems look surprisingly similar.

A most striking feature of Fig. 10.3 is that one finds new information repositories and new information processing activities on every new subsystem level. The same patterns of information repositories and information processing are found again and again. This indicates that there are numerous production and control levels in a large system. The control activities on one level are perceived as part of the production activities when viewed from the control system one level above.

Nevertheless, this hierarchical structure would be perfectly suited for partial automation if it were unique and stable. However, the structure is not unique. This kind of system stratification is always man-made. Therefore, it is not always quite clear what is production and what is control on each level. Production tasks and control tasks tend to be performed by the same persons, and it is not always easy to distinguish between the two. In non-computerized manual systems, this mixing of tasks is only of local consequence; thus, it does not affect the overall functioning of the system. However, as soon as computers are involved in the operations, it is very tempting to collect information from all over the system into one large database management structure. Then the mixing of production and control not only occurs within each local subsystem in the hierarchy, but among subsystems both on different levels and on the same level.

This could be harmful because of the global interdependencies created among the tasks. Changes in one part of the system might easily require modifications in other parts of the system, when one would expect the two system parts to be well insulated from each other.

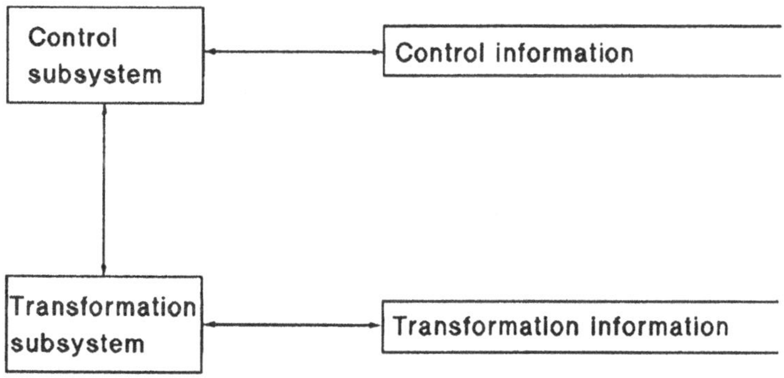

Fig. 10.4. A production system decomposed into subsystems for control and transformation (of products)

This feature is even more pronounced if we reformulate our systems model somewhat. In Fig. 10.4 the transformation subsystem consists of the production activity and product repository of the model in Fig. 10.2. The control subsystem acts like the control activity of the previous model. The information repository decomposes into control information and transformation information. The control information component is assumed to contain information about the state and goals of the transformation subsystem. The transformation information component is assumed to contain information which is relevant for performing the transformations themselves without regard to coordination issues.

A consequence of the idea of having several levels of control is depicted in Fig. 10.5. The transformation subsystem on each level is perceived to consist of one control part and one transformation part on the next lower level. This leads to a model of boxes within boxes. The control information component on each level refers to phenomena in the "inside box" on the next lower level of control. From an information processing point of view this might have serious consequences. It means that the structure of the control information on one level reflects the structure of the system on the lower levels.

Consequently, if changes in the lower-level system's structures are implemented, we can expect that higher-level information system constructs will also require changes. Thus, it becomes very difficult to insulate one part of the information system from the effects of modifying another part of the system. Local changes have a tendency to propagate into the neighboring subsystems.

In most practical projects where human beings are involved, it is not possible to produce nice, clean system descriptions like those depicted in Fig. 10.5. The channels of control, influence and interaction are usually not

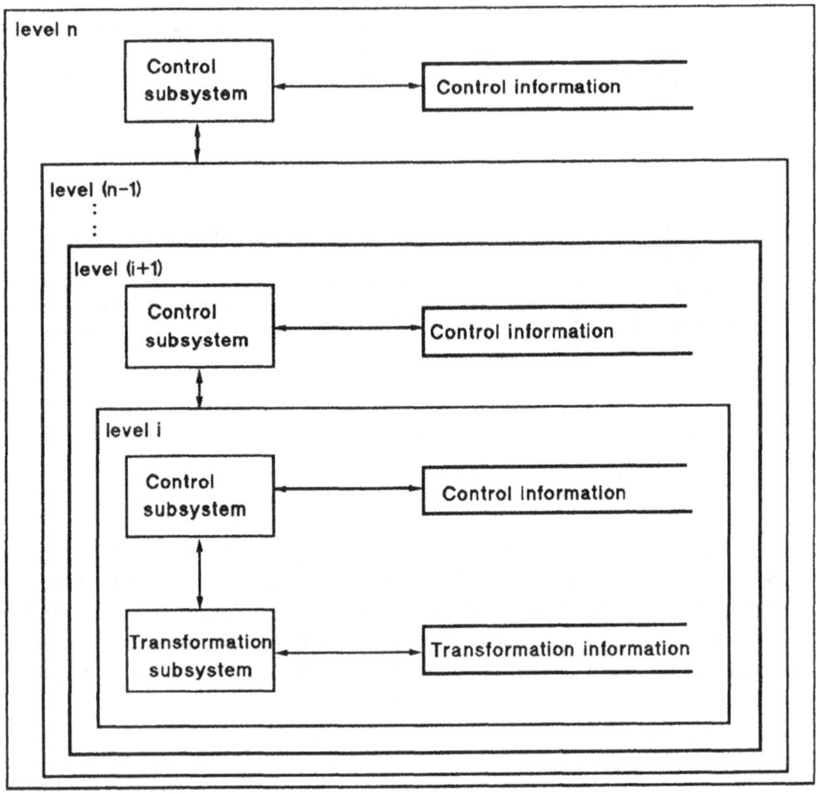

Fig. 10.5. Production systems have many levels of controls

simple or static. The consequence for the information system designer is that the information structures usually contain information elements which refer to several levels of control. The designer can easily find himself in a situation where his design decisions influence the behavior of those parts of the system to which the design refers. Viewing the relationship between control and production, this appears to be an unavoidable consequence.

To be able to exercise control, one needs to have proper knowledge about the state of the production activities. Different control systems will generally require different state information. So the production system and the control system are designed as two distinctly different parts of an integrated system design. The control activities refer to the production activities and vice versa. This is the usual case and it is perfectly all right to do it this way as long as the interactions can be kept locally, so that the designer can perceive the consequences of his decisions. However, as soon as we start to group together information and activities that belong to different levels of control and transformation, we may introduce interaction among subsystems and activities which are fairly distant from each other as regards cause-effect relationships.

10.6 Self-Referential Information Systems

An information system conveys knowledge about situations and behavior of a world that is separate from itself. The relevant parts of the world must be abstracted in order to provide a basis for the design of the information system. So the world is abstracted into a model of the world. The world model may be represented in terms of data, and by rules that represent the legal states of the world. Information systems contain models of the relevant data, and rules for the manipulation of this data. Information systems communicate data with their environment, which is also part of the "external" world.

Relationships between "the world" and its associated data processing system are depicted in Fig. 10.6. The behavior of the world may be abstracted in a set of possible world-states, and in a set of possible transitions between states. The states and the state-transitions constitute a model of the world. They may be represented in a database and in data processing rules. The database represents the state of the world, while the rules represent state transitions. The rules transform the effect of state transitions into modifications of the database that conform to the relevant state-changes. The structure of data and data processing rules are represented in database schemas and computer programs, and are stored in data dictionaries and software libraries.

When computerized information systems are put into operation, they interact with the world that they model and they soon become parts of that world (Fig. 10.7). Information systems consequentially are self-referential systems. They may have to contain a model of themselves, and of their own interaction with their environments. This property of information systems also has consequences for how computer programs should be viewed in relation to the interactions with their environments.

A computer program is a statement of data and algorithm that defines an automatic procedure which can be executed by a computer. If we employ the production system view of the previous section, the computer programs can be viewed as being the production machines that collect, store, transform, and distribute the information, i.e., the products of the information system.

In information systems engineering, one is concerned with providing methods for the creation and maintenance of functionally satisfactory and cost-effective programs. Functionally satisfactory programs must be useful for the organization to which they belong. However, judging a program's usefulness is a rather subjective matter and judging one program to be more useful than another is even more subjective.

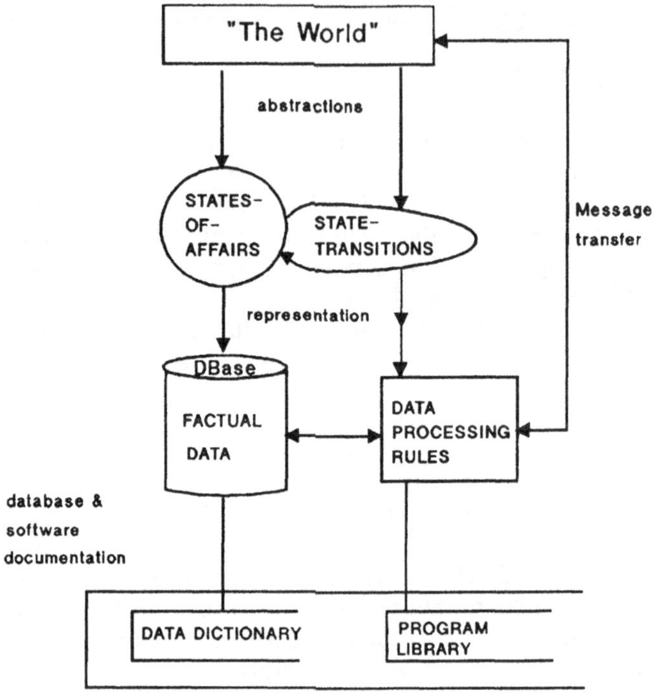

Fig. 10.6. Relationships between "the world" and associated data processing system

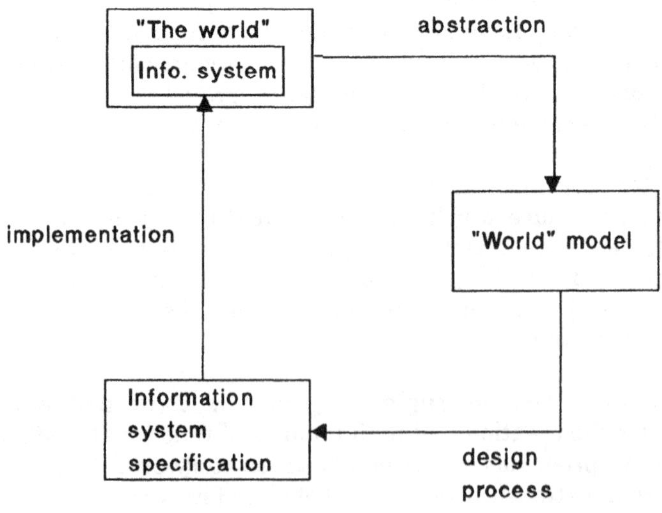

Fig. 10.7. Information systems are parts of the world that they model

The usefulness of most programs will change with time as the needs of the organization change. The consequence of organizational change is the manifestation of new and different requirements for program functions. We are led into a never-ending process of changing both the program and the program's environment.

Most natural and artificial systems evolve. However, because of the close coupling of software with humans, society, and their activities, and because it seems so simple to change a computer program, the rate of software evolution is very high, both in absolute terms and relative to the professional lifetime of computer programmers.

Lehman has proposed a program classification scheme based on the degree of change a program is expected to undergo during its lifetime [LEHM80]. A program is classified to be either an S-program, a P-program, or an E-program:

- S-programs are static, in the sense that one will never need to change them once they have been built.
- P-programs are applied to problems in a changing environment, and will possibly require changes.
- E-programs are evolutionary, in the sense that they will definitely undergo changes in the course of their lifetime.

10.6.1 Static Programs

A static program is complete if it satisfies a given specification. An example of an S-program is a program that can calculate the function y = sin(x) with some pre-specified precision. For a program to be of class S, a problem statement must be possible from which a complete and authoritative specification of a program for its solution can be derived. The traditional approach to the development of S-programs is illustrated in Fig. 10.8. The S-problem is stated relative to a universe of discourse. The S-problem is describable by a formal statement of the type: calculate sin(x), with precision 10^{-7}, for x-values given in radians. From this formal problem statement, a program specification is worked out and a program is subsequently written in some programming language.

The traditional approach to determining if the program is correct is to have the program undergo extensive testing. The usual way to test programs is to perform several test executions so that the computational results can be precisely predicted. If discrepancies are found between the output and the expected results, the program will be modified until it displays the wanted behavior.

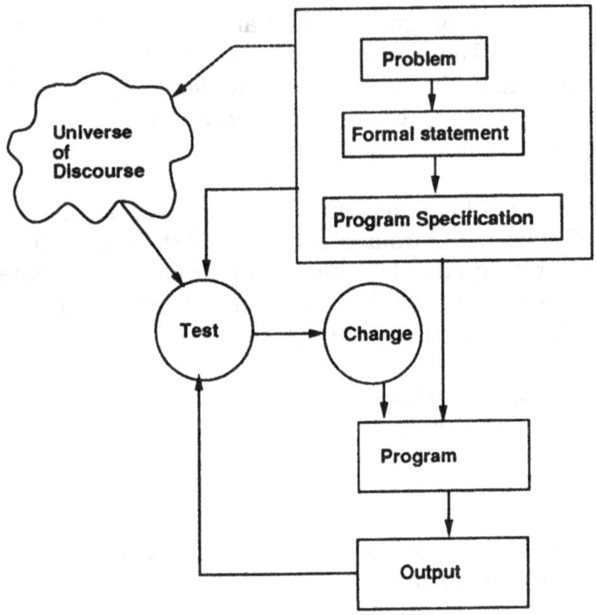

Fig. 10.8. S-program development: traditional approach (from [LEHM80])

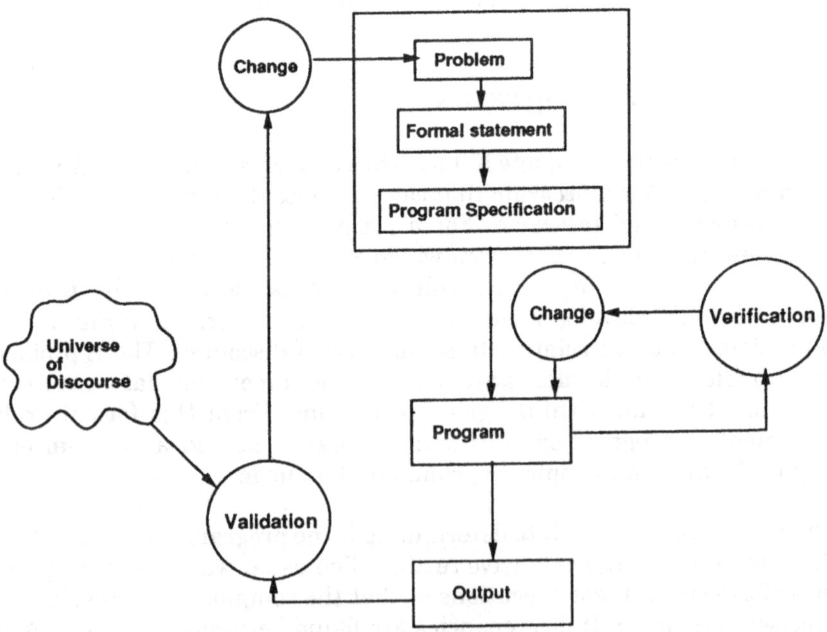

Fig. 10.9. S-program validation (from [LEHM80])

Industry in general and the majority of programmers rely primarily on such procedures to determine the acceptability of programs. In practice, program specifications often do not exist at all. If a specification exists it is mostly used to determine the set of tests that are to be applied. Testing is generally viewed as an unsatisfactory basis for demonstrating the veracity of programs. Testing can only show the presence of faults, never their absence. The number of different paths which one can follow through a typical program execution is usually so large that it becomes practically impossible to test every possibility. So it is quite possible that undetected errors remain in a program even after the most extensive testing.

Program validation is concerned with evaluating a program in relation to the needs of the program users (Fig. 10.9). It may happen that the S-problem relates to some real-world phenomenon. In this event the program output may not be satisfactory even after the S-program has been proven to be correct with respect to its specification. When this happens either some part of the problem statement must be incorrect, or some aspect of the program specification is inappropriate. Thus, a new problem statement and/or a new program specification must be worked out. A new program must be written, tested and/or verified and its result be subsequently validated. It may be that the changes are achieved by extensions of the original versions. In the S-program situation they nevertheless represent the implicit definition of a new program.

Program verification proofs demonstrate equivalence between a program and its specification. Since by definition an S-program may have a specification, S-programs can be formally proven correct. We may not know the proof, but we know that there is a proof. So we can replace the test procedure with a verification procedure, by which we prove if the program text is equivalent to the authoritative program specification (Fig. 10.9). If equivalence can be disproved, we change the program text until we get one which is equivalent to the specification.

10.6.2 Problem Oriented Programs

A problem oriented program (P-program) can be deemed to be complete only if it satisfies the needs of its users. A P-program's satisfaction of a program specification is relevant only if the specification reflects the needs of the environment. The P-program development process is illustrated in Fig. 10.10. The basic difference between the S-program situation and the P-program situation is that in the latter, the program designers are assumed to gain a better understanding of the real problem to be solved as they test and validate their programs. Changes in the problem statement will suggest themselves as the program is executed. The universe of discourse to which the problem statement refers is itself likely to change. This change too is likely to demand or provide opportunity for change to the problem statement, and hence to the program.

The basic problems to which the P-programs are related are problems that relate to actual needs as ultimately experienced in practice, rather than those that are primarily of intellectual interest like mathematical problems. Therefore, it is realistic to see the P-program process as a gradual approach to the identification and formulation of the problem and its solution. The resultant P-program must then be viewed as one evolving program, rather than as a sequence of new programs as is the case with S-programs.

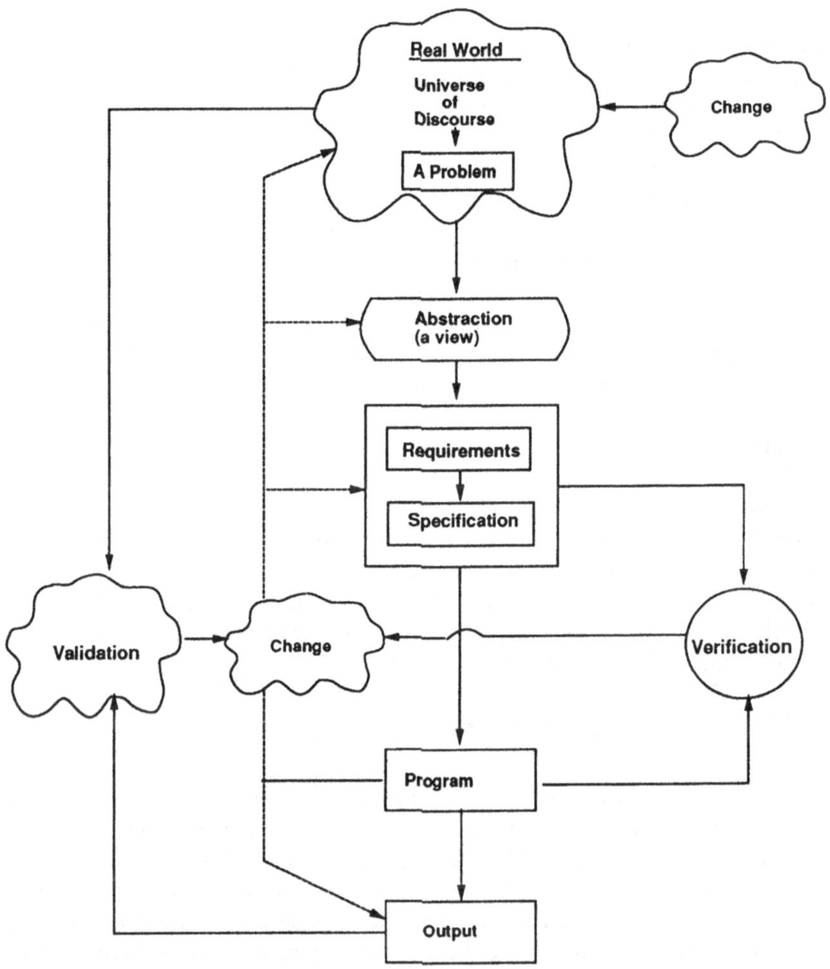

Fig. 10.10. P-program development process (from [LEHM80])

10.6.3 Evolutionary Programs

Evolutionary programs (E-programs) implement applications that control activities and/or events in the environments in which they are embedded and executed. That is, an E-program is itself a part of the universe that it both models and controls. The E-program contains a model of its own interaction with its operational environment.

The basic E-program development loop is illustrated in Fig. 10.11. In order to specify the properties of an E-program to be developed or modified, one must predict the E-program's impact on its operational environment and on the people who are to work with it.

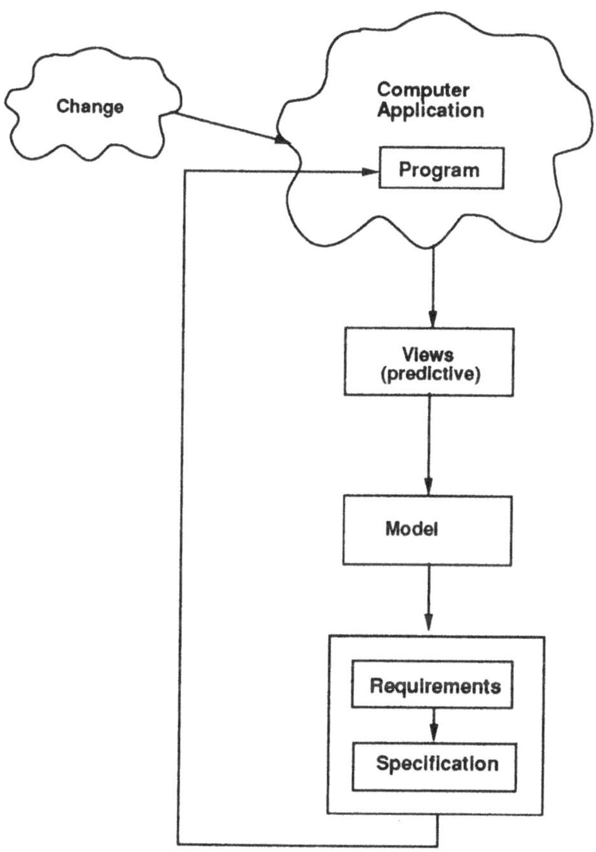

Fig. 10.11. E-Program basic development loop (from [LEHM80])

The designer must perceive the combined behavior of the program and its environment as it will be once the program has been implemented and installed. This prediction cannot be perfect, and system validation will inevitably result in a series of changes to the systems (Fig. 10.12).

So E-programs are true evolutionary programs. They change the operational environment because they are embedded in the environment. That change in turn generates further evolutionary pressure. The process of change and evolution is never-ending, the alternative being increasing ineffectiveness and ultimate obsolescence.

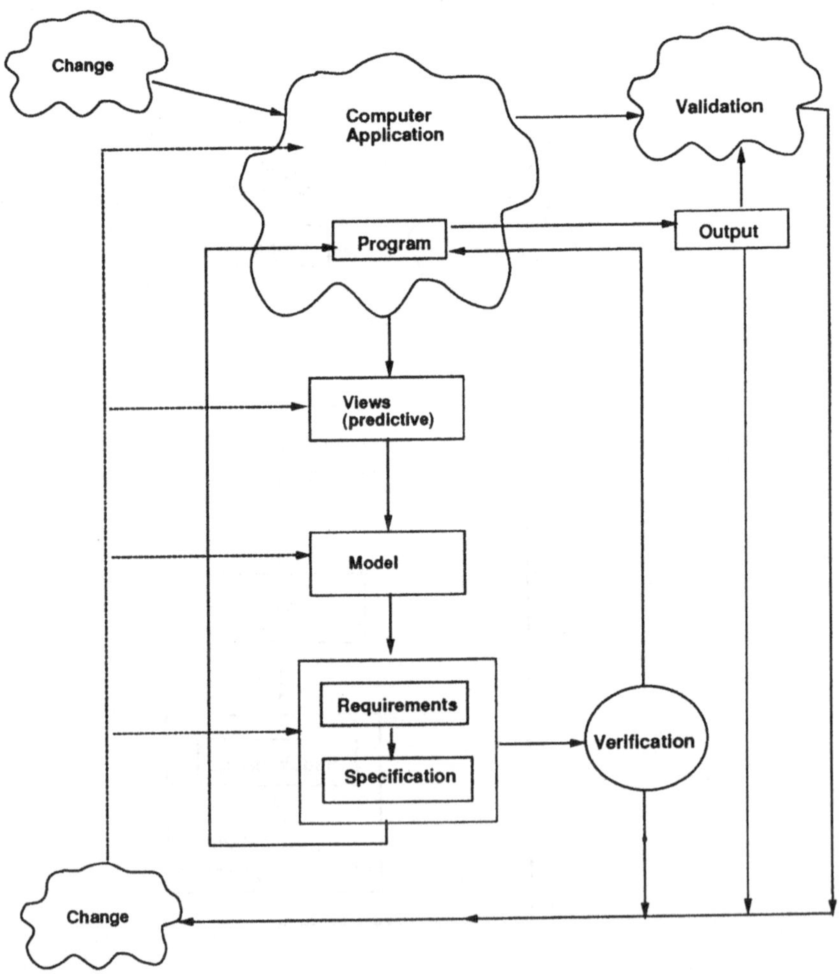

Fig. 10.12. Full E-program (from [LEHM80])

Exercises

1. Define the terms *data, information* and *knowledge* and state their relationships.

2. Consider the following phrase

   A Course in Information System Engineering

   Classify the phrase as either data, information or knowledge according to

   a) someone who does not have any knowledge of the English language or computer literacy,
   b) someone who is British but does not have any idea about computers,
   c) a British lecturer in computer science.

3. Give some examples of different types of information and knowledge.

4. Classify the following into S-programs, P-programs or E-programs:

   a) a program to find the square root of a positive real number by the Newton-Raphson method,
   b) programs related to the accounting subsystem of an organization,
   c) the first prototype of an accounting subsystem in an organization,
   d) the operating system MS-DOS, version 1.0.

Chapter 11

Three Domains of Information Systems Modeling – and the Object-Oriented Approach

We may distinguish three domains in information systems modeling: *the subject domain, the interaction domain* and *the implementation domain* [WINO79].

The subject domain consists of people, physical entities, ideas, goals, actors, and activities that exist independently from the information systems and data systems that support the activities in the subject domain. Knowledge representation techniques belong to the subject domain. So do simulation techniques.

The interaction domain consists of messages, transactions, information transformations, and archives that participate in the communication between the information system and its environment, and in the communication among different parts of the information system. Conventional systems analysis techniques, e.g., data flow analysis, belong to the interaction domain.

The implementation domain consists of computer-related entities like files, subroutines, datafields, databases, communication protocols, and operating systems. Conventional programming belongs to the implementation domain. The implementation domain also comprises the organization which supports the operation of information systems, including the appropriate human and technical resources.

The three modeling domains represent three different points of view on an information system. The three views are, nevertheless, intertwined as well as separated.

Let us distinguish the functional aspect, the technological aspect and the operational aspect of an information system. The three aspects may be characterized by the words *what, how* and *who*:

what shall the system do: the functional aspect,

how shall the solution be: the technological aspect,

who shall operate the system: the operational aspect.

At first glance one would easily claim that the subject domain includes the *who*-aspect of the information systems environment. The interaction domain is seen to include both the *what*-aspect and parts of the *who*-aspect of the information processing system, while the implementation domain includes the *who*- and *how*-aspects of the information processing system. We have previously argued that an implemented information system becomes part of its environment as soon as it is put into operation. The operational aspects of an information system are, in principle, no different from the operational aspects of its environment. A natural conclusion is therefore that the modeling concepts of the three domains of subject modeling, interaction modeling and implementation modeling must have some commonality.

Most contemporary information-system modeling approaches do not distinguish between the three modeling domains. The commonest approach is to start the analysis/design work in one of the three domains, and treat the various aspects of *what, how* and *who,* from the dominant point of view of the starting domain. This approach leads to a number of quite different information-system development methods. Some examples are:

- the subject domain may be viewed from an operational perspective, as a set of interacting entities, and represented by a computational model, e.g., a simulation model, or an object-oriented model.

- the subject domain may be viewed from a functional perspective, and represented by rules that reflect relevant domain properties, e.g., as the knowledge base of an expert system.

- the interaction domain may be viewed from a functional perspective, and represented by, e.g., dataflow diagrams.

- the interaction domain may be viewed from an operational perspective, and represented by a prototype of a human-computer interface.

- the implementation domain may be viewed from a technological perspective and represented by, e.g., organizational charts and computer configuration descriptions.

- the implementation domain may be viewed from an operational perspective and represented by, e.g., computer programs.

Common to contemporary information-system development methods is that none of them offer adequate support for developers in all phases and for all aspects of the system development process. We shall discuss modeling in the three domains in more detail in the sequel. Our main emphasis is on the clarification of modeling issues. Our main objective is to show commonality among modeling approaches in the three different domains of modeling. The object-oriented approach to systems analysis and design has been proposed for dealing with all three modeling domains. The same modeling constructs are to be used in all three domains. The object-oriented approach is explained in some detail at the end of the chapter.

11.1 Subject Domain Modeling

A number of subject-domain modeling techniques have been proposed. Simulation techniques are modeling methods that have been known for a long time. Conceptual modeling originated in the database field in the early 1970s. Knowledge representation techniques like semantic networks, so-called frame-based approaches, production rule techniques, and logic programming have been developed within the artificial intelligence community in the 1970s and 1980s. The subject domain models are often called *reality models*, or *world models*. The purpose of a subject-domain modeling exercise is to create a model of some part of the world in such a way that the model can serve as a common reference framework to support discourse about world properties.

World models have both structural and behavioral aspects, representing static and dynamic properties, respectively. As an example, consider a world of persons, managers and employees. The structural aspect is dealt with by defining entity sets PERSON, MANAGER and EMPLOYEE, where MANAGER and EMPLOYEE are subsets of the PERSON set. This definition implies that every employee is a person, and every manager is a person.

We may formally express this as follows, in an Entity-Relationship fashion:

1) **entityset** PERSON, MANAGER, EMPLOYEE;

2) MANAGER $\subseteq$ PERSON

3) EMPLOYEE $\subseteq$ PERSON

An additional static property would involve the clarification of the relationship between managers and employees. Is a manager an employee, or are managers and employees considered to be different species? If managers are always considered to be employees, then the set of managers is a subset of the set of employees. This is formally expressed as:

4) MANAGER $\subseteq$ EMPLOYEE

If no manager is considered to be an employee, then the MANAGER entity set and the EMPLOYEE entity set have to be disjunct sets. This is formally expressed as

5) MANAGER $\cap$ EMPLOYEE = $\emptyset$

where $\emptyset$ denotes the empty set. Although this might seem somewhat artificial, one might envisage a situation where only owners of a business are permitted to be managers.

It is usually required of the modeling techniques that also subtle properties can be described rather than only the apparent properties such as all employees are persons. We need to have modeling tools which enable us to specify that employees have names, addresses, eye colors, skills, and ages. There are several ways of expressing such features. One will usually classify names, addresses, etc., as different types of data, and, might also say that employees have the property of having names, ages, etc. A complementary statement would be to say that the attributes of employees are name, age, address, and so on.

A formal expression might look like:

6) **datatype** NAME, ADDRESS, AGE, SKILL;

7) **attribute** (EMPLOYEE) : NAME, ADDRESS, AGE, SKILL;

The feature of *property inheritance* is implicit in such a reality model. *Property inheritance* means that if the members of a set have some property, then the property must be shared by the members of all subsets of the set; e.g., if every person has a name, then every employee must have a name, provided that every employee is a person. Thus, the statements

8) **attribute** (PERSON) : NAME, AGE;

9) **subset** (PERSON) : MANAGER;

imply that MANAGER has attributes NAME and AGE.

Some of the dynamic aspects are shown through modeling the circumstances by which an individual becomes a member of a set, or loses its membership in the set. In the case of employees and managers, one way of modeling some of the dynamic aspects is to declare that an individual person has to be hired prior to becoming an employee, and that he or she stop being an employee after being fired. This knowledge could be formally expressed as

10) $S_i \models x \in \text{PERSON} \wedge x \notin \text{EMPLOYEE} \wedge \text{hire}(x) \rightarrow S_{i+1} \models x \in \text{EMPLOYEE}$

11) $S_j \models x \in \text{EMPLOYEE} \wedge \text{fire}(x) \rightarrow S_{j+1} \models x \notin \text{EMPLOYEE}$

which is to be read:

if, in state S_i, it is true that x is a person but not an employee, and x is hired,

then, in the next state of the system, S_{i+1}, it is true that x is an employee;

and,

if, in state S_j, it is true that x is an employee and x is fired,

then, in the next state S_{j+1} of the system, it is true that x is not an employee.

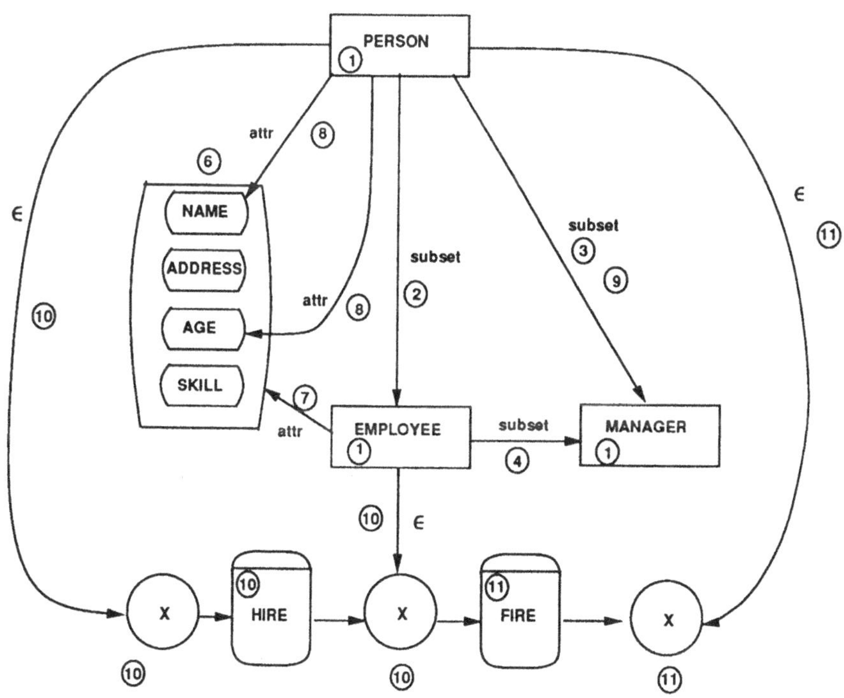

Fig. 11.1. Graphical representation of subject-domain model

Modeling in the subject domain is usually done in an informal fashion using graphical notations for sketching entities, their relationship, and their behavior. Nevertheless, there is a close relationship between the mathematical formulas and graphical notations. One example is shown in Fig. 11.1 which depicts a graphical representation of the formulas in the previous part of this section. The numbers in the picture refer to the appropriate formula in the text that introduced the corresponding system's feature. For example, the number 1 in the square boxes PERSON, MANAGER, EMPLOYEE means that these boxes represent concepts which were introduced in formula 1. Please note that formula 5 is not shown in the drawing, because it is in conflict with formula 4 (a manager cannot both be an employee and not be an employee at the same time).

So we have specified the circumstances that lead an individual to enter/leave the set of employees. In a similar way we can describe promotion of employees to managers, and subsequent demotions of managers. Reality modeling is, of course, more than what is shown here; nevertheless, the main issues of structure and behavior are always to be found in subject-domain modeling. Different ways of treating these issues will be described in later chapters.

11.2 Interaction Domain Modeling

When working in the domain of interaction, we are concerned with developing suitable information services to satisfy requirements which originate in the information systems environment. Let us expand a little on the employee-manager example. Normally, employees are hired and fired as a result of a decision making process and not on an ad hoc basis. There is hopefully someone somewhere who makes considered decisions based on the best possible information about employee's potentials and track records. We would expect to find a personnel management department, a personnel archive, and procedures for hiring and firing. To clarify what happens in such a system, and what should happen in a future system, we will normally have to analyze a rather complex pattern of interaction among entity classes, information repositories, decision makers, and procedures which enable the decision makers to handle the different situations which might arise.

It is important to be aware that there are no "correct" solutions to these problems. There is no such thing as a universally correct way of hiring and firing employees. Each enterprise has to establish its own set of procedures which define what the correct way of hiring/firing should be in that environment. Establishing solutions to these kinds of problems must be based upon a consensus process within the company. Several persons have

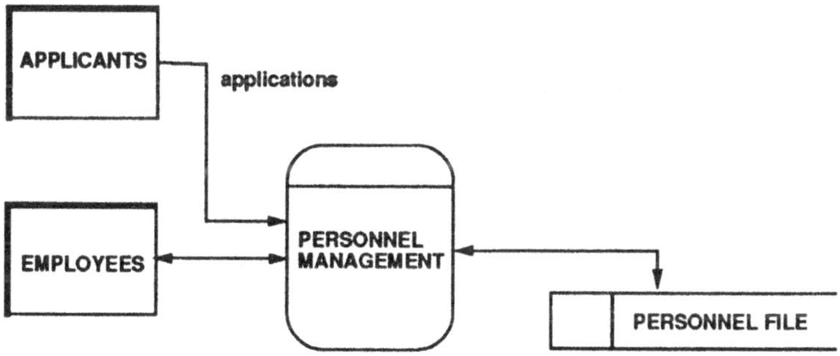

Fig. 11.2. Personnel management – interaction with applicants, employees and personnel file

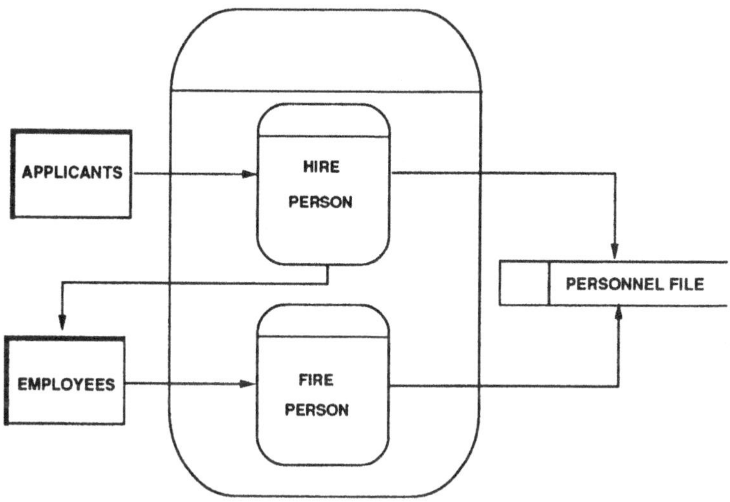

Fig. 11.3. Personnel management: hiring and firing

to agree upon a "correct" procedure for the hiring/firing of employees. The persons who are involved in the consensus process have to develop a common understanding of the problem so that they can agree on a solution. Consequently, they need a common language for describing information processing problems. Natural languages are used to some degree but they suffer from being rather imprecise so that ambiguities often appear. Diagram-based languages are quite common in all branches of engineering. There are several proposals for diagrammatic languages. We shall use dataflow diagrams. One possible solution of the hiring/firing example is depicted in Figs. 11.2 – 4.

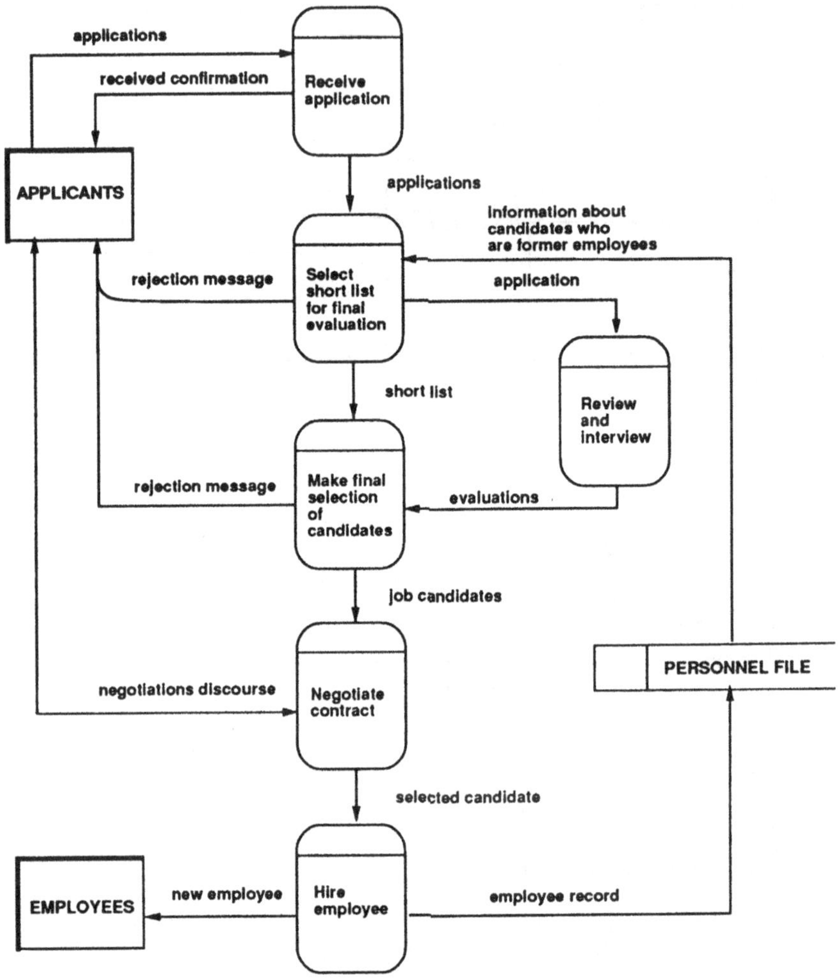

Fig. 11.4. Details of personnel hiring activity

Looking at the three diagrams of Figs. 11.2 – 4, we see that there is an increasing number of details in each picture. We start out with a very simple diagram stating that there is personnel management activity, which interacts with applicants, employees, and an archive containing personnel information. In this diagram we express the view on a very high level of abstraction. In the next diagram we expand the detail somewhat, stating that personnel management consists of one activity for hiring and one activity for firing personnel. In the third diagram (Fig. 11.4) we suddenly explode into a number of new details, showing what personnel hiring is all about. We state that when applicants send in their applications, they will receive a confirmation that their application has been received. We further state that only a few applicants (the "short-listed" ones) are selected for interview and closer review. It is indicated what information each sub-activity is expected to produce, and what information each sub-activity is expected to need. The result of a successful hiring is stated to be a new employee and an update of the personnel file.

The three diagrams show an application of the integral principle of hierarchical systems decomposition. We begin by depicting an imprecise statement of the system, adding to it more and more detail as we gain understanding of the system by decomposing it into smaller and smaller parts.

A drawback of these kinds of diagrammatic language is their lack of expressiveness. For example, algorithms become very clumsy if we try to specify them using diagrammatic languages. We therefore must enhance diagrammatic specification languages with some other language for specifying algorithms. There is a wide choice available, ranging from stylized natural languages, decision tables, and programming languages, to mathematical logic. Various alternatives are treated elsewhere in this text.

In addition to diagram-based specification languages and languages for specifying algorithms there are languages for specifying the content and structure of messages, documents, and archives. There are also methods and techniques for specifying man/machine interface designs.

In particular, one should note that it is impossible in practice to distinguish between the subject domain and the interaction domain in a strict sense. The reason for this is that the information system becomes intertwined with its environment. Unless the system is analyzed very thoroughly, it becomes impossible to determine if agents of the subject domain exist independently of the information system, or if they exist because of the information system.

A consequence of this observation is that the subject domain and the interaction domain should be represented by similar modeling concepts. The modeling situation is exemplified in Fig. 11.5. In the diagram EMPLOYEE is the set of present and former employees.

The PERSONNEL_FILE is supposed to contain information about employees, present and former. We have indicated this by explicitly relating the PERSONNEL_FILE to the EMPLOYEE entity set, through a **reference**-relation. This should be understood to mean that PERSONNEL_FILE contains one record for each member of the EMPLOYEE-set. We have further indicated in the figure that PERSONNEL_FILE can be interpreted as being an entity-set consisting of record-entities, that is, y is a record entity and y is a member of PERSONNEL_FILE. By viewing the system in this way, there is no longer any big difference between subject-domain modeling and interaction-domain modeling. An information entity in the interaction domain is just another entity of a particular kind, as is a

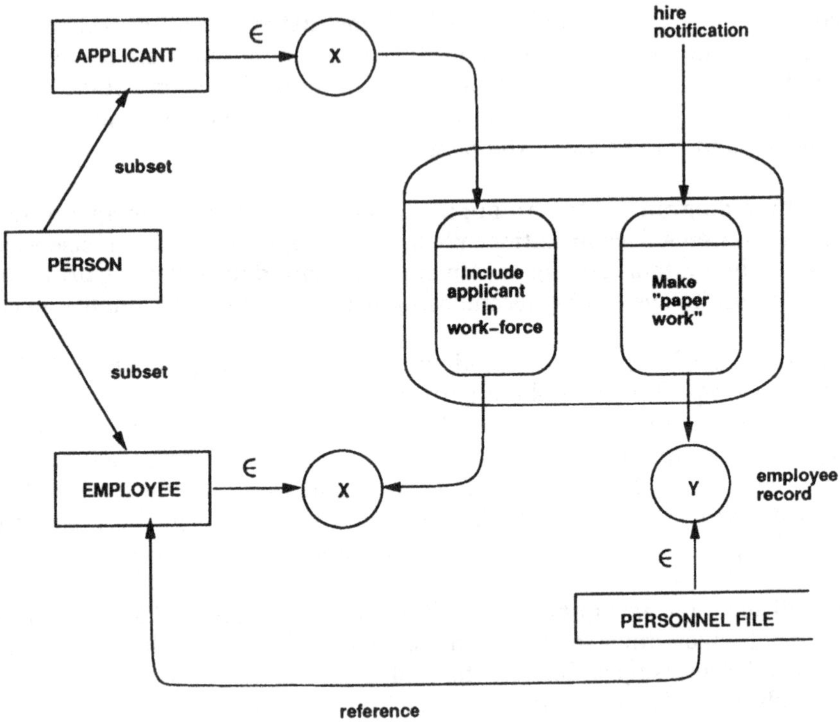

Fig. 11.5. Details of process of hiring employee

person-entity or a football-entity. If the arguments given here are accepted, it has consequences for the choice of systems description methods for the interaction domain.

Most of the contemporary methods for information-system modeling have been invented to satisfy needs from the software design end of development projects. From this point of view the "worldly" processing is one thing, the data processing is another thing, and the two are naturally separated. It has been perfectly acceptable that the programming-language type of modeling construct should be used for modeling in the interaction domain.

There is a need for new ways of modeling in the interaction domain, so as to bring about an integrated way of viewing the computer programs' interaction with their environments. We are faced with the challenge of representing the behavior of software by using modeling concepts which can also be used for modeling the software's environment. A possible answer may be found in the object-oriented approach.

11.3 Implementation Domain Modeling

The major challenges of the implementation domain are to choose technical solutions that can satisfy stated functional requirements, and that can minimize operational costs. We are concerned with software and databases, data communication and operating systems. Because our point of view is the implementer's, we prefer to state our specifications in implementation oriented terms. Our tools are programming languages and database schema languages, rather than non-executable graphical sketches.

In the implementation domain we would describe the hiring of an employee in terms of a hire-transaction, which collects the relevant employee data from a computer operator, and subsequently creates an employee-record which is stored in the personnel file. The natural way of starting the specification would be to write a pseudo-program:

```
program hire (hire-screen, personnel-file);
setupscreen (hire-screen);
fillinscreen (hire-screen);
createrecord (hire-screen,employee-record);
store (employee-record, personnel-file);
comment write some message in hire-screen;
end hire-program;
```

This would be followed by detailed program specifications written in an appropriate programming language for each of the components in the pseudo-program above.

Programming languages are not very effective when used for documentation. Additional tools have had to be used to cure this deficiency. These tools are usually based on informal diagramming techniques, and are used to describe functional aspects of software, and the interaction between the software and its environment. This is exemplified in Fig. 11.6.

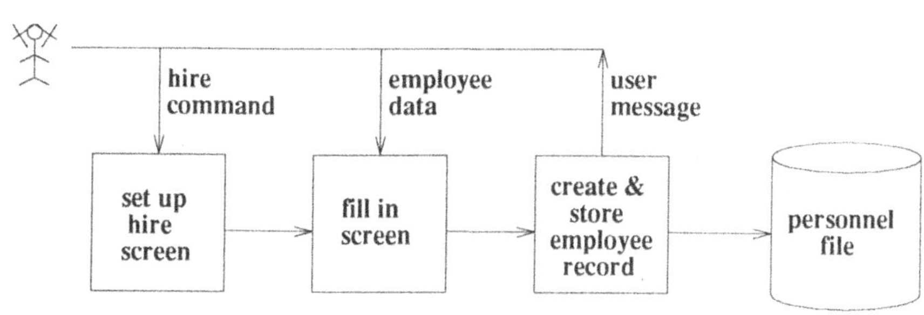

Fig. 11.6. Overview of a hire-transaction

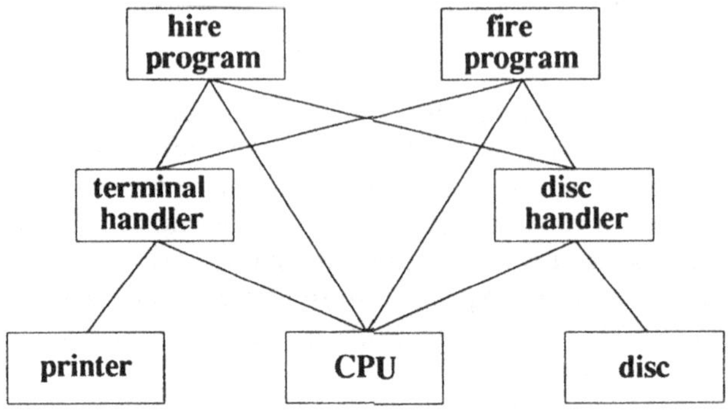

Fig. 11.7. A computer resource model of the hire/fire system

In the implementation domain we also concentrate on performance issues. The application programs require computer resources in order to execute. The computer resources are offered by, e.g., terminal handlers, and disc handlers. One possible computer resource model of a hire/fire system is depicted in Fig. 11.7. We see from the figure that the hire-program (as well as the fire-program) uses terminal, cpu, and disc resources.

The diagrams show very distinctly two typical features of those documentation techniques that are designed as additions to programming languages:

- There is no formal relationship between the diagrams and the software that the diagrams represent;

- The diagrams describe systems functionality in an implementation oriented style.

These two features lead to severe difficulties during software maintenance. When software is modified, there is no automatic updating of the diagrammatic documentation, because of the lack of direct relationship between code and documentation. Furthermore, modifications of systems functionality have to be expressed in terms of the current implementation. These are severe drawbacks, and decrease the usefulness of informal documentation.

The two weaknesses mentioned above are present in most contemporary systems development methods to a lesser or larger extent. In the structured analysis/structured design approach the lack of formal relationships between the diagrams and the resulting software leads to difficulties during maintenance. The state-of-the-art in information systems engineering has still not reached a stage which permits the integrated modeling of the three domains. Currently the best-known approach to achieving this end is the object-oriented approach, which will be surveyed in the next section.

11.4 The Basic Concepts of the Object-Oriented Approach

The object-oriented (OO) paradigm is a new approach for software development. In this paradigm, the real world is viewed as consisting of autonomous, concurrent objects interacting with each other. Each object has its own states and behavior, resembling their counterparts in the real world.

The object-oriented paradigm sprang from programming, has matured into design, and has recently moved into analysis [KORS90]. Object-oriented programming can be traced back to more than two decades ago when Simula 67 was invented. Later at the end of 1970s, the pure object-oriented language Smalltalk [GOLD84] was introduced and, with its success, inspired substantial interest in the object-oriented approach.

The basic concepts are *object, object class, encapsulation, inheritance, polymorphism*, and *dynamic binding*.

11.4.1 Objects

Objects are the basic building blocks of an object-oriented system. In the OO paradigm, the real world is viewed as consisting of objects, and hence many real world applications can be considered object-oriented systems, which may or may not be computerized. This provides unified modeling in the real world realm, the software design realm, and the programming realm. It allows seamless transition from one realm to another.

The notion of an object includes the following:

- An object models an entity or thing in the application domain. For example, books, employees, etc., in the real world can all be modeled by objects.

- An object has a set of attribute values that define a state of the object. For example, the status attribute of a library book may have as its values "available", "checkout", "on reserve", "missing", and "removed". These values may be used to determine the state of a book object at any time.

- An object has a set of operations that the object is capable of performing to change its attribute values, and may cause changes to attribute values of other objects. For example, filling an order in a retail company may cause the following changes: 1) the order changes its state from "new order" to "filled order"; 2) the customer's balance is changed to reflect the additional amount that is charged to the customer; and 3) the inventory level or quantities-on-hand of the merchandise is updated to reflect the amount sold to the customer.

- An object encapsulates both its attributes and operations; this means that the attributes and operations of an object are modeled and stored together with the object. In the function-oriented and data-oriented paradigms, the attributes and the operations of an object are modeled and stored separately as described in Chaps. 2 and 3.

- An object has an identity that can be used to uniquely identify the object, or distinguish the object from other similar objects. Each object has its own identity so that even if two objects have the same attribute values, they can still be identified by using their identities.

11.4.2 Object Class

An object class defines the structure, i.e., the attributes and their types, and the operations. The definitions of the operations differ substantially between the authors. The differences lie in the level of abstraction and the mechanism used to specify the operations. An object class can be interpreted in two different ways: 1) it defines the intention, i.e., constraints on the attribute values that an object of the class can have, and constraints on the invocation of the operations; 2) it defines the Herbrand universe, i.e., all possible objects of the class. To distinguish between these two interpretations, some authors use object type to refer to the first interpretation, and object class to refer to the second interpretation. However, the term "object class" or simply "class" is most often used to refer to the type of a class of objects.

An object class is a subclass of another object class if every object of the former is also an object of the latter. The latter is called a superclass of the former. An object class may have more than one superclass. For example, graduate teaching assistants are a subclass of both students and employees. The notions of object class and object subclass are closely related to the notion of inheritance.

11.4.3 Encapsulation

Encapsulation means modeling and storing, with an object, the attribute values and the operations the object is capable of performing. In a conventional paradigm, the modeling of these two aspects is done separately. For example, in structured analysis, the operations that can be performed on book objects are modeled using DFDs, while the attributes of book objects are specified in a data dictionary.

Encapsulation relates closely to the notion of information hiding [PARN72B], which suggests that a software module designer should try to hide or localize the internal linkage of data structures, and implementation details of the procedures. Encapsulation provides an effective way to enforce information hiding, because the data aspect of an object may be made private and access to these private data can only be achieved through invoking the operations. Thus, the ripple-effect of changes to data structures and algorithm implementation may be minimized.

The advantages of an integrated treatment of attributes and operations of an object are:

1. That the semantics of an object is more complete because it also describes which attributes are changed by which operations, and what are the effects of such changes.

2. That it supports information hiding; and hence, it may facilitate software testing and maintenance.

3. That it facilitates software reuse because: a) the semantics of an object is clear and easy to understand; b) objects are loosely coupled and can be added or removed more easily; and c) tailoring of reusable objects would create less ripple-effect due to information hiding.

11.4.4 Inheritance

Inheritance means *properties defined for an object class are automatically defined for all of its subclasses* (unless selective, or overriding inheritance is specified).

Inheritance may be considered along two orthogonal dimensions. Along the vertical dimension, it is concerned with how properties of a superclass are inherited by a subclass. At least two types of this inheritance are described in the literature: *class inheritance*, and *instance inheritance*. Class inheritance is also called type inheritance, that is, a more specific object type (i.e., a subtype) inherits the attributes (values) and operations defined for a more general object type (i.e., a supertype). The more specific object type, however, may have additional attributes and operations of its own. Instance inheritance is per object inheritance and is based on the axiom that objects of a subclass are objects of any superclass of the subclass.

Suppose book is a subclass of document and my_book is an object of book. Then by this axiom my_book also is an object of document. The title of my_book as a document also is the title of my_book as a book. In this sense my_book as a book inherits the title (whose value is of type string) from the document superclass. Strictly speaking, this cannot be considered as inheritance because my_book remains one and the same object – it does not inherit anything from any other object. It does, however, play two different roles in this example: as a document, and as a book. In this sense, some authors prefer to call object subclasses roles.

Along the horizontal dimension, it is concerned with inheritance from one, or more than one superclass, i.e., multiple inheritance. If the superclasses use the same name for different properties, then from which superclass shall the subclass inherit the property? The common solution is to define a total order that linearizes the superclasses so that there is a unique selection for each property defined in the superclass.

The most promising benefit of inheritance is software reuse, which has been utilized widely by software engineers. Since properties of a superclass are automatically defined for all of its subclasses, software that implements the operations of the superclass can be re-used by the subclasses. For example, a `print` operation for a document object class may be re-used to print a technical report if the printing of a technical report requires the same actions to be performed. Methods inherited from a superclass must be retested in the context of the subclass, because the testing using the superclass context may not include all the cases that may occur in the context of the subclass [PERR90], [WEYU86].

11.4.5 Polymorphism

Polymorphism means the ability to take more than one form: an attribute may have more than one type of values, and an operation may be implemented by more than one method. A simple example is a list object, where the type of its elements may change. If `sort` is an operation on list, then polymorphism means: 1) the element type of the list to be sorted may be different at different times; and 2) the `sort` operation may be implemented by different algorithms (and the appropriate implementation will be chosen by the compiler at run time). Another commonly used example is from computer graphics, where a graphics operation, such as `draw`, is implemented for different graphic objects, such as arc, rectangle, line, point, etc. Depending on the type of the object to which the operation is to be applied (i.e., the target object), the appropriate implementation is selected. Thus, instead of `draw_arc`, `draw_rectangle`, `draw_line`, `draw_point`, etc., only one operation name is sufficient. In this sense, the `draw` operation is said to be semantically overloaded because the operation name has been used to mean the drawing of more than one type of graphic object.

Dynamic binding is an effective mechanism to implement polymorphism. As discussed in the last section, an operation may have more than one implementation. The choice of which implementation to use when an operation is invoked is determined at run time, by the types and number of the arguments, and the function pointed to by a function pointer.

11.5 Object-Oriented Analysis

The objective of object-oriented analysis is to produce a software requirements specification, to be used as basis for object-oriented design and object-oriented programming. The software requirement specification states the capabilities the software system must provide, and the constraints on the implementation. The analysis task is to identify and specify key application objects (including their relationships and behaviors), and requirements and constraints on their evolution. Object-oriented analysis is an important activity since the quality of the software requirement specification largely determines the extent to which the software system would satisfy the user's needs.

In the object-oriented analysis phase, the analyst studies existing documentation and communicates extensively with users to identify their business goals, current problems, and from these to derive their needs. An object-oriented analysis model of the current application is developed during this process. Based on the needs and the analysis model, the software requirements and constraints are derived. Current research has focused on developing better analysis models for representing objects, object classes, and their interactions.

In this section, we briefly describe two object-oriented analysis methods: Coad and Yourdon's object-oriented analysis OOA, Rumbaugh et al.'s object modeling technique OMT.

11.5.1 Coad and Yourdon's Approach

Coad and Yourdon's object-oriented analysis [COAD90] is a five-step method:

1. *Finding Class-&-Objects*. An object is an abstraction of something in an application domain that the system needs to keep information about or interact with. An object encapsulates the attribute values and services (i.e., operations). A class is a collection of objects.

2. *Identifying Structures*. The subclass/superclass relationships, called Gen-Spec Structures, and Whole-Part Structures are identified and specified. A Whole-Part Structure specifies the relationships between objects such that one object has several other objects as its parts or components. For example, door, tire, engine, windshield, etc., are parts of a car.

3. *Identifying Subjects.* A subject is a mechanism for guiding a reader, e.g., an analyst, domain expert, manager, or client, through a large, complex structure (Gen-Spec, or Whole-Part Structure). Class-&-Objects that belong to a subject of conversation are grouped together to facilitate communication.

4. *Defining Attributes.* Attributes are values that describe the states of an object and are exclusively manipulated by the services of that object. Attributes for the object classes are identified and specified using the modeling symbols.

5. *Defining Services.* A service is an object behavior required by an application. The services provided by each class are identified and specified using state transition diagrams and service description symbols.

Part of an analysis model in Coad and Yourdon's notation is shown in Fig. 11.8. In particular, Fig. 11.8a shows that Document is a superclass of Book, and Reserved_Book. Fig. 11.8b is a description of the Document Class, and Fig. 11.8c describes a document check_out procedure (i.e., a service).

Coad and Yourdon's modeling method is also capable of describing Whole-Part Structures, connections between objects, and message passing, but these are not presented in Fig. 11.8. The constructs of Coad and Yourdon's approach are summarized in Fig. 11.9.

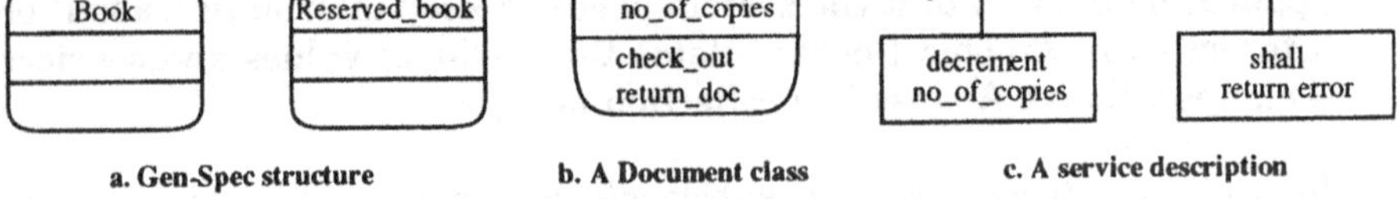

Fig. 11.8. An example in Coad and Yourdon's notation

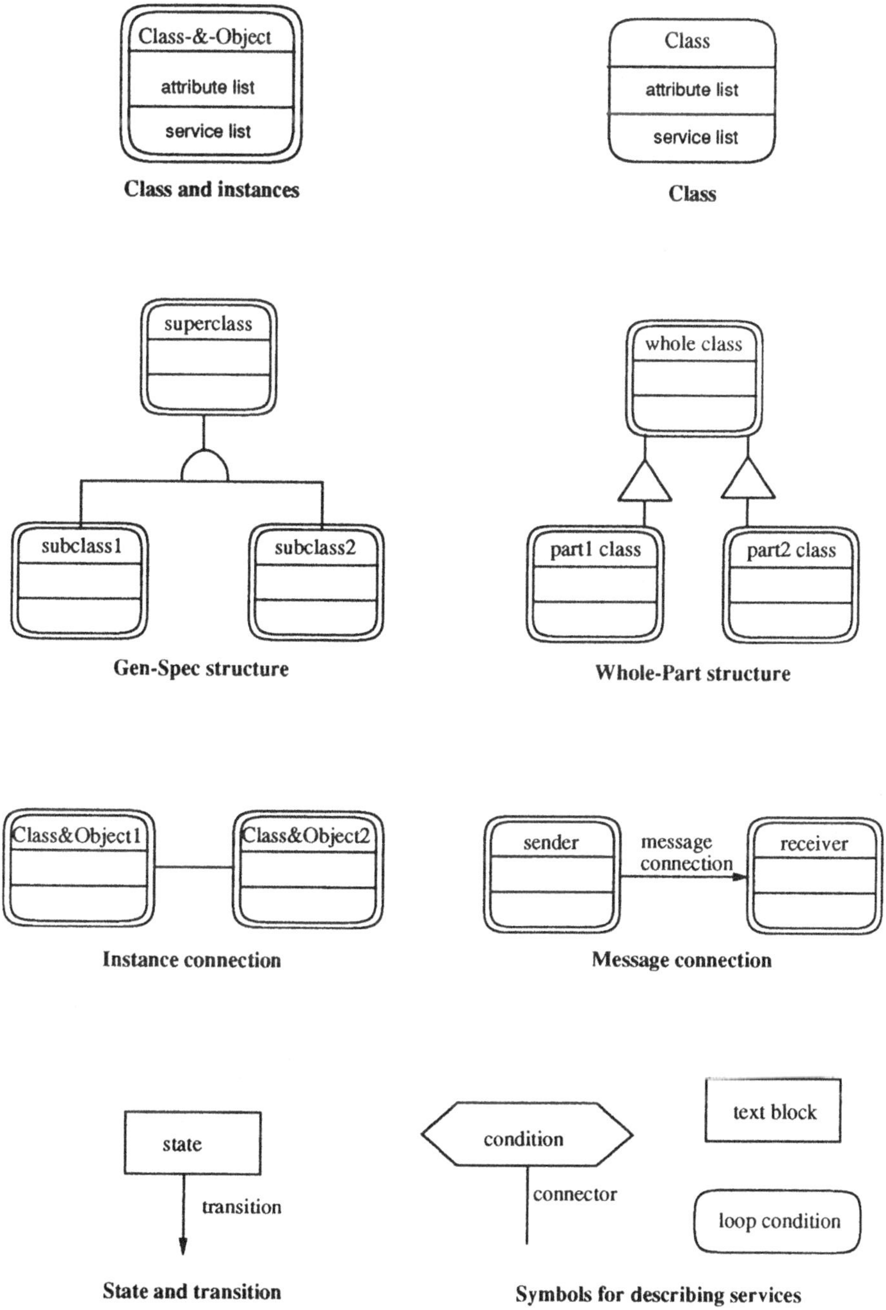

Fig. 11.9. Coad and Yourdon's notations for OOA

11.5.2 The Object Modeling Technique

The Object Modeling Technique (OMT) [RUMB91] is based on three major models: the object model, the dynamic model, and the functional model. The object model describes the static structure of the objects and their relationships. It can be seen as an extended Entity-Relationship model. The dynamic model describes the state transitions of the system being modeled. It consists of a set of concurrent state transition diagrams. The functional model describes the transformations of data values within a system. It is described using data flow diagrams. The OMT analysis methodology consists of the following steps:

1. *Identifying objects and classes*. This is done by studying a problem statement and other documentation to identify the possible objects and classes. The result is documented using OMT symbols.

2. *Preparing a data dictionary*. This step involves preparation of a short description for each object and class identified in the last step.

3. *Identifying associations between objects*. Associations are derived from stative verbs or verb phrases, such as *next to, part of, contained in, drives, writes, has,* etc. They describe a relationship between two or more objects.

4. *Identifying object attributes and link attributes*. Attributes usually correspond to nouns followed by possessive phrases, such as "the title of a document", "the number of copies of a book", etc. Adjectives may correspond to enumerated attribute values, such as "available", "unavailable", "missing", and "removed" may represent the values of a library book's status.

5. *Organizing and simplifying object classes using inheritance*. This is achieved by generalizing several subclasses to yield a superclass so that the attributes for the subclasses may be moved up to the superclass. A class may also be "divided" into several simpler subclasses by recognizing adjective-noun phrases, such as "graduate student" and "undergraduate student". These two phrases suggest that the class of students might be divided into two subclasses, corresponding to graduate and undergraduate students.

6. *Verifying that access paths exist for likely queries*. This is done by navigating through the object model(s), guided by the queries to be answered to ensure that the information needed to answer the queries can be derived from the model.

7. *Iterating and refining the model.*

8. *Grouping classes into modules.*

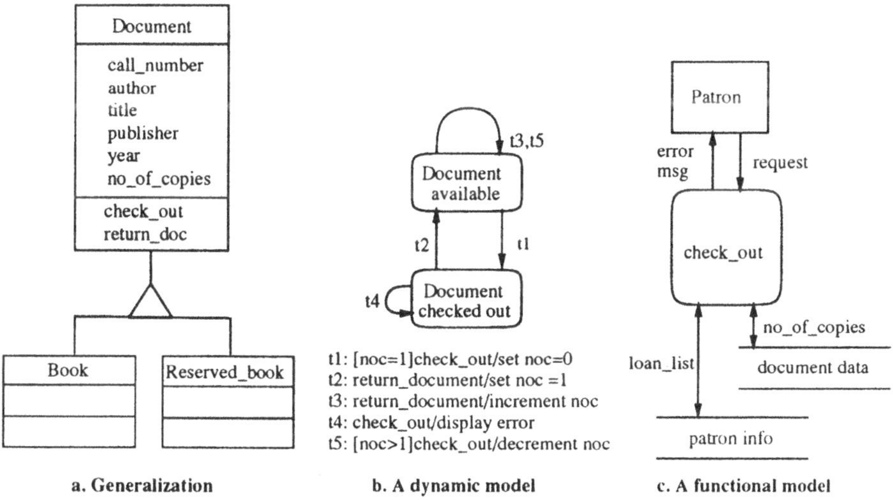

Fig. 11.10. An example in OMT notation

To illustrate this, Fig. 11.10 shows part of an OMT model for a library application. In particular, Fig. 11.10a specifies an inheritance hierarchy and the attributes and operations for the Document class, Fig. 11.10b describes the state dependent behavior of document, and Fig. 11.10c shows what are the input and output of the check_out operation.

OMT also provides constructs for modeling various types of aggregation, association, instance of a class, derived attribute and class, constraint, and many other features. These are summarized in Fig. 11.11. Note that Fig. 11.11 only depicts the basic modeling constructs for object modeling. Advanced constructs, constructs for dynamic modeling, and function modeling are to be found in [RUMB91].

11.6 Object-Oriented Design

Unlike in the structured paradigm, the distinction between object-oriented analysis and object-oriented design is not obvious and it is difficult to draw a clear line between them. This has been referred to as the seamless transition from object-oriented analysis to object-oriented design. In our opinion, the distinction is in orientation: the analysis phase is application or problem oriented, while the design phase is software oriented. This means that the focus in analysis is to identify and specify the application and application requirements, while in design the focus is to derive, from the analysis model, software and object structures that best satisfy the chosen architecture and software design principles.

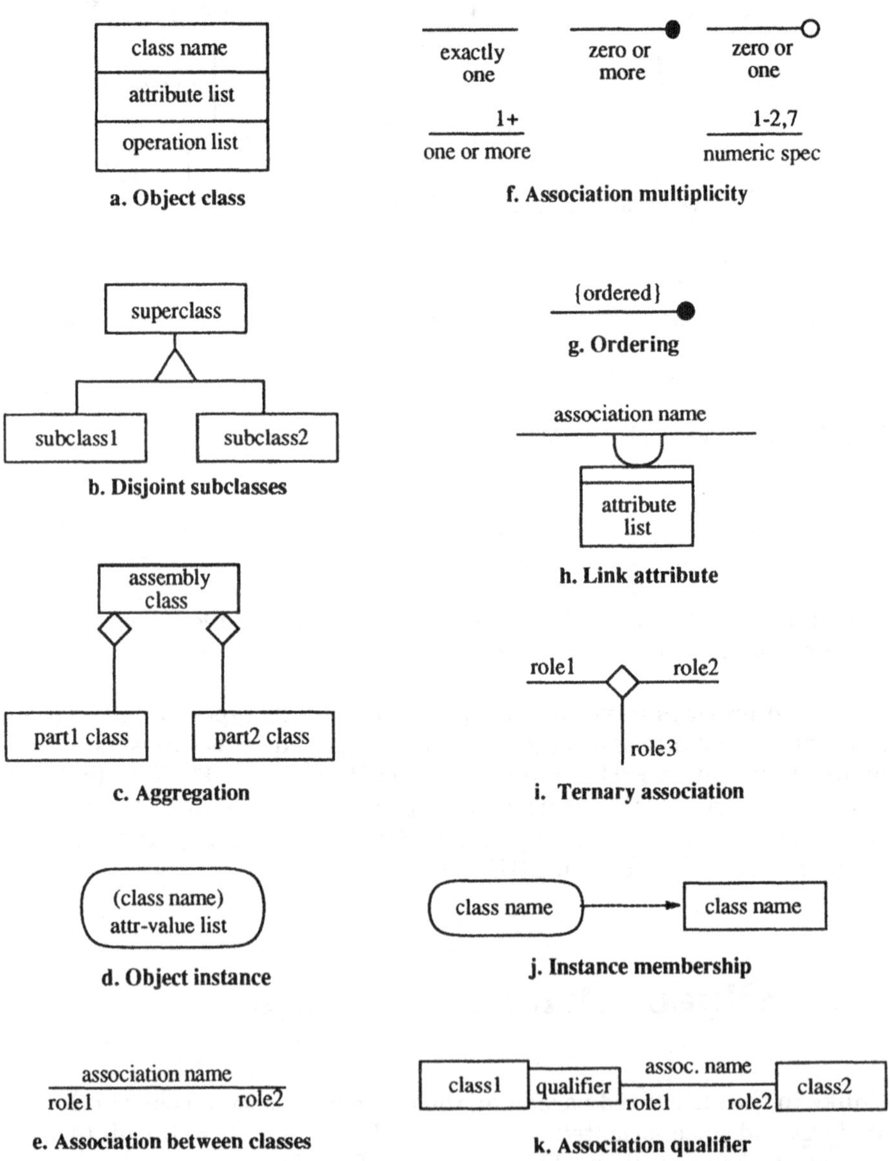

Fig. 11.11. Rumbaugh et al.'s basic notations for object modeling

Object-oriented design is a methodology dependent activity. This means that different object-oriented analysis methods would lead to different design methods. Moreover, design methods tend to involve considerations of various design decisions, which depend on the application on hand and the target system. To avoid getting into these details, we choose to give a general presentation of software design.

Object-oriented design consists of software architectural design and object design. In architectural design, the overall software architecture, consisting of interacting software subsystems, is derived. In object design, the logical data structures, algorithms, and the interfaces between the objects are determined.

11.6.1 Architectural Design

The object-oriented paradigm is basically a bottom-up approach for software development, because the object-oriented analysis phase is focused on specifying the application domain objects, their relationships, and behavior, while the overall software architecture is derived from the analysis model in the design phase.

It is commonly agreed that an object-oriented system consists of a set of subsystems. According to Rumbaugh et al. [RUMB91], a subsystem is a package of classes, associations, operations, events, and constraints that are interrelated and have a reasonably well-defined, small interface with other subsystems. A subsystem may be identified by the services it provides.

Unfortunately, different authors propose different ways to organize the subsystems. In our opinion, the subsystems may be organized in a tree or lattice structure, as in structured design. The root of the tree represents the system, which consists of several subsystems, each of which in turn consists of lower level subsystems, and so on. A system or subsystem can be viewed as an object and their relationships may be modeled by the Whole-Part [COAD90], or aggregate [RUMB91] constructs. The information processed by a system or subsystem may be modeled as attributes and the services provided may be modeled as operations.

At this level, the object classes identified during the analysis phase are of secondary importance. They are associated with the subsystems according to the functions they provide. For example, we may consider that a library system consists of four subsystems: *circulation, cataloging, purchasing,* and *patron assistance*. Since the main processing of documents is done by circulation in a library system, we can associate the document class with the circulation subsystem. When a new document is purchased and catalogued, the cataloging subsystem may request the circulation subsystem to create an instance for this new document by sending the necessary information to the circulation subsystem.

11.6.2 Object Design

Object design is based on the analysis results. The goal is to determine the data structures to be used to implement the attributes of the object classes, the algorithms to implement the operations, and the interfaces between the objects. Software design principles such as abstraction, information hiding, and modularity are applied to obtain better software.

Object designs may be described using Brunet's notations (see Fig. 11.12) [BRUN91]. Other notations, such as Booch's notation [BOOC86] and Wasserman's notation [WASS90], may also be used to represent object designs.

In illustration, Fig. 11.13 shows how these symbols are used to describe two modules that implement two object classes and their interactions. The diagram shows a patron object and a document object. The patron object is

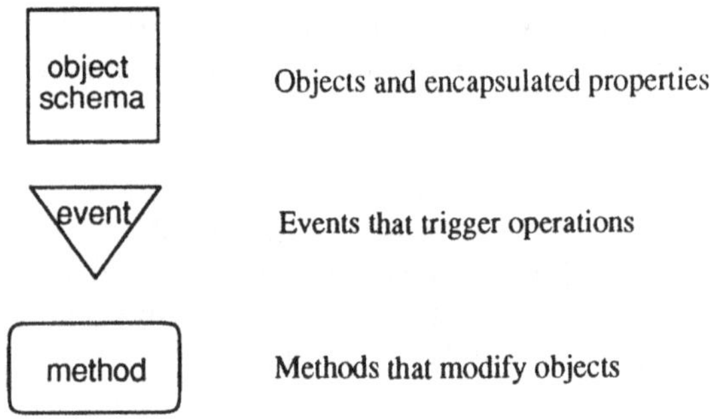

Fig. 11.12. Brunet's symbols used for object design

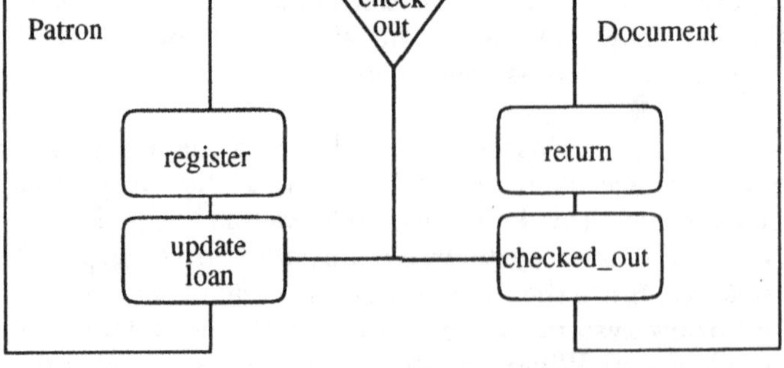

Fig. 11.13. Diagram illustrating interaction between objects

assumed to have two methods, `register` (a patron) and `update loan` (for a patron). The document object has two methods, `return` (a document) and `check_out` (a document). A check_out event invokes both the `check_out` method and the `update` method to update the information about the particular patron and the document being checked out.

As described above, object design also must determine the data structures and algorithms to implement the object classes. This is done as in traditional software design and is omitted in this book.

11.7 Object-Oriented, Function-Oriented, and Data-Oriented Approaches

The development of the object-oriented paradigm has also been motivated by problems associated with the function-oriented and data-oriented approaches of software development. The structured analysis and structured design method [DEMA78], [GANE78], [YOUR89] has been the most recognized function-oriented approach, and semantic data models [CHEN76], [HULL87], [PECK88] are often associated with the data-oriented paradigm.

In a function-oriented approach, software development begins with identifying and specifying the functions or transformations that manipulate the data stored in a system. Data flow diagrams (DFD) are the most often used tool for this purpose. A DFD is a directed graph, in which nodes represent functions and directed edges represent interaction between the functions through data flows. The contents and logical structures of the data, being stored in or passed throughout a system, are specified separately in a data dictionary. The resulting DFD analysis model is then converted into a tree or lattice structure, representing the invocation relationships between the software modules. This software architectural design is then used to guide the programming and testing activities (i.e., top-down, or bottom-up programming and integration testing).

Industrial experience has shown that the functional approach has several drawbacks:

- consistency between the DFD and the data dictionary is difficult to maintain;

- since the functions are tightly coupled through the data flows, it is difficult to eliminate or minimize the so-called "ripple effect";

- the difference in point of view between following data flows in structured analysis and building hierarchies of tasks in structured design has always been a major problem [COAD90], [KORS90];

- the tight coupling between the functions, and treating data and functions separately makes software reuse difficult.

In a data-oriented approach, the starting point is modeling of the data structures. The specification of the functions and procedures is done according to the data structures. Usually, these two aspects are specified in two separate models: a snapshot model and a process model. The drawbacks of this approach are:

- consistency between the snapshot model and the process model is difficult to maintain;

- depending on the process or transaction specification language, interaction between the functions might not be explicitly modeled, and hence their effect is difficult to comprehend;

- how to convert an analysis model into a good software design has seldom been addressed;

- since functions and data are treated separately, software reuse is still difficult.

The object-oriented paradigm recognizes the drawbacks of the other paradigms and attempts to provide an integrated and balanced treatment of the static and the behavioral aspects of a system. The starting point is objects, which encapsulates both the static and the behavioral aspects of an application. Based on objects, a new set of concepts are introduced into system modeling, design, and implementation.

Exercises

1. In the development of a Payroll subsystem in an organization, give examples of elements in the

   a) subject domain,
   b) interaction domain,
   c) implementation domain.

2. State the advantages and disadvantages of the DFD technique used in interaction domain modeling.

3. Model the interaction domain of the problem given in Figs. 11.2–5 by using structured English.

4. Develop a complete object oriented model for the Car Rental Agency (CRA) problem introduced in Exercise 6 in Chap. 2.

5. Explain how object oriented analysis helps in maintenance and reuse.

6. Compare object oriented analysis with other modeling techniques.

7. Give examples of

   a) an object,
   b) a class,
   c) a subclass,
   d) encapsulation,
   e) inheritance,
   f) polymorphism.

Chapter 12

Model Integration with Executable Specifications

Most contemporary approaches to information systems development are based on the idea that a system is built through a succession of development phases. Separate issues are dealt with in different phases. Functional issues are central to the first phase, while issues of software and database design are treated in later phases. This is explained in some detail in previous chapters of this book.

Iterative and explorative approaches are weak on providing simple user-oriented system abstractions

There are other approaches. In particular there is an approach which is known as the *iterative development* approach. The main idea is that the most important operational functions of an information system are designed, implemented, installed and put into operation as quickly as possible. The system's evolution is seen as a sequence of additions and modifications to the specifications and to the software, as users' operational experiences are forcing system changes. In the artificial intelligence community a similar approach is known as *explorative programming*. System specifications are expressed in mathematically founded languages, e.g., LISP, and are consequently directly executable. System evolution appears in the form of modifications to the executable specifications.

The major weakness of these two approaches is that system modifications are discussed relative to software details of an operational software system. It is well known that it is difficult to lead a discussion about principles when the basis for the discussion contains many details. One gets bogged down by detail: it becomes difficult to see the forest because of the trees. It is necessary to remove unnecessary detail from the system description in order to achieve a better basis for human beings to understand the system. Alas, there is no effective technique for abstracting away irrelevant specificational detail in the iterative and explorative approaches.

Specification statements may be viewed as constraints on a solution space

There is another way of viewing systems development that emerges naturally from our previous discussion on the constructivity issue. We may view a specification statement as a *constraint on a solution space*. Assume that we state that an information system should support the ticketing of an airline. That statement will exclude all solutions that do not have the functional property of supporting the ticketing of the airline. We may later on state that the response time for a request for flight information about any route that the airline serves should be less than 0.5 second for 80% of the requests, and not more than 2.5 seconds for the remaining 20%. This statement will constrain the solution space even more.

Information system development may consequently be seen as a process of gradually constraining a solution space until only one solution remains. This view has important consequences for information system modeling. It implies that in order to control the system development process we must be able to find out whether a proposed solution is within the solution space or not. This can only be done effectively if we have access to a specification method which is constructive with respect to the relevant system properties.

Constructivity in information systems modeling implies that systems specifications are executable

Constructivity in a modeling approach is independent of which system level is being described. It does not matter whether a system is described on the level of algorithms or on the level of data flow. A property which is constructive on one level must be constructive on every level of specificational detail. A consequence of this is that the modeling constructs of a constructive approach must be the same regardless of which system level is being described. This means that the modeling constructs that are being used for abstract descriptions of a system must be included in the constructs that are being applied for the most detailed algorithmic descriptions. System specifications are therefore, in principle, executable specifications provided that enough detail is supplied.

In the remainder of this chapter we shall discuss some of the features of a modeling approach which is based on executable specifications, and which aims at overcoming two of the major deficiencies of most contemporary modeling approaches: lack of constructivity and lack of abstraction facilities. The modeling approach is called PPP, which stands for Phenomena, Processes, and Programs [GULL91].

12.1 Constructivity in Information Systems Modeling

The major feature of a constructive subsystem structure is that the system's properties can be formally derived from statements of components' structural relationships and components' properties. It is commonly found that for some properties it is possible to find constructive modeling techniques, while for other properties it is not possible to model constructively. We will show three levels of modeling detail in information systems specification and discuss them from the point of view of obtaining the property of constructivity. Our example is the process of finding the solution of the 2nd-degree equation $Ax^2 + Bx + C = 0$.

The list of input and output elements may be constructed from the component structure of a system

It is well known among first-year high-school students that this equation has the two solutions

$$x_1 = -\frac{B}{2A} + \sqrt{\left(\frac{B}{2A}\right)^2 - \frac{C}{A}}$$

$$x_2 = -\frac{B}{2A} - \sqrt{\left(\frac{B}{2A}\right)^2 - \frac{C}{A}}$$

This knowledge is depicted in Fig. 12.1.a. Our next step is to work out the details of the solution. We choose first to calculate

$$p = -\frac{B}{2A} \qquad \text{and then} \qquad q = \sqrt{\left(\frac{B}{2A}\right)^2 - \frac{C}{A}}$$

But what happens if q is negative? Second-year high-school students might have heard about complex numbers, and will know that the first-year students have a simplified view of the harsh realities of the world of mathematics.

The more realistic solution process is depicted in Fig. 12.1.b. We distinguish between three cases: the equation having two real solutions, one real solution, and two complex solutions. There is an inconsistency between the external properties of the simplistic solution and the external properties of the realistic solution. We see that it is easy to derive the list of output elements x_1, x_2, x, z_1, and z_2 which constitute the output element property of the system (Fig. 12.1.c). The specification model used is constructive with respect to finding the list of output elements. However, the specification model is too limited to carry knowledge about the interdependence among the output elements. We know that the equation cannot simultaneously have both real and complex solutions.

Relationships among input elements and among output elements constitute a constructive property in the process-port model

We have to extend the specification model if we want to represent that feature. The simple input/output specification model of Fig. 12.1 is extended with facilities to state relationships between the output elements. We use the logical AND-operator and the XOR-operator (XOR = exclusive OR). If the AND-operator is applied to two or more output elements of a process, all of the output elements are required to be produced before the process terminates. If the XOR-operator is applied to two or more output elements, one and only one of the output elements is required to be produced before process termination. The corresponding graphical notation is shown in Fig. 12.2.a. The symbols for the logical operators are called ports, and the specification model is called the process-port model. It should be noted that the full model is more detailed than what is shown here [GULL91]. Further details will be introduced in the next section.

In Fig. 12.2.b the process port model is applied for the solution of the 2nd-degree equation. In the first step both p and q are calculated, while in the second step indicates that only one of the cases $q < 0$, $q = 0$, $q > 0$ can happen, so that in the third step we get either x_1 and x_2, or x, or z_1 and z_2. The specification model is constructive with respect to the relationship among the output elements. We can therefore derive the relationships among the output elements on the "supersystem" level if we know the specification of the components' relationships and components' output port properties. In Fig. 12.2.c we have depicted the results of having non-constructive or constructive specification models. In the first case, we cannot derive any relationship among the output elements. In the second case, the derived relationship among output elements is represented by a compound port, which is to be read XOR $((x_1\ x_2), x, (z_1\ z_2))$.

Can the algorithmic relationship between input and output elements be made to be a constructive property?

But we are still not satisfied. If we look at Fig. 12.2.b, we see that we need very little additional specificational detail to be able to relate the results of the system x_1, x_2, etc., to the input A, B, C, of the system.

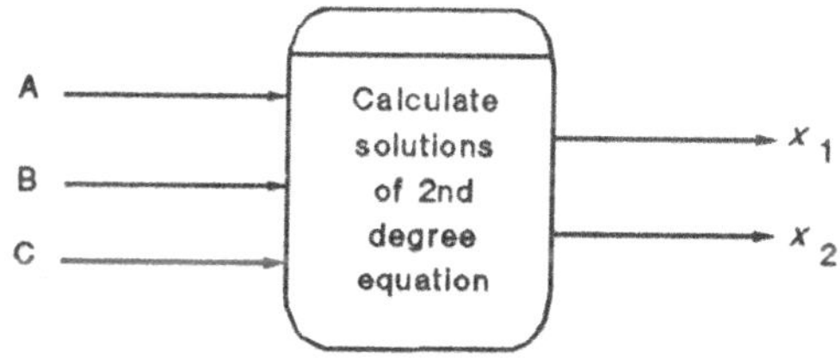

(a) Solutions of 2nd-degree equation, simplistic view

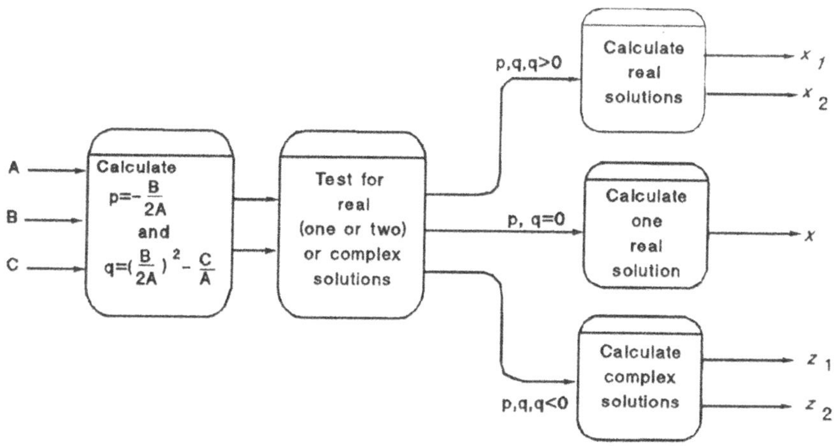

(b) Solutions of 2nd-degree equation, details of realistic view

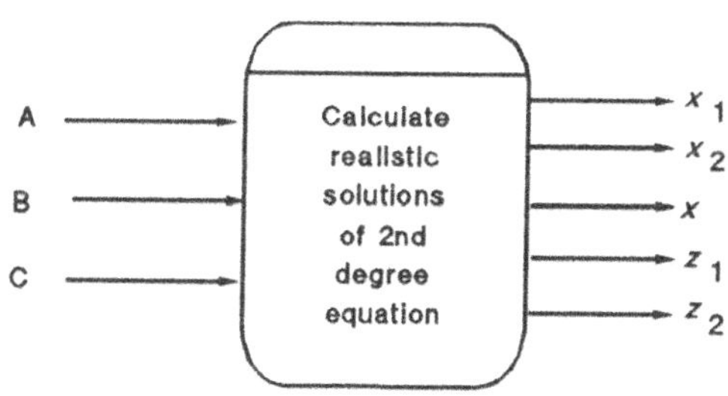

c) "Super system" of system in (b)

Fig. 12.1. Solution of 2nd-degree equation, simple I/O model

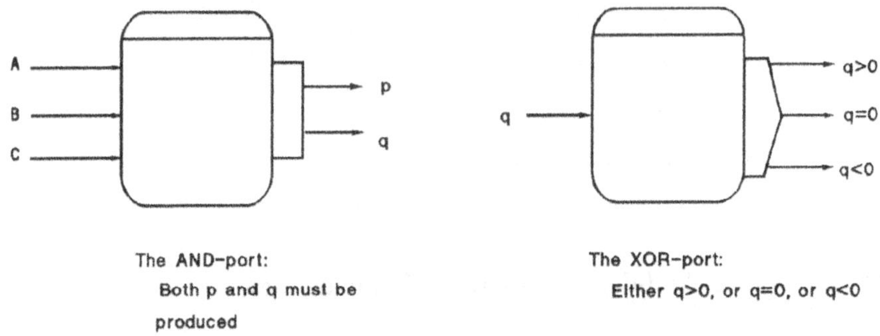

a) Information processes with AND-port and XOR-port

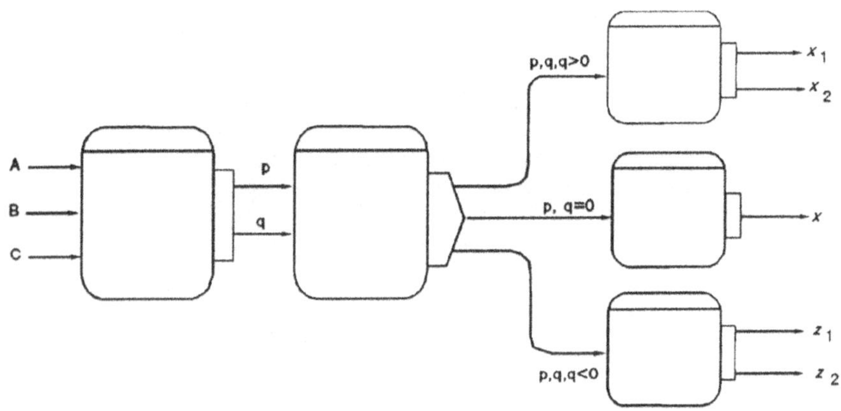

(b) Solution of 2nd-degree equation, specified with process-port model

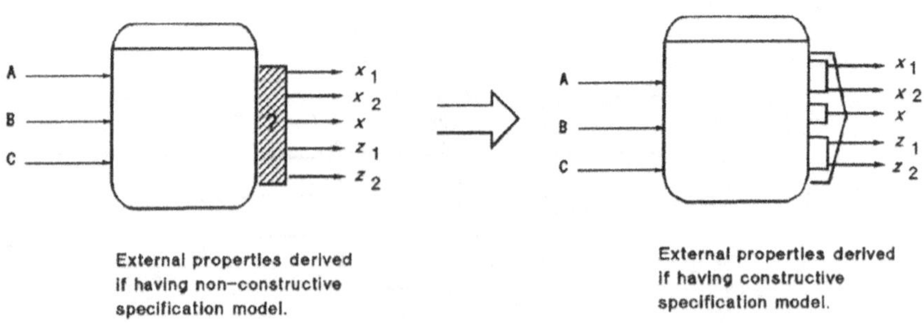

(c) Compound port, constructed from subsystem in (b)

Fig. 12.2. Solution of 2nd-degree equation, specified with process-port model

Let us assume that we can functionally relate the input and output of each system component to each other so that we have

process 1: $p = -\dfrac{B}{2A}$; $q = p^2 - \dfrac{C}{A}$

process 2: if $(q > 0)$ go to 3
if $(q = 0)$ go to 4
if $(q < 0)$ go to 5

process 3: $x_1 = p + \sqrt{q}$

$x_2 = p - \sqrt{q}$

process 4: $x = p$

process 5: $z_1 = p + \mathrm{j}\sqrt{-q}$

$z_2 = p - \mathrm{j}\sqrt{q}$

where $\mathrm{j} = \sqrt{-1}$

The crucial question, then, is whether it is possible, by using standard mathematical techniques, to derive the derivation rule

$$\textbf{if} \left(\left(\frac{B}{2}\right)^2 - AC\right) > 0 \qquad \textbf{then}\ x_1 = -\frac{B}{2A} + \sqrt{\left(\frac{B}{2A}\right)^2 - \frac{C}{A}}$$

$$\textbf{and}\ x_2 = -\frac{B}{2A} - \sqrt{\left(\frac{B}{2A}\right)^2 - \frac{C}{A}}$$

and so on for all of the output elements based on the system specifications given so far. It has been demonstrated that it is possible to achieve this for systems of standard mathematical formulas by applying symbolic evaluation techniques. If this is also made possible in general, we have a very potent specification model because we can then achieve constructivity on the functional level, meaning that we can derive external functional properties of a computer program in terms of a set of explicit derivation rules. If this type of constructivity was available, it would mean that we would be able to find out if two solution proposals exhibited different external functionality. We could also determine whether a solution proposal was consistent with requirements specifications expressed on the level of functional constraints. This problem is treated in more detail in Chap. 14.

12.2 The PPP Approach

The PPP approach is still in the research realm. PPP stands for Phenomena, Processes and Programs. It consists of three submodels: Process Model (PrM), Phenomenon Model (PhM), and Process Life Description (PLD) which serves as the program model. A user interface description language is in the planning stage [GULL91]. PPP aims at model integration through all phases in a system's life. Rather than creating a completely new model in all aspects, the starting point for choosing modeling constructs in PPP has been the well-known approaches of Entity-Relationship, structured analysis, and graphical program specifications. The three sets of modeling constructs are extended and integrated into an executable specification model which covers all life-cycles phases.

The three submodels of PPP are described in the remainder of this section, followed by an example where the integration of the three models is shown. The presentation is based on a paper by Gulla, Lindland and Willumsen [GULL91].

12.2.1 The Phenomenon Model – PhM

In the realm of phenomenon modeling the intention is to describe the properties of the problem domain as they exist independently of the information system. The purpose is to describe real world objects and their relationships as well as to describe their informational counterparts. A PhM model should give a picture of the problem's characteristica as well as serving as a basis for database design. An object oriented approach to PhM would seem natural. The PhM of the present version of PPP is, however, an extension of the ER-model. The formal parts of the extension is described in the set-theoretic information model of Chap. 14.

An ER-diagram in PhM notation is shown in Fig. 12.3. Entity classes and relationship classes appear as in the classical ER-model, except for the diagrammatic symbol for relationship which is different. Attribute names appear on the arrows between ER-classes and datatypes, which are indicated by ellipsoids.

To improve the expressive power of PhM a few modeling constructs have been added to the original ER-model. The attribute relation has been extended. Covering specifications have been added to the cardinality specifications of the relationships. Subclassifications of entity classes have been made possible. The extensions are depicted in Fig. 12.4.

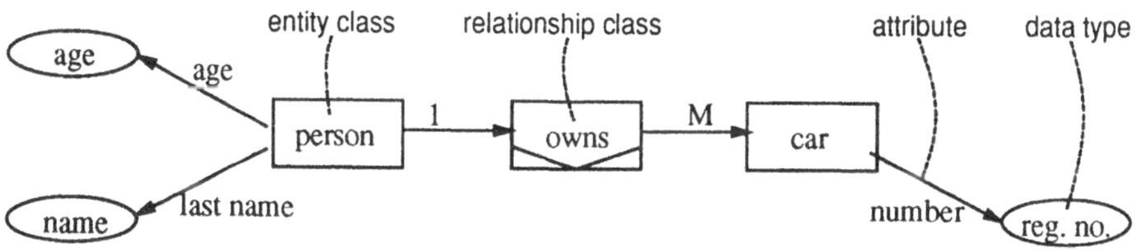

Fig. 12.3. Simple ER model in PhM notation

Attributes, identifiers, repeating groups **and** ***qualities*** **relate data types and entity/relationship classes**

In the original ER-model there is only one type of relationship between entity sets and data types, and no relationship at all between a relationship set and a data type. In PhM this has been extended. Both entity classes and relationship classes may be related to data types. The relations are:

— *identifier* which uniquely determine instances of an entity class. That is, the identifiers of two instances of the same entity class must have different values. Relationship classes are not allowed to have identifiers. The diagrammatic symbol for identifier is **id**.

— *attribute* which is used to denote single-value properties of instances of classes. Several instances of a class may share the same attribute value, e.g., several persons may be of the same age. The diagrammatic symbol for attribute is **att**.

— *repeating group* which is used denote multiple-value properties of instances of a class, e.g., a person may have several children. Several instances of a class may share the same attribute value, e.g., two persons may share the same children. The diagrammatic symbol for repeating group is **rep**.

— *quality* which is used to denote properties that are characteristic of a class rather than of the elements of the class, e.g., the number of instances in the class. The diagrammatic symbol for quality is **qual**.

Covering and subclassification may be expressed for entity classes and relationship classes

If all instances of an entity class have to participate in a relationship then this entity class is *fully related* to that relationship class. Otherwise the entity class is only *partially related* to the relationship class, meaning that not every instance of the entity class has to be related to an instance of the

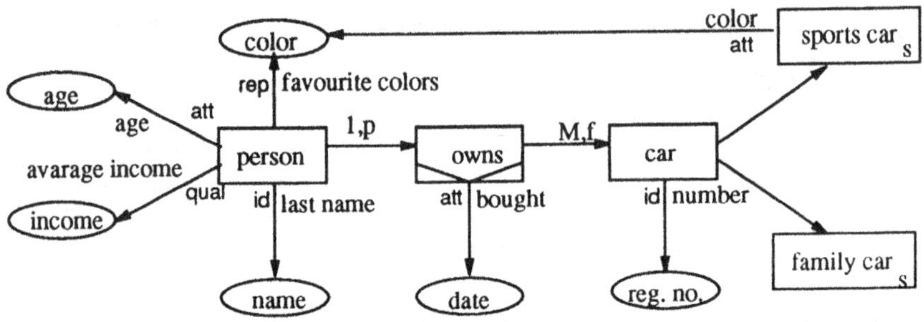

Fig. 12.4. A more sophisticated PhM model

relationship class. The diagrammatic symbols are **f** for *full* and **p** for *partial*. In the example of Fig. 12.4 we see that *not every person owns a car, but every car is owned by someone.*

A class may have several subclasses. If a class has a subclass then all members of the subclass are also members of the class. In the example of Fig. 12.4 the entity class *car* has subclasses *sports car* and *family car*. Subclasses inherit the properties of their superclasses, e.g., sports cars and family cars have registration numbers for cars.

12.2.2 The Process Model — PrM

The process model has already been partly introduced in Sect. 12.1. PrM may be seen as an extension of the ordinary dataflow model of structured analysis. One may also reverse this and view dataflow diagrams to be abstracted PrM specifications. Which of the two views is adopted does not matter very much. We shall anyhow choose to explain the PrM modeling constructs as additions to the well-known dataflow technique.

PrM incorporates the structured analysis constructs of process, datastore, dataflow and external entity. The latter is replaced with the *agent* concept. The PrM modeling constructs which are additional to the dataflow model are concerned with *resources* (and *agents*), *timers*, *triggering* and *termination* of processes, and with *ports* (Fig. 12.5, 12.6).

Resources contain items that are necessary for a process to run, e.g., computing resources or communication resources of various kinds.

Agents are external entities which have dynamic properties. Agents are considered to be external to the information system being considered, and are parts of that system's environment.

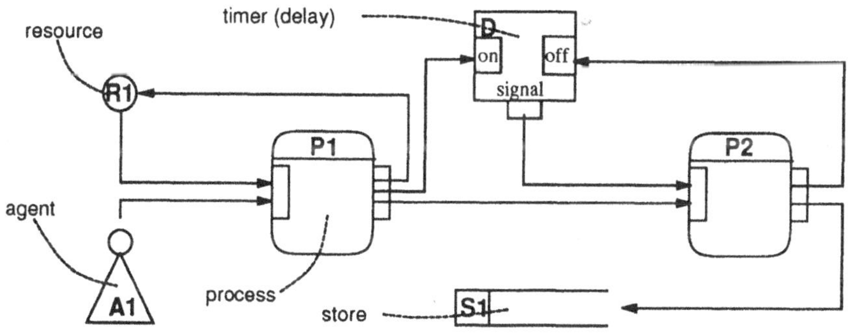

Fig. 12.5. Basic concept of the PrM language

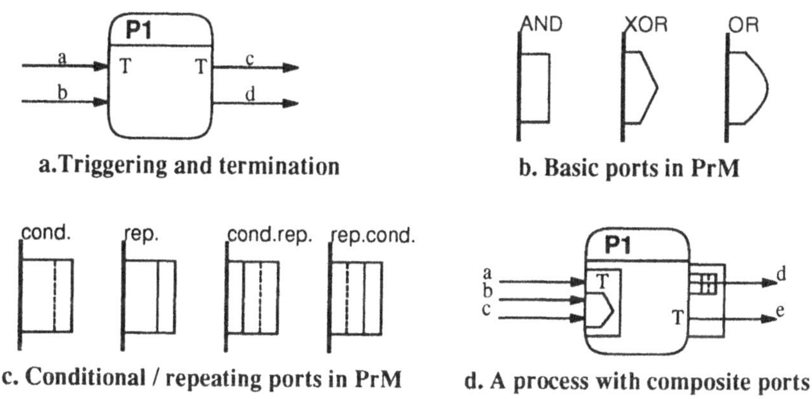

Fig. 12.6. Auxiliary concepts of PrM

Timers are *clocks* or *delays*. Clocks are used to model events that occur at particular points in time, e.g., at a particular date. Delays are used to model events that take place at some point in time relative to (after) some other event taking place.

Resources, agents, and timers are depicted in Fig. 12.5.

Triggering and *termination* of processes are indicated by denoting the respective dataflows by a T-symbol in the processes where flows originate or end (see Fig. 12.6a). The notions of triggering and termination introduce the modeling of control flow. A triggering input flow can only be received when a process is idle, and the receipt of a triggering flow will cause a process to change its state from idle to active. Similarly, a terminating flow is caused by the process changing its state from active to idle.

Consider the process description of Fig. 12.6a. When a data item is received in flow **a** the process is triggered, while the process causes a data item to be sent in flow **c** when the process activity is terminated. Other flows may be sent and/or received anytime between triggering and termination of the process. The sequence of actions in Fig. 12.6a becomes the following: first receive **a**, then receive **b** and send **d** in any order, and finally send **c**.

Ports relate input flows to each other (*input ports*), and they relate output flows to each other (*output ports*). Ports have been introduced in the previous section. There are three basic kinds of ports (Fig. 12.6b):

- AND ports, meaning that all flows are to be received/sent
- XOR ports, meaning that only one of the flows of the port is to be received/sent, and
- OR ports, meaning that one or more of the flows are to be received/sent.

A port may have the additional property of being

- *conditional,* meaning that the flows may or may not be received/sent, and/or
- *repeating,* meaning that the flows may be received/sent several times during a process execution.

Conditionality is added to a port by means of a dotted line, while repetition is added by means of an unbroken line. In Fig. 12.6c the conditional, repeating, conditional repeating, and repeating conditional AND ports have been depicted. The style of the corresponding XOR and OR constructs is similar.

Composite ports are constructed by putting ports inside of each other as shown in Fig. 12.6d, which illustrates the following: When a triggering flow **a** is received the P1-process also receives either **b** or **c**, but not both, as indicated by the XOR-port inside the AND port. On the output side the flow **d** may be (COND) sent several times (REP). Consequently the **d**-output may occur zero, one or several times. The **e**-flow will be sent on termination of the P1-process, as indicated by the termination symbol (T) in the diagram's output port.

Process decomposition is achieved by applying the PPP modeling constructs all through to the algorithmic level. At the most detailed level it is convenient to change the mode of expression, because the graphical language becomes impractical. For this purpose the PLD language is introduced.

12.2.3 The Process Life Description (PLD)

The PLD language is a graphical programming language [TRIP76]. The modeling concepts and the graphical notation differs only slightly from other graphical programming languages; see, e.g., [BROW83]. The major primitive concepts of PLD are the *start*, *send* and *receive* concepts, as well as *sequence*, *assignment*, *iteration* and *choice*. These primitive concepts provide the smallest meaningful semantic units of the PLD language.

Each PLD concept is associated with a graphical symbol. A PLD statement consequently appears as a collection of interrelated graphical symbols. The graphical symbols may be related in two different ways. A symbol *succeeds* another symbol if it is drawn immediately below the other symbol, or if there is a link between the lower left corner of the other symbol and the upper left corner of the symbol which succeeds it. A symbol is *right-connected* to another symbol if there is a link from its upper left corner to the upper right corner of the other symbol. Examples of succeeding symbols and right-connected symbols are shown in Fig. 12.7.

The primitive concepts of PLD comprise the basic concepts of structured programming

The primitive PLD concepts are explained below, and exemplified in Fig. 12.7.

Start indicates the beginning of a PLD specification. It is only used at the top of the PLD program, and is followed by a block of PLD statements.

Assignment symbolizes a block of program statements, or a subprogram call. The graphical symbol is a rectangle containing written assignment statements. An assignment construct may succeed any other primitive construct, and be succeeded by any other construct (except a start construct).

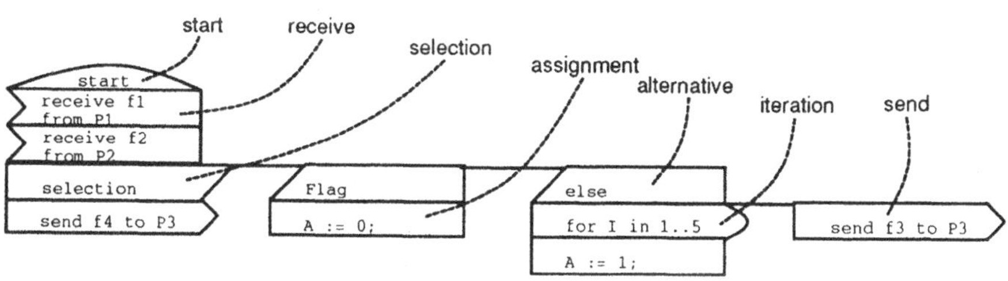

Fig. 12.7. A simple PLD model

Choice specifies a selection among alternatives. A choice construct consists of one selection part, whose only role is to indicate the choice situation, and a set of alternative parts. The selection part may be succeeded by any primitive construct (except a start construct). The alternative parts are right-connected, and each contains a condition succeeded by a PLD block. The alternatives must be mutually exclusive.

Iteration indicates a program loop, and must contain a condition telling when to stop the iteration. A block of PLD statements must be right-connected to an iteration construct. The execution of this PLD block constitutes one iteration. An iteration construct may be succeeded by any other PLD construct (except a start construct).

Send is used to specify what data are to be sent to other PLD programs, or are to be sent to data stores, agents, timers, and resources. PLD programs that receive data must contain a corresponding *receive* construct. The text inside the graphical symbol specifies what data to send where. A send construct may be succeeded by any other PLD construct (except a start construct).

Receive is used to specify which data are input to a PLD program. The text inside the graphical symbol specifies which data are received from where. A receive construct may be succeeded by any other PLD construct (except a start construct).

A complete PLD program consists of a start construct succeeded by a PLD block, which specifies the behavior of the process. The graphical relationships among the symbols provides the basis for the interpretation of a PLD program. Each primitive concept denotes the execution of one or more program statements. The general rule is that *the flow of control is from left to right and from top to bottom.*

PLD program skeletons may be automatically derived from PrM processes

The PrM and PLD languages are partly overlapping. The external properties of a PrM process, as expressed by the port structures, can be represented in a PLD diagram. So process ports may be transformed to PLD format. An example is shown in Fig. 12.8. Process P1.1 `Verify transaction` receives a triggering Transaction from `Customer`, and requires Previous balance from the Account data store, before processing can start. We see how the input flows are transformed to *receive* constructs in the PLD diagram. The output port consist of nested XOR ports. We see how these are transformed to *selection* constructs in the PLD diagram, and how the outgoing data flows are transformed to *send* constructs.

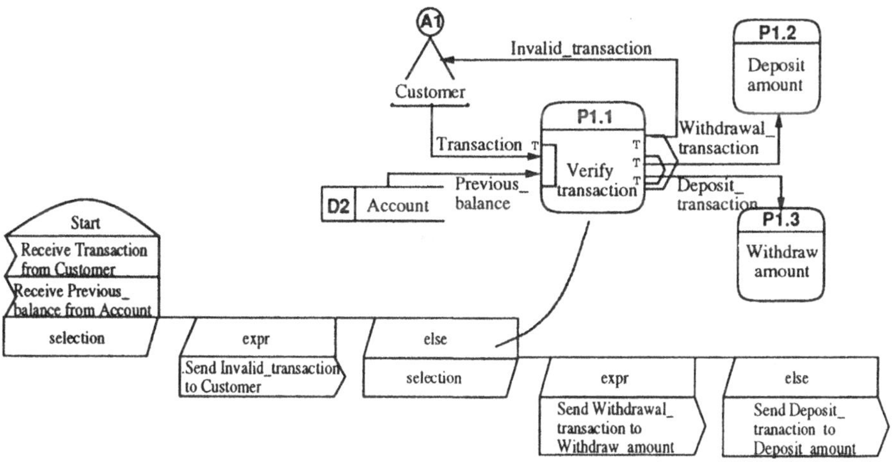

Fig. 12.8. Consistency checks between PrM and PLD

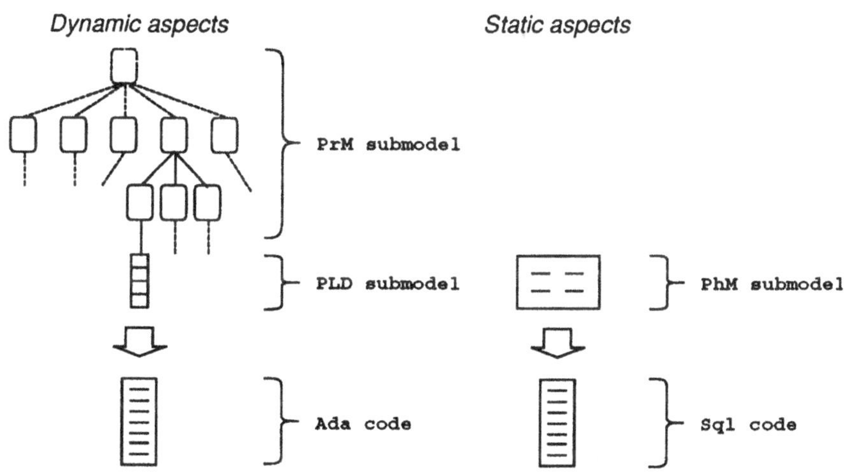

Fig. 12.9. Dynamic aspects

Examples of transformation rules from PrM to PLD are:

- Input flows map to *receive* constructs,
- Output flows map to *send* constructs ,
- Non-triggering *receive* constructs succeed triggering *receive* constructs,
- A terminating *send* construct succeeds non-terminating *send* constructs.

Different port constructs map to different PLD constructs. We shall not delve into detail here.

PLD specifications may be transformed into executable code in various different programming languages. An example of transforming from PLD to Ada is shown in the next section. Transformations from PLD to, e.g., C or C++ is also possible. Furthermore, the PhM constructs may be transformed to, e.g., relational schemas, with associated SQL code as indicated in Fig. 12.9.

12.2.4 An Example of Applying the PPP Model

The following example is taken from [GULL91]. Assume that the problem domain is banking, and that we have four different kinds of transactions: `Add customer and account`, `Remove customer and account`, `Make deposit`, and `Make withdrawal.`

Data such as customer name and address, account number and deposit are associated with the transactions, together with processing rules of different kinds. The top level PrM diagram is shown in Fig. 12.10. The corresponding phenomenon model of the problem domain is shown in Fig. 12.11, and contains the three entity classes Customer, Account, and Transaction, together with their respective attributes. The relationships indicate that several transactions may be performed on each account, and that the relations between accounts and transactions are timestamped.

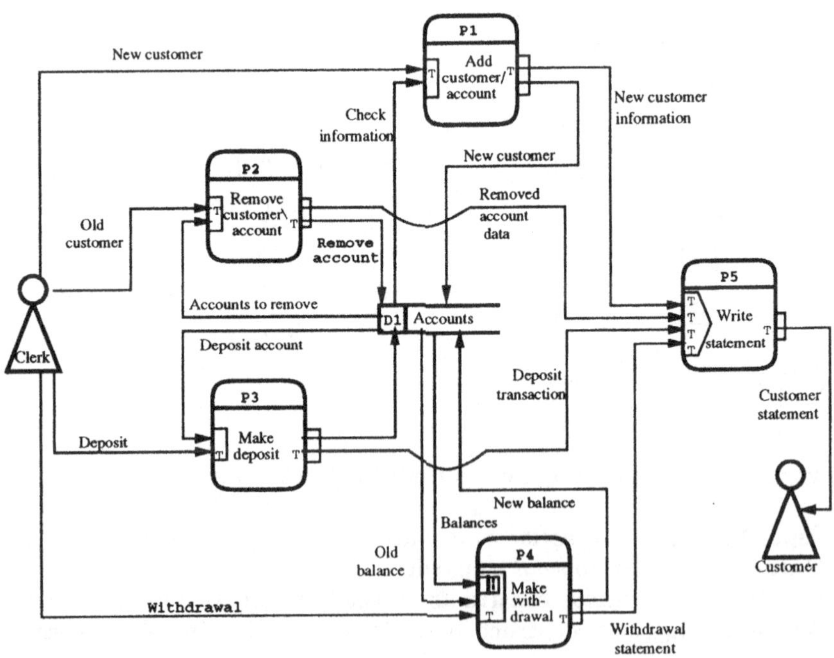

Fig. 12.10. The top level PrM diagram of the bank example

The four banking transactions appear in the PrM model of Fig. 12.10. One of them, P4 **Make withdrawal**, has been further decomposed in Fig. 12.12. Three processes are used to describe the behavior of P4. They are P4.1 **Accept withdrawal**, P4.2 **Register transaction**, and P4.3 **Make w_statement**. The **Accept withdrawal** process checks whether the transaction can be accepted with respect to the given processing rules. **Register transaction** stores the transaction data and computes the new balance, while **Make w_statement** issues a statement to the customer.

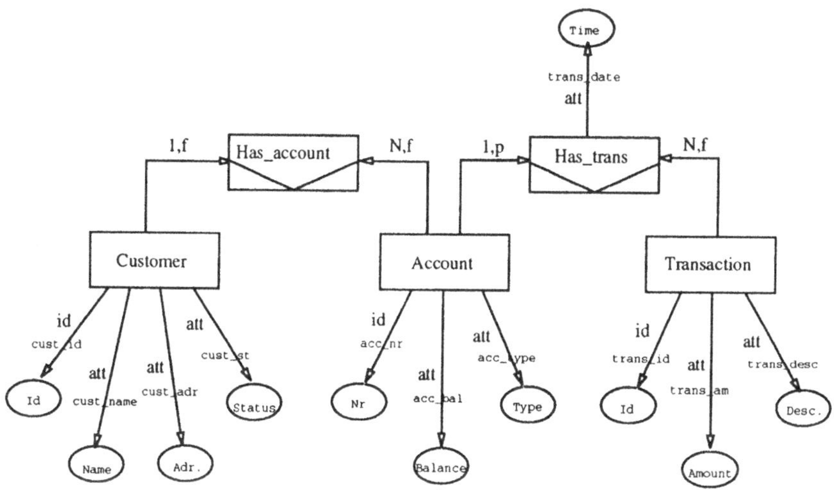

Fig. 12.11. The PhM-diagram for the bank example

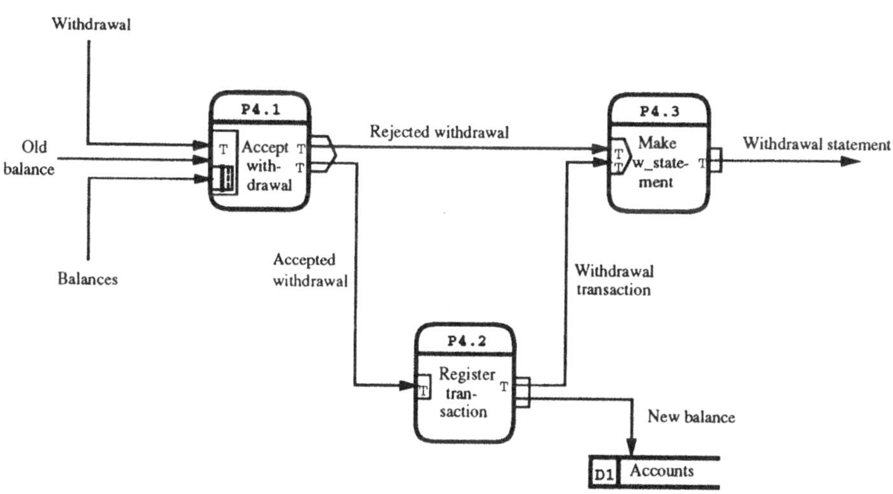

Fig. 12.12. The decomposed PrM-diagram P4: Make withdrawal

The behavior of P4.1 **Accept withdrawal** is specified in a PLD diagram of Fig. 12.13. The process is triggered when a withdrawal message is received from the bank clerk. The triggering message contains an account number, a date and an amount to be withdrawn. The process first checks to find whether the account contains enough money to satisfy the withdrawal request. If not, it is checked whether the customer has other accounts with enough money (the conditional repeat port). If the transaction is accepted it is sent on to P4.2 **Register transaction**. Otherwise the transaction is rejected and sent to P4.3 **Make w_statement**.

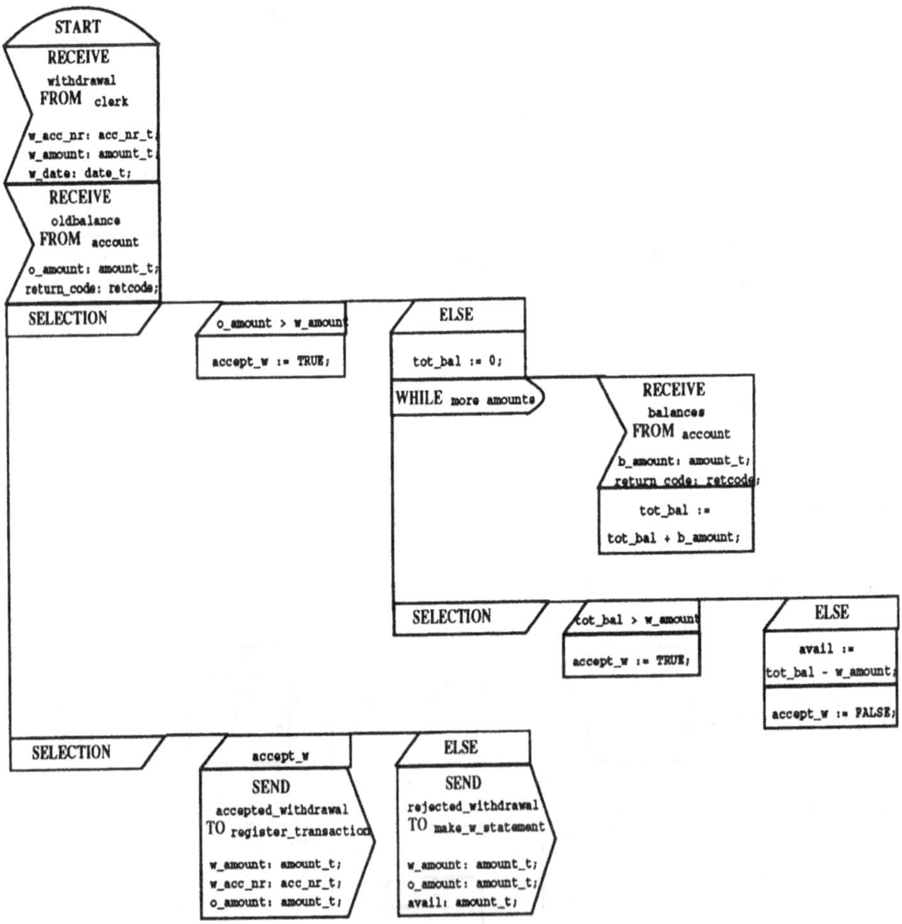

Fig. 12.13. The PLD-diagram for process P4.1 Accept withdrawal

We have now developed a basis for generating code. We may generate a database schema from the phenomenon model, as shown below:

```
create table customer
   (cust_id      int         not null,
    acc_nr       int         not null,
    cust_name    char(40)    null,
    cust_adr     char(40)    null,
    cust_st      char(10)    null)

create table account
   (acc_nr.      int         not null,
    acc-bal      float       null,
    acc_type     char(40)    null)

create table transaction
   (trans_id     int         not null,
    trans_am     float       null,
    trans_desc   char(40)    null)

create table has_trans
   (acc_nr       int         not null,
    trans_id     int         not null,
    trans_date   date        not null)
```

We have chosen to generate Ada code from the PLD specification of process P4.1 `Accept withdrawal`, as shown below:

```
-- process    : P4.1
-- status     : nonterminating
-- task       : accpt_withdrawal
-- input flow to process:
--    from A1 : withdrawal
--    from D1 : old_balance
--    from D1 : balances
-- output flow from process:
--    to P4.2 : accepted_withdrawal
--    to P4.3 : rejected_withdrawal
------------------------------------------
TASK accept_withdrawal is
   ENTRY withdrawal ( var1 : in acc_nr-t;
                      var2 : in amount_t;
                      var3 : in date_5;
   ENTRY SHUT_DOWN
END accept_withdrawal;

TASK BODY ACCEPT_WITHDRAWAL IS
accept_w : BOOLEAN;
avail, b_amount, o_amount, tot_bal, w_amount : amount_t;
return_code : retcode;
w_acc_nr : acc_nr_t;
w_date : date_t;
```

```
BEGIN
   TASK_CYCLE: LOOP
      SELECT
         ACCEPT withdrawal(   var1 : in acc_nr-t;
                              var2 : in amount_t;
                              var3 : in date_t) DO
            w_acc_nr := var1;
            w_amount := var2;
            w_date := var3;
         END withdrawal;
      OR
         ACCEPT SHUT_DOWN;
         EXIT TASK_CYCLE;
      END SELECT;
      ADBCMD( DBPROC, "SELECT BALANCE FROM ACCOUNT ");
      ADBCMD( DBPROC, "where nr = w_acc_nr");
      ADBSQLEXEC( DBPROC);
      return_code := ADBRESULTS( DBPROC);
      IF return_code := ADBNEXTROW( DBPROC);
         ADBBIND( DBPROC, 1, o-amount);
         return_code := ADBNEXTROW( DBPROC);
      END IF;
      IF o_amount > w_amount THEN
         accept_w := TRUE;
      ELSE
         tot_bal := 0;
         ADBCMD( DBPROC, "select balance ");
         ADBCMD( DBPROC, "from customer, has_account, account ");
         ADBCMD( DBPROC, "where customer.name = w_name, ");
         ADBCMD( DBPROC, "has_account.user_id = customer.id, ");
         ADBCMD( DBPROC, "has_account.acc_nr = account.nr ");
         ADBSQLEXEC( DBPROC);
         return_code := ADBRESULTS( DBPROC);
         IF return_code = SUCCEED THEN
            ADBBIND( DBPROC, 1, b_amount);
         END IF;
         WHILE return_code /= NO_MORE_ROWS
            return_code := ACBNEXTROW( DBPROC);
            tot_bal := tot_bal + b_amount;
         END WHILE
         IF tot_bal > w_amount THEN
            accept_w := TRUE;
         ELSE
            avail := tot_bal - w_amount;
            accept_w := FALSE;
         END IF;
      END IF;
      IF accept_w THEN
         register_transaction.accepted_withdrawal(w_amount,
                                                  w_acc_nr,o_amount);
      ELSE
         make_w_statement.rejected_withdrawal(w_amount,
                                                   o_amount,avail);
      END IF,
   END LOOP TASK-CYCLE;
   EXCEPTION
      WHEN TASKING_ERROR => TERMINATE;
      WHEN SYBASE_ERROR => ADBPERROR(ADBERRNO),
END accept_withdrawal;
```

12.3 The Problem of Removing Irrelevant Specificational Detail

The number of specification details increases rapidly as we approach the implementation level. As the amount of detail increases, so the possibility decreases for humans to understand the implications of the specifications. The specification may become too large to be grasped in a short time. It may include detail that is of little interest to some of the actors involved, e.g., in deliberations that may lead to changes in the specification.

Irrelevant details must be removed so that the human's mind can be more easily focused on relevant issues. We face the problem of deciding which parts of a specification are important, of deciding how to show specificational detail, in which format, and at which level of abstraction. The abstraction problem has not been solved in information systems engineering. So we shall try not more than to present the problem in this text.

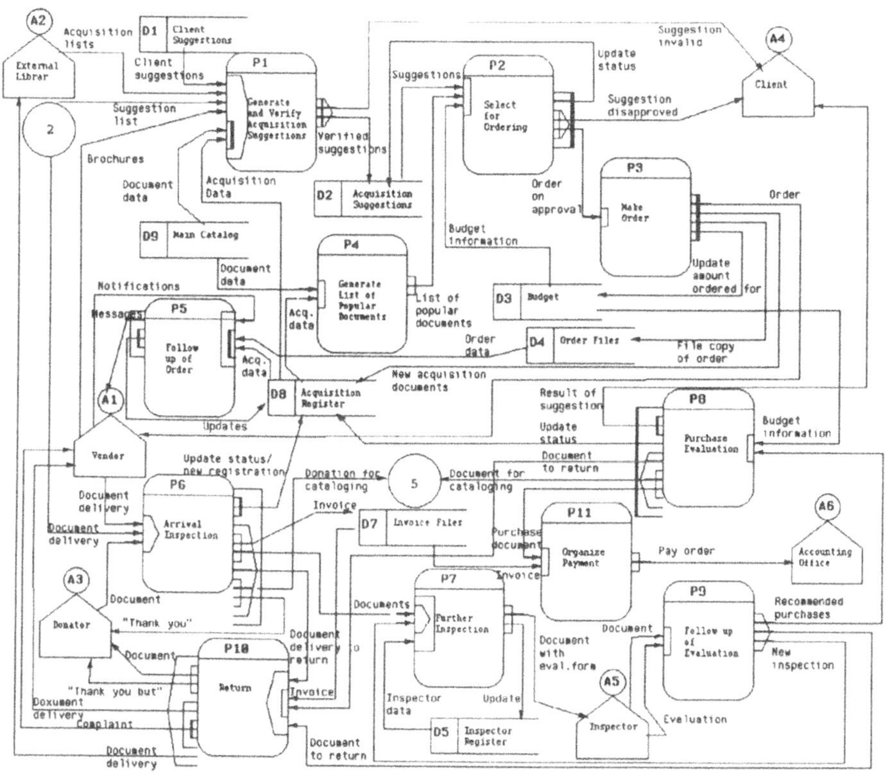

Fig. 12.14. PrM specification of the acquisition activity [LOVS90]

We shall take PrM diagrams as examples. In the next section we shall discuss an approach to abstraction in the context of simple flowcharts.

PrM diagrams contain more detail than dataflow diagrams. When going beyond the level of toy-examples, the number of details becomes large, and the diagrams become difficult to draw and difficult to understand. An example of a library information system is shown in Fig. 12.14. Because of the large number of details the diagram becomes ugly, and consequently becomes difficult to read. The amount of details is, however, inherent in the information system problem itself. If the details are not specified properly the system can not be executed. So we either have to live with the large number of details in the PrM specification or we have to express the details in some other formalism, e.g., some programming language. The latter solution does not remove the problem of excessive detail. It only hides the problem in incomprehensible code which nobody expects to be understandable anyway.

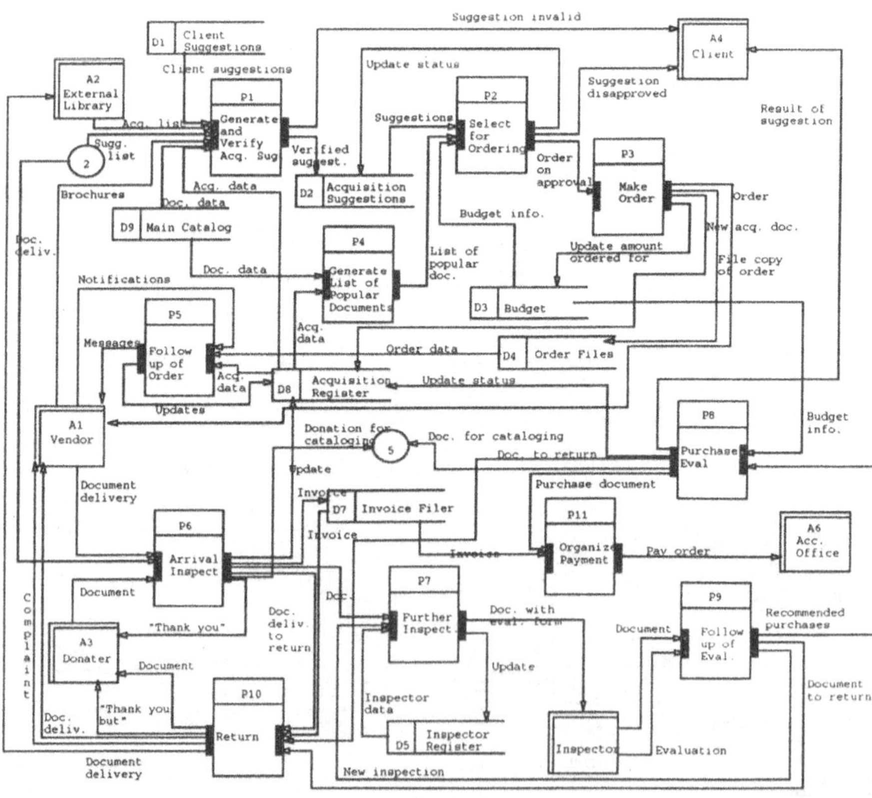

Fig. 12.15. PrM specification where the ports are abstracted away [SELT93]

In order to make the diagrams easier to understand they must be simplified. The simplifications should be done according to preset criteria, so that they may be done by computer. Simplifications done by humans in ad hoc ways become impossible to maintain, and too cumbersome to produce. So we need standard abstraction mechanisms. One such mechanism may be the removal of ports in PrM diagrams. The example system with ports abstracted away is depicted in Fig. 12.15.

We see that removal of ports is certainly not enough to create an understandable picture of the system. Because the ports enforce a certain lay-out of the diagrams one may achieve a simpler diagram through reformatting. Another alternative may be to simplify the dataflow structure of the diagram. An example is shown in Fig. 12.16. All dataflows that can be collapsed have been be collapsed. We shall not explain in detail which rules that have been applied to achieve this remarkably simplified diagram. We shall be content to indicate that it indeed seems to be possible to abstract away specificational detail to enhance the readability of complex information systems specifications, without trading away formality of expression.

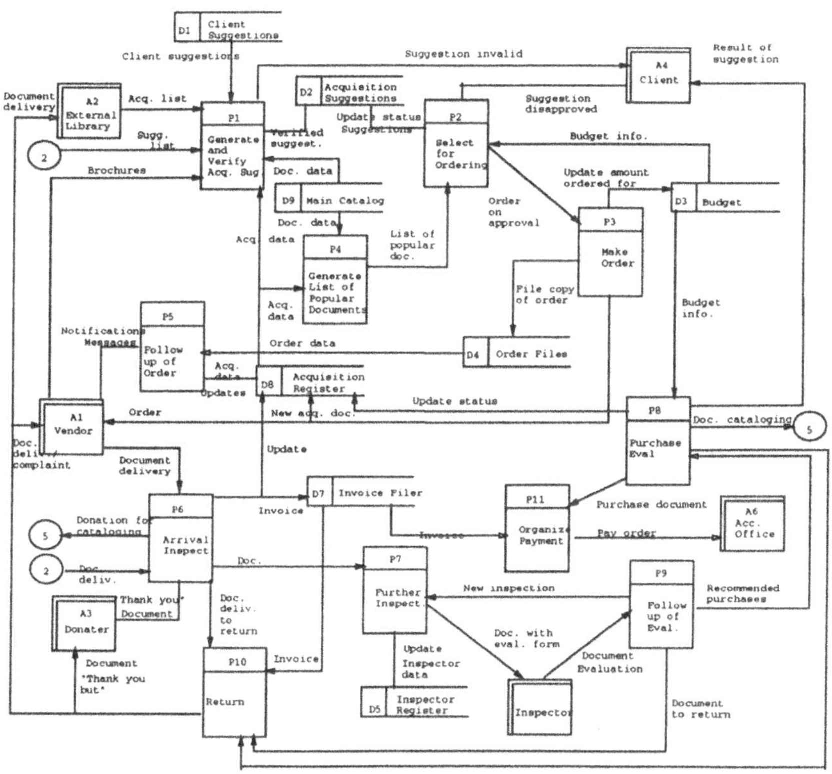

Fig. 12.16. PrM specification where all possible flows are collapsed [SELT93]

12.4 A Simple Method for Abstracting Away Modeling Detail

The discussion of the previous section indicates that specification models may become overloaded with specificational detail if they are to have sufficient constructive properties. This will be counter to the equally important requirement that specifications be easily understandable. The solution to these conflicting requirements is found in suitable methods for abstracting away specificational detail when the details are not needed. We shall demonstrate how abstraction algorithms may be used to explain how various types of conventional flowcharts relate to each other.

Flowcharts are diagrams that present properties of software systems in graphical form. Software systems often have high structural complexity, and consist of many details. It therefore becomes impossible to depict all of the relevant properties of a piece of software in one single graphical picture. Various diagramming techniques have been proposed for the presentation of different systems properties. The process diagram is the most commonly used diagram for describing computer programs (Fig. 12.17). The process sequences are clearly described in such a diagram, while the input/output aspects are overlooked. Another diagram type presents the relationships among the inputs and outputs of processes (Fig. 12.18). The data precedence diagram depicts the precedence relationships between the data objects which appear in a computer program. For example, if D_9 may be produced on the basis of D_4 alone, then D_4 is the only precedent data of D_9.

There are also other diagram types of the flowchart kind. At first glance the various diagram types appear to be formally unrelated to each other. Each of them is used for presenting some particular feature of a software system. We may say that each diagram type represents one particular point of view of a software system. However, there is no commonly accepted technique for relating the various points of view to each other. We shall show how this can be done, by using a simple formal model for representing those basic systems properties that are commonly depicted in flowcharts. We shall show how various flowchart types may be formally derived from a basic systems model [AUGL75], as shown in Fig. 12.19.

A data processing system may be described by the basic object types

D – data objects,

P – process objects,

S – signal objects, representing sequencing relations between process objects,

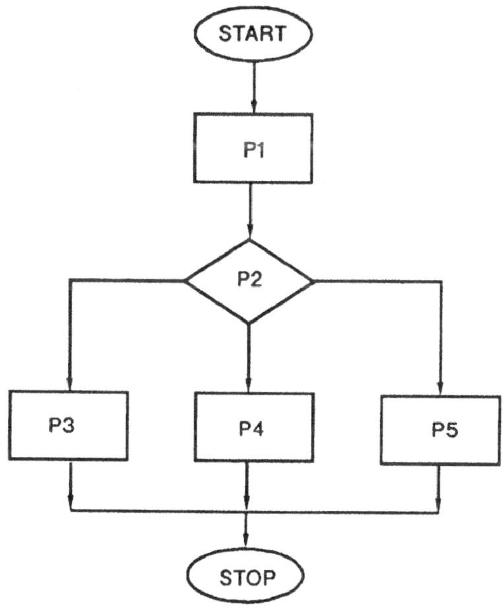

Fig. 12.17. A process diagram

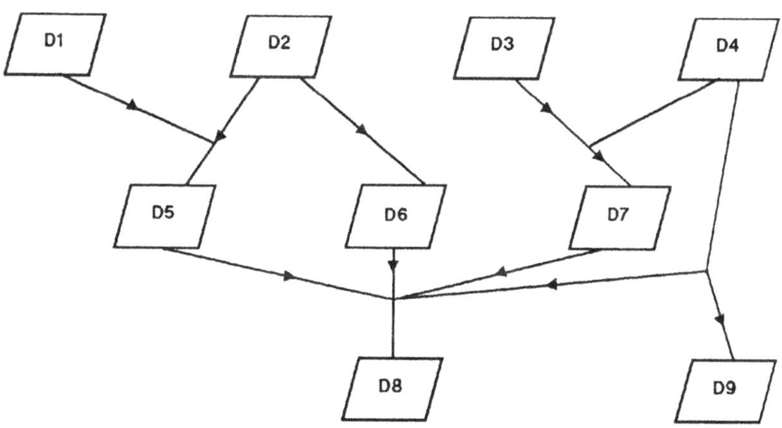

Fig. 12.18. A data precedence diagram

and the basic relationship types

$I \subseteq D \times P$ – the input relation, which connects a data object to a process object,

$O \subseteq P \times D$ – the output relation, which connects a process object to a data object,

$N \subseteq S \times P$ – the entry relation, which connects a signal object to a process object,

$X \subseteq P \times S$ – the exit relation, which connects a process object to a signal object.

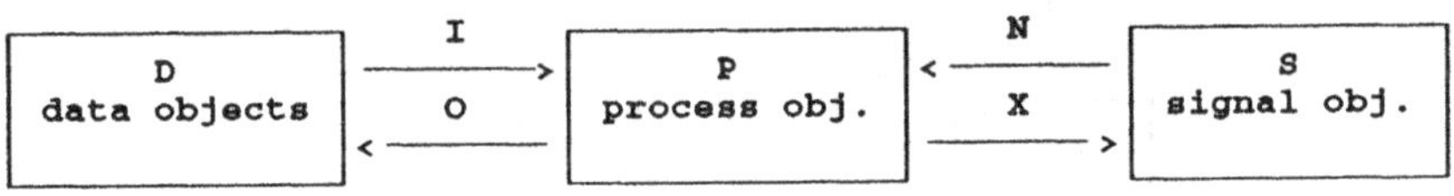

Fig. 12.19. Object types and relationship types in a model for comparing flowcharts

We shall now show how six different diagram types may be derived from the basic form. Diagrams may be designed either to present a specific property of a system or to give an overall view of it. The relations that are presented in the diagrams are not always the primary relations that have been defined in the system model. This will be pointed out in the definitions of the various diagram types. Where the relations are derived, the derivation rule will be described. The diagrams will be defined by defining the set OBJ which is the set of objects, and the set REL which is the set of relationships that are presented in the diagrams.

System diagrams (Fig. 12.20) give an overall view of a system by presenting all objects and all primary relations between them, in terms of the basic representation model.

Diagram definition: $OBJ = D \cup S \cup P$

$REL = I \cup O \cup N \cup X$

An example is shown in Fig. 12.20.

Data objects are represented by rhomboids, signal objects by ellipses, and process objects by rectangles and diamonds (for branching processes). We see immediately that the complexity of this diagram type will quickly become overwhelmingly large. The next few flowchart types show some possible simplifications.

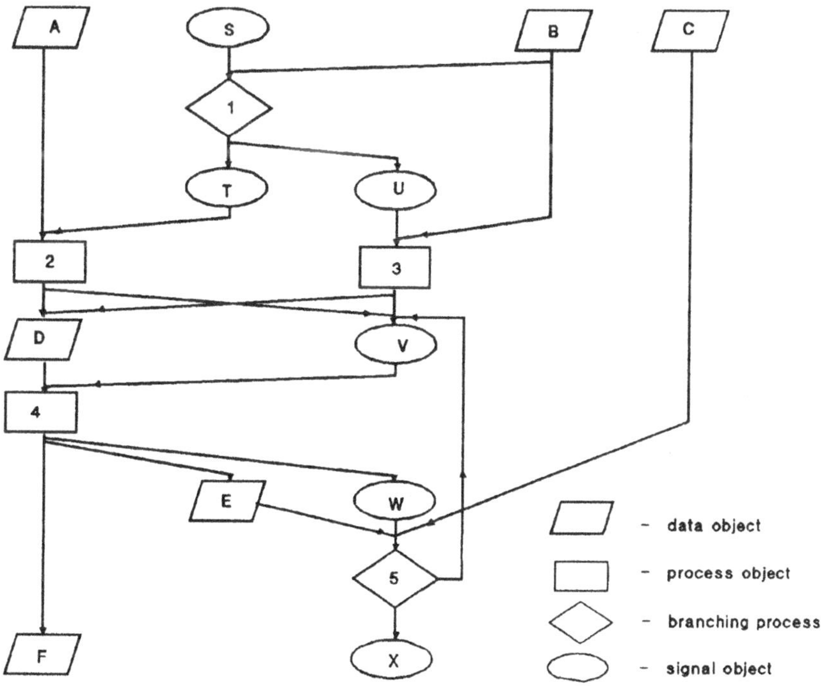

Fig. 12.20. Example of system diagram

DPD diagrams (Fig. 12.21) present the primary relations in the system as they are given between data- and process-objects (DPD: Data-Process-Data).

Diagram definition: $OBJ = D \cup P$

$$REL = I \cup O$$

The signal objects, that is, the sequencing relations between processes, are not shown in the DPD diagrams. Only data exchange between processes is shown, not the control flows.

We see that branching processes 1 and 5 appear with no outputs. This is because the two processes only have control functions in the system. They appear in the diagram because all process objects are required to appear. A more subtle definition of the diagram may suppress processes that have control functions alone.

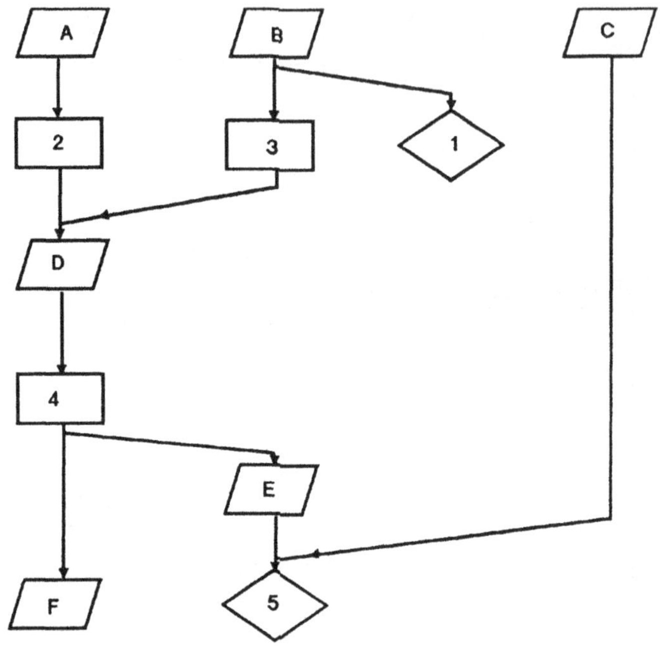

Fig. 12.21. Example of the DPD diagram of the system in Fig. 12.20

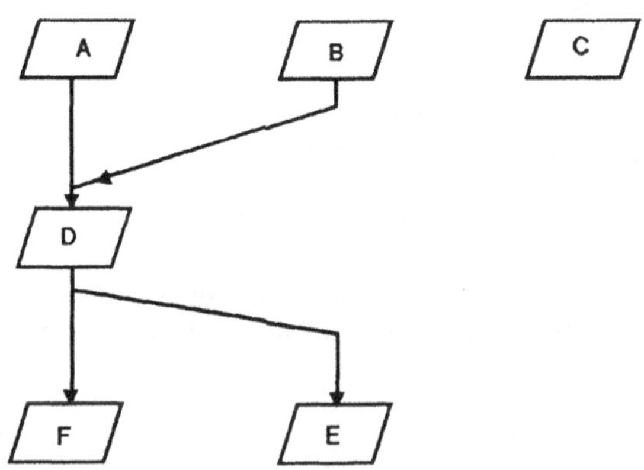

Fig. 12.22. Example of the DD diagram of the system in Fig. 12.20

DD diagrams (Fig. 12.22) present precedence relationships between data objects.

Assume $\exists P_j \in P$ and $\exists D_i, D_k \in D$, then

by definition: OBJ = D

$$REL = \{(D_i, D_k) \mid (D_i, P_j) \in I \wedge (P_j, D_k) \in O\}$$

If some data object D is the output of a process, which require inputs A and B, then A and B are said to be the precedents of D. That is, A and B must be known before D can be produced. The DD diagram is a data precedence diagram. The DD-diagram comprises all of the data-objects of the system, as well as the derived precedence relationships between data-objects.

PP/D diagrams (Fig. 12.23) present the processes and their data connections (process to process via data relationship). If there is a relation from process 2 to process 4, this means that process 4 needs data generated by process 2.

Assume $\exists D_j \in D$ and $\exists P_i, P_k \in P$, then

by definition: OBJ = P

$$REL = \{(P_i, P_k) \mid (P_i, D_j) \in O \wedge (D_j, P_k) \in I\}$$

This example diagram shows one branching process that is not connected to any other process. Also in this case the reason is that the process (no. 1) has a process control function only. It does not interact with any other process by means of data, but only through signals. Therefore it appears isolated in a PP/D diagram. Also in this case one may obtain nicer diagrams by applying a more complicated algorithm for selecting objects and relations from the comprehensive systems model.

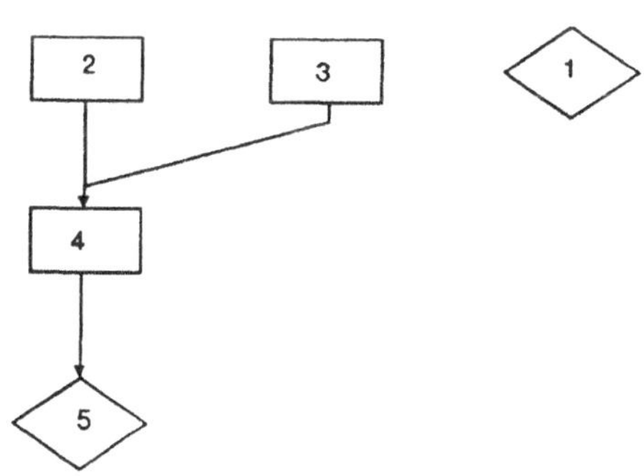

Fig. 12.23. Example of the PP/D diagram of the system in Fig. 12.20

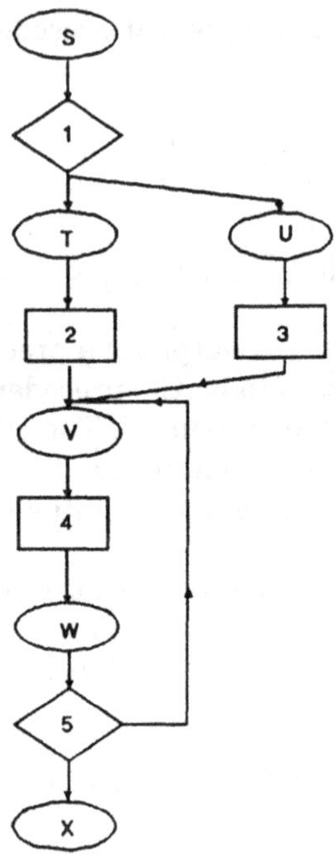

Fig. 12.24. Example of the SPS diagram of the system in Fig. 12.20

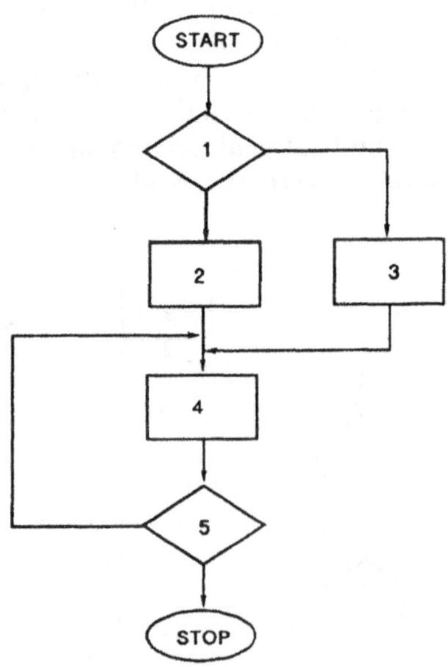

Fig. 12.25. Example of the a PP/S diagram of the system in Fig. 12.20

SPS diagrams (Fig. 12.24) present the dynamical structure, (signal-process-signal diagrams). They show the primary dynamic relations between signal- and process-objects together with the signal- and process-objects themselves.

Diagram definition: $OBJ = S \cup P$
$REL = N \cup X$

PP/S diagrams (Fig. 12.25) present the dynamic structure in a compressed form (process to process via signal relationships). The objects are all of the processes, as well as the start- and stop-signals on the system's boundary. The relations are the primary relations between boundary-signals and processes, together with the derived dynamic relations between processes.

Assume $\exists S_j \in S$ and $\exists P_i, P_k \in P$ and further let P_l be any element in P, then

by definition:
$$OBJ = P \cup \{S_j \mid (S_j, P_i) \in N \wedge (P_l, S_j) \notin X\}$$
$$\cup\ S_j \mid (P_i, S_j) \in X \wedge (S_j, P_l) \notin N\}$$
$$REL = \{(P_i, P_k) \mid (P_i, S_j) \in X \wedge (S_j, P_k) \in N\}$$
$$\cup\ \{(S_j, P_i) \mid (S_j, P_i) \in N \wedge (P_l, S_j) \notin X\}$$
$$\cup\ \{(P_i, S_j) \mid (P_i, S_j) \in X \wedge (S_j, P_l) \notin N\}$$

A relation from process A to process B means that process A is executed immediately before process B. A relation from a signal to a process means system execution may start with this process, and a relation from a process to a signal means execution may terminate with this process. To make this last interpretation more explicitly visible on the diagram, the texts within signal-boxes are (in contradiction with the semantic definition) changed to the words "start" and "stop" respectively. The reader will easily see that this diagram type is commonly known as a "process flowchart"!

Exercises

1. State the advantages of using a model which allows executable specifications, in the development of an information system.

2. Explain the terms of *constructivity* and *constraints*. Give examples.

3. Explain the differences between the ER model and PrM.

4. Compare the PrM with the DFD technique, stating the advantages and disadvantages of both.

5. Compare PLD with structured English. State the advantages and disadvantages.

6. Apply the PPP model to the CRA problem which is introduced in Exercise 6 in Chap. 2.

7. Discuss the importance of "abstracting the specifications".

8. Given the problem below:

   "Find all possible roots of the quadratic equation

   $$Ax^2 + Bx + C = 0$$

   where A, B and C are real numbers,"

   a) draw a system diagram,
   b) draw a DPD diagram,
   c) draw a DD diagram,
   d) draw a PP/D diagram,
   e) draw a SPS diagram,
   f) draw a PPS diagram,

   by identifying the relevant objects (i.e., P-, D-, or S-objects) associated with each diagram.

Chapter 13

An Example of Comparing Information Systems Analysis Approaches

We shall compare two approaches to information systems analysis through an example. The example system is a one-bit window protocol for message transmission. One approach is oriented towards object descriptions. The other one is a stimulus-response approach (see Sect. 1.6).

One of the specification models is object oriented. The other is process oriented. We have not tried to evaluate the two approaches, but let the examples speak for themselves. The approaches are discussed at length elsewhere in this text.

13.1 The Example: A One-Bit Window Protocol

A one-bit window protocol transmits messages from a SENDER to a RECEIVER via a transmission channel. Messages may be lost or distorted during the transmission, but the order in which they were transmitted may not be interchanged. Both the SENDER and the RECEIVER may use, e.g., check-sum methods to determine if a message has been distorted during transmission. Distorted messages are ignored, and are treated as if they were lost during transmission.

A basic problem is that the RECEIVER cannot know whether the SENDER has sent a message until it is received by the RECEIVER. Therefore the RECEIVER cannot notify the SENDER if a message has been lost during transmission. The SENDER must be able to find out whether a message has been lost, without having to interrogate the RECEIVER. For this purpose the RECEIVER is required to return an acknowledgement to the SENDER that its message has been received.

If the SENDER does not receive, within some preset time interval (=*TIMEOUT*), an acknowledgement that the message has been received by the intended RECEIVER, it will repeatedly retransmit the message (until a positive acknowledgement has been received).

To enable both SENDER and RECEIVER to keep track of the order in which the messages have been transmitted, each message is associated with a one-bit sequence number. The value of the sequence number for a sequence of correctly transmitted messages will therefore alternate between 0 and 1. The "window" will be of size one, because there will at any time be at most one message lacking acknowledgement.

The SENDER will transmit a new message (from its queue of messages to be sent) as soon as it receives an acknowledgement that the previous message has been received. The sequence number of the new message will be the opposite of the sequence number of the previously sent message. If an acknowledgement has not been received within some preset time interval (= *TIMEOUT*), the sender will re-transmit the previously sent message, together with its original sequence number.

The RECEIVER will send an acknowledgement to the SENDER whenever it receives a message that is recognized as not having been distorted. The RECEIVER keeps its own local sequence number, whose value is the expected value of the sequence number of the next message to be received.

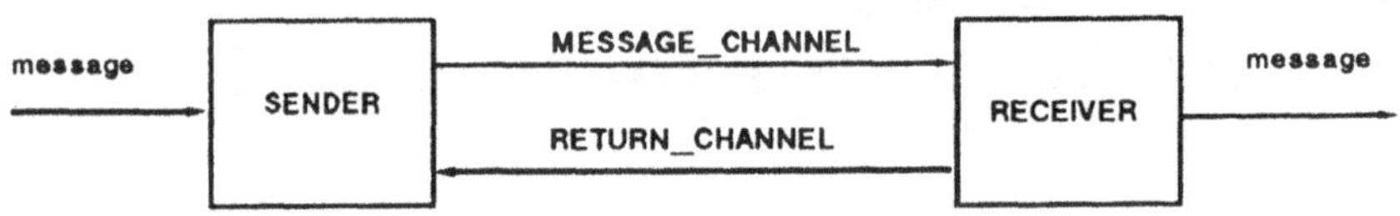

Fig. 13.1. Simplified message transmission system

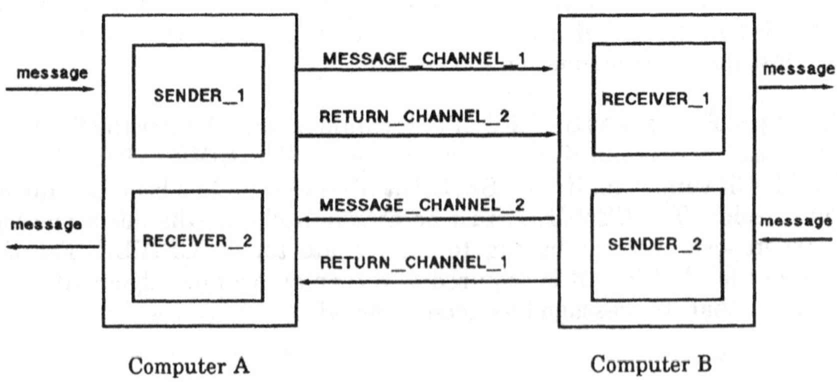

Fig. 13.2. Computerized data communication system

The acknowledgement is the sequence number of the received message, and is sent back to the sender of that message. When the acknowledgement has been sent, the RECEIVER changes its local sequence number, in order to prepare for the next message to be received. Whenever a transmission is initiated, the SENDER and the RECEIVER have to share information about the value of the sequence number of the first message to be transmitted.

To simplify the modeling of the protocol, we introduce some further rules and constraints. We assume that a SENDER and a RECEIVER are related by one MESSAGE_CHANNEL, and one RETURN_CHANNEL. We also assume that one channel can only transmit one message at a time. Acknowledgements are viewed as "empty" messages, that is, only the sequence number is transmitted over the (return) channel (Fig. 13.1).

In practice, computers will play the roles of both senders and receivers at the same time (Fig. 13.2). Acknowledgements are "piggybacked", that is, they are added to messages that are sent in the opposite direction of the message that is to be acknowledged. The message channel of one sender-receiver pair occupies the same communication channel as the return channel of another sender-receiver pair. However, we shall not complicate the problem further, and will therefore stick to the problem as explained above (Fig. 13.1).

13.2 Object-Oriented Analysis of the Communication Protocol

An object-oriented systems specification consists of one description of the "static" properties of the system, and one description of the "dynamic" properties of the system.

The static properties are specified by assigning "attributes" to the various "objects" of the system, assuming that each attribute may take on some value, and that an object is characterized by the values of its attributes.

The dynamic properties may be specified through a description of the "events" that are associated with the changes of the system. Each event may be described by two rules, one rule that specifies the conditions that must be satisfied prior to the event taking place, and one rule that specifies the conditions after the event has taken place. These two rules are called the precondition and the postcondition rules.

The attributes we have chosen for our modeling exercise are [HOVE84]:

seqno – the sequence number (= 0 or 1) that is attached to messages that are transmitted through the one-bit communication protocol (We adopt the notational convention that we may negate *seqno* by writing ~*seqno*, that is, ~*seqno* = 0 if *seqno* = 1, and ~*seqno* = 1 if *seqno* = 0),

state – which for the SENDER may have the values *READY_TO_SEND* or *READY_TO_RECEIVE*, and for the channels may have the values *ACTIVE* or *PASSIVE*,

time – which represents the elapsed time since the SENDER transmitted a message that has not been acknowledged over the RETURN_CHANNEL. The time-attribute may take on one of the values 0 or *TIMEOUT*. The last value indicates that either the message or the acknowledgement has been lost during transmission.

A channel may be in an *ACTIVE* state, or in a *PASSIVE* state. In its *ACTIVE* state a channel is transmitting a message, which can be read on its receiving side. In its *PASSIVE* state nothing is happening on the channel, and no message may be read on its receiving side.

Only the SENDER has a time-attribute for keeping track of the elapsed time since it sent its last message out into the MESSAGE_CHANNEL. If no acknowledgement has been received from the RECEIVER within the time *TIMEOUT* from when the message was sent, some action will be taken by the SENDER (e.g., retransmission).

The properties of the one-bit window protocol may be specified by sentences of the forms

SENDER(*seqno,state,time*)
RECEIVER(*seqno*)
MESSAGE_CHANNEL(*seqno,state*)
RETURN_CHANNEL(*seqno,state*)

where the attributes *seqno, state,* and *time,* may hold values as explained above (Fig. 13.3).

The behavior of the communication protocol may now be described by specifying the pre- and post-conditions of the events that can take place in the system. There are three classes of events: those that are associated with the behavior of the SENDER of messages, those that are associated with the behavior of the RECEIVER of messages, and those that are associated with transmission errors on the CHANNELs.

There are three SENDER events. They are called **SEND_MESSAGE**, **RECEIVE_ACKNOWLEDGEMENT**, and **RETRANSMIT_MESSAGE**. The SENDER will be set in the *READY_TO_RECEIVE* state after it has sent (or re-

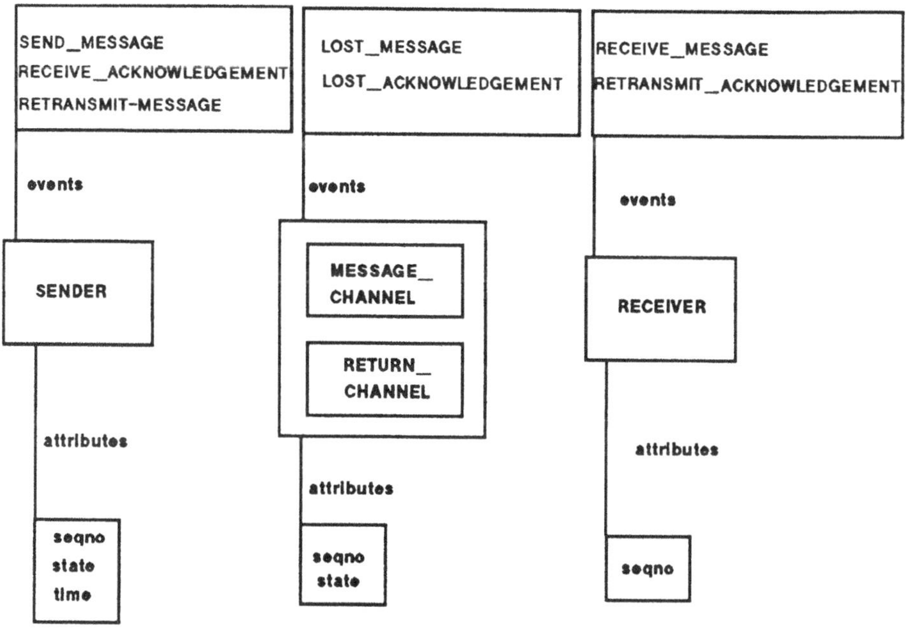

Fig. 13.3. Objects, attributes and events of the one-bit window protocol

transmitted) a message, and it will be set in the *READY_TO_SEND* state after it has successfully received an acknowledgement of reception of the previously sent message.

There are two transmission channel events that are both concerned with faulty transmission. They are **LOST_MESSAGE** and **LOST_ACKNOWLEDGEMENT**. Both of the event-descriptions mark the loss of information from the transmission channels by setting *time=TIMEOUT* for the SENDER, thereby indicating that the SENDER will never get an acknowledgement from the RECEIVER. The two RECEIVER events have to do with returning acknowledgements to the SENDER for messages received. They are called **RECEIVE_MESSAGE and RETRANSMIT_ACKNOWLEDGEMENT**. Both of them return an appropriate sequence number (*seqno*) to the SENDER (Fig. 13.3).

The SENDER events are: **SEND_MESSAGE, RECEIVE_ACKNOWLEDGEMENT** and **RETRANSMIT_MESSAGE:**

i) **SEND_MESSAGE:**
   precondition SENDER(*seqno, READY_TO_SEND, time*=0)
   postcondition SENDER(*seqno, READY_TO_RECEIVE, time*=0)
   $\wedge$ MESSAGE_CHANNEL(*seqno, ACTIVE*);

Remark: The precondition rule expresses that a SENDER must be in the state = *READY_TO_SEND*, prior to sending a message, and that it must have received acknowledgement of its previously sent message (*time*=0). The postcondition rule expresses that the SENDER will now start to wait for an acknowledgement (*READY_TO_RECEIVE*), and that the message (together with the appropriate sequence number) has been sent out into the transmission channel, which is now *ACTIVE*.

ii) **RECEIVE_ACKNOWLEDGEMENT**:

precondition	SENDER(*seqno, READY_TO_RECEIVE, time*=0)
$\wedge$	RETURN_CHANNEL(*seqno, ACTIVE*)
postcondition	SENDER(*~Seqno, READY_TO_SEND, time*=0)
$\wedge$	RETURN_CHANNEL(*seqno, PASSIVE*);

Remark: If the sequence number held by the SENDER (which is the sequence number associated with the previously sent message), matches the sequence number returned from the RETURN_CHANNEL, the sequence number changes (from 0 to 1 or vice versa), the SENDER may send a new message, and the RETURN_CHANNEL is put in the *PASSIVE* state.

iii) **RETRANSMIT_MESSAGE**:

precondition	SENDER(*seqno, READY_TO_RECEIVE, time=TIMEOUT*)
postcondition	SENDER(*seqno, READY_TO_RECEIVE, time*=0)
$\wedge$	MESSAGE_CHANNEL(*seqno, ACTIVE*);

Remark: If no acknowledgement has been received (*time* = *TIMEOUT*), the previous message and the previous sequence number is re-transmitted.

The CHANNEL events are: **LOST_MESSAGE** and **LOST_ACKNOWLEDGEMENT**:

iv) **LOST_MESSAGE**:

precondition	SENDER(*seqno, state, time*=0)
$\wedge$	MESSAGE_CHANNEL(*seqno, ACTIVE*)
postcondition	SENDER(*seqno, state, time=TIMEOUT*)
$\wedge$	MESSAGE_CHANNEL(*seqno, PASSIVE*);

Remark: This represents the consequences of a message disappearing.

v) **LOST_ACKNOWLEDGEMENT**:

precondition	SENDER(*seqno, state, time*=0)
$\wedge$	RETURN_CHANNEL(*s, ACTIVE*)
postcondition	SENDER(*seqno, state, time=TIMEOUT*)
$\wedge$	RETURN_CHANNEL(*s, PASSIVE*);

Remark: Note that a sequence number of the acknowledgement on the RETURN_CHANNEL that disappears does not have to be equal to the sequence number of the SENDER. It simply does not matter, because it disappears anyway, and it is irrelevant which value it has.

The RECEIVER events are **RECEIVE_MESSAGE** and **RETRANSMIT_ACKNOWLEDGEMENT**:

vi) **RECEIVE_MESSAGE**:

precondition RECEIVER(*seqno*)
∧ MESSAGE_CHANNEL(*seqno, ACTIVE*)
postcondition RECEIVER(*~seqno*)
∧ RETURN_CHANNEL(*seqno, ACTIVE*)
∧ MESSAGE_CHANNEL(*seqno, PASSIVE*);

Remark: The RECEIVER changes the value of its sequence number, as it makes itself ready to receive the next message.

vii) **RETRANSMIT_ACKNOWLEDGEMENT**:

precondition RECEIVER(*seqno*)
∧ MESSAGE_CHANNEL(*~seqno, ACTIVE*)
postcondition RECEIVER(*seqno*)
∧ RETURN_CHANNEL(*~seqno, ACTIVE*)
∧ MESSAGE_CHANNEL(*~seqno, PASSIVE*);

Remark: The RECEIVER receives a message with a the opposite sequence number of what was expected. The only way that this can happen is as a result of a retransmission of a previously sent message, because the associated acknowledgement was lost during transmission. Therefore, the appropriate action is to retransmit the lost acknowledgement, that is, to send the previous sequence number out into the RETURN_CHANNEL once more.

In order to complete the specification, we must state the initial conditions for the communication protocol. They are

SENDER(*seqno=0, state=READY_TO_SEND, time=0*)
∧ RECEIVER(*seqno*=0)
∧ MESSAGE_CHANNEL(*s, state=PASSIVE*)
∧ RETURN_CHANNEL(*s, state=PASSIVE*);

13.3 The Communication Protocol Modeled as a State-Transition Machine

It is interesting to note that we may develop a state-transition diagram for the one-bit window protocol system by transforming the object-oriented specification from the individual level to the systems level. We can construct the state transition diagram by starting at the initial state and applying the specifications of the various events in order to find consecutive systems states. The procedure is shown in the sequel.

The initial condition corresponds to

System state 0:

SENDER(*seqno=0, state=READY_TO_SEND, time=0*)
∧ RECEIVER(*seqno=0*)
∧ MESSAGE_CHANNEL(*s, state=PASSIVE*)
∧ RETURN_CHANNEL(*s, state=PASSIVE*);

Only the precondition SENDER(*seqno, READY_TO_SEND, time=0*) of the **SEND_MESSAGE** event is consistent with the initial conditions of system state 0. In order to specify the system state that will become the result if a **SEND_MESSAGE** event takes place, we have to substitute the **SEND_MESSAGE** event's postcondition

SENDER(*seqno, READY_TO_RECEIVE, time=0*)
∧ MESSAGE_CHANNEL(*seqno, ACTIVE*);

for the appropriate sentences in the specification of system state 0. When we apply the substitution (Fig. 13.4), we get

System state 1:

SENDER(*seqno=0, READY_TO_RECEIVE, time=0*)
∧ RECEIVER(*seqno=0*)
∧ MESSAGE_CHANNEL(*seqno=0, ACTIVE*);
∧ RETURN_CHANNEL(*s, state=PASSIVE*);

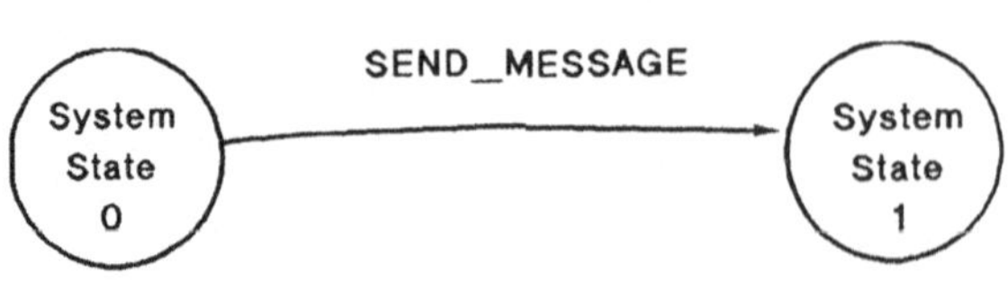

Fig. 13.4. State-transition diagram, states 0 and 1

There are two events that have preconditions that are compatible with system state 1, and that may therefore take place when the system is in state 1. They are **RECEIVE_MESSAGE**, with the specification

precondition	RECEIVER(*seqno*)
∧	MESSAGE_CHANNEL(*seqno, ACTIVE*)
postcondition	RECEIVER(*~seqno*)
∧	RETURN_CHANNEL(*seqno, ACTIVE*)
∧	MESSAGE_CHANNEL(*seqno, PASSIVE*);

and **LOST_MESSAGE**, with the specification

precondition	SENDER(*seqno, state, time=0*)
∧	MESSAGE_CHANNEL(*seqno, ACTIVE*)
postcondition	SENDER(*seqno, state, time=TIMEOUT*)
∧	MESSAGE_CHANNEL(*seqno, PASSIVE*);

When we apply the **LOST_MESSAGE** event, we get into system state 2, and when we apply the **RECEIVE_MESSAGE** event, we get into system state 3 (Fig. 13.5).

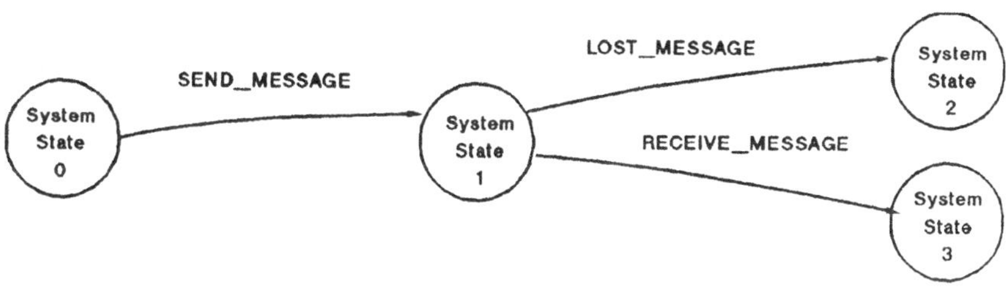

Fig. 13.5. State-transition diagram, states 0, 1, 2 and 3

System state 2:

This state is the result of a **LOST_MESSAGE** event taking place in system state 1, and has the specification:

SENDER(*seqno=0, READY_TO_RECEIVE, time=TIMEOUT*2)
∧ RECEIVER(*seqno=0*)
∧ MESSAGE_CHANNEL(*seqno=0, PASSIVE*);
∧ RETURN_CHANNEL(*s, state=PASSIVE*);

The only event which is compatible with state 2 is **RETRANSMIT_MESSAGE** which has the following specification:

precondition	SENDER(*seqno, READY_TO_RECEIVE, time = TIMEOUT*)

postcondition SENDER(*seqno, READY_TO_RECEIVE, time*=0)
$\wedge$ MESSAGE_CHANNEL(*seqno, ACTIVE*);

By substituting the postcondition into the state 2 specification, we get

SENDER(*seqno=0, READY_TO_RECEIVE, time*=0)
$\wedge$ RECEIVER(*seqno*=0)
$\wedge$ MESSAGE_CHANNEL(*seqno=0, ACTIVE*);
$\wedge$ RETURN_CHANNEL(*s, state=PASSIVE*);

which is exactly equal to the specification of system state 1 (see Fig. 13.6).

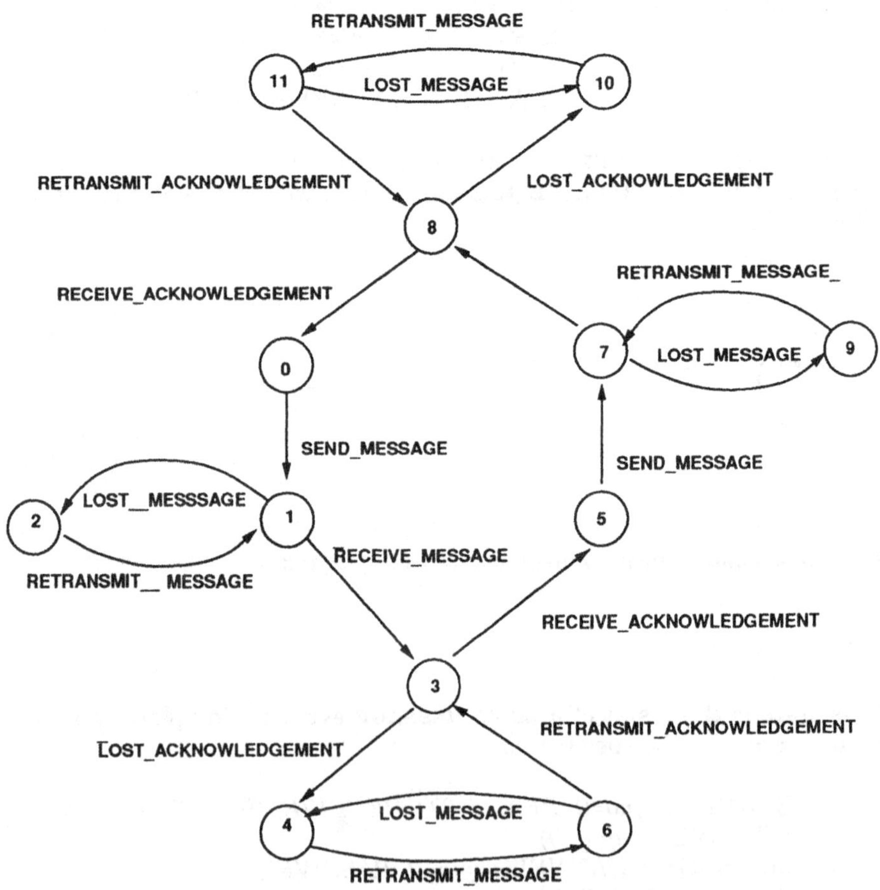

Fig. 13.6. State-transition diagram of a one-bit window protocol

System state 3:

This state is the result of a *RECEIVE_MESSAGE* event taking place in system state 1, and has the specification

SENDER(*seqno=0, READY_TO_RECEIVE, time*=0)
∧ RECEIVER(*seqno*=1)
∧ MESSAGE_CHANNEL(*seqno=0, state=PASSIVE*);
∧ RETURN_CHANNEL(*seqno=0, state=ACTIVE*);

By inspection we find that the events **RECEIVE_ACKNOWLEDGEMENT** and **LOST_ACKNOWLEDGEMENT** are both compatible with state 3 (see Fig. 13.6).

So far we have constructed a state-transition diagram with three states in addition to the initial state. If we continue the construction, we end up with a diagram as shown in Fig. 13.6. There are 11 relevant states in addition to the initial state, altogether 12 states. These 12 systems states are considerably fewer than the total number of states that might have been possible if we did not take the events into consideration.

Looking at the attribute descriptions of the four objects whose properties we have described above, we note that each of the attributes may take one of only two values. The seqno-attribute may be 0 or 1, the SENDER state may take one of two values, and so on. This means that the number of possible states for each system-part may be calculated. They are for the

SENDER	2x2x2 = 8 states
RECEIVER	2 states
MESSAGE_CHANNEL	2x2 = 4 states
RETURN_CHANNEL	2x2 = 4 states

The protocol system may therefore be in any of 8x2x4x4 = 256 states, if all combinations of the individual states of the systems parts are permitted.

The permitted state transitions are constrained by the events that can take place in the system. From the specifications of the events one may find which of the individual state combinations are permitted on the systems level. For example, system states that are unreachable from other systems states are without interest for us, and are not permitted. In the state-transition diagram of Fig. 13.6 we have reduced the system's state space from 256 possible combinations of individual states, to the 12 systems states that are permitted by the system's events.

13.4 Stimulus-Response Analysis of the Communication Protocol

Our analysis, and our specifications, have so far been directed towards describing the properties of the individual parts of the system, and of their interactions. We have oriented ourselves towards analysis of the "objects" that constitute the system. We shall now repeat the analysis of the communication protocol, using an information-processing oriented stimulus-response approach.

The stimulus in this case is the appearance of one (or more) messages to be transmitted from a sender of the messages data, via a transmission channel, to a receiver of the data. So the message(s) will trigger a SEND_DATA process, that will send the data out into the transmission channel, to be received at the other end of the channel by a RECEIVE_DATA process. In order to provide for the sender to check that the data transmission has taken place as expected, the receiver sends an acknowledgement back to the sender. The sender will need the results of a process ANALYZE_ACKNOWLEDGEMENT in order to decide if a new message may be transmitted, or if a retransmission has to be performed because of transmission errors. The (stimulus-response) process that results from this analysis is depicted in Fig. 13.7 as a data flow diagram (DFD). Note that we use the terms SEND_DATA and RECEIVE_DATA, instead of SEND_MESSAGE and RECEIVE_MESSAGE, which might have been more appropriate, in order to avoid confusion with the use of these terms in the previous sections of this text.

This analysis is, so far, pretty sketchy. Our next step is to look at what the appropriate response should be if no acknowledgement is received of the data that has been transmitted, and at what the method should be to ensure the appropriate response. The description of the one-bit window protocol determines that a message should be retransmitted if an acknowledgement has not been received within a time = TIMEOUT after it was sent into the transmission channel. We obviously need a time measuring device, a TIMER.

A modification of the previous DFD is shown in Fig. 13.8. The DFD has been amended with a TIMER, which is a time measuring device that has the following properties:

- it starts counting time when it receives an ON-signal,
- it stops counting time when it receives an OFF-signal,
- it sends a signal when a time interval TIMEOUT has passed since it received an ON-signal, if it has not received an OFF-signal in the meantime.

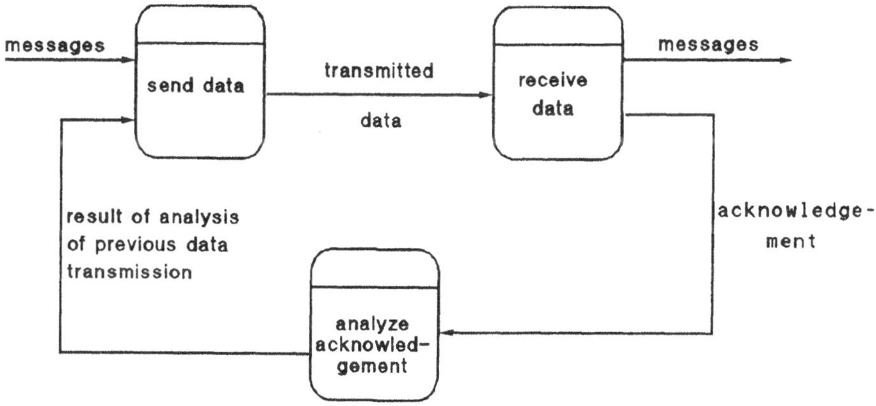

Fig. 13.7. A high level DFD for the communication protocol

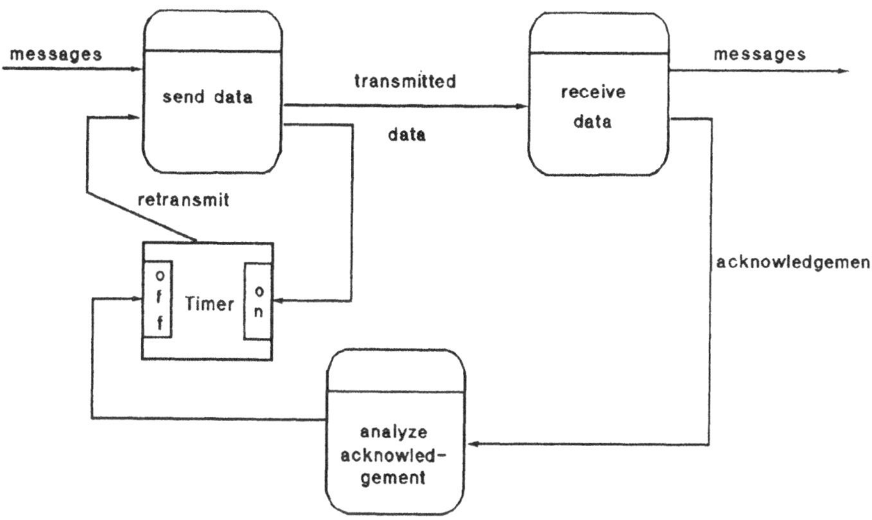

Fig. 13.8. A DFD with a TIMER, for the communication protocol

The diagram in Fig. 13.8 should be interpreted as follows:

The TIMER starts to count when the SEND_DATA process sends its data into the transmission channel. If an acknowledgement has not been received within the time interval TIMEOUT, a retransmission signal is forwarded to the SEND_DATA process. When an acknowledgement is received, an OFF-signal is sent to the TIMER, and information is sent to SEND_DATA about the successful transmission, so that the next message may be transmitted.

However, there are more pitfalls in this problem than may be seen at first glance. How would this system react if the appropriate data was received by RECEIVE_DATA, but the acknowledgement disappeared? According to the diagram in Fig. 13.8, the message would be retransmitted, and received by a RECEIVE_DATA process that did not know that it was receiving a retransmission. The sequence number – the alternating bit – in the one-bit window protocol takes care of this problem for us.

A more complete picture of the data flows and control flows of a one-bit window protocol is depicted in the diagram of Fig. 13.9. Loss of messages are indicated by a "sink" symbol. The "sinking" of a flow is to be interpreted as the loss of the data in that flow, and as a subsequent termination of the associated process. For example, in Fig. 13.9 the "sinking" of the message-flow from the RECEIVE_DATA process is to be interpreted as the loss of that data, followed by a termination of the process. This will be clarified further in the next diagram (Fig. 13.10).

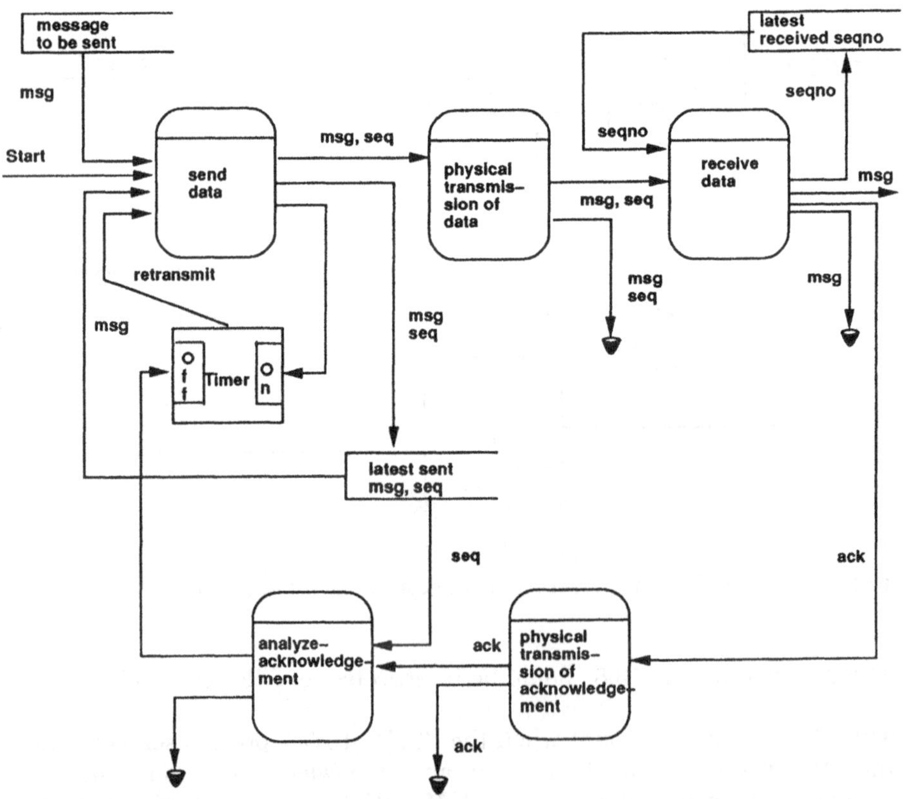

Fig. 13.9. Data flows and control signals for the one-bit window protocol

We shall first explain the system as it is depicted in Fig. 13.9. We have modeled the channel transmission as two DFD-processes, with inputs and outputs.

We have furthermore introduced three data stores, one for the messages to be sent, one that always contains the last message that was sent by the SEND_DATA process, and its associated sequence number, and one data store for the sequence number of the last message that was correctly received by the RECEIVE_DATA process.

If we now follow the response of an external start signal, we observe that a message *msg* together with a sequence number *seq* is delivered from the SEND_DATA process to the PHYSICAL_TRANSMISSION_OF_DATA process, where it is either lost or sent on to the RECEIVE_DATA process. The SEND_DATA process also starts the TIMER, and stores the transmitted data in the appropriate data store.

The RECEIVE_DATA process compares the received data's sequence number with the sequence number that is stored in the data store. It may discard the message (send it into the"sink"). It may also accept it, that is, send it out of the system into the environment, and it may send an acknowledgement *ack* for physical transmission back to be subject to analysis by the ANALYZE_ACKNOWLEDGEMENT process. If *ack* is not lost (sent into "sink") by the PHYSICAL_TRANSMISSION_OF_ACKNOWLEDGEMENT process, it is compared with the latest sent sequence number. The acknowledgement may be discarded (sent into "sink"), but the TIMER may also be shut off, and the transmission of a new message may be started. A retransmission of the latest sent message may also be ordered, in case the TIMER is not shut off by the ANALYZE_ACKNOWLEDGEMENT process.

In the explanation above we have relied heavily on interpreting the diagram by using our knowledge of the problem at hand. Extensive interpretation is necessary because of the sketchy nature of the DFD-like diagrams that we have used. This is partly due to the fact that some details are supposed to appear at a more detailed level of the (hierarchically organized) specification, e.g., the relationship between *ack* and *seq*. But interpretation is also necessary because DFD-style diagrams lack the capability to express simple logical relationships between different flows of data and/or control, for example, one may not express the fact that data is either lost or not lost during physical transmission.

We may improve the specificational precision of the diagram by simple means (Fig. 13.10). We introduce logical operators in order to represent logical relationships between flows. The graphical symbol of an XOR-operator is shown for the PHYSICAL_TRANSMISSION_OF_DATA process. We see from the diagrammatic representation that when the (*msg,seq*)-data is sent to physical transmission, it is either lost (into the "sink"), or it is

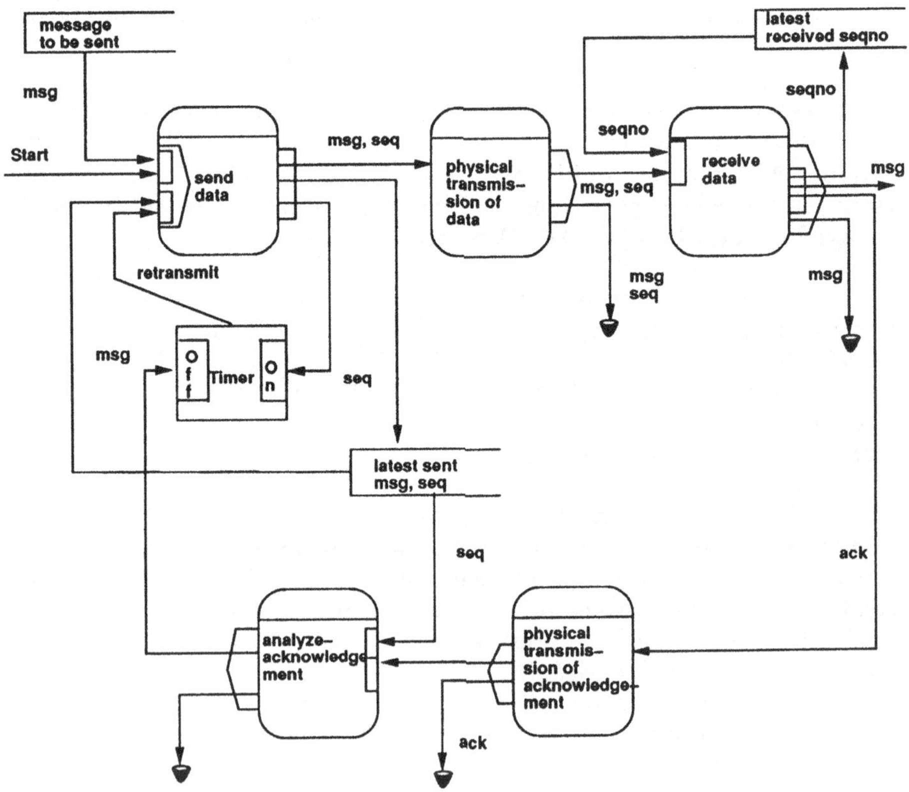

Fig. 13.10. Data flows, control signals and process ports for the one-bit window protocol

sent to the RECEIVE_DATA process. The graphical symbol for a logical AND is a rectangle, e.g., the RECEIVE_DATA process requires as its input both the physically transmitted (*msg,seq*)-data, AND the latest received sequence number from the appropriate data store. Nested logical formulas may also be graphically represented, e.g., the output of RECEIVE_DATA is either the loss of a message (sent into "sink"), or (XOR) it is all of *ack* AND *msg* AND *seqno*. The graphical representations of the logical formulas that relate the flows, may be viewed as "ports" through which input data and output data have to flow. The word "port" or "process port" may be used in the sequel to denote these logical formulae.

We may now interpret the diagram without having to depend as much on knowledge not contained in the diagram as for a "normal" DFD. The *start*-stimulus triggers the SEND-DATA process, which takes a message *msg* from the datastore, generates a sequence number *seq*, stores the (*msg,seq*)-

pair in the *latest-sent-message*-datastore, starts the TIMER, and terminates when sending the (*msg,seq*)-pair to PHYSICAL_TRANSMISSION_OF_DATA.

We may follow the response pattern quite easily, until exhaustion. When analyzing the response, we find that the specification of the start and the end of the response is incomplete. It is indicated in the diagram that the terminating flow of the ANALYZE_ACKNOWLEDGEMENT process merges with the external *start* stimulus, thus initiating the whole message-transmission process once more. The details of this of course have to be worked out. Furthermore, there is no specification of how the first message shall be treated, in order to let the sender and the receiver share the initial value of the sequence number. There is also no specification of how the last message in the queue of incoming messages shall be treated, so that the system may terminate itself when there are no more messages to be transmitted.

All of the lacking details have to be worked out. They may appear either as additions to the diagram already shown, or in the detailing of the various processes, as a result of the usual hierarchical decomposition of the processes.

When following the response paths of the diagram of Fig. 13.10, we note that whenever it is discovered that a message has been lost or distorted, one terminates the process and relies on the time-out mechanism to take care of the retransmission. For example, if a distorted message is discovered by the RECEIVE_MESSAGE process, it terminates and leaves it to the time-out mechanism to order a retransmission. This introduces (unnecessary) delays, because we have to wait until TIMEOUT for a retransmission, even if the system "knows" long before that a retransmission is necessary. How to improve our solution, in order to speed up the solution, is left as an exercise to the reader.

Exercises

1. Explain why state transition diagrams are used in the example in Sect. 13.3. What other type of technique could be used instead?

2. Modify the example given in the chapter with the constraint that there exists only one channel between the sender and the receiver, to be used for both messages and acknowledgements.

3. State the advantages and disadvantages of enriching the conventional DFD technique by ports and timers.

Chapter 14

Formal Modeling Approaches

We shall round off this text with a chapter on modeling approaches which aim at a higher degree of formality than most of the methods that we have covered so far. We shall give examples of *static modeling approaches* and *dynamic modeling approaches*, and also give a short introductory discussion on *temporal modeling approaches*.

Formal modeling in information systems engineering is the subject of much research. We do not intend to give more than a taste of some of the results. To give a fair review of all of the work that is being done in this research area goes far beyond the scope of an introductory text like ours. Therefore many significant research contributions and researchers have not been mentioned in this chapter.

Static modeling approaches view the application discourse as a snapshot of interrelated classes of entities and a set of processes operating upon the snapshot model. Characteristics of this kind of approach are simplicity, straightforwardness, and low development cost. Due to these properties, static modeling approaches are widely used in practice. However, using processes to model the behavior or dynamic properties of an application has a number of drawbacks. A process description consists of a sequence of instructions. Therefore, the description is imperative rather than declarative. The imperative style implies a prescription for software design; therefore, the design space of the later development stages is reduced. Secondly, the real-world semantics is hidden in sequences of instructions; therefore, the conceptual model description may be difficult to understand and change.

We shall discuss various types of information, their properties, and their capture. Furthermore, we shall explain two of the most common approaches to the practical representation of information in information systems engineering. They are the set-theoretic approach and the semantic network approach.

In the set-theoretic approach, the world is viewed as consisting of entities and relationships among these entities. Events, situations, occurrences, etc., are not properly described by using the set-theoretic approach. This approach is therefore said to be a static modeling approach, as opposed to dynamic modeling approaches which can also capture behavioral properties of the world. Two well-known examples of set-theoretic models are the entity-relationship model [CHEN75] and the socalled NIAM-model [NIJS77], which is a binary data model. The entity-relationship model was originally proposed for the modeling of data. The entity-relationship approach (ER-approach) may also be used for the modeling of static properties of the world, independently of the way that the world is represented by data in a database [SOLV80].

Another static modeling approach is that of "semantic networks". The idea of a semantic network does not stem from information processing but from a completely different area, namely psychology. It grew from the attempt to model human verbal memory. Semantic networks were introduced into information processing by Quillian [QUIL68]. He attempted to find a "data structure" that could be used to represent information in a human-like manner. Quillian's model is organized in such a way that concepts are explained by other concepts, which in turn can be explained by other concepts. The semantic network language is as powerful as first-order logic in expressing relationships among concepts.

Dynamic modeling approaches provide more suitable and adequate mechanisms for modeling the behavior aspects of an application. One of the most important features of a dynamic modeling approach is that state transitions are described explicitly and abstractly, without considering in full detail the mechanisms which achieve them. Instead of specifying sequences of instructions, the behavior aspects are modelled by a pair consisting of a precondition and a postcondition. Thus, how the postcondition can be established is independent of the conceptual model description. The system designer is free to choose a mechanism to implement the specification so as to achieve higher system performance. Moreover, the conceptual model description will be easier to understand because details are suppressed.

We will present two dynamic modeling approaches: ACM/PCM and the Behavior Network Model (BNM). The ACM/PCM modeling approach was proposed by Brodie and Silva for database modeling [BROD82]. It stands for "Active Component Modeling and Passive Component Modeling". The passive component modeling is the modeling of the static aspects of the application, i.e., entities and their relationships. The active component modeling is the modeling of the behavioral aspects, i.e., operations, actions and transactions. The BNM is an extension of the set-theoretic information model described in Sect. 14.1 [KUNG86]. The behavioral aspects of the application are accommodated by extending the information model with a modified Petri net component which describes the state transitions of the application.

14.1 The Set-Theoretic Approach to Information Modeling

The set-theoretic approach views the world as consisting of interrelated phenomena. These phenomena are classified into entity classes and relationship classes. An entity class is a collection of entities. A relationship class is a set of relations or associations between the individuals of two entity classes. Although n-ary relationships may be described in the set-theoretic models, we will only present binary relationships in this text. This presentation is based on a generalization and formalization of Chen's original entity-relationship model [CHEN75], [SOLV80].

14.1.1 Classification

The formation of the entity or relationship classes is based on the recognition of similarities between individual phenomena. For example, the entity class PERSON is an abstraction of a collection of individual entities which have the similarity that they are human beings. Therefore, the worldly phenomena are classified according to similarities among the phenomena. Elements of an entity class may or may not have a physical existence. For example, lectures, seminars, etc., can be regarded as entities that do not physically exist. That is, an entity class may represent concrete as well as abstract things that we want to talk about.

By using set notation, we may define entity and relationship classes in the following way:

Examples:

$\text{PERSON} = \{ x \mid x \text{ is a person} \}$
$\text{MEN} = \{ y \mid y \in \text{PERSON and the sex of } y \text{ is male} \}$
$\text{WOMEN} = \{ z \mid z \in \text{PERSON and the sex of } z \text{ is female} \}$
$\text{BOAT} = \{ w \mid w \text{ is a small vessel for travelling across water} \}$
$\text{MARRIAGE} = \{ \langle u,v \rangle \mid u \in \text{MEN and } v \in \text{WOMEN and } u, v \text{ are married} \}$

To facilitate communication, a graphic language is used. In this language, an entity class is represented by a rectangle and a relationship class is represented by a rectangle with double bottom lines. For example, the fact that men and women are married can be depicted as follows:

However, the above picture is not precise enough. For instance, it is not clear whether one man is married to one or several women. Another ambiguous point is whether every man or woman must be married. These problems are resolved by introducing additional constructs. That is, one may specify whether a subset, or the entirety of an entity class is involved in a relationship. The property that *one* or *many* elements of an entity class can be related to elements of another entity class can be specified by using **1** or **M**. For example, the figure below expresses that not every man and not every woman is married, as indicated by **p**, which stands for partial involvement. Further, a man or woman can marry only one woman or man.

A textual specification is also possible. For instance, the graphic representation above can be specified by the following text:

EntityClass :	MEN, WOMEN
Relationship:	MARRIAGE (**p**:**p**) (**1**:**1**)
	from MEN **to** WOMEN

If a full involvement is required, we can use **f** (= full) instead of **p** (= partial). The figure below illustrates an example where it is required to express that each scientific paper must be written by some author(s) and each author must write some paper(s). Further, a paper may have several authors and an author may write several papers; this information is displayed by the two **M**'s labelling the two arrows.

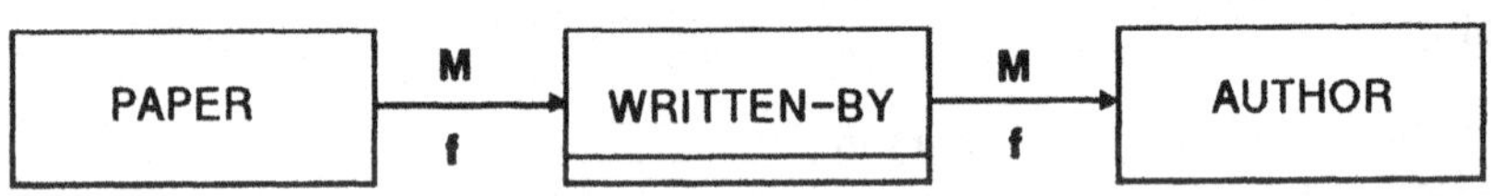

Of course, on can combine the above introduced constructs in various ways in order to specify different situations. It is easily seen that altogether 16 different combinations can be constructed. These are sufficient for specifying most situations.

14.1.2 Derived Relationships

Some relationships can be constructed from other relationships. This is done by using the relation composition operator $\circ$. Let R be a relationship from entity class X to entity class Y and S a relationship from Y to entity

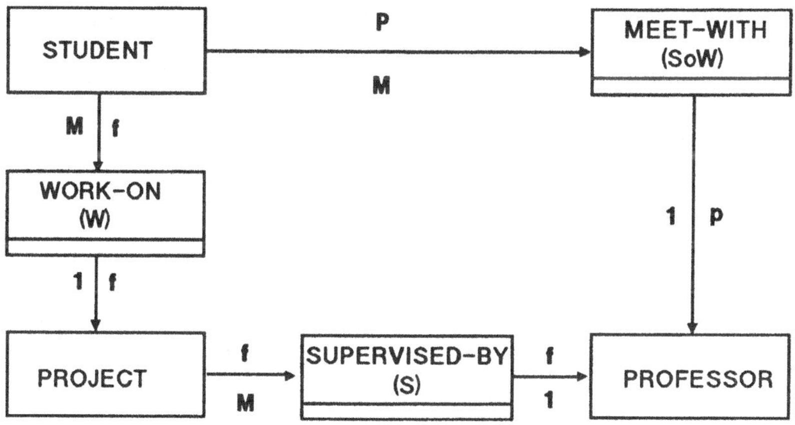

Fig. 14.1. A relationship composition example

class Z. Then the relationship $S \circ R$ is called a composite relationship of R and S, which maps X into Z,where

$$S \circ R \;=\; \{ <x,z> \mid x \in X \wedge z \in Z \wedge (\exists y)\,(y \in Y \wedge <x,y> \in R \wedge <y,z> \in S)\}$$

The operation of obtaining $S \circ R$ is called composition. Note that $S \circ R$ is empty if $(\exists y)(y \in Y \wedge <x,y> \in R \wedge <y,z> \in S)$ is not true. An example of relationship composition is shown in Fig. 14.1, where the MEET-WITH relationship is a composition of the other two relationships.

In the figure, the relationship MEET-WITH is defined in terms of the WORK-ON and SUPERVISED-BY relationships, as indicated by the relationship composition SUPERVISED-BY o WORK-ON.

The notation MEET-WITH = SUPERVISED-BY o WORK-ON seems to read backward. This results from the mathematicians' custom of denoting the value of a function f at argument a by $f(a)$. The value of g at the argument $f(a)$ is then written $g(f(a))$. By definition of composition, $g(f(a))$ and $(g \circ f)(a)$ denote the same value.

To see how the properties of the MEET-WITH relationship can be derived, we introduce some notation. The fact that WORK-ON relates a student to only one project and a project to several students, i.e., a many-to-one relationship, is denoted by **card** (WORK-ON) = (**M**:**1**). The fact that this relationship relates a subset of students to a subset of projects, i.e., (**p**, **p**), is denoted by **func**(WORK-ON) = (**p**, **p**). That is, **card** is the cardinality property and **func** is the functional property of the relationship. In general, let R be a relationship, then the cardinality property of R can be denoted by **card**(R) = (x,y), where x, y is either *1* or *M*. Likewise, the functional property of R can be denoted by **func**(R) = (θ,γ), where θ, is either **p** or **f**. If $R = R_1 \circ R_2$ is a relationship obtained by composition, then the cardinality and functional properties of R can be derived from those of R_1 and R_2. The cardinality property is calculated by using the operator **ö**, which has operands $x_1, x_2 \in \{1, M\}$. The result of the operation is defined as follows:

1 ö 1 = **1**
1 ö M = **M**
M ö 1 = **M**
M ö M = **M**

The reader may view **1** as **false**, **M** as **true**, and the **ö** operator as the logical connective ∨. Then the same rule applies here as for ∨. That is, the **ö** and the ∨ operations are isomorphic.

Suppose that **card**(R_1) = (x_1,y_1) and **card**(R_2) = (x_2,y_2). Then

card(R) = **card**($R_1 \circ R_2$) = (x_1**ö**x_2, y_1**ö**y_2).

For example, in Fig. 14.1,

card(WORK-ON) = (**M**:**1**) and **card**(SUPERVISED-BY) = (**M**:**1**);

therefore,

card(MEET-WITH) = **card**(SUPERVISED-BY o WORK-ON) = (**MöM**,**1ö1**) = (**M**:**1**).

Similarly, the functional property is computed by using the ⊞ operator which is also a binary operation. Its operands are $\chi, \chi_2 \in$ {**p**, **f**}. The rules of the operation are defined as follows:

p ⊞ **p** = **p**
p ⊞ **f** = **p**
f ⊞ **p** = **p**
f ⊞ **f** = **f**

The reader may assume that **p** and **f** correspond to **false** and **true** respectively. Then the ⊞ operation is isomorphic to the logical connective ∧.

In Fig. 14.1, we have

func(WORK-ON) = (**p**, **p**) and

func(SUPERVISED-BY) = (**f**, **p**), therefore,

func(MEET-WITH) = **func**(SUPERVISED-BY o WORK-ON) = (**p⊞f**,**p⊞p**) = (**p**,**p**).

A composite relationship is a derived relationship. Therefore, we can express such a relationship by using a logical formula. For instance, the MEET-WITH relationship can be expressed by the equation

$$\text{WORK-ON}(u,v) \wedge \text{SUPERVISED-BY}(v,w) \rightarrow \text{MEET-WITH}(u,w)$$

where the variables u, v, w are assumed to be universally quantified. The composite relationship is the logical conjunction of the component relationships over common elements.

14.1.3 Subclassification of Entities

An entity class A may be a subset of another entity class B. In this case, we say that B is a generalization of A. For example, we may wish to express that managers and secretaries are employees, and employees are persons. This can be displayed by using the **subset** construct (Fig. 14.2).

The graphic representation of Fig. 14.2 may be specified by the text

EntityClass: PERSON, EMPLOYEE, MANAGER, SECRETARY
subset (PERSON): EMPLOYEE
subset (EMPLOYEE): MANAGER, SECRETARY

The equivalent expressions in predicate logic appear as follows:

$$(\forall x)(\ x \in \text{MANAGER} \rightarrow x \in \text{EMPLOYEE}\ )$$
$$(\forall x)(\ x \in \text{SECRETARY} \rightarrow x \in \text{EMPLOYEE}\ )$$
$$(\forall x)(\ x \in \text{EMPLOYEE} \rightarrow x \in \text{PERSON}\ )$$

That is, each arc labelled by **subset** corresponds to a universally quantified well-formed formula, which states that any individual belonging to the subclass belongs to the superclass. By using the transitivity property of the implication sign, we can easily derive the following well formed formulas:

$$(\forall x)(\ x \in \text{MANAGER} \rightarrow x \in \text{PERSON}\ )$$
$$(\forall x)(\ x \in \text{SECRETARY} \rightarrow x \in \text{PERSON}\ )$$

Therefore, the subset hierarchy has the transitivity property.

The **subset** construct can also be applied to relationship classes. For instance, the relationship of being classmates implies the relationship of being schoolmates. Another example is the father relationship, which is a subset of the ancestor relationship. Further, the ancestor relationship can be defined recursively. Figure 14.3 illustrates how recursive definitions can be displayed.

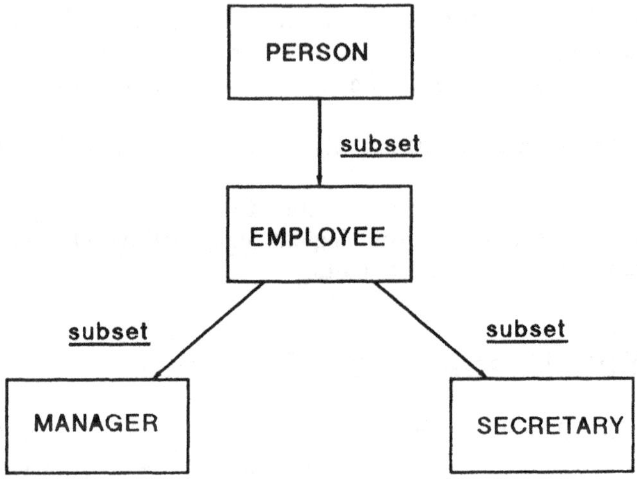

Fig. 14.2. The subset construct

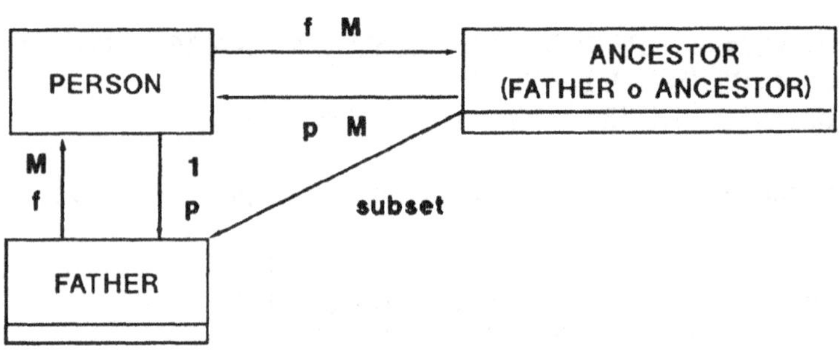

Fig. 14.3. A recursive relationship

We see that the FATHER relationship is a subset of the ANCESTOR relationship. This corresponds to the logical formula

$$\text{FATHER}(u,v) \rightarrow \text{ANCESTOR}(u,v)$$

The composite relationship ANCESTOR corresponds to the formula.

$$\text{FATHER}(u,v) \wedge \text{ANCESTOR}(v,w) \rightarrow \text{ANCESTOR}(u,w)$$

14.1.4 Attributes of Entities

The recognition of the similarities among phenomena implies a recognition of common gross properties of these phenomena. Persons normally have hair of a certain color, and have certain height, weight, addresses, and so on. Properties as such are denoted by symbols. In our common practice, we know that different sets of symbols are used to describe different properties. For instance, character strings beginning with capital letters are used to denote the names of persons as well as countries. Positive integers are used to denote the age of persons as well as the number of seats in an air-plane. Sets of symbols like these are called *value sets*. The meaning of the elements of a value set is defined by an association of the value set to an entity or relationship class. The association is called an attribute of the entity or relationship class. For example, the attributes: SSN, NAME, and ADDR associate three different value sets to the entity class PERSON, as illustrated in Fig. 14.4.

The value set of an attribute can be a complex data structure, as shown in Fig. 14.4. An identifier of an entity or relationship class is a subset of attributes which can be used to identify uniquely the individuals of the entity or relationship class, e.g., the SSN in Fig. 14.4.

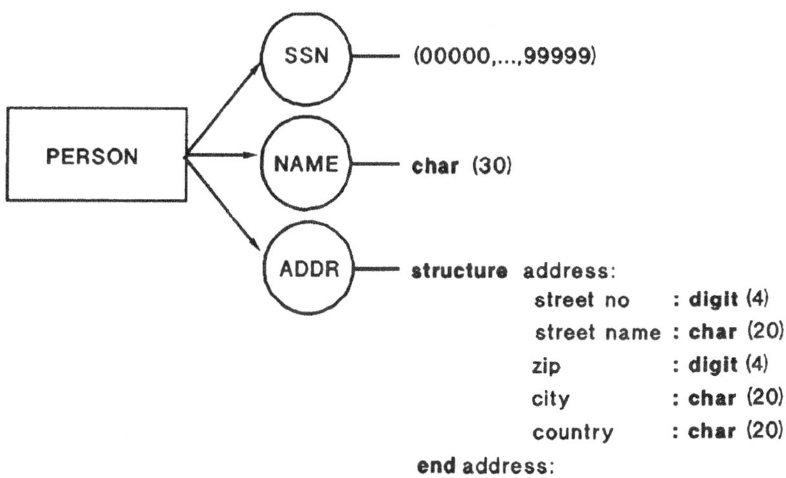


```
EntityClass: PERSON
Attribute(PERSON): SSN,NAME, ADDR
Identifier(PERSON): SSN
ValueSet (SSN): (0000,....,99999)
ValueSet (NAME): char (30)
ValueSet (ADDR): structure address:
                       street no   : digit (4)
                       street name : char (20)
                       zip         : digit (4)
                       city        : char (20)
                       country     : char (20)
                 end address:
```

Fig. 14.4. Entities and attributes

The way in which the phenomena are classified is dependent on the application in question. That is, whether a phenomenon such as marriage should be viewed as an entity or as a relation between two persons depends on the view of the user's community. Similarly, a concept such as color can be classified as an entity set as well as an attribute of an entity set. Conflicting views like this must be resolved during the system analysis phase. Unfortunately, there is no easy solution to such problems. It often depends on the analyst's experience, knowledge, and the art of compromise. A cookbook approach to providing a set of guidelines is not adequate.

14.1.5 Inheritance of Attributes

The attributes of an entity class are inherited by all of its subclasses. A subclass also inherits the relationships of its superclasses. Since an identifier is also an attribute, the identifier(s) of the superclass is also inherited by its subclasses. The inheritance rule gives a convenient way for specifying a model of the relevant concepts. Figure 14.5 illustrates this advantage.

In Fig. 14.5, since MANAGER is a subset of EMPLOYEE, MANAGER will also have E#(employee number), NAME, and SALARY. Further, the relationship WORK-IN is also inherited by MANAGER.

Figure 14.5 also shows that AVERAGE-SALARY is a quality related to MANAGER. This means that AVERAGE-SALARY applies to the entity class MANAGER as a whole rather than an individual as an attribute does. Those who are familiar with knowledge representation tools may recognize that the attribute construct corresponds to the member slot and the quality construct corresponds to the own slot in, say, the knowledge representation tool KEE (KEE = Knowledge Engineering Environment) [KUNZ84].

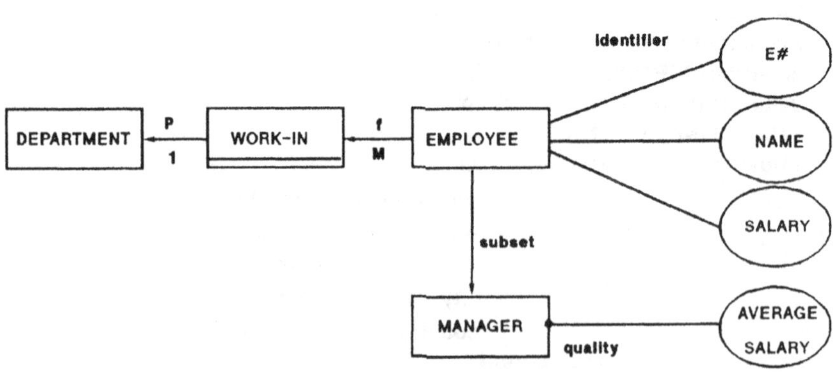

Fig. 14.5. Property inheritance

14.2 The Semantic Network Approach to Information Modeling

A semantic network consists of a set of nodes interconnected by a set of arcs. Each node of the net denotes an object, a situation, or a subnet, which is a semantic net in itself. Each arc denotes an instance of a binary relation between the two nodes connected. Figure 14.6 depicts a simple example, taken from the organization of a scientific conference, as explained in Sect. 1.1.1 (the IFIP conference example).

The potential participants at such a conference may also play other roles, that is, they may belong to other entity sets. They may be authors of papers, referees of papers, or members of the program committee or of some other interest group that is invited to participate at the conference.

The arc pointing from node AUTHOR to node POTENTIAL PARTICIPANT is labelled by **s**. This indicates that AUTHOR is a subset of POTENTIAL PARTICIPANT. Similarly, REFEREE, PROGRAM_COMMITTEE_MEMBER, and INTEREST_GROUP_MEMBER are all subsets of POTENTIAL PARTICIPANT. Note that these subsets are not necessarily disjoint. For instance, a program committee member may be an author. An example of this is also shown in the figure, i.e., JOHN is an element of AUTHOR, and he is also an element of PROGRAM_COMMITTEE_

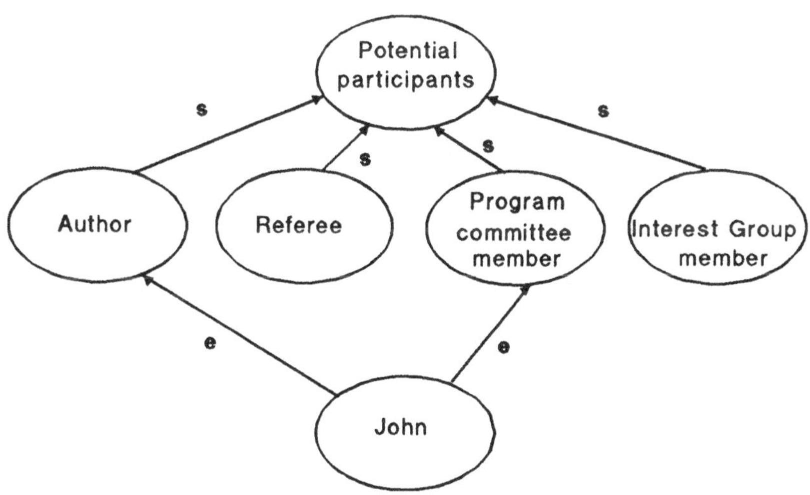

Fig. 14.6. A semantic network example

MEMBER. This is graphically represented by the two arcs pointing from JOHN to AUTHOR and to PROGRAM_COMMITTEE_MEMBER respectively. Note that these two arcs are labelled by **e** to suggest the 'element of' relation, that is, JOHN is an element of the set of AUTHORS, as well as of the set of PROGRAM_COMMITTEE_MEMBERS.

Note that the information model in this case captures a conference policy rule. Some scientific conferences will not permit program committee members to submit papers to the conference, in order to avoid conflicts of interest when papers are to be selected for presentation. Other conferences are so large that the program committee is split into several subgroups. A paper which has been authored by a program committee member can then be evaluated by some other part of the program committee, to avoid conflicts of interest. It is the latter situation that is depicted in Fig. 14.6.

14.2.1 Disjoint Subset and Distinct Element

Many practical circumstances require that disjoint subsets and distinct elements of a set can be represented. In such cases, the labels of the arcs will be **ds** for *disjoint subset* and **de** for *distinct element*. Figure 10.7 gives an example of this kind of situation.

Figure 14.7 says that ACCEPTED PAPER and REJECTED PAPER are two disjoint subsets of the set PAPER, that is, a paper cannot be an accepted paper as well as a rejected paper at the same conference.

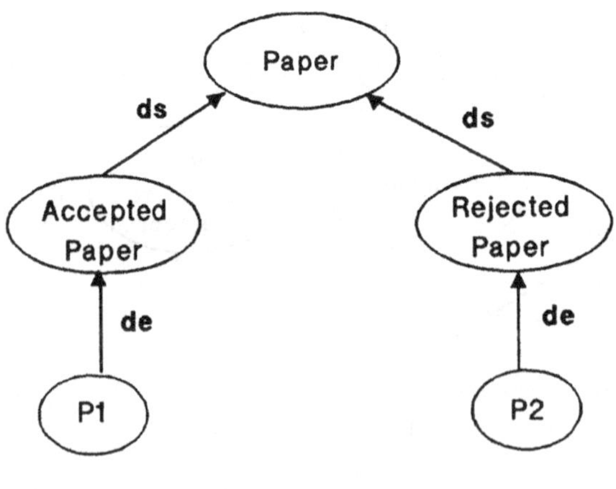

Fig. 14.7. Disjoint subset and distinct element

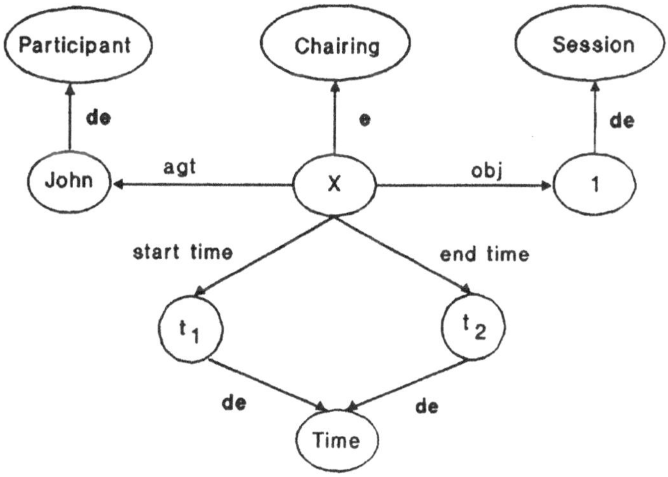

Fig. 14.8. A deep case in a semantic net

14.2.2 Deep Cases

In order to describe general relationships between the nodes, the semantic network model provides a construct called a *deep case* description. An example is the expression:

Participant John chairs session 1 during the time interval from t_1 to t_2.

Figure 14.8 shows the description of such a situation in a semantic net.

A deep case can be described by a *situation* node, e.g., CHAIRING in Fig. 14.8. An instance of that situation is represented by an element node, say x, which has arcs leading to other nodes in the network. These arcs and nodes together depict the situation instance at a more detailed level. The **agt** arc indicates that John is the *agent* of the chairing *action*, the **obj** arc indicates that session 1 is the *object* of that action and the start-time and the end-time of the action is t_1 and t_2.

14.2.3 Spaces

Abstraction is a useful mechanism in conceptual modeling. Abstractions can be achieved in semantic networks by using *spaces* and *compound spaces*. The central idea is to allow groups of nodes and arcs to be bundled together

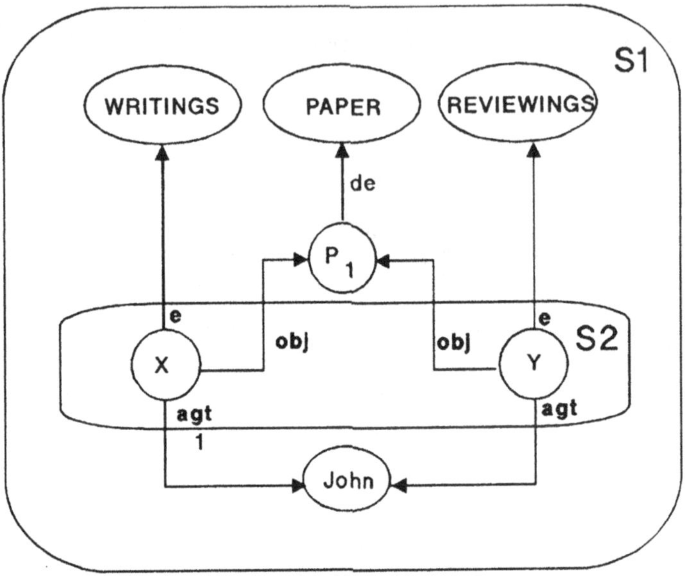

Fig. 14.9. A conjunction situation in spaces

into units called *spaces*. A space is then treated at the same level as nodes and arcs. A compound space is also called a *vista*, which is made up of spaces and compound spaces. Thus compound spaces will be called spaces in the sequel. A space is represented by a rectangle encompassing the group of nodes and arcs. Figure 14.9 illustrates this mechanism.

The semantic net in Fig. 14.9 states that John writes and reviews paper P1. We see that there are two spaces, S1 and S2. The space S2 is a (simple) space since it contains only nodes and arcs. S1 is a compound space since it contains another space as well, i.e., S2. By definition, the internal structure of S2 is not visible from the view of S1, and in this sense S2 can be regarded as an abstraction of some situation. Note that the coexistence of the two instances of WRITINGS and REVIEWINGS is represented by using a space S2. This means that the information about John reviewing his own paper is contained wholly within space S2.

14.2.4 Constraints

In refereeing practice, it is not allowed for someone to review a paper written by himself. For simplicity, we temporarily ignore the tense and express this requirement in the following well-formed formula:

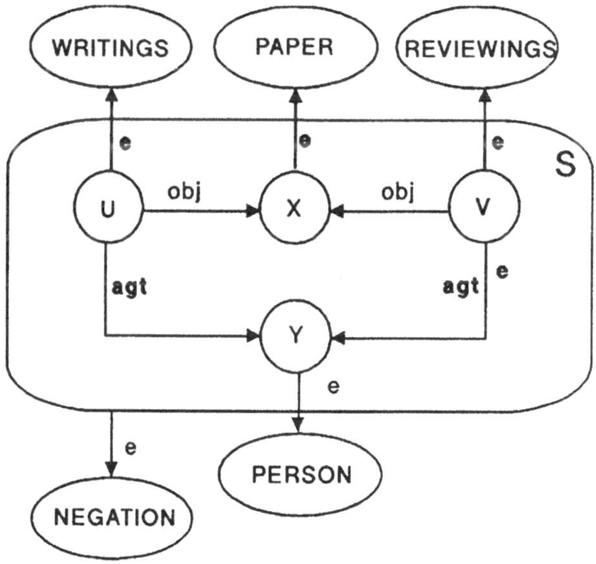

Fig. 14.10. A semantic constraint in semantic-net form

$$\sim(\exists x)(\exists y)(\ \text{PAPER}(x) \wedge \text{PERSON}(y) \wedge \text{WRITE}(y,x) \wedge \text{REVIEW}(y,x)\ )$$

where PAPER(x) is true iff x is a paper and PERSON(y) is true iff y is a person, the predicate WRITE(y,x) says that y writes x and REVIEW(y,x) means that y reviews x. The above formula states that nobody writes and reviews the same paper. Such a statement can be viewed in two ways:

1) It states the fact that nobody writes and reviews the same paper.

2) It requires that the statement must always be true. In this case, we are using the formula as a semantic constraint specifying that in no state of the information system can anyone write and review the same paper.

The above semantic constraint can be represented in semantic network notation as in Fig. 14.10, where we only use one space S as opposed to the semantic net of Fig. 14.9 where we used a compound space to express the same thing in the particular case of the person John.

The semantic net states that the situation represented by space S is not true. Space S states that there exists some person Y and some paper X such that Y writes and reviews X. Note that existential quantification is achieved in a semantic network through explicit indication of the existence of an element (or situation, etc.) of some kind. So far, we have incidentally discussed how semantic network constructs can be used to express logical connectives such as conjunction and negation. This implies that the semantic network language is as powerful as the propositional logic language.

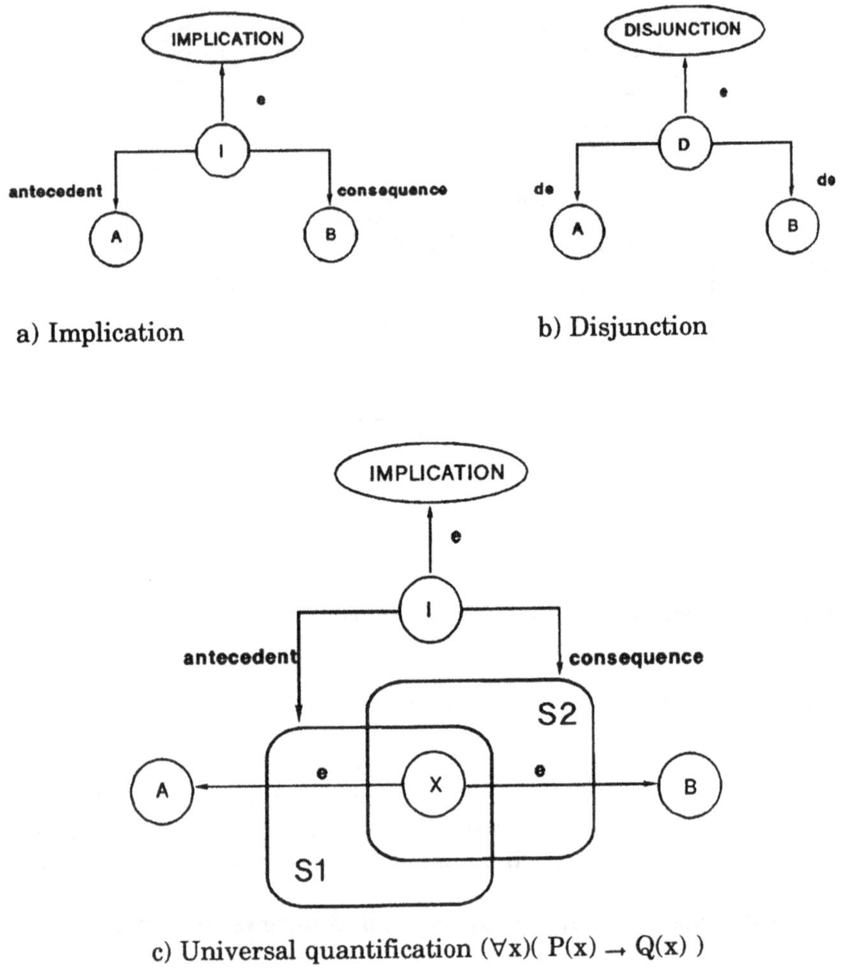

Fig. 14.11. Other logical connectives in semantic-network form

Moreover, we have also shown that the existential quantifier can be expressed in semantic network notation. Since the universal quantifier can be replaced by the existential quantifier and the negation connective, we may then regard the semantic network language as being as powerful as the predicate logic language. However, it is very useful and desirable to express in semantic-network form the implication and disjunction connectives, and universal quantification. Figure 14.11 illustrates how these can be expressed.

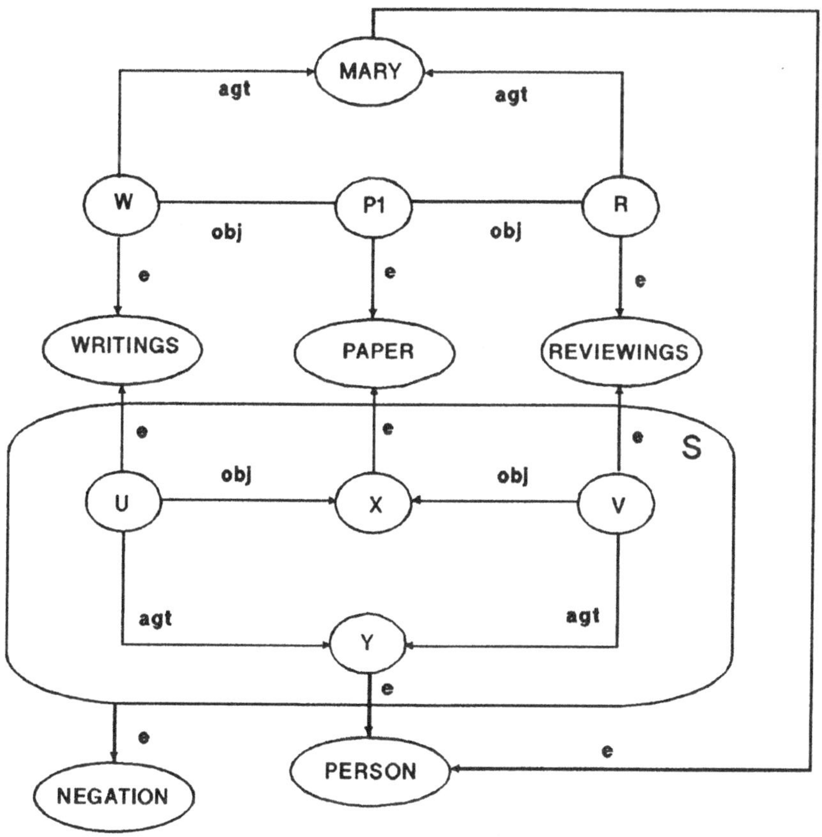

Fig. 14.12. An inconsistent description in a semantic network

Note that inconsistency may occur in a semantic network model. Again, consider the semantic constraint that nobody may write and review the same paper. If we extend the network in Fig. 14.10 by adding a few nodes and arcs, as shown in Fig. 14.12, the resulting network becomes inconsistent. That is, its counterpart in first-order logic is inconsistent. In fact, Fig. 14.12 says that

> Nobody can write and review the same paper, and
> Mary writes and reviews paper P1.

which is obviously a contradiction.

14.3 The ACM/PCM Modeling Approach

ACM/PCM is a modeling methodology for the design and development of moderate-to-large-scale data-intensive systems. The main principles of ACM/PCM are abstraction and localization. The principle of abstraction is the suppression of irrelevant details in order to concentrate on the more important issues at various modeling stages. The structure (i.e., static) and behavior aspects are modeled by data abstractions and abstract data types. Data abstraction is a set of composition/decomposition rules which can be used to form higher-level objects from lower-level constituents, and vice versa. The principle of localization enables the designer to model individual properties of application objects separately and then integrate these descriptions to produce a complete design.

In ACM/PCM, a database application is described in three levels: the transaction level, the conceptual level and the database level (see Tab. 14.1). These three levels correspond roughly to the three-level architecture of a relational database system [DATE82]. The transaction level is designed for specific applications and the end-user needs. It provides mechanisms for modeling application-oriented transactions, queries and reports. The queries and reports are constructed by using application objects and transactions are described in terms of actions. The application objects and actions are described in the conceptual level, which integrates the local views described in the transaction level. The application objects and actions in turn are described in terms of database objects and operations. The difference between an action and a database operation is that an operation is performed on a single instance of a database object, e.g., the insertion of an employee number, and the update of a salary value. An action can be performed on an application object, e.g., to insert or delete a hotel reservation. An action as such may invoke several database operations in a certain manner. For example, deleting a hotel reservation consists of deleting the reservation number, the room, the client, and the arrival and departure dates of the reservation.

	Structure modeling	Behavior modeling
Transaction level	Queries, reports	Transactions
Conceptual level	Application objects	Actions
Database level	(Database) objects	Database operations

Table 14.1. Three levels in ACM/PCM

The structure and behavior modeling in ACM/PCM is done by using the Extended Semantic Hierarchy Model (SHM+). In the rest of this section, we will first briefly present the structure modeling concepts and constructs and then give a more thorough description of behavior modeling.

14.3.1 Structure Modeling in SHM+

SHM+ provides one structural concept, the object, and four abstraction mechanisms: classification, aggregation, generalization, and association. For convenience of explanation, we will refer to object classes as objects unless objects and object classes must be distinguished.

Real-world entities are classified into object classes in much the same way as described for the set-theoretic information model. This is called the classification abstraction of the ACM/PCM approach. Classification represents an "instance-of" relationship between an object class in a schema and an object in a database. For instance, a relation tuple

<26374, Hsiao , $40,000>

is an instance of the object class EMPLOYEE, which has employee number, name and salary as the common properties of individual employees.

Aggregation is the abstraction mechanism in which a relationship between component objects is formed as a higher-level aggregate object. Aggregation represents the "part-of" relationship. For example, when building a university information system, we may describe an enrollment as an aggregation of a student, a class, and a grade. A class can in turn be regarded as an aggregation of a course, a semester, an instructor, and a room. These two aggregates may be graphically represented as in Fig. 14.13, where the underscored components take part in the key of the aggregate. In mathematical terms an aggregate is a subset of the cartesian product of its components, i.e., an aggregate $A \subseteq C_1 \times C_2 \times \ldots C_i \ldots \times C_n$, where C_i is a part of A.

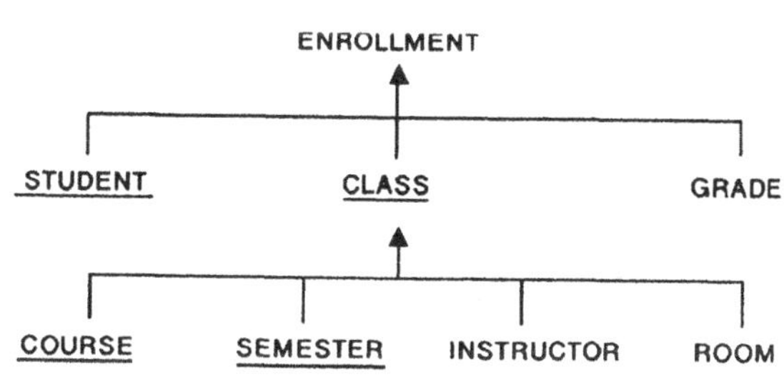

Fig. 14.13. Aggregates in SHM+

Generalization is the abstraction mechanism in which a relationship between class objects is considered as a higher-level generic object. It is the "is-a" relationship. Such a relationship is often encountered in conceptual modeling. For example, EMPLOYEE may be considered as a generic object class of MANAGER and SECRETARY. Therefore, we have that manager is-a employee and secretary is-a employee. In mathematical terms: manager $\subseteq$ employee, secretary $\subseteq$ employee.

An important feature of SHM+ is that it supports the "principle of relativism". By this principle, an object can be considered both independently and in terms of any relationship in which it takes part. For instance, EMPLOYEE can be considered as an independent object which is an aggregation of a set of properties. On the other hand, EMPLOYEE can also be considered as either the generalization of MANAGER and SECRETARY or a subset of PERSON. These views coexist. Figure 14.14 illustrates how multiple views can be represented in SHM+.

Association is the abstraction mechanism for representing the "member-of" relationship. For example, MAN and WOMAN are two members of SEX-GROUP. Further, MAN and WOMAN may be regarded as subsets of PERSON. These are depicted in Fig. 14.15.

In summary, we show the graphic symbols for representing the abstraction mechanisms in Fig. 14.16.

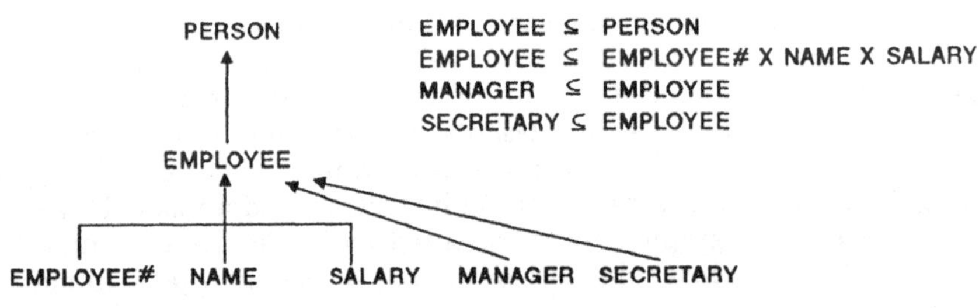

Fig. 14.14. An example of semantic relativism

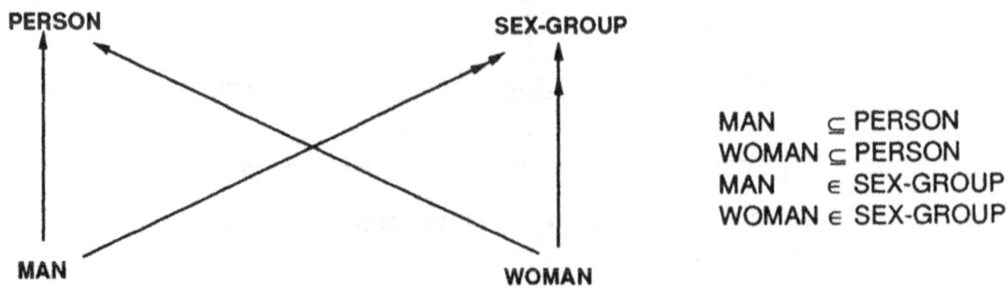

Fig. 14.15. An association example

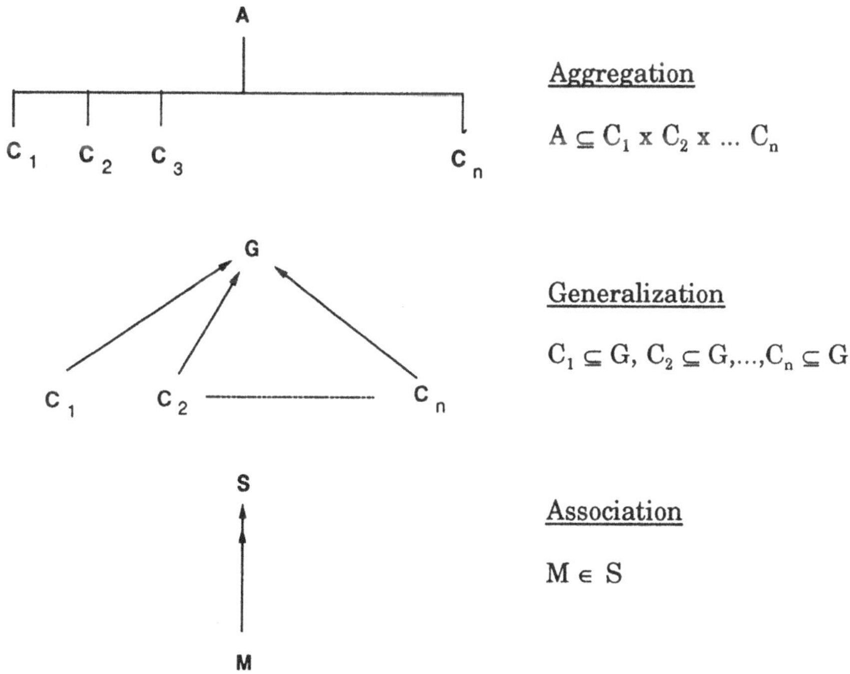

Fig. 14.16. Graphical symbols for structure modeling in SHM+

14.3.2 Behavior Modeling in SHM+

Behavior modeling is achieved by specifying the flow of control in applying the database operations and actions. SHM+ provides three control abstraction mechanisms which are analogues of aggregation, generalization, and association. As shown in Table 14.1, the evolution of the information system is specified at three different levels. These three levels correspond to the database level, the conceptual level, and the transaction level.

In the database level, a database operation either updates or retrieves a single instance of a database object. A set of database operations are provided by SHM+, e.g., INSERT, DELETE, UPDATE, FIND, CREATE, and REQUEST. Control abstractions are used to construct higher-level "operations" from database operations. The three control mechanisms of SHM+ are called sequence, choice, and repetition and are the three basic control flow constructs of structured programming. Figure 14.17 shows the graphic symbols of these three control abstractions, which are analogous to the graphic symbols of their structure modeling counterparts.

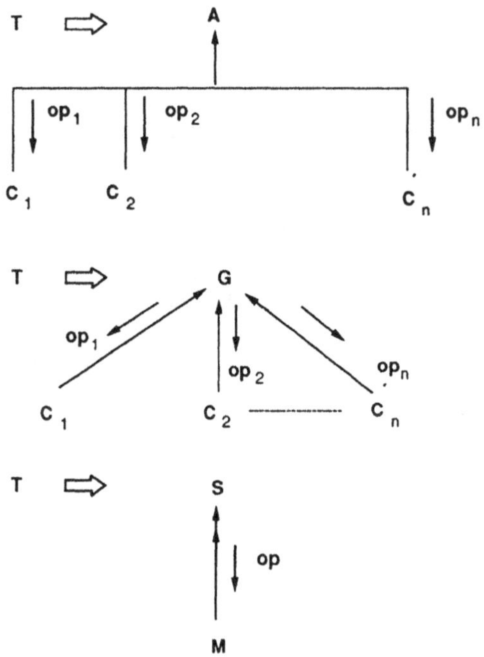

Sequence:; T is a sequence of operations op_i on c_i,i=1,...,n.

Choice:; T is a selection of operations op_i on c_i, i=1,...,n.

Repetition:; T is a repetition of operation op on each member of M.

Fig. 14.17. Graphic symbols for behavior modeling

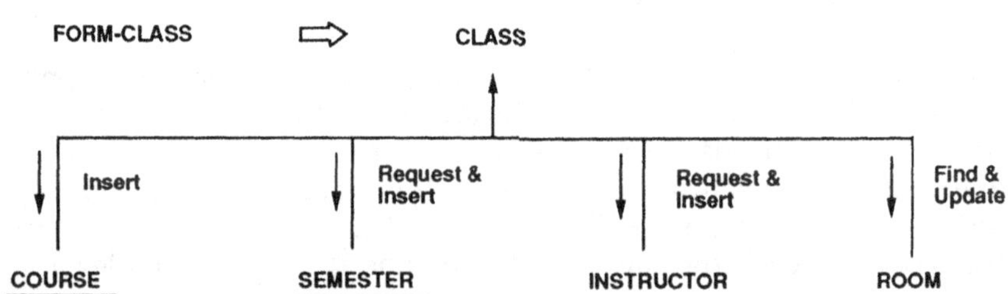

Fig. 14.18. The "FORM-CLASS" action schema

In SHM+, an action can be defined by using the three control abstractions. For example, the action FORM-CLASS can be considered as a sequence of operations as shown in Fig. 14.18.

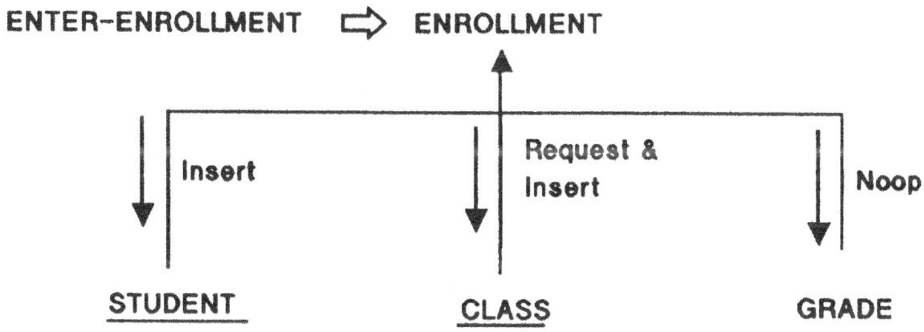

Fig. 14.19. The ENTER-ENROLLMENT action schema

The schema for the action ENTER-ENROLLMENT is shown in Fig. 14.19. Note that the modeling of the ENTER-ENROLLMENT need not consider the operations that can be performed on the aggregate CLASS. Neither does it have to consider the operations on the properties of CLASS. We also notice that when an enrollment is entered, we do not have a grade for the student enrolling in a class. Therefore, "noop" (no operation) is applied to GRADE.

An action schema can be specified by an action specification, which is more precise and suitable for analysis and verification. An action specification involves a precondition, a postcondition, and possibly a database operation. The precondition ensures that the action can be invoked, while the postcondition checks the success of the action execution. As an example, consider the specification of the ENTER-ENROLLMENT action schema:

```
ACTION          : ENTER-ENROLLMENT(s)
IN              : (s : STUDENT, c : CLASS, g : GRADE)
OUT             : (ENROLLMENT)
PRECONDITION    : s is-a STUDENT? AND c is-a existing CLASS?
POSTCONDITION   : ENROLLMENT(s, c, null) exist?
DB-OPERATION    : INSERT ENROLLMENT(s, c, null)
```

We have chosen to express the pre- and postconditions in a slightly different way from the original proposal in [BROD82]. However, this will not distort the original idea of SHM+).

At the transaction level, end-user requirements are modeled and specified. Modeling at the transaction level involves the design and specification of transaction schemas. Suppose that some end-user requirements include retrieving the average class grade. A transaction for this is shown in Fig. 14.20, where AVERAGE-CLASS-GRADE is an aggregation of CLASS and AVERAGE-GRADE.

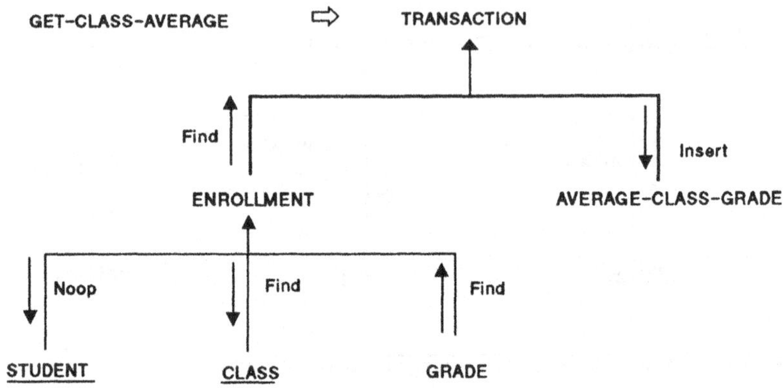

Fig. 14.20. A transaction schema in SHM+

The transaction schema shows that in order to compute the class average, the class in question must be located and the set of grades for the class found (and retrieved, indicated by the upward arrows). The computed AVERAGE-CLASS-GRADE is then inserted. The specification of the transaction becomes:

TRANSACTION	: GET-CLASS-AVERAGE(c)
SCOPE	: (c : CLASS, e : ENROLLMENT, a : AVERAGE-GRADE)
PRECONDITION	: c is-a existing CLASS?
	GRADES of c available?
POSTCONDITION	: AVERAGE-CLASS-GRADE(c, a) exists?

14.3.3 Action and Transaction Programs

Action and transaction programs are constructed by introducing active roles into behavior specifications. This is done by analyzing the behavior schemas and specifications. For example, the action program for ENTER-ENROLLMENT can be specified as:

```
ACTION      : ENTER-ENROLLMENT(s)
IN          : (s : STUDENT, c : CLASS, g : GRADE)
OUT         : (ENROLLMENT)
              if s is-a STUDENT then REQUEST-CLASS(c)
              if c is-a existing CLASS
              then INSERT ENROLLMENT(s, c, null)
```

The example shows that an action program can be easily constructed by transforming the precondition and postcondition into conditions of the IF-statement. The active roles are the database operations specified in the action schema. In a similar way, we can transform a transaction schema and specification into a transaction program:

```
TRANSACTION : GET-CLASS-AVERAGE(c)
SCOPE       : (c : CLASS, e : ENROLLMENT, a : AVERAGE-GRADE)
              if c is-a existing CLASS and GRADES of c available
              then RETRIEVE-GRADES(c)
                   COMPUTE-AVERAGE(c, a)
                   INSERT-AVERAGE-CLASS-GRADE(c, a)
              check AVERAGE-CLASS-GRADE(c, a) exists
```

The active roles are a number of action invocations. For instance, if the test of "s is-a STUDENT" is true, then an interactive action REQUEST-CLASS(c) is performed which requests the user for a class in which the student is to be enrolled. The postcondition of the action schema is translated into the condition expression of the second **if**-statement. This conditional expression checks if the requested class c is an existing CLASS. When the test is successful, the object <s, c, null> is inserted into ENROLLMENT. A similar explanation can be given for the transaction program above. In these program specifications, the designer needs to know only what the actions do, but not how the actions do their tasks.

ACM/PCM is a design methodology that covers the whole process of database design, specification, and implementation. Both structural and behavioral aspects are modeled and integrated. The constructs for structure and behavior modeling have a similar structure. Three levels of modeling and specifications are proposed. The experience of Brodie and Silva with SHM+ has shown that neither transaction modeling nor conceptual modeling should be completed independently and that design is an interactive process. That is, a preliminary design of transaction schemes facilitates modeling at the conceptual level. After a more thorough modeling at the conceptual level, transaction modeling is then refined. Although not presented, ACM/PCM also provides methods for specifying exception handling through the use of the preconditions and postconditions. This should also be considered as an important feature of any modeling approach.

14.4 Petri Nets

In this section we introduce some of the basic concepts and constructs of a Petri net. Many interesting Petri net topics cannot be covered here. The interested reader is referred to [PETE77] or [REIS85] for more about Petri nets and their applications.

14.4.1 Basic Concepts

A Petri net is a formal model of information and control flows in a system. In particular, it provides mechanisms for describing systems that may exhibit asynchronous and concurrent activities. Figure 14.21a shows a simple Petri net.

The Petri net in Fig. 14.21a has four transitions t_1, ..., t_4 and four places P_1, ..., P_4. A transition is represented by a bar and a place is represented by a circle. A place P_i is said to be an input place of a transition t_j if there is an arrow going from P_i to t_j. For example, P_1 and P_2 are the input places of t_2, while P_3 is the input place of t_3. Note that transition t_1 has no input places and is called a source. A place P_i; is said to be an output place of a transition t_j if there is an arrow going from t_j to P_i. For instance, P_1 is the output place of t_1 and P_3 is the output place of t_2. Note that transition t_3 has two output places P_2 and P_4. Transition t_4 has no output places and is called a sink.

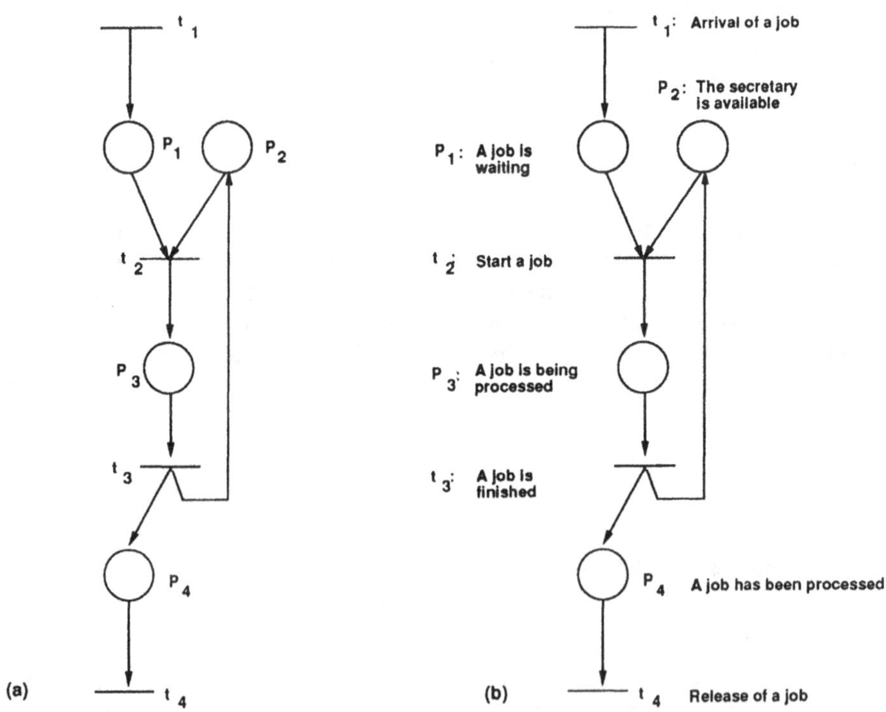

Fig. 14.21. A Petri net and an interpretation

A Petri net can also be defined by a four-tuple N = <P, T, I, O>, where P and T are the sets of places and transitions, respectively, I is the input function which defines the set of input places of each transition, O is the output function which defines the set of output places of each transition. For instance, the Petri net of Fig. 14.21.a can be specified as:

$P = \{ P_1, P_2, P_3, P_4 \}$
$T = \{ t_1, t_2, t_3, t_4 \}$

$I(t_1) = \{ \}$,	$I(t_2) = \{ P_1, P_2 \}$
$I(t_3) = \{ P_3 \}$,	$I(t_4) = \{ P_4 \}$
$O(t_1) = \{ P_1 \}$,	$O(t_2) = \{ P_3 \}$
$O(t_3) = \{ P_2, P_4 \}$	$O(t_4) = \{ \}$

The graphic notation is more intuitive and easy to explain. Therefore, in this book we will use the graphic representation of a Petri net.

The Petri net in Fig. 14.21a can be regarded as a statement in a formal language. It is analogous to a statement in first-order logic. The Petri net can be interpreted by associating meanings to the places and transitions. Figure 14.21b illustrates a possible interpretation of the net in Fig. 14.21a. Thus, the net in Fig. 14.21b can be used to model the behavior of a secretary in an office. Associating different meanings to the places and transitions will result in a different interpretation which models a different piece of reality. For instance, if we interpret P_2 as "the CPU is idle", then the resultant net may be a model of a unit-processor computer system.

14.4.2 Markings and Execution

An interesting feature of a Petri net is that it is an executable model. The execution of a Petri net is achieved by assigning tokens to places. A token is usually represented by a dot residing in a place and an arbitrary number of tokens can be assigned to a place. Figure 14.22a illustrates an assignment of tokens to the places of Fig. 14.21b. A Petri net with tokens is a marked Petri net. The distribution of tokens in a marked Petri net is called a marking, which in effect defines a state of the system at some point in time.

The marked Petri net in Fig. 14.22a expresses that the secretary is processing a job and there are two jobs waiting. These are illustrated by the token in P_3 and the two tokens in P_1. Since P_2 and P_4 contain no token at all, we know that the secretary is not available and no job has been processed. A Petri net as such is known as an uninterpreted Petri net since the tokens of the net are not interpreted. That is, a token is nothing more than an "indicator", the occurrence of which indicates that the condition or situation represented by the place is "on". For example, P_3 and the token in P_3 together indicate that "a job is being processed". Similarly, the occurrence of the two tokens in P_1 indicates that two jobs are waiting; however, the two waiting jobs are assumed to be described elsewhere by some other means, e.g., one of the jobs may be a letter to be typed and the other may be a meeting to be scheduled. The occurrence of the two tokens clearly is not adequate for these purposes. One modification of the original definition of a Petri net results in so-called "interpreted Petri nets", in which tokens have types. The BNM model we are going to study in the next section is a kind of interpreted Petri net.

In a marked Petri net, a transition is said to be enabled if each of its input places contains a token. In Fig. 14.22a, two transitions are enabled, i.e., t_1 and t_3. Transition t_3 is enabled since its only input place contains a token. Transition t_1 is enabled since it has no input place; hence, each of its input places contains a token. Transition t_2 is not enabled since one of its input places does not contain a token.

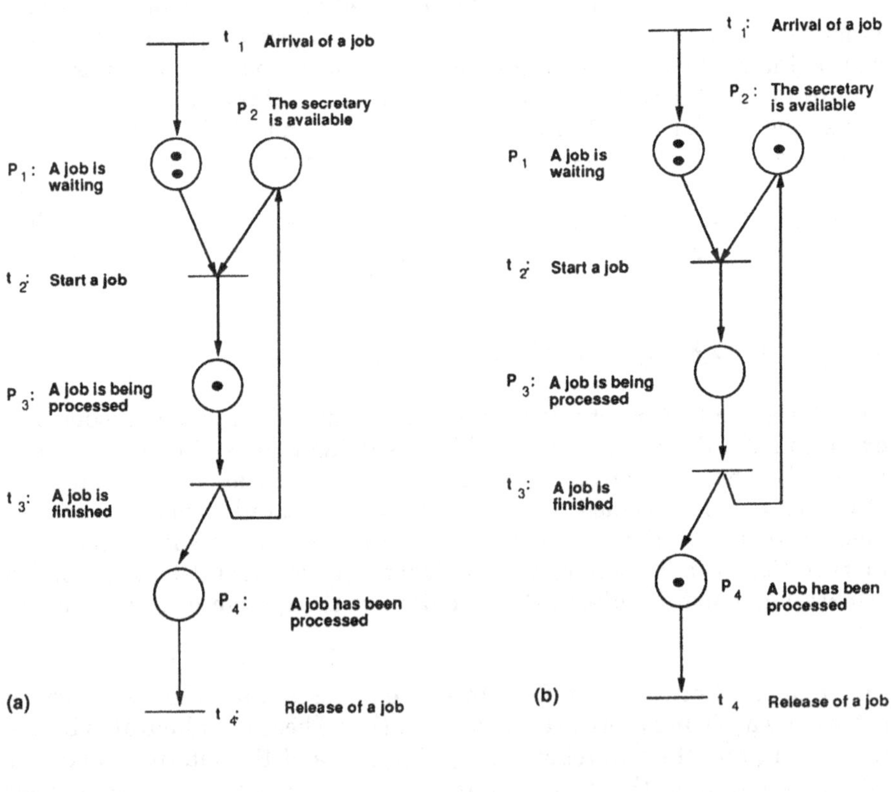

Fig. 14.22. Marked Petri nets

If a transition is enabled, then it can be fired eventually, but the time at which it is fired is nondeterministic. Further, the time taken to fire a transition is assumed to be zero. Therefore, only one transition can be fired at one time. The firing of a transition results in the removal of one token from each of its input places and the placing of one token into each of its output places. For example, in Fig. 14.22a, transition t_3 is enabled; therefore, this transition can be fired at some time. The firing of t_3 has the result that the token in P_3 is removed and a token is placed into P_4 and P_2, respectively. The resultant Petri net is shown in Fig. 14.22b, which says that the secretary is available and a job has been processed. In this new state, t_2 is enabled because the requirement to start a job is fulfilled. The firing of t_2 takes place eventually. This nondeterminism may exist since the secretary may not start processing a job right away. The nondeterminism may also be viewed as some sort of system overhead in scheduling.

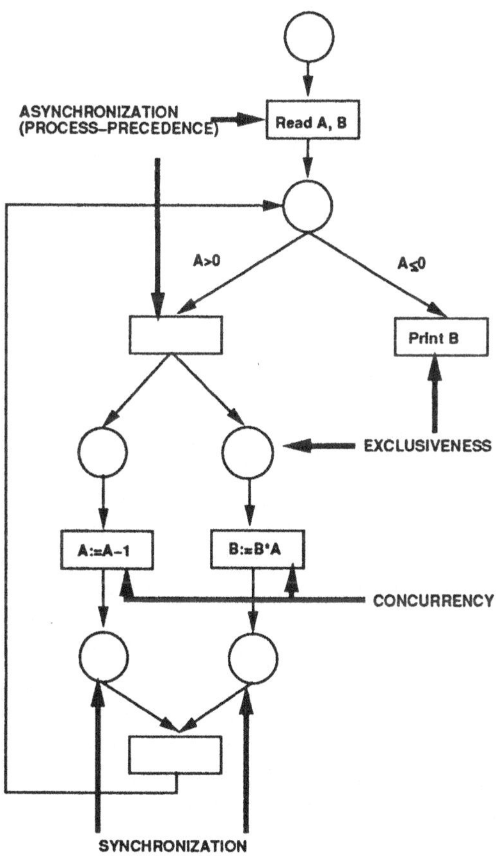

Fig. 14.23. System properties modeled by a Petri net

Another kind of nondeterminism in Petri nets is that when more than one transition is enabled, then any one of the enabled transitions may fire. The choice as to which transition fires is made in a nondeterministic manner by random selection or external force such as priority selection. In Fig. 14.22a, both t_1 and t_3 are enabled. Therefore, either t_1 or t_3 can fire, although we have chosen to fire t_3 in the above. If t_1 were fired instead of t_3, then the resultant net would have three tokens in P_1 and the other places would remain the same. That is, a new job arrived while a job was being processed. This nondeterministic feature resembles real-world situations in which several events may happen concurrently and the pattern of these event happenings is not unique.

A Petri net can be used to model a number of properties of a system. Figure 14.23 illustrates an example in which asynchronization (or precedence), as well as synchronization, exclusiveness, and concurrency are modelled. Note that we have displayed a transition by a box containing the meaning of the transition. The net in Fig. 14.23 models the computation of B:=B*A in terms of additions, where A≥1.

14.5 The Behavior Network Model

An information system can be modeled by a Petri net (or a set of Petri nets). However, a Petri net is good for describing the behavior properties but not so suitable for modeling the static properties of a system. For example, we have pointed out earlier that information about a job must be described elsewhere since an uninterpreted token is just an indicator of the holding of some condition. In conceptual modeling, we also need to describe the structural (static) aspects of a piece of reality. The objective of the BNM (Behavior Net Model) is to combine the static (set-theoretic) model and the Petri net model to support both structural and behavioral modeling. To support structural modeling, a "type" is associated with each of the places. This is achieved by introducing some links between a static model and a Petri net model. The behavioral aspects are supported by augmenting each transition with a precondition and a postcondition. The firing of an enabled transition requires that the precondition must be true when the transition is fired. After firing, the postcondition becomes true. Figure 14.24 gives a BNM model for a payroll processing system. The system computes the payment for each of the employees according to the hours the employee has worked. As for any system of this kind, the process must stop if the number of employees is finite. We will analyze the behavior net in Fig. 14.24 and see whether these requirements are satisfied.

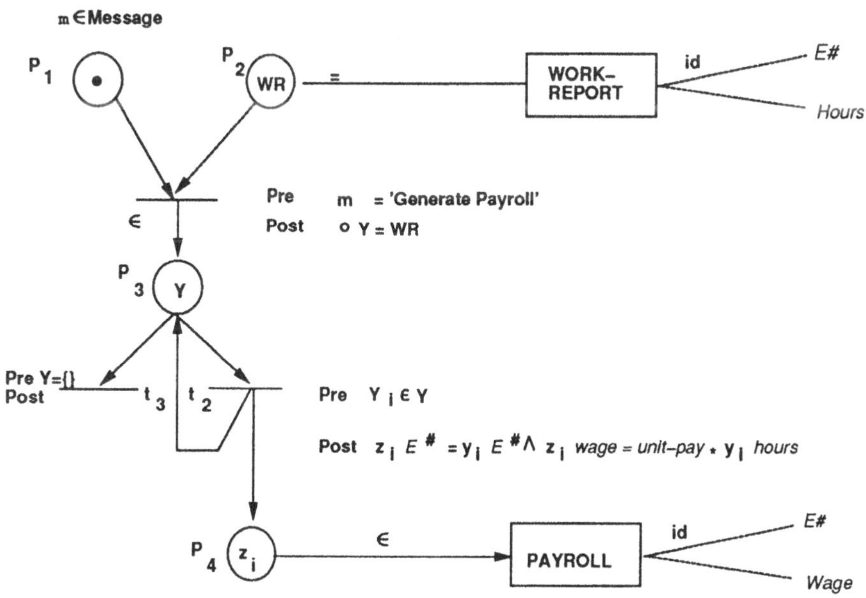

Fig. 14.24. A BNM model for a payroll processing system

In Fig. 14.24, two entity classes are defined, i.e., WORK-REPORT and PAYROLL. These two entity classes define the structural aspects of the payroll processing reality. The behavioral aspects are described by three transitions. Transition t_1 has two input places, P_1 and P_2. Place P_1 contains a set of tokens or elements. A typical element of P_1 is referred to by m, which belongs to a pool MESSAGE of messages being specified elsewhere. An alternative way of doing this is shown by the connection between place P_4 and the entity class PAYROLL where the structure of the elements in P_4 is explicitly displayed. Therefore, we know that each token of P_4 has an $E^{\#}$; and a WAGE component.

Place P_2 is linked to the entity class WORK-REPORT by an arc with label '='; this means that tokens of P_2 denote the entire entity class WORK-REPORT. Since the entity class WORK-REPORT is unique at any given time, it implies that there is only one element in P_2 at any given time. The typical element of P_2 is referred to by WR, which is a set in its own right.

The above discussion indicates that there are two kind of tokens: tokens representing individuals and tokens representing an entire class. As a convention, tokens representing individuals are referred to by strings of lowercase letters, e.g., m for the tokens in P_1 and z_i for the tokens in P_4. Tokens representing entire classes are referred to by strings of capital letters, e.g., WR and Y.

The tokens of an input place are either consumed or referenced by a transition. For simplicity, we will say that the input place is consumed or referenced. If it is consumed, then a token will be removed from the place when the transition is fired. If it is referenced, then no token will be removed when the transition is fired. In Fig. 14.24, a solid line from a place to a transition represents that the place is consumed by the transition while a dashed line from a place to a transition represents that the place is referenced by the transition. Thus, place P1 is consumed by t_1 and P_2 is referenced by t_1.

We now explain the firing of a transition. Suppose that, in Fig. 14.24, place P_1 contains some messages. Since place P_2 denotes the entire entity class WORK-REPORT, the place must contain one element, which is a (possibly empty) set. Therefore, t_1 is enabled. t_1 can be fired if its precondition is true, i.e., for some m $\in$ MESSAGE such that m = 'generate payroll'. In this case, t_1 can be fired. The firing of t_1 removes the token m = 'generate payroll' from P_1 and places one token, which is referred to by Y, in place P_3. Y is a copy of WR, as specified by the postcondition of the transition.

Now, both t_2 and t_3 are enabled. However, t_3 can be fired if Y = { }, as specified by the precondition. On the other hand, t_2 can be fired if there exists a $y_i \in Y$. The preconditions of t_2 and t_3 are made exclusive such that one and only one of them will be true at any time. Which of the preconditions is true depends on whether Y is empty or not. If Y is empty, then t_3 will be fired sooner or later which will remove the only token from place Y; at this point, the whole process stops, since no transition can be fired.

The preconditions and postconditions are specified in some predicate language. In particular, the precondition specifies the requirement on the input tokens for the firing of the transition. As mentioned above, the input individuals are referred to by strings of lowercase letters. The occurrence of a lowercase string in a precondition stands for the existence of an individual. For example, the precondition of t_1 is: m='generate payroll', which expresses that "there exists an m such that m='generate payroll'. A upper case string in a precondition stands for the entire entity or relationship class. Therefore, it refers to the entire class at a particular time. For example, the precondition of t_2 states that "there exists a y_i such that $y_i \in Y$".

The postcondition specifies the anticipated properties of the output tokens. The specification of a postcondition may involve the next operator '∘'. For instance, the postcondition of t_1 is '∘Y = WR' asserts that in the next state, the value of Y is WR. This corresponds to an assignment statement in a programming language. If an assertion or subexpression of an assertion does not involve the next operator, then it is a general assertion. For instance, the subexpression

$$z_i.E^{\#}=y_i.E^{\#} \wedge z_i.wage=unit\text{-}pay*y_i.hours$$

of the postcondition of t_2 asserts that "there exists a z_i such that ...".

14.6 The Retail Company Example

As an application of the behavior network model, we show in Fig. 14.25 the description of a retail company. It is assumed that the following six events may affect the status of the entity classes:

1) The event of placing an order by an individual on a certain date *d*. This states that a customer with customer number *cn* wants to order some quantity *qty* of part with part number *pn*. The variables *d*, *cn* and *qty* are treated as parameters of the event.

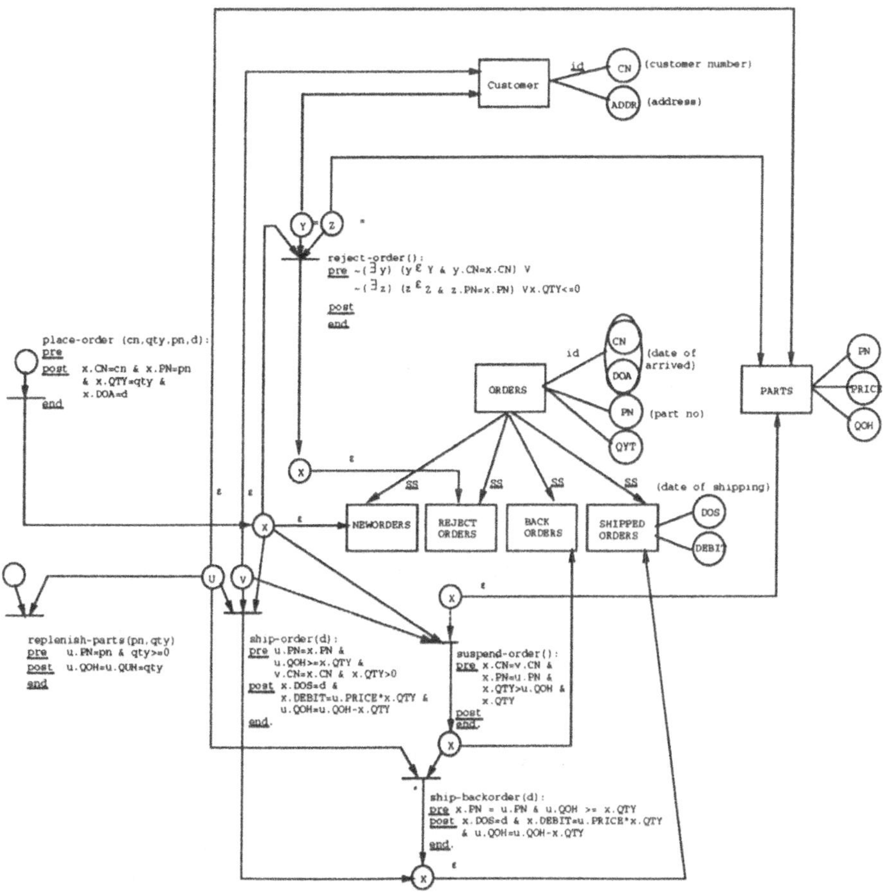

Fig. 14.25. The retail company example

2) The event of rejecting a new order. This may happen when one of the following situations exists: the customer number is invalid, the part ordered is not dealt with by the retail company, or the quantity ordered is zero or less.

3) The event of shipping an order. This happens when the quantities on hand (*qoh*) is sufficient for filling the order.

4) The event of suspending an order. This happens when the stock level is not enough for filling the order.

5) The event of filling a back order. This may happen when the stock level is increased by a replenishment.

6) The event of replenishing parts which causes the stock level to increase.

These six events are described by the six transitions in Fig. 14.25. In conceptual modeling, we often need to specify some semantic constraints. These constraints define other properties of the entities and relationships of the piece of reality that cannot be described by the graphic symbols conveniently. Another reason for specifying semantic constraints is that we want to state such constraints explicitly. Consider, for example the behavior net model in Fig. 14.24, one might want to specify explicitly that no employee can have a wage of more than $7,500. Such a constraint may be specified in two different ways: either by constraining the value set of *wage* of PAYROLL or by a semantic constraint as follows:

$$\sim(\exists x)(\, x \in \text{PAYROLL} \wedge x.\mathit{wage} > 7{,}500 \,) \tag{14.1}$$

To satisfy this constraint, the behavior net in Fig. 14.24 should be modified. A possible modification may be that the precondition of t_2 includes the checking that the wage of z_i is less than $7,500. The modification may also introduce an additional transition to handle the exceptional case when the wage of an employee exceeds $7,500. In this sense, the semantic constraints can serve as a set of correctness criteria for a conceptual model. We will return to this point when we discuss the analysis of a behavior net.

Suppose that $Y \neq \{\,\}$, then there must be some element $y_i \in Y$. In this case, transition t_2 can be fired. Firing t_2 results in that the token Y, which is a set, is removed and a new token is generated and placed into the same place, P_3. In particular, the new token denotes the set $Y-\{y_i\}$. Moreover, a token z_i is placed in the output place P_4. The properties of the elements being placed in the output places are specified by the postcondition of the transition. The computation of the wages for the employees is repeated until $Y=\{\,\}$, in which case, t_3 is fired and the process stops.

14.7 Simulation in the Behavior Network

A behavior net model can be used as a simulation model of the future system to be constructed. We may use two methods for the purpose of simulation. One involves the assignment of concrete instances to the entity and relationship classes and uses the behavior net model to simulate the behavior of the system. In this case, prototypes of the transitions must be implemented. When a transition is fired, the control is passed to the prototype procedure which implements the transition. When the procedure terminates, the control is returned to the simulation monitor. This approach may be used at the later stage of system development where most of the transitions have been implemented. The simulation results can be used to validate the conceptual model, identify system performance bottlenecks, etc.

Another method is based upon abstract simulation or symbolic execution. In this approach, we do not use concrete instances of the entity and relationship classes. Further, we assume that a place is in two states: it either contains an individual satisfying the precondition of a transition, or it does not contain such an individual. Thus, a marking of a behavior network can be defined as a vector of q components, where q is the number of places of the behavior net. Each entry of the vector is either 0 or 1. A 0-entry indicates that the corresponding place contains no tokens while a 1-entry indicates that the place contains at least one token. For example, a marking of the behavior net in Fig. 14.24 is a vector of four components:

$$\mu = [\mu_1, \mu_2, \mu_3, \mu_4]$$

where μ_i, $i = 1,\ldots,4$, is either 0 or 1. If place P_i contains one or more elements, then $\mu_i = 1$; otherwise, $\mu_i = 0$.

The initial state of the system is denoted by an initial marking. For example, the initial state of the payroll processing system can be $\mu = [1,1,0,0]$, which means that there are some messages in the message pool; no copy of WR has been produced, and no element of PAYROLL has been generated.

Starting in the initial state, we can execute the behavior network and study the dynamic property of the model. The execution involves the construction of a tree. The nodes of the tree are denoted by markings and the edges are denoted by transitions. In particular, the root of the tree is denoted by the initial state. In the development of the tree, we do not take into account the preconditions and the postconditions of the transitions. Consider, for example, the execution of the net in Fig. 14.24 starting in the initial state $\mu = [1,1,0,0]$. The only transition which is enabled is t_1. The new marking

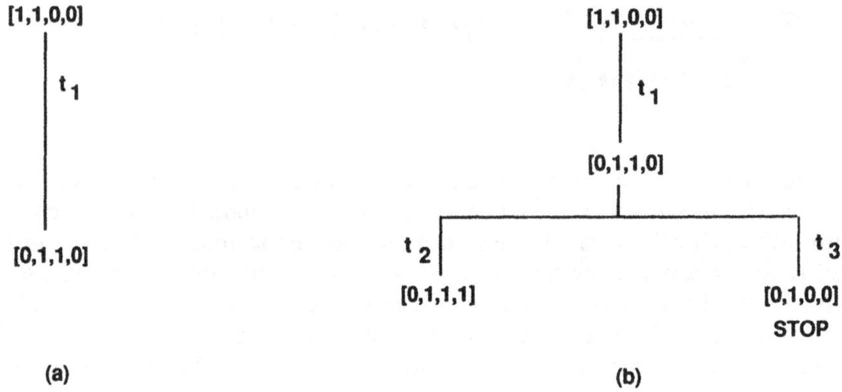

Fig. 14.26. Tree construction for the behavior net

resulting from firing t_1 will be $\mu' = [0,1,1,0]$. We may interpret this result as firing t_1 a sufficient number of times so that all the messages in P_1 have been consumed. The tree that is constructed so far is shown in Fig. 14.26.

In marking [0,1,1,0], both t_2 and t_3 are enabled. The firing of these two transitions add two branches to the tree, as shown in Fig. 14.26b. In the marking [0,1,0,1], no transition is enabled; hence, no progress can be made. This fact is indicated by the "STOP" beneath the right-most branch. If the construction is continued along the left branch, then we will finally construct the tree for the behavior net.

The tree construction process can be made more systematic by using two matrices

Given a marking, the enabled transitions can be identified by using a Boolean matrix M that has q rows and r columns, where q is the number of places and r is the number of transitions. If place P_i is an input place of transition t_j, then $M_{ij} = 0$, otherwise, $M_{ij} = 1$, where $i = 1,\ldots, q$ and $j = 1,\ldots,r$. For example, the M matrix for the behavior net in Fig. 14.24 is

$$M = \begin{array}{c} \\ \end{array}\begin{array}{ccc} t_1 & t_2 & t_3 \end{array}$$

$$M = \begin{bmatrix} & 1 & 1 \\ & 1 & 1 \\ 1 & & \\ 1 & 1 & 1 \end{bmatrix} \begin{matrix} P_1 \\ P_2 \\ P_3 \\ P_4 \end{matrix}$$

Given a marking μ, the enabled transitions can be identified by calculating $V = \mu \, \Delta \, M$, which is a 1xr vector, where

$$V_j \;=\; \bigwedge_{i=1}^{q} (\mu_i \vee M_{ij})$$

If $V_j = 1$, then, transition t_j is enabled.

Given the initial marking $\mu_0 = [1,1,0,0]$, the calculation of $V = \mu_0 \, \Delta \, M$ becomes:

$$V_1 = (1 \vee 0) \wedge (1 \vee 0) \wedge (0 \vee 1) \wedge (0 \vee 1) = 1 \wedge 1 \wedge 1 \wedge 1 = 1$$
$$V_2 = (1 \vee 1) \wedge (1 \vee 1) \wedge (0 \vee 0) \wedge (0 \vee 1) = 1 \wedge 1 \wedge 0 \wedge 1 = 0$$
$$V_3 = (1 \vee 1) \wedge (1 \vee 1) \wedge (0 \vee 0) \wedge (0 \vee 1) = 1 \wedge 1 \wedge 0 \wedge 1 = 0$$

Since $V_1 = 1$ and $V_2 = V_3 = 0$, only t_1 can be fired in μ_0.

After firing a transition, a new marking must be produced. The computation of the new marking can also be done by using a matrix, denoted by A. If the behavior net has q places and r transitions, then A is a r x q matrix.

Each entry of A is either −1, 0, or 1.
A_{ij} = −1 means that the firing of transition t_i consumes one token of place P_j.
A_{ij} = 0 means that firing t_i does not affect P_j.
A_{ij} = 1 means that t_i places one token into P_j when it is fired.

Given a marking μ and a matrix A, the marking μ' resulting from firing t_i is the sum of μ' and the i-th row of A.

The matrix A for the behavior network in Fig. 14.24 is

$$A = \begin{array}{c} \\ t_1 \\ t_2 \\ t_3 \end{array} \begin{array}{c} \begin{array}{cccc} P_1 & P_2 & P_3 & P_4 \end{array} \\ \begin{bmatrix} -1 & 0 & 1 & 0 \\ 0 & 0 & 0 & 1 \\ 0 & 0 & -1 & 0 \end{bmatrix} \end{array}$$

We explain how to obtain the entries of the first row. Consider transition t_1 in Fig. 14.24. Since P_1 emits a solid line to t_1, this means that when t_1 is fired, one token will be consumed, therefore, $A_{11} = -1$. However, P_2 emits a dashed line to t_1, which means that t_1 only refers to the tokens in P_2; therefore, the firing of t_1 will not remove any token from P_2. Thus, $A_{12} = 0$. Furthermore $A_{14} = 0$, since there is no connection between t_1 and P_4.

Finally, $A_{13} = 1$ since P_3 is an output place but not an input place of t_1. Note that if a place P_j is both an output and a consumed input place of a transition t_i, e.g., P_3 with respect to t_2, then $A_{ij} = 0$.

Consider the example in Fig. 14.24. The initial marking is [1,1,0,0]. The matrix A is as shown above. Suppose that t_1 is fired. Then the resultant marking is

$$\mu_1 = [1,1,0,0] + [-1,0,1,0] = [0,1,1,0]$$

The firing of t_2 in μ_1 results in

$$\mu_2 = [0,1,1,0] + [0,0,0,1] = [0,1,1,1]$$

These two new markings are as shown in Fig. 14.26b.

By using M and A, we can easily construct the tree for the example in Fig. 14.24. Since there are finitely many places and each component of a marking can only take two different values, there is a finite number of markings. In particular, if a behavior net has q places, then there are at most 2^q different markings. The number of nodes of the tree can be reduced by the following rule: if a node n_2 is identical to a node n_1 along the path leading from the root to n_2, then we will not expand below n_2, since all nodes which can be generated from n_2 can also be generated from n_1. Figure 14.27 displays the complete tree for the net in Fig. 14.24.

The leftmost branch is closed since the tip of the branch is identical to its parent. This is indicated by a cross 'X' beneath the tip. The other two branches are stopped since no transition can be fired in the tip markings.

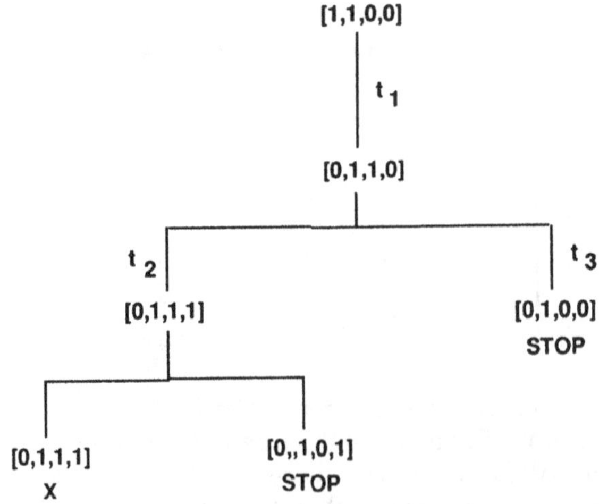

Fig. 14.27. The complete tree for a behavior network

14.8 Derivation of External Systems Properties Using Path Analysis

By path analysis, we mean the analysis of the paths of the tree constructed above when the preconditions and postconditions of the transitions are taken into account. We will use Fig. 14.27 as the running example. First we see that the leftmost branch of the tree has two identical nodes, i.e., the tip and its parent. This means that t_2 can be fired indefinitely many times and each firing will result in the same marking or state. Thus, we can replace the last edge by a loop going from [0,1,1,1] to itself. This results in Fig. 14.28 below. The second thing to do is to fill in the preconditions and the postconditions of the transitions. This is also illustrated in Fig. 14.28.

In Fig. 14.28, we have abbreviated the postcondition of t_2 by

$Q_k = z_k \cdot E^{\#} = y_k \cdot E^{\#} \wedge z_k.\mathit{wage} = \mathit{unit\text{-}pay} * y_k.\mathit{hours} \wedge oY = Y - \{y_k\}$

where k = i, i+1. The subscript i+1 is used to distinguish the first firing of t_2 and the other firings.

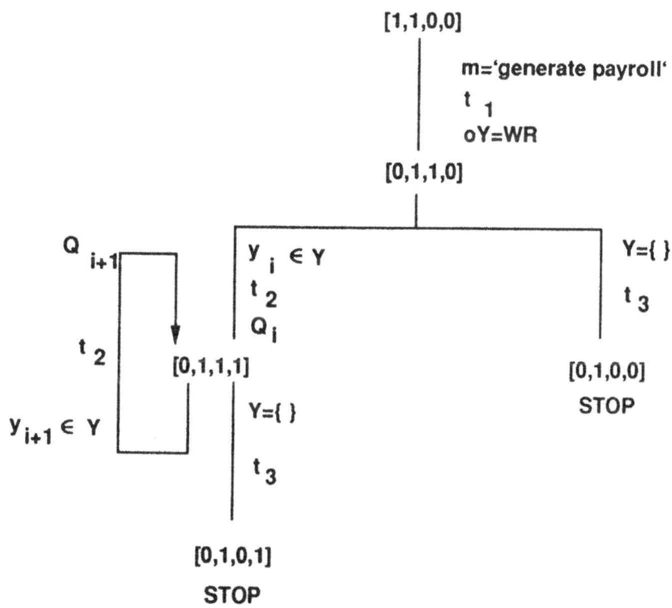

Fig. 14.28. Path analysis with pre- and postconditions

Figure 14.28 shows that there are two independent paths, representing two independent executions of the behavior net. The left path involves a loop as a result of repeatedly firing t_2. The right path represents a special case, that is, WR = { }. The path analysis is to construct the path condition for which one of the two paths will be executed. Further, the analysis will also produce the final assertion when the execution of the path is terminated.

We analyze the right path first. For convenience of explanation, we show the right path and the analysis in Fig. 14.29.

```
                              Path             State        Path
                              Condition        Change       Assertion
------------------------------------------------------------------------
  [1,1,0,0]
       |
       |  m='generate      m='generate
  t1   |     payroll'      payroll' ∧
       |
       |  OY=WR                                Y:=WR
       |
  [0,1,1,0]
       |  Y={}             WR={}
  t3   |
       |
       |
  [0,1,0,0]
   STOP                                                     STOP
------------------------------------------------------------------------
```

Fig. 14.29. Analysis of the right path

The initial path condition is assumed to be true. We then travel along the right path. When a precondition is encountered, it is put into conjunction with the existing path condition. When a postcondition is encountered, it is separated into two parts: one that involves the next operator and another that does not. The part involving the next operator in effect changes some values of the variables. These changes are shown under the "State Change" column. The other part is an assertion, which is collected in the "Path Assertion" column. When the terminal of the path is reached, the special path assertion STOP is entered. The assertion STOP stands for "the process stops".

Figure 14.29 can be used to state some general properties of the path. For example, the last marking [0,1,0,0] indicates that place P_4 is empty. Further, we know that elements of P_4 are elements of the entity class PAYROLL. This implies that PAYROLL is also empty. The conjunction of the path conditions above the last marking is

$$m=\text{'generate payroll'} \wedge WR=\{\ \} \tag{14.2}$$

Therefore, we obtain the following property

$$\text{m='generate payroll'} \wedge \text{WR}=\{\,\} \Rightarrow \text{PAYROLL} = \{\,\} \qquad (14.3)$$

which states that

"if m='generate payroll' and WR is empty then PAYROLL is empty".

Similarly, we may formulate the following assertion for the path assertion STOP:

$$\text{m='generate payroll'} \wedge \text{WR}=\{\,\} \Rightarrow \text{STOP} \qquad (14.4)$$

Expressions (14.3) and (14.4) can be combined into the assertion

$$\text{m='generate payroll'} \wedge \text{WR}=\{\,\} \Rightarrow (\text{PAYROLL} = \{\,\} \wedge \text{STOP}) \qquad (14.5)$$

which states that

if m='generate payroll' and WR is empty, then PAYROLL is empty, and the process stops.

		Path Condition	State Change	Path Assertion
[1,1,0,0]				
t_1	m='generate payroll'	m='generate payroll' ∧		
	Y=WR		Y:=WR	
[0,1,1,0]				
	$y_1 \in Y$	$y_1 \in WR$ ∧		
t_2	Q_1		Y:=WR-{y_1}	$z_1.E'=y_1.E'$ ∧ z_1.*wage=unit-pay** y_1.*hours*∧
[0,1,1,1]				
:				
:				
[0,1,1,1]				
	$y_{1+k} \in Y$	$y_{1+k} \in$ WR-{y_1}-... -...- {y_{1+k-1}} ∧		
t_2	Q_{1+k}		Y:=WR-{y_1}	$z_{1+k}.E'=y_{1+k}.E'$ ∧ -...-{y_{1+k}} z_{1+k}.*wage=unit-pay** y_{1+k}.*hours* ∧
[0,1,1,1]				
	Y={}	WR-{y_1}-... -...-{y_{1+k}}={}		
[0,1 0,1] STOP				STOP

Fig. 14.30. Analysis of the left branch

Properties as expressed by (14.3) – (14.5) do not completely describe the properties of the behavior net in Fig. 14.24. To obtain the complete specification, we need to analyze the left path. In the following discussion, we will first analyze the left path and formulate a complete set of properties for the behavior net.

The analysis of the left path is shown in Fig. 14.30. Note that we have stretched the loop into a sequence of indefinite length.

After the first firing of t_2, we can formulate the following two rules in the same manner as we formulated (14.3) and (14.4):

$$m=\text{'generate payroll'} \wedge y_i \in WR \Rightarrow PAYROLL \neq \{\ \} \qquad (14.6)$$

$$m=\text{'generate payroll'} \wedge y_i \in WR \Rightarrow z_i.E^{\#}=y_i.E^{\#} \wedge z_i.\mathit{wage}=\mathit{unit\text{-}pay}*y_i.\mathit{hours} \qquad (14.7)$$

By $(A \wedge B) \rightarrow C \Leftrightarrow A \rightarrow (B \rightarrow C)$, equation (14.7) can be rewritten as:

$$m=\text{'generate payroll'} \Rightarrow y_i \in WR \Rightarrow z_i.E^{\#}=y_i.E^{\#} \wedge z_i.\mathit{wage}=\mathit{unit\text{-}pay}*y_i.\mathit{hours} \qquad (14.8)$$

The sequence of t_2 firings has the same form as the first firing, therefore, (14.8) can be generalized to:

$$m=\text{'generate payroll'} \Rightarrow (\forall y)(\ y \in WR \Rightarrow (\exists z)(\ z.E^{\#}=y.E^{\#} \wedge z.\mathit{wage}=\mathit{unit\text{-}pay}*y.\mathit{hours}\ ) \qquad (14.9)$$

Finally, when the end of the path is reached, we have

$$m=\text{'generate payroll'} \wedge |WR| \leq N \Rightarrow STOP \qquad (14.10)$$

where $|WR| \leq N$ means the cardinality of WR is at most N, where N is a constant. In other words, WR is finite. This is derived from the condition that

$$WR - \ldots - \{y_{i+k}\} = \{\ \} \qquad (14.11)$$

Combining the results of both of the branches, we have from (14.3) and (14.6):

$$m=\text{'generate payroll'} \Rightarrow (\ WR = \{\ \} \text{ iff } PAYROLL = \{\ \}\ ) \qquad (14.12)$$

From (14.4) and (14.10), we have

$$m=\text{'generate payroll'} \Rightarrow (\ |WR| \leq N \Rightarrow STOP) \qquad (14.13)$$

Putting (14.9), (14.12) and (14.13) together, we obtain

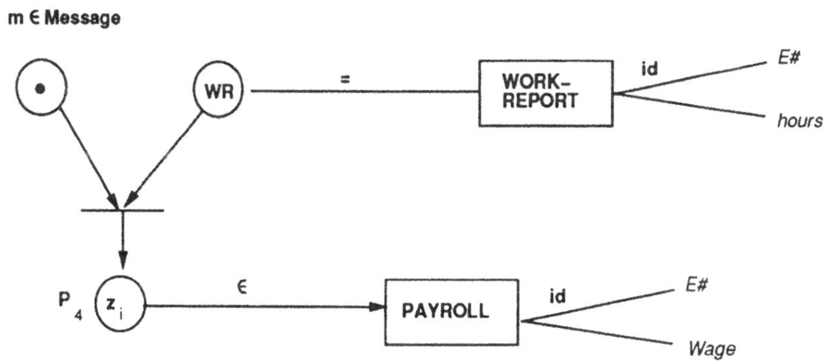

Fig. 14.31. A higher-level behavior network

$$m=\text{'generate payroll'} \Rightarrow \tag{14.14}$$

$$(\forall y)(\ y \in WR \Rightarrow (\exists z)(\ z.E^{\#}=y.E^{\#} \wedge z.\mathit{wage}=\text{unit-pay}*y.\mathit{hours}\ ))$$
$$\wedge\ (\ WR = \{\ \}\ \text{iff}\ PAYROLL = \{\ \}\ )$$
$$\wedge\ (|WR| \leq N \Rightarrow STOP)$$

Expression (14.14) is the complete specification of the general property of the behavior net in Fig. 14.24. From this, we can obtain a higher-level behavior net as shown in Fig. 14.31, where the precondition of the transition t is m='generate payroll', and the postcondition is similar to the consequent part of (14.10).

14.9 The Temporal Dimension of Information System Modeling

The dynamic and temporal aspects of systems are commonly considered in engineering and science. The behavior of a system is conceptually modeled by a set of equations where time plays a fundamental role as an independent variable. Numerical approximation requires that we transform the model to a discrete model that only considers a finite set of points on the time axis. Numerical simulation can then be used to produce a "history" of system states. We can then study the chronology for model correctness, performance, etc.

Conceptual information modeling has similarities with numerical modeling of physical systems. In both cases, we wish to maintain a conceptual model of some system of interest. The difference is that in information modeling, we have to deal with a world which is difficult to approximate by traditional mathematical equations. As a matter of fact, our knowledge of conceptual information modeling is still in its infant stage. However, it has been recognized for a long time that an information model should describe the temporal dimension of the system. For instance, Young and Kent have included the concept of time in abstract formulation of data processing problems [YOUN58]. Unfortunately, the temporal dimension did not receive enough attention until the mid-1970s. In recent years, a handful of disciplines have been concerned with the temporal dimension of information modeling. Designers of computerized information systems have to deal with historic information. Researchers in Artificial Intelligence have to include in knowledge representation not only a snapshot description but also the evolution of the world. Many logicians have regarded classical logic as not being adequate for capturing the semantics involving time or temporal reference. In natural language processing, we have to face and provide mechanisms for the machine to understand tensed sentences.

An information system is a model of some application. When the state of affairs for the application changes, the state of the information system must also change. This requires that the information system must be able to reflect the evolution of the application. The evolution of the application can be regarded as a result of event happenings, where the concept of event includes the occurrence of time points. Figure 14.32 illustrates the relationship between an information system and a piece of reality.

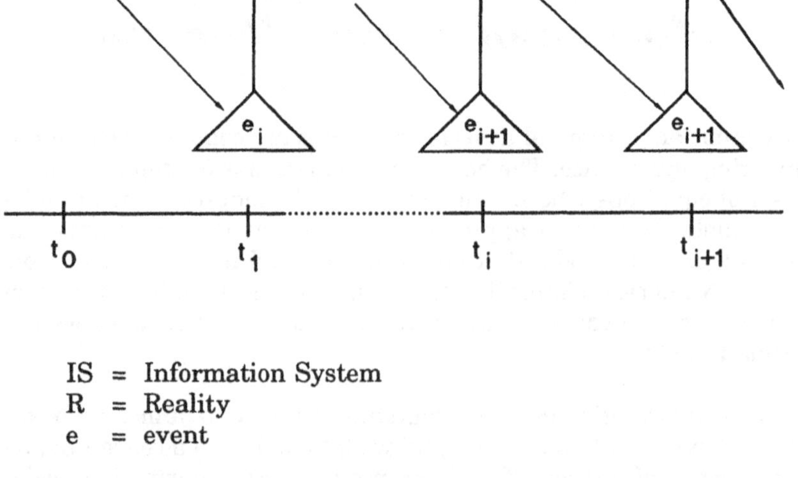

IS = Information System
R = Reality
e = event

Fig. 14.32. Information system and reality

At time t_i the contents of the information system represent the state of affairs of a piece of reality at t_i. At time t_{i+1}, some event happens, e.g., hire an employee or issue a ticket; thus the state of affairs changes. The contents of the information system must reflect this change by updating the contents. An event can be either an internal or an external event. An internal event may happen when the information system state fulfills certain conditions. For example, when the inventory level drops below certain limit, a purchase order is issued. An internal event may also happen by prescheduling. For example, a Ph.D. dissertation is scheduled at 10:00 am on May 15; therefore, the event took place at that time. An external event usually takes place because of some external decisions, such as firing or hiring of an employee.

The operational and the deductive approaches to describing the evolution of knowledge bases

Figure 14.32 reveals the evolution of the information system according to the evolution of a piece of reality. If we regard the information system as a knowledge base **K**, then the evolution of the knowledge base can be described by two different approaches: the operational and the deductive approach. Both approaches can be viewed as a transformation of the form

$$\mathbf{K}_{i+1} = \delta(e_i, \mathbf{K}_i) \tag{14.15}$$

where δ is a mapping from one knowledge base $\mathbf{K}_i$ and an event e_i to another knowledge base $\mathbf{K}_{i+1}$.

In the operational approach, the effect of an event e_i is modelled by an operation which consists of some added knowledge K_{ai} and some deleted knowledge K_{di}. Thus, (14.15) becomes:

$$\mathbf{K}_{i+1} = \mathbf{K}_i + K_{ai} - K_{di} \tag{14.16}$$

The deductive approach assumes the "vector addition principle". That is, the transition from $\mathbf{K}_i$ to $\mathbf{K}_{i+1}$ is viewed as the sum of the knowledge about the event e_i and the knowledge about the consequence of the event e_i (see Fig. 14.33). The knowledge of the consequence of e_i is usually expressed in terms of deductive laws, e.g., in a deductive database. Let K_{ei} be the knowledge of e_i and C_{ei} be the knowledge of the consequence of e_i. Then (14.15) becomes:

$$\mathbf{K}_{i+1} = \mathbf{K}_i + K_{ei} + C_{ei} \tag{14.17}$$

Note that the deductive approach assumes an accumulative memory such that knowledge of the real world is never forgotten once obtained. Conceptual modeling along the temporal dimension assumes one of these two approaches, as we will see in the sequel.

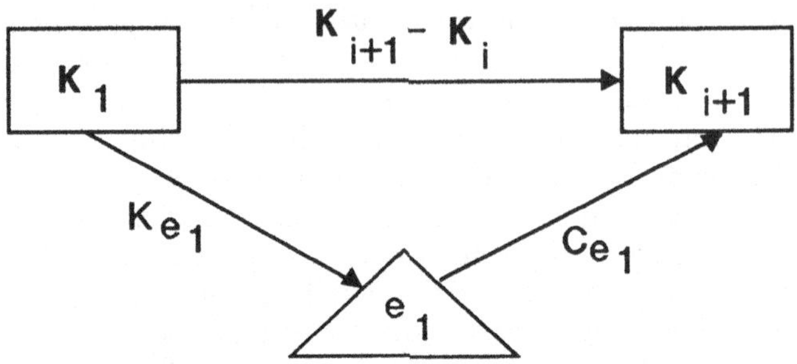

Fig. 14.33. The vector addition principle

Several time concepts may be involved in the specification of information systems

The evolutionary aspect is only one of the issues along the temporal dimension. Another issue concerns the specification and manipulation of time in information modeling. Such a requirement may arise from information needs like:

a) Quantities-on-hand two weeks ago.

b) The average sales figures for 1986.

c) Productivity trends after the computer system has been installed.

d) It was reported that by January 1986 one in three Norwegians had a car.

e) The total sales figures in December for the last three years.

These requirements involve time in one way or another. For example, statement a) involves two instances of time: the time of assertion and the time of reference. The time of assertion is the time at which the statement is made which in this example is not given. It depends on the time of conversation, e.g., now, today or yesterday. The time of reference is the time which participates in the definition of a piece of information. It is sometimes dependent on the time of assertion. For example, the time of reference for statement a) is two weeks ago and the exact value of this depends on the time of assertion. The time of reference of statement b) is 1986, which is explicitly given; hence, its value is independent of the time of assertion. This example partly depends on the time of reference, i.e., it must be later than 1986.

Both assertion time and reference time are important in information modeling. Without time of reference, a piece of information may be "incomplete" or meaningless. Since the time of reference participates in the definition of information, it is also called the intrinsic time. On the other hand, the time at which a piece of information is obtained is called the extrinsic time. Obviously, the assertion time must be greater than or equal to the extrinsic time. Note that a piece of information might not contain an intrinsic time component. For example, the amount of order #30345 is $2,000.

Note also that there are situations in which the concept of time is involved but explicit reference to time is not required or not possible. For instance, a book that has been reserved by someone cannot be borrowed by others. In a university administrative information system, it may require specifying that no student can have credit for a course that the student has previously taken for credit. These statements refer to some states in the past, but the time points associated with these states are either not required or cannot be determined due to incomplete knowledge. In these cases, we have so-called topological time which is temporally indefinite.

The above discussion indicates that two kinds of time reference are required. One requires the specification of and reasoning through information involving particular points in time. On the other hand, topological time deals with uncertainty in time modeling. Obviously, conceptual modeling requires both of these. Thus, by the temporal dimension of information modeling, we mean modeling of evolution that takes into account topological, as well as explicit time reference.

14.10 Modeling With Explicit Time Reference

Modeling with explicit time reference can be achieved in many different ways. Indeed, a great deal of literature on this topic has been published (see, e.g., the comprehensive survey by Bolour et al. [BOLO82]). The simplest way of incorporating time into a conceptual model is to give a time stamp to each of the time-varying attributes of an entity. This method may be called the single-time stamping method. For convenience, let us assume that predicate calculus is used as the modeling language. Then we may define a predicate SALARY(*e, s, t*) for representing that employee *e* has salary *s* at time *t*. In many cases, we may reason through the extension of the predicate to obtain the salary for a given employee. For example, if we know that

SALARY(John Smith, $40,000, t_1) and
SALARY(John Smith, $50,000, t_3)

then we might derive SALARY(John Smith, \$40,000, t_2), where $t_1 \le t_2 < t_3$. However, when we are doing so, we have already assumed that there exists no other salary value for Smith at times between t_1 and t_2. To be precise, we may express this assumption as:

$$\sim(\exists t)\,(\exists s)\,(\ t_1 < t \le t_2 \wedge s \ne \$40{,}000\ )$$

The single-time stamping method is not a good device for conceptual modeling. It has to "split" an entity description into parts describing the time-varying attributes of the entity separately. The reason behind this is that the instances of the entity type can no longer be uniquely identified without involving a time stamp. Usually, an entity type will have more than one time-varying attribute. In such cases, it is impossible to select one of the time stamps to be involved in the identification of an instance, unless all these attributes change at the same times and same rate. The drawback of having to describe entities in pieces contradicts the principle of object-orientation, where modular or cohesive specification of real-world objects is required.

A serious problem arises if John Smith happened to be fired some time between t_1 and t_2 and rehired after t_2. This is illustrated in Fig. 14.3, where Smith did not have a salary at t_2. Without special care, the single-time stamping method as described above would assert that the salary at t_2 was \$40,000; an incorrect statement.

The problem illustrated in Fig. 14.34 can be solved by using double (or dual) time stamping: one for the starting time and the other for the end time. By this method, the salary property of employees may be defined by the following predicate:

$$\text{SALARY}'(e, s, t_s, t_e)$$

where t_s and t_e are the starting time and the end time, respectively. If employee e has salary s since t_s, then t_e is "now", a special time stamp. For example,

SALARY'(John Smith, \$40,000, 1984, now) (14.18)

states that Smith has salary \$40,000 since 1984 (and his current salary is still \$40,000). Suppose that "now" is 1986, then the salary of Smith in 1985

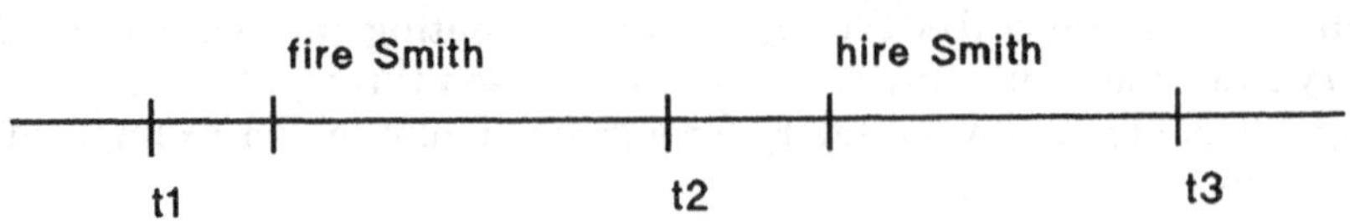

Fig. 14.34. A problem with single time stamping

was $40,000, because 1984 < 1985 < now. If Smith is fired, then (14.18) is updated and the new tuple is

SALARY'(John Smith, $40,000, 1984, 1986) (14.19)

Suppose that Smith is rehired in 1988, and "now" is 1991 then we have:

SALARY'(John Smith, $40,000, 1984, 1986) and (14.20)
SALARY'(John Smith, $50,000, 1988, now) (14.21)

Thus, we know that Smith has no salary in 1987.

The dual-time stamping method assumes a discrete time axis, although the time points on the axis can be infinite. Without this assumption, updating of a time-varying attribute will cause problem. Suppose that we have (14.18) and the salary of Smith was raised to $50,000 in 1985. How can we specify the end time for the current tuple and the starting time for the new tuple? If we adopted the above mechanism, then we would have a contradictory assignment of salaries to Smith, i.e., Smith having a salary of $40,000 and $50,000 for the same time stamp 1985. A straightforward solution for this problem would be:

- Refine the time scale to avoid having two different attribute values for the same time stamp, e.g., we may assume that the time scale for salary is months rather than years;

- Assume that the two consecutive intervals defined by the two pairs of time stamps are adjacent and disjoint. Further, each interval includes the starting point but excludes the end point. This is shown in Fig. 14.4. However, the assumption is valid only if the time axis is supposed to be discrete.

The dual-time stamping method still requires that the description of an entity class (or entity type) be split into parts, as required by the single-time stamping method. Many modeling methods use either the single- or the dual-time stamping mechanism.

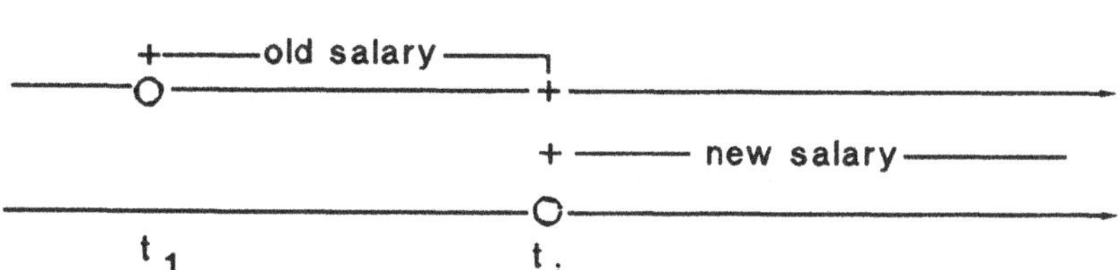

+ = time point not included;
0 = time point included.

Fig. 14.35. Two adjacent intervals in dual-time stamping

14.11 Modeling With Topological Time

Modeling with topological time is usually done by using some form of temporal logic. The most commonly used temporal logic is the one adapted from modal logic. Modal logic can be seen as an extension of classical logic including two modal operators: $\Box$(necessarily) and $\Diamond$ (possibly). For instance, $\Box$A means that A is necessarily true and $\Diamond$A means that A is possibly true. These two operators can be interpreted in terms of the temporal dimension as follows:

$\Box$A: A is always true, denoted by $\underline{\text{always}}$ A
$\Diamond$A : A is sometimes true, denoted by $\underline{\text{sometime}}$ A

It is easy to verify that $\underline{\text{sometime}}$ A if and only if ~$\underline{\text{always}}$~ A. Thus, $\underline{\text{always}}$ and $\underline{\text{sometime}}$ are dual to each other. It is sometimes useful to indicate the directions of such temporal statements. This means that we may wish to express statements such as "always in the past A is true", or "sometimes in the future B is true", etc. Therefore, we may introduce '←' (in the past) and '→' (in the future) for these purposes. Figure 14.36 illustrates all the possible combinations between the temporal operators and the two direction signs.

As an application of this simple temporal language, we show how the property of the behavior net in Fig. 14.31 can be expressed in a more proper way. Before introducing temporal logic, we expressed this property in (14.13). However, expression (14.13) does not indicate the time needed to produce the PAYROLL items. Therefore, it is not a statement along the temporal dimension. We may now reexpress (14.13) as:

$$
\begin{aligned}
& m=\text{'generate payroll'} \Rightarrow \\
& \quad (\underline{\text{sometime}}^{\rightarrow} (\forall y)(y \in WR \Rightarrow \\
& \quad (\exists z \in PAYROLL)(z.E^{\#}=y.E^{\#} \wedge \mathit{wage}=\mathit{unit\text{-}pay} * y.\mathit{hours}))) \\
& \wedge (WR = \{\ \} \text{ iff } PAYROLL = \{\ \}) \\
& \wedge (|WR| \leq N \Rightarrow \\
& \quad \underline{\text{sometime}}^{\rightarrow} \text{ STOP })
\end{aligned}
\qquad (14.22)
$$

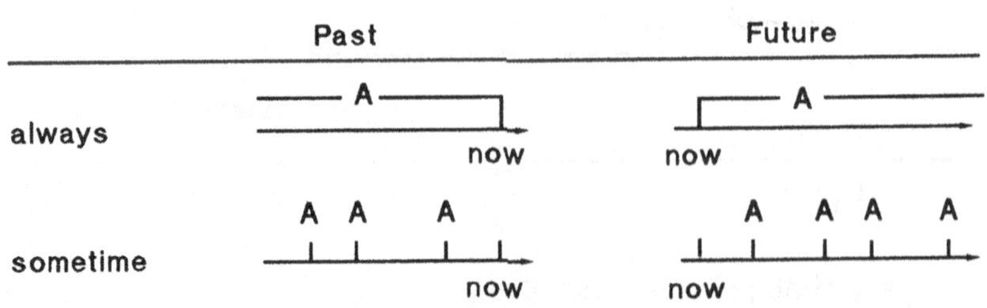

Fig. 14.36. Combination of temporal operators and direction signs

Expression (14.22) states that if m='generate payroll', then sometime in the future, a payroll entry will be produced for each of the entries in WORK-REPORT, and WORK-REPORT is empty if and only if PAYROLL is empty, and if WORK-REPORT is finite, then sometime in the future, the process stops.

We introduce some more constructs. One of the constructs is the "next" operator, o;, which we have already encountered in the behavior network model. Another construct is the "inclusive", denoted by a prime. The inclusive construct can be used together with the temporal operators to express e.g. $\underline{\text{always}}^{\leftarrow'}$A, which means always in the past including the present, A is true.

In temporal logic, we may talk about individuals in several states. This means that we need a way to distinguish individuals of a particular state and individuals ranging over several states. An arbitrary individual of a specific state will still be referred to by a variable, as in the first-order language. An arbitrary individual which ranges over several states will be called a parameter, which is denoted by a variable with a prime, e.g., x', y', etc. We explain this by two examples.

Let CUSTOMER(x,y) express customer x with balance y. Let ORDER(x,y,z) denote that customer x orders part y in quantity z, and PART(x,y) denotes the stock level of part x is y. Further, let SHIPMENT(x,y,z) denote shipment of part y in quantity z to customer x. (Note that we have made this example as simple as possible so that it can be read easily.) The temporal constraint

> Give a customer priority, if his balance is always less than \$1,000, where priority means filling the order immediately whenever possible.

can be expressed by:

$$\underline{\text{always}}^{\leftarrow'}(\exists b)(\text{CUSTOMER}(c',b) \wedge b<1{,}000) \rightarrow$$
$$(\underline{\text{always}}^{\rightarrow'}(\text{ORDER}(c',p',z') \wedge (\exists q)(\text{PART}(p',q) \wedge q \geq z') \rightarrow \circ\text{SHIPMENT}(c',p',z'))$$

Let E(x,y) denote that employee x has salary y. Then the temporal constraint stating that salary never decreases can be expressed by:

$$E(x',y') \rightarrow \underline{\text{always}}^{\rightarrow'} (\forall z)(\, E(x',z) \rightarrow z \geq y' \,)$$

Note the difference between variables and parameters. In the first example above, variable b denotes some balance in a past state or the present state. The balance denoted by b may differ from state to state. Similarly, the quantities-on-hand q may differ from state to state but it must be greater than or equal to the quantity ordered. On the other hand, parameter c' denotes the same customer over sequences of states. This means that priority is given to customers who always have a balance of less than \$1,000.

Exercises

1. Discuss the advantages and disadvantages of formalizing the modeling approaches.

2. Model the Car Rental Agency problem of Chap. 2, Exercise 6, by
   a) E-R,
   b) Semantic Networks,
   c) ACM/PCM,
   d) BNM.

3. Are Petri nets completely sufficient to model information systems? If so, why? If not, why not?

4. Give a classification of formal modeling approaches.

5. Compare the E-R approach with the semantic network approach.

6. Compare ACM/PCM with BNM.

7. Is ACM/PCM an object oriented approach ? If so, why? If not, why not?

8. Define the interfaces between the three levels of the ACM/PCM modeling approach.

References

[ANDE85] Andersen R., Hodneland E., Knarbakk R.: KOKO - Konfigurasjonskontroll i TINA. Dept. Comp. Science, NTH, Univ. Trondheim, Tech.Rep. 14/85, 1985

[ANDE90] Andersen R.: Integration of CASE tools. In: NIK'90, Norsk Informatikk Konferanse, Bergen, Norway, November 1990, pp. 71-80

[ANSI77] Jardine D.A.: The ANSI/SPARC DBMS model. North-Holland, Amsterdam 1977

[AUGL75] Auglænd K., Sølvberg A., Tørlen R.: Proposals to standard flowcharts in systems design. In: M. Lundeberg, J. Bubenko (eds.): Systemering. Studentlitteratur, Lund 1975

[BAEC87] Baecker R.M., Buxton W.A.S.: Readings in human-computer interaction. Morgan Kaufmann, San Mateo CA 1987

[BELA76] Belady L.A., Lehman M.M.: A model of large program development, IBM Syst. J. 3, 225-252 (1976)

[BERG81] Bergland G.D., Gordon R.D.: Structured design methodologies. In: B.D. Bergland, R.D. Gordon (eds.): Software design strategies. IEEE Computer Society Press 1981, pp. 297-315

[BIRT73] Birtwistle G.M., Dahl O.J., Nygaard K.: Simula begin. Auerbach/Studentlitteratur, Lund 1973

[BOLO83] Bolour A., Dekeyser L.J.: Abstractions in temporal information. Inf. Syst. 8:1, 41-49 (1983)

[BOOC86] Booch G.: Object-oriented development. IEEE Trans. Softw. Eng., SE-12:2, 211-221 (1986)

[BOOC91] Booch G.: Object oriented design with applications, Benjamin/ Cummings, Redwood City CA 1991

[BROD82] Brodie M.L., Silva E.: Active and passive component modelling: ACM/PCM. In [OLLE82], pp. 41-92

[BROO75] Brooks F.P.: The mythical man-month. Addison-Wesley, Reading MA 1975

[BROO82] Brookes C.H.P., Grouse P.J., Jeffery D.R., Lawrence M.J.: Information systems design. Prentice-Hall, Englewood Cliffs 1982

[BROW83] Brown E.J.: On the application of Rothon diagrams to data abstractions. SIGPLAN Notices 18:12, 17-24 (1983)

[BRUN91] Brunet J.: Modeling the world with semantic objects. Proc. of the IFIP conference, Quebec, Canada, October 1991

[CARD85] Cardelli L., Wegner P.: On understanding types, data abstracting and polymorphism. ACM Comput. Surv. 17:4, 471-522 (1985)

[CARR82] Carroll J.M.: Learning, using and designing command paradigms. Human Learning 1:1, 31-62 (1982)

[CHEN76] Chen P.P.: The entity-relationship model: towards a unified view of data. ACM Trans. Database Syst. 1:1, 9-36 (1976)

[CHEN85] Chen P.P.: Database design based on entity and relationship. In: S.B. Yao (ed.): Principles of database design. Prentice-Hall, Englewood Cliffs NJ 1985, pp. 174-210

[COAD90] Coad P., Yourdon E.: Object-oriented analysis. Yourdon, New York 1990

[CODD70] Codd E.F.: A relational model for large shared data bases. Commun. ACM 13:6, 377-387 (1970)

[DATE77] Date C.J.: An introduction to database systems. 2nd ed., Addison-Wesley, Reading MA 1977

[DBTG74] National Bureau of Standards Handbook 113, CODASYL Data Description Language Journal of Development. Government Printing Office, Washington DC 1974

[DEMA78] DeMarco T.: Structured analysis and system specification. Yourdon, New York, 1978

[DITT91] Dittrich K.R., Dayal U., Buchmann A.P. (eds.): On object-oriented database systems. Springer-Verlag, Berlin 1991

[EHRI84] Ehrich H.D., Lipeck U.W., Gogolla M.: Specifications, semantics, and enforcement of dynamic database constraints. In: Proc. 10th Int. Conf. on VLDB, Singapore, August 1984, pp. 301-308

[ELMA89] Elmasri R., Navathe S.B.: Fundamentals of database systems. Benjamin/Cummings, Redwood City CA 1989

[FELD79] Feldman S.I.: MAKE-A program for maintaining computer programs. Software Management & Experiences 9:3 (1979)

[FOLL70] Føllesdal D., Hilpinen R.: Deontic logic: an introduction. In: R. Hilpinen (ed.): Deontic logic: introductory and systematic readings. Reidel, Dordrecht 1970

[FRAT80] Frates J., Moldrup W.: Introduction to the computer. Prentice-Hall, Englewood Cliffs NJ 1980

[FROS84] Frost R.A. (ed.): Database management systems. McGraw-Hill, New York 1984

[GANE78] Gane C., Sarson T.: Structured systems analysis: Tools and techniques. Prentice-Hall, Englewood Cliffs NJ 1978

[GARD87] Gardner, H.: The mind's new science. Basic Books, New York 1987

[GAYD88] Gaydasch A.: Effective database management. Prentice-Hall, Englewood Cliffs N.J. 1988

[GOLD83] Goldberg A., Robson D.: Smalltalk-80: the language and its implementation. Addison-Wesley, Reading MA 1983

[GOLD84] Goldberg A.: Smalltalk-80: the interactive programming environment. Addison-Wesley, Reading MA 1984

[GRØN88] Grønli R., Pedersen K.: RAPVERC. Dept. Computer Science, NTH, Univ. Trondheim, Techn. Rep. 13/88, 1988

[GULL91] Gulla J.A., Lindland O.I., Willumsen G.: PPP - an integrated CASE environment. In: R. Andersen et al. (eds.): Proc. CAiSE'91, Lecture Notes in Computer Science, 498, Springer-Verlag, Berlin 1991, pp 194-221

[HART68] Hartmann W., Matthes H., Proeme A.: Management information systems handbook. McGraw Hill, New York 1968

[HEND91] Henderson-Sellers B., Constantine L.L.: Object-oriented development and functional decomposition. J. Object-Oriented Program. 33:9, 11-17 (1991)

[HENR84] Henry S., Kafura D.: The evaluation of software systems' structure using quantitative software metric. Softw. Pract. Exper. 14:6, 561-573 (1984)

[HONE82] Honeywell, Alsys: Reference manual for the ADA programming language. Washington DC, 1982

[HOVE84] Hove J.O.: Kung's method for consistency proof and model construction applied to communication protocols (in Norwegian). Technical report # 22/84, Dept. Computer Science, The University of Trondheim, Norway

[HULL87] Hull R., King R.: Semantic data modeling: survey, applications, and research issues. ACM Comput. Surv. 19:3, 201 - 260 (1987)

[IBM 89] IBM: AD/Cycle concepts. IBM Corporation 1989

[ICHB79] Ichbiah J.D. et al.: Rationale for the design of the ADA programming language. SIGPLAN Notices 14 B (1979)

[IFIP66] IFIP: Vocabulary of information processing. International Federation for Information Processing and International Computation Centre, 1. English Language Edition, Amsterdam 1966

[IMS 74] IMS/VS General information manual, GH20-1260. IBM Corporation, White Plains, NY 1974

[JACK75] Jackson M.A.: Principles of program design. Academic Press, London 1975

[JENS80] Jensen R.W., Tonies C.C.: Software engineering. Prentice-Hall, Englewood Cliffs NJ 1980

[KEEN81] Keen P.G.W.: Information systems and organizational change. Commun. ACM 24:1, 24-33 (1981)

[KEEN82] Keen P.G.W., Bronsema G.S., Zuboff S.: Implementing common systems: One organization's experience. Systems, Objectives, Solutions No. 2, 125-142 (1982)

[KORS90] Korson T., McGregor J.D.: Understanding object-oriented: a unifying paradigm. Commun. ACM 33:9, 40-60 (1990)

[KUNG83] Kung C.H.: An analysis of three conceptual models with time perspective. In: T.W. Olle et al. (eds.): Information systems design methodologies: a feature analysis. North-Holland, Amsterdam 1983, pp. 141-168

[KUNG86] Kung C.H., Sølvberg A.: Activity and behavior modeling. In: [OLLE86]

[KUNZ84] Kunz J.C., Kehler T.P., Williams M.D.: Applications development using a hybrid AI development system. AI Mag. 5:3 (1984)

[LANG66] Langefors B.: Theoretical analysis of information systems. Studentlitteratur, Lund 1966

[LAWR82] Lawrence M.J.: An examination of evolution dynamics. In: Proc. Conf. Software Eng., IEEE 1982, pp. 188-196

[LEHM80] Lehman M.M.: Programs, life cycles and laws of software evolution. In: Proc. IEEE Spec. Iss. on Software Engineering, September 1980, pp. 1060-1076

[LIND77] Lindsay P.H., Norman D.A.: Human information processing: An introduction to psychology. Academic Press, New York 1977

[LOND72] London K.R.: Decision tables. Auerbach Publishers, Princeton PA 1972

[LOVS90] Løvseth K.W.: TEMOCCA - Temporal modeling concepts - A case study. Dept. Electrical Eng. and Comp. Sc., NTH, University of Trondheim 1990

[LOWE73] Lowe T.C.: Analysis of an information system model with transfer penalties. IEEE Trans. Comput. C-22, 469-480 (1973)

[MARC77] March S.T., Severance D.G.: The determination of efficient record segmentations and blocking factors for shared data files. ACM Trans. Database Syst. 2:3, 279-296 (1977)

[MCCA89] McCabe T., Butler C.W.: Design complexity measurement and testing. Commun. ACM 32:12, 308-320 (1989)

[METZ73] Metzger P.W.: Managing a programming project. Prentice-Hall, Englewood Cliffs NJ 1973

[MIDA88] MIDAS. Modell for systemutvikling. Versjon 1, STATOIL, Stavanger, Norway 1988

[MONA92] Monarchi D.E., Puhr G.I.: A research typology for object-oriented analysis and design. Commun. ACM 35:9, 35-47 (1992)

[MUMF78] Mumford E., Land F., Hawgood J.: A participative approach to the design of computer systems. Impact of Science on Society 28:3, 235-53 (1978)

[MYER78] Myers G.J.: Composite/structured design. Van Nostrand Reinhold, New York 1978

[NIJS77] Nijssen G.M.: On the gross architecture for the next generation data base management systems. In: Proc. IFIP Congress 1977, North-Holland, Amsterdam 1977, pp. 327-335

[OFTE81] Oftedal H., Sølvberg A.: Data base design constrained by traffic load estimates. Inf. Syst. 6:4, 267-282 (1981)

[OLIV82] Olive A.: DADES A methodology for specification and design of information systems. In: [OLLE82], pp. 285-334

[OLLE82] Olle T.W., Sol H.G., Verrijn-Stuart A.: Information systems methodologies: A comparative review. North-Holland, Amsterdam 1982

[OLLE86] Olle T.W., Sol H.G., Verrijn-Stuart A.: CRIS III - improving the practice. North-Holland, Amsterdam 1986

[ORRK77] Orr K.T.: Structured systems development. Yourdon, New York 1977

[PAGE80] Page-Jones M.: The practical guide to structured systems design. Yourdon, New York 1980

[PARN72] Parnas D.L.: On criteria to be used in decomposing systems into modules. Commun. ACM 14:12, 1053-1058 (1972)

[PECK88] Peckham J., Maryanski F.: Semantic data models. ACM Comput. Surv. 20:3, 153-190 (1988)

[PERL85] Perlman, G.: Making the right choices with menus. Human-Computer Interaction – Interact '84, North-Holland, Amsterdam 1985

[PERR90] Perry D.E., Kaiser G.E.: Adequate testing and object-oriented programming. J. Object-Oriented Program. 2, 13-19 (1990)

[PETE77] Peterson J.L.: Petri nets. ACM Comput. Surv. 9:3, 224-252 (1977)

[PRES82] Pressman R.S.: Software engineering: a practitioner's approach. McGraw-Hill, New York 1982

[QUIL68] Quillian M. Ross: Semantic memory. In: M. Minsky (ed): Semantic Information Processing. MIT Press, Cambridge 1968

[REIS85] Reisig W.: Petri nets. An introduction. EATCS Monographs on Theoretical Computer Science, Vol. 4. Springer-Verlag, Berlin 1985

[RITT72] Rittel H.: On the planning crisis: systems analysis of the 'first and second generations'. Proc. Systems Analysis Seminar, European Association of National Productivity Centres, 1971, and in Bedriftsøkonomen nr. 8, 1972, Norway

[RUBI92] Rubin K.S., Goldberg A.: Object behavior analysis. Commun. ACM 35:9, 48-62 (1992)

[RUMB91] Rumbaugh J. et al.: Object-oriented modeling and design. Prentice-Hall, Englewood Cliffs NJ 1991

[SELT93] Seltveit A.H.: An approach to large-scale IS development based on systematic complexity reduction. Ph.D.Thesis, Univ. Trondheim 1993

[SFIN87] SFINX Consortium: SFINX Software factory integration and experimentation, Software Factory Concepts, Esprit Project 1262

[SHNE92] Shneiderman B.: Designing the user interface. 2nd ed., Addison-Wesley, Reading MA 1992

[SIME73] Sime M.E., Green T.R.G., Guest D.J.: Psychological evaluation of two conditional structures used in computer languages. Int. J. Man-Mach. Stud. 5 (1973)

[SMIT90] Smith C.U.: Performance engineering of software systems. Addison-Wesley, Reading MA 1990

[SOLV80] Sølvberg A.: A contribution of the definition of concepts in expressing user's information systems requirements. In: P.P. Chen (ed): Entity relationship approach to systems analysis and design. North-Holland, Amsterdam 1980, pp 359-380

[STAM73] Stamper R.: Information in business and administrative systems. B.T. Batsford, London 1973

[STEV74] Stevens W., Myers G., Constantine L.: Structured design. IBM Syst. J. 13:2, 115-139 (1974)

[TANG86] Tang C.S.: A temporal logic langua modeling of information and expert systems. In: Proc. IFIP TC2 Working Conf. on Knowledge and Data, North-Holland, Amsterdam 1986

[TICH85] Tichy W.F.: RCS - A System for version control. Software Management & Experience 15:7 (1985)

[TICH88] Tichy W.F.: Tools for software configuration management. In: J. Winkler (ed.): Proc. Int. Workshop on Software Version and Configuration Control, B.G. Teubner, Stuttgart 1988

[TRIP76] Tripp L.L., Peters L.J.: Design representation schemes. In: Proc. MRI symposium on computer software engineering, 1976, pp. 31-56

[VESS86] Vessey I., Weber R.: Structured tools and conditional logic: an empirical investigation. Commun. ACM 29:1 (1986)

[WALD84] Walden K.: Automatic generation of MAKE dependencies. Software Management & Experiences 14:6 (1984)

[WARN76] Warnier J.D.: The logical construction of programs. 3rd edn., Van Nostrand Reinhold, New York 1976

[WASS90] Wasserman A.I. et al.: The object-oriented structured design notation for software design representation. Computer 23:3, 50-63 (1990)

[WAYN86] Wayne A.B.: Software configuration management. Addison-Wesley, Reading MA 1986

[WEYU86] Weyuker E.J.: Axiomatizing software test data adequacy. IEEE Trans. Softw. Eng. SE-12:12, 1128-1138 (1986)

[WINO79] Winograd T.: Beyond programming languages. Commun. ACM 22:7, 391-401 (1979)

[WIRT76] Wirth N.: MODULA: a language for modular multi-programming. Zürich 1976

[WIRT81] Wirth N.: Lilith: a personal computer for the software engineer. In: Proceedings 5th International Conf. on Software Engineering. IEEE Computer Society Press 1981, pp.2-16

[YAOS85] Yao S.B. (ed.): Principles of data base design. Prentice-Hall, Englewood Cliffs NJ 1985

[YOUR89] Yourdon E.: Modern structured analysis. Prentice-Hall, Englewood Cliffs NJ 1989

Index

Springer-Verlag and the Environment

We at Springer-Verlag firmly believe that an international science publisher has a special obligation to the environment, and our corporate policies consistently reflect this conviction.

We also expect our business partners – paper mills, printers, packaging manufacturers, etc. – to commit themselves to using environmentally friendly materials and production processes.

The paper in this book is made from low- or no-chlorine pulp and is acid free, in conformance with international standards for paper permanency.

Made in the USA
Monee, IL
13 September 2026

62973755R00308